ACADEMIC LABORATORY CHEMICAL HAZARDS GUIDEBOOK

ACADEMIC LABORATORY CHEMICAL HAZARDS GUIDEBOOK

WILLIAM J. MAHN

VAN NOSTRAND REINHOLD
New York

*To my wife, Carol Wickell,
and
N. Irving Sax, mentor and good friend*

Extreme care has been taken in preparation of this work. However, author and publisher make no warranty or representation, expressed or implied, with respect to accuracy, completeness, or utility of the information contained in this document; nor do the author or publisher assume any liability with respect to the use of or reliance upon, or for damages resulting from the use of or reliance upon, any information, procedure, conclusion, or opinion contained in this document.

Copyright © 1991 by Van Nostrand Reinhold

Library of Congress Catalog Number 90-12821
ISBN 0-442-00165-7

All rights reserved. No part of this work covered by the copyright hereon may be reproduced or used in any form by any means—graphic, electronic, or mechanical, including photocopying, recording, taping, or information storage and retrieval systems—without written permission of the publisher.

Printed in the United States of America

Van Nostrand Reinhold
115 Fifth Avenue
New York, New York 10003

Van Nostrand Reinhold International Company Limited
11 New Fetter Lane
London EC4P 4EE, England

Van Nostrand Reinhold
102 Dodds Street
South Melbourne 3205, Victoria, Australia

Nelson Canada
1120 Birchmount Road
Scarborough, Ontario M1K 5G4, Canada

16 15 14 13 12 11 10 9 8 7 6 5 4 3 2 1

Library of Congress Cataloging in Publication Data
Mahn, William J.
  Academic laboratory chemical hazards guidebook/William J. Mahn.
  p. cm.
  Includes index.
  ISBN 0-442-00165-7
  1. Chemical laboratories—Safety measures. 2. Chemicals—safety measures. I. Title.
QD51.M29 1990
642.028′9—dc20

CONTENTS

Preface ix

PART 1. MANAGING CHEMICAL HAZARDS IN THE LABORATORY   **1**

1. REACTIVE HAZARDS, UNDERSTANDING AND CONTROL   3

General Rules for Handling Chemicals   3
Types of Reactivity Hazards   4
   Flammable Compounds   4
     Properties of Flammable Substances
     Sources of Ignition
     Basic Rules for Using Flammable Substances
     Highly Flammable Solvents
     Flammable or Explosive Gases and Liquefied Gases
   Explosives   7
     Peroxides
     Per Salts
     Oxidizing Agents
     Detonators
     Handling Explosive Materials
   Incompatible Chemicals   15
   Corrosives—Including Acids and Alkalies   15
   Exothermic and Other Reactions   17
   Operations Involving Hazardous Reactions   19
Suggested Readings   19

2. TOXIC HAZARDS   20

Basic Rules of Toxicology   20
   Duration of Exposure   21
   Acute and Chronic Toxicity   21
   Expressions of Toxicity and Dose Response   21
   Routes of Entry   22
     Inhalation
     Contact with Skin or Eyes
     Ingestion
     Injection
     Other Routes
Types of Toxic Chemicals   25
   Carcinogens   25
   Poisons   25
   Irritants   26
     Corrosive Chemicals
     Strong Acids
     Strong Bases
     Dehydrating Agents
   Asphyxiants   28
   Anesthetics   28

Embryotoxins	28
Allergens	29
Particulates	29
Understanding Toxic Effects	**29**
Cancer	30
Nervous System Damage	32
Liver Damage	33
Kidney Damage	33
Reproductive System Effects	33
Mutagenesis	33
How Toxic Is "Toxic"?	**34**
Dose Response and Threshold	34
Variables to the Dose-Response Relationship	35
Practicality of Using Available Information	35
Extrapolating from Animal Experiments to Humans	36
Correcting for Toxicokinetic Differences	
The Surface-Area Concept	
Differences in Species' Sensitivity	38
Differences Between Individuals	38
Interactions and Synergisms	40
Approaches in Dealing with Uncertainty	40
Suggested Readings	**40**
3. PROCEDURES FOR HANDLING TOXIC MATERIALS	**42**
General Control Measures	**42**
Other Safety Precautions	42
Potential Problem Areas	43
Specific Procedures for Toxic Chemicals	**44**
Procedure B: For Substances of Moderate Chronic or High Acute Toxicity	44
Procedure A: For Substances of High Chronic Toxicity	45
A Specific Example for Polycyclic Aromatic Hydrocarbons	47
Animal Experiments with Substances of High Chronic Toxicity	48
Other Specific Hazards	48
Beryllium	
Carbonyls	
Chromium Solutions	
Corrosives	
Cyanides	
Epoxy Resins	
Gases and Vapors	
Heavy Metals	
Welding Vapors and Ozone	
Other Rules for Handling Toxic Materials	49

Quality Control—Evaluation of the Efficiency of the Exposure Control Program	50
Material Safety Data Sheets	50
Suggested Readings	**50**
4. CHEMICAL LABELING AND STORAGE	**51**
Labeling	**51**
Information on Labels	52
A Model Labeling System	52
Interpreting Labels	53
Examples	54
Ordering and Receiving Chemicals	**54**
Purchasing Philosophies: Cost Versus Safety	54
Chemical Grades	54
Ordering and Procuring Chemicals	55
Receiving Ordered Chemicals	55
Recordkeeping	**56**
Inventory Control	56
Dated Receiving System	56
Recordkeeping Detail	56
Chemical Storage	**57**
General Rules for Chemical Storage	57
Chemical Storerooms and Stockrooms	57
Central Storerooms	
Stockrooms	
Chemical Storage Patterns	
Flammable Liquids	
Toxic Substances	
Compressed Gases	
Transport of Chemicals from Stockrooms to Laboratories	60
Solid and Liquid Chemicals	
Cylinders of Compressed Gases	
Chemical Storage in Laboratories	61
General Considerations	
Flammable Liquids	
Toxic Substances	
Compressed Gases	
Radioactive Materials	
Poor Storage Practices	**63**
Reagents in Hoods	63
Flammable Reagents in Standard Refrigerators	63
Food in Refrigerators	63
Storage and Handling of Specific Materials	**63**
Oxidizing Materials	63
Spontaneously Flammable or Easily Oxidizable Substances	64

Peroxide Formers	64		
Corrosive Chemicals	64		
Toxic and Corrosive Materials	64		
Materials That React Dangerously with Water	64		
Finely Divided Materials	64		
Suggested Readings	65		

5. LABORATORY WASTES — 66

- **Legal Aspects** — 66
  - RCRA — 66
    - *Definitions*
  - State Laws Governing Waste Disposal — 67
- **Types of Waste** — 67
  - Nonhazardous Waste — 68
    - *Ordinary Trash*
    - *Glass*
    - *Nonhazardous Biological Wastes*
  - Hazardous Waste — 68
    - *Toxic Waste*
    - *Ignitable Waste*
    - *Corrosive Materials*
    - *Reactive Waste*
- **General Considerations for Disposal** — 70
  - The First Step — 71
  - Disposal Choices — 71
  - Recycling — 71
  - Return to Supplier or Give to Another Organization — 71
  - Incineration and Solvent Burners — 71
  - Disposal to the Sewer System — 72
  - Conversion of Hazardous Materials to a Less Hazardous Form — 73
  - Consolidation of Hazardous Materials Before Disposal — 73
  - Disposal of Unlabeled Chemicals — 73
  - Landfills — 73
  - Contractor Disposal — 74
- **Specific Procedures** — 74
  - Disposal of Liquid Chemical Wastes — 74
  - Lab Packs — 75
  - Disposal of Hazardous Biological Wastes — 75
- **Storage of Wastes** — 76
- **Transportation of Chemical Wastes** — 76
- **Minimization of Waste** — 76
- **Recordkeeping** — 76
- **Suggested Readings** — 77

PART 2. CHEMICAL ENTRIES — 79

PART 3. APPENDICES — 281

APPENDIX A: HAZARDOUS RATINGS AND CLASSIFICATIONS — 283

- Toxicity Rating Scale — 283
- Fire Hazard Ratings and Classes — 283
- Underwriters Laboratories Scale of Relative Hazard of Various Flammable Liquids — 283

APPENDIX B: NATIONAL FIRE PROTECTION ASSOCIATION (NFPA) LABELS — 284

- Dimensions of NFPA Labels — 285
- NFPA Hazard Ratings and Signal Words — 285

APPENDIX C: HAZARDOUS MATERIALS WARNING LABELS — 286

- General Guidelines on the Use of Labels — 286
- UN Class Numbers — 286
  - Explosives (UN Class 1) — 286
  - Compressed Gases (UN Class 2) — 287
  - Flammable Liquids (UN Class 3) — 287
  - Flammable Solids (UN Class 4) — 288
  - Dangerous When Wet (UN Class 4) — 288
  - Oxidizing Materials (UN Class 5) — 288
  - Poisonous Materials (Class B Poisons, UN Class 6) — 289
  - Irritating Materials (UN Class 6) — 289
  - Biomedical Materials (UN Class 6) — 289
  - Radioactive Materials (UN Class 7) — 290
  - Corrosives (UN Class 8) — 290
  - Other Regulated Materials (ORMs) (49 CFR, Part 172.316) — 291
- Special Handling — 291
- Special-Item Markings — 292
- Manufacturers' Labels — 292
- International Labeling — 293
  - International Labeling for Explosives — 293
- Suggested Reading — 293

APPENDIX D: HAZARDOUS LABORATORY SUBSTANCES — 294

APPENDIX E: STANDARD CHEMICAL LABEL STATEMENTS — 306

- Health-Related Statements — 306
- Health Hazard Action Statements — 307
- Fire Hazard Amplifying Statements — 307
- Fire Hazard Action Statements — 308

viii CONTENTS

Reactivity Hazard Amplifying Statements	308
Reactivity Hazard Action Statements	308
First-Aid Statements	309
Statements Specifying Specific Disposal Instructions	310
Sample Label	310

APPENDIX F: TOXICOLOGY GLOSSARY   311

APPENDIX G: SCIENCE INVENTORY AND STORAGE HAZARDS   315

Suggested Shelf Storage Patterns	**326**
Organics	326
Inorganics	327
Compressed Gases	328

INDICES   **329**

Subject Index   329

Chemical Index   332

CAS Number Index   341

PREFACE

Academic Laboratory Chemical Hazards Guidebook contains the information most often needed by laboratory workers. It is designed to be the first-choice reference for those who require information on the hazards and handling procedures of the chemicals with which they work. Intended for use in universities, colleges, high schools, and middle schools, the book is also suitable for individuals handling hazardous chemicals and laboratory equipment in engineering, research and development, clinical, quality control, and food science laboratories. In general, the book is an attempt to promote and improve safety in the laboratory.

With chapters on reactivity, toxicity, handling, storage and labeling, and disposal, Part 1 of this title describes the management of chemical hazards. Part 2 outlines the hazardous properties and correct handling of over 200 chemicals commonly found in laboratories. The extensive cross referencing allows the reader to easily look up a particular chemical by its Chemical Abstracts Service (CAS) number or chemical name (most synonyms are given). The appendices, Part 3, provide information on toxicity and fire hazards, NFPA labels, DOT warning labels, how to generate your own labels, a toxicology glossary, and chemical storage plans.

This book and a companion volume, *Fundamentals of Laboratory Safety: Physical Hazards in the Academic Laboratory,* allow the laboratory worker (students, technicians, instructors, and scientists alike) to quickly find information on specific laboratory hazards and their management by use of the extensive table of contents and indices. Both books are beneficial to workers not only in an academic science laboratory (both physical and biological), but also those in the engineering lab and even a machine shop. All these areas involve the use of hazardous materials and equipment. The goal has been to provide a volume that is more likely to contain the required information than any other single source. If the needed information is not in this book or its companion title, I have tried to direct the reader to the next best reference.

I wish to thank Gene Falken who was very instrumental on the start of this project.

William J. Mahn

Part 1
MANAGING CHEMICAL HAZARDS IN THE LABORATORY

1
REACTIVE HAZARDS, UNDERSTANDING AND CONTROL

INTRODUCTION

Reactive hazards in the laboratory are the result of both chemical properties and unsafe operations. Most reactive hazards in the laboratory are the result of uncontrolled energy release. There are several variables in experiments and other operations which can increase the possibility of accident. In order to control reactive hazards it is necessary to control these variables. This control requires some preplanning and knowledge of the properties of the chemicals involved.

Temperature is the most important factor in controlling hazardous reactions in the laboratory. An accepted rule of thumb is that reaction rates double for every 10°C increase in temperature. Thus a reaction may be slowed and made more controllable by keeping the temperature low. Results of high temperature may include an increase in corrosion rates or excessive pressure.

An uncontrolled exothermic reaction can cause a temperature increase that leads to a runaway reaction. One of the factors contributing to a runaway reaction is a pressure increase which also increases the reaction rate and may cause an explosion. Many explosions of laboratory and pilot plant equipment have occurred because of runaway reactions.

The second most important controllable factor is chemical concentration. An increase in concentration increases the rate of a reaction. It is always better to work with small amounts of a chemical to minimize both reactive and toxic hazards. Reagents should normally be used in concentrations of about 10% (if they are soluble at that level) to minimize hazards. If a reagent is particularly reactive, however, then concentrations should be reduced to 5% or even 2%. If a reaction is being tried for the first time, it is even more important to keep concentrations as low as possible.

The presence of a catalyst may contribute greatly to the hazard. Catalysts vary in behavior, sometimes speeding up a reaction, sometimes slowing down or controlling a reaction. A full understanding of the characteristics of a catalyst is essential to safe use of it.

Another factor in hazardous reactions is improper operation. Improper operation can include many things—the addition of incorrect materials, adding in the wrong order, failure to measure amounts accurately, failure to mix adequately, or failure to provide adequate safety measures. An example is the combining of two reactive materials in substantial quantities without stirring; when mixing is initiated, the reaction can proceed with great vigor and sometimes destructive violence.

Even when reactive hazards are understood and controlled, emergency preplanning should be considered essential for any laboratory operation. There should be specified people who will take action in case of an injury, performing rescue and giving first aid to the injured. There must be trained and knowledgeable people available to size up the situation and handle the necessary fire protection equipment. In case of a release of gases or a spill of hazardous materials, there must be evacuation procedures to get personnel away from the problem and to a safe area. General procedures are discussed in detail in *Fundamentals of Laboratory Safety: Physical Hazards in the Academic Laboratory* (Chap. 12).

Specific emergency procedures should be developed for specific problems. For example, with some highly reactive systems, provision should be made for dumping the contents of the reacting vessel into a neutralizing or deactivating solution.

GENERAL RULES FOR HANDLING CHEMICALS

There are important general rules for the proper handling of reagents. These rules protect the reagent from contamination and the user from problems caused by incompatibility.

1. Never return a reagent to the storage bottle. An incompatible mixture may be created by mistake.
2. Keep reagent containers closed. Dust and vapors may escape from an open container, and gaseous or suspended material may enter, changing the nature of the reagent.
3. Never use improper or unlabeled reagents. If you are not certain of the nature of the substance, don't use it.
4. Never insert a spatula, stirring rod, or other device into a storage reagent bottle. Remove contents by tilting and rolling the jar, dumping contents onto a dish, watchglass, beaker, filter paper, or glassine paper. Spatulas can be used with care in laboratory reagent bottles.
5. Screw-top bottle caps must be placed on clean filter paper with the top down when they are removed from the bottle.
6. Special care should be taken when handling concentrated sulfuric and other acids. Dilution of the acid should be done by pouring the acid into water with vigorous stirring.

TYPES OF REACTIVITY HAZARDS

Flammable Compounds

Flammable substances are among the most common of the hazardous materials found in laboratories. However, the ability to vaporize, ignite, and burn or to explode varies with the specific type or class of substance. Prevention of fires and explosions requires knowledge of the flammability characteristics (limits of flammability, ignition requirements, and burning rates) of combustible materials likely to be encountered under various conditions of use (or misuse) and of the appropriate procedures to use in handling such substances.

Properties of Flammable Substances. Flammable substances are those that readily catch fire and burn in air. A flammable liquid does not itself burn; it is the vapors from the liquid that burn. The rate at which different liquids produce flammable vapors depends on their vapor pressure, which increases with temperature. The degree of fire hazard also depends on the ability to form combustible or explosive mixtures with air, the ease of ignition of these mixtures, and the relative densities of the liquid with respect to water and of the gas with respect to air. These concepts can be evaluated and compared in terms of a number of properties.

Flash Point. An open beaker of diethyl ether set on the laboratory bench next to a Bunsen burner will ignite, whereas a similar beaker of diethyl phthalate will not. The difference in behavior is due to the fact that the ether has a much lower flash point. The flash point is the lowest temperature, as determined by standard tests, at which a liquid gives off vapor in sufficient concentration to form an ignitable mixture with air near the surface of the liquid within the test vessel. The concentration required is the lower explosive limit (LEL). The flash point is determined by the Tagliabue open-cup method (TOC), the Cleveland open-cup method (COC), or the Tagliabue closed-cup method (TCC). The TCC and COC are generally used for higher flash points. The open-cup method most closely approximates real-world conditions. These acronyms are common in listings of chemical hazards as found in Material Safety Data Sheets (MSDS) and the Chemical Entries in this volume.

Many common laboratory solvents and chemicals have flash points that are lower than room temperature.

Ignition Temperature. The ignition temperature or autoignition temperature of a substance, whether solid, liquid, or gaseous, is the minimum temperature required to initiate or cause self-sustained combustion independent of the heat source. It differs from the flash point in that a flame, spark, or other ignition source is not required to start the fire. A steam line or a glowing lightbulb may ignite carbon disulfide (ignition temperature 80°C/176°F). Diethyl ether (ignition temperature 160°C/320°F) can be ignited by the surface of a hot plate. Because of its extreme flammability, ether is available for laboratory use only in metal containers.

These physical properties can be understood by considering the properties of gasoline and engine oil. No one needs to be reminded that gasoline vapor mixed with air is explosive. Gasoline (*not* listed in Table 1–1) is explosive because of its low flash point. In winter a grade of gasoline of higher volatility and lower flash point is provided in order to facilitate ignition (starting). (Some chemists prime their car's carburetor with ethyl ether, flash point −45°C/−49°F, or petroleum ether, flash point −56°C/−69°F, to start in extremely cold weather, but this practice should not be condoned.) Whereas gasoline spilled on a hot engine will ignite, the crankcase oil which bathes the crankshaft and hot pistons does not do so, due to the relatively high ignition temperature of the motor oil (versus the low ignition temperature of gasoline). The circulating engine coolant also helps to keep the temperature of the engine far below the ignition temperature of the crankcase oil.

TABLE 1-1. Hazardous Properties of Some Common Laboratory Chemicals

Chemical	Class	Flash Point (°C)	Boiling Point (°C)	Ignition Temp. (°C)	Flammable Limit (% by volume in air)		Vapor Density	U.L. Flammability Rating
					Lower	Upper		
Acetaldehyde	1A	−37.8	21.1	175.0	4.0	60.0		
Acetic acid		42.8			5.4			
Acetone	1B	−17.8	56.7	465.0	2.6	12.8	2.0	90
Amyl acetate		25.		399.	1.1		4.49	55–60
Benzene	1B	−11.1	80.0	560.0	1.3	7.1	2.77	95–100
Butyl alcohol		100.		371.	1.4		2.55	40
Carbon disulfide	1B	−30.0	46.1	80.0	1.3	50.0	2.64	110
Carbon tetrachloride		None		None	None		5.32	0
Cyclohexane	1B	−20.0	81.7	245.0	1.3	8.0	2.9	
Ethyl acetate		−4.		427	2.2		3.04	High
Ethyl alcohol	1B	12.8	78.3	365.0	3.3	19.0	1.59	70
Ethyl ether		−45.		180.	1.9		2.56	100
Ethylene dichloride		−7.		429.	6.2		3.4	
n-Heptane	1B	−3.9	98.3	215.0	1.05	6.7		
n-Hexane	1B	−21.7	68.9	225.0	1.1	7.5		
Isopropyl alcohol	1B	11.7	82.8	398.9	2.0	12.0		
Methane (gas)		−7.		Low	5.0		0.55	High
Methyl alcohol	1B	11.1	64.9	385.0	6.7	36.0	1.11	70
Methylene chloride		None					5.37	
Methyl ethyl ketone	1B	−6.1	80.0	515.6	1.8	10.0		
Pentane	1A	−40.0	36.1	260.0	1.5	7.8		
Petroleum ether		−56.		246.	1.4		2.5	95–100
Styrene	1B	32.2	146.1	490.0	1.1	6.1		
Tetrachloroethane		None		None	None			0
Toluene	1B	4.4	110.6	480.0	1.2	7.1	3.14	75–80
p-Xylene	1C	27.2	138.3	530.0	1.1	7.0		

Limits of Flammability. It is possible for a flammable liquid to be above its flash point and yet not ignite in the presence of an adequate energy source. The explanation for this phenomenon lies in the composition of a fuel–air mixture that may be too lean or too rich for combustion.

Each flammable gas or liquid (as a vapor) has two fairly definite limits defining the range of concentrations in mixtures with air that will propagate flame and explode. The lower flammable limit or lower explosive limit (LEL) is the minimum concentration (percent by volume) of the vapor in air below which a flame is not propagated when an ignition source is present. Below this concentration, the mixture is too lean to burn. The upper flammable limit or upper explosive limit (UEL) is the maximum concentration (percent by volume) of the vapor in air above which a flame is not propagated. Above this concentration, the mixture is too rich to burn. The flammable range or explosive range consists of all concentrations between the LEL and the UEL. This range becomes wider with increasing temperature and in oxygen-rich atmospheres. The limitations of the flammability range, however, provide little margin of safety from a practical point of view because, when a solvent is spilled in the presence of an energy source, the LEL is reached very quickly and a fire or explosion will ensue before the UEL can be reached.

Vapor Density. A property of most flammable solvents that is not widely appreciated is that their vapor densities are higher than air (see Table 1–1) and as a result their vapors have a tendency to creep along benches into sinks and along floors. This property is well known for ethyl ether, but ethyl acetate and benzene have vapor densities even greater than that of ether. Special care should be taken to eliminate ignition sources at a lower level than that at which a flammable substance is being used. Flammable vapors from massive sources such as spillages have been known to descend into stairwell and elevator shafts and ignite on a lower story. If the path of vapor within the flammable range is continuous, the flame will propagate itself from the point of ignition back to its source.

The flash points, boiling points, flammable limits, ignition temperatures, and vapor densities of a number of common laboratory chemicals are given in

Table 1–1 and in the Hazardous Chemical Entries. It should be remembered, however, that tabulations of properties of flammable substances are based on standard test methods, for which the conditions may be very different from those encountered in practical use. Large safety factors should be applied. For example, the published flammable limits of vapors are for uniform mixtures with air. In a real situation, point concentrations may exist that are much higher than the average. Thus it is good practice to set the maximum allowable concentration for safe working conditions at some fraction of the tabulated lower LEL; 20% is a commonly accepted value.

Spontaneous Ignition. Spontaneous ignition or combustion takes place when a substance reaches its ignition temperature without the application of external heat. The possibility of spontaneous combustion should always be considered, especially when materials are stored or disposed of. Materials that are susceptible to spontaneous combustion include oily rags, dust accumulations, organic materials mixed with strong oxidizing agents (such as nitric acid, chlorates, permanganates, peroxides, and persulfates), alkali metals such as sodium and potassium, finely divided pyrophoric metals, and phosphorus.

Sources of Ignition. For a fire to occur, three distinct conditions must exist simultaneously: a concentration of flammable gas or vapor that is within the flammability limits of the substance; an oxidizing atmosphere, usually air; and a source of ignition. Removing any of the three will prevent the start of a fire or extinguish an existing fire. In most situations air cannot be excluded. The problem, therefore, usually resolves itself into preventing the coexistence of flammable vapors and an ignition source. Because spillage of a flammable liquid is always a possibility, strict control of ignition sources is mandatory.

Many sources—electrical equipment, open flames, static electricity, burning tobacco, lighted matches, and hot surfaces—can cause ignition of flammable substances. When these materials are used in the laboratory, close attention should be given to all potential sources of ignition in the vicinity. The vapors of all flammable liquids are heavier than air and are capable of traveling considerable distances along the floor or ground. This possibility should be recognized, and special note should be taken of ignition sources at a lower level than that at which the substance is being used.

The act of pouring a flammable substance from one container to another can cause a static discharge. A fire which destroyed a large part of the chemistry building of the University of Minnesota started when benzene vapors self-ignited (due to static charges) while being transferred from a drum. Metal containers, funnels, and tubing used during transfer of flammable substances should be properly bonded and grounded to discharge static electricity. Solvents with low flash points should always be transferred by pouring through a chemically resistant, stainless steel funnel to which ground-wire leads have been clamped. The grounded metal funnel serves as a discharger. This technique is especially useful when nonmetallic containers (especially plastic) are used; the bonding is made to the liquid rather than to the container. When no solution to the static problem can be found, then all processes should be carried out as slowly as possible to give the accumulated charge time to disperse.

Basic Rules for Using Flammable Substances. Basic precautions for safe handling of flammable materials include the following:

1. Flammable substances should be handled only in areas free of ignition sources.
2. Flammable substances should never be heated by using an open flame. Preferred heat sources include steam baths, water baths, oil baths, heating mantles, and hot-air baths.
3. When transferring flammable liquids in metal equipment, static-generated sparks should be avoided by bonding and the use of ground straps.
4. Ventilation is one of the most effective ways to prevent the formation of flammable mixtures. An exhaust hood should be used whenever appreciable quantities of flammable substances are transferred from one container to another, allowed to stand in open containers, heated in open containers, or handled in any other way.

Highly Flammable Solvents. Among the most hazardous liquids are those that have flash points at room temperature or lower, particularly if their range of flammability is broad. Materials that have flash points higher than the maximum ambient summer temperature do not ordinarily form ignitable mixtures with air under normal (unheated) conditions, but, as shown in Table 1–1, many commonly used substances are potentially very hazardous, even under relatively cool conditions.

In Table 1–1, note the high flammability rating of acetone, benzene, carbon disulfide, ethyl acetate, ethyl alcohol, ethyl ether, petroleum ether, and toluene—solvents which can be found in any laboratory—and methane gas, which can be generated at the laboratory bench or can escape from gas cylinders. High flammability usually occurs with materials that have low flash points and low ignition temperatures. Carbon disulfide (CS_2) is notorious for having an extremely low ignition temperature (100°C) and a low flash point (−30°C). The consequence of this is that CS_2 can ignite even on the steam bath, a fact that should always be kept foremost in mind when working with this solvent.

Flammable or Explosive Gases and Liquefied Gases. Compressed or liquefied gases present hazards in the event of fire because the heat will cause the pressure to increase and may rupture the container. Leakage or escape of flammable gases can produce an explosive atmosphere in the laboratory. Acetylene, hydrogen, ammonia, hydrogen sulfide, and carbon monoxide are especially hazardous. Acetylene and hydrogen have very wide flammability limits, which adds greatly to their potential fire and explosion hazard. (See *Fundamentals of Laboratory Safety: Physical Hazards in the Academic Laboratory,* Chap. 8, for specific precautions for the use of flammable gases.)

Even if it is not under pressure, a substance is more concentrated in the form of a liquefied gas than in the vapor phase and may evaporate extremely rapidly. Oxygen, in particular, is an extreme hazard; liquefied air is almost as dangerous [if allowed to boil freely, it will have an increasing concentration of oxygen (boiling point −183°C/−297.4°F) because the nitrogen (boiling point −196°C/−320°F) will boil away first]; and even liquid nitrogen, if it has been standing around for some time, will have absorbed enough oxygen to require careful handling. When a liquefied gas is used in a closed system, pressure may build up, so that adequate venting is required. If the liquid is flammable (e.g., hydrogen), explosive concentrations may develop. Any, or all, of the three problems—flammability, toxicity, and pressure buildup—may become serious.

Explosives

Some chemicals decompose when heated. Slow decomposition may not be noticeable on a small scale, but on a large scale, or if the evolved heat and gases are confined, an explosive situation can develop. The heat-initiated decomposition of some substances, such as certain peroxides, is almost instantaneous. Light, mechanical shock, and certain catalysts are also initiators of explosive reactions. Hydrogen and chlorine react explosively in the presence of light. In general, compounds containing the following functional groups tend to be sensitive to heat and shock: acetylide, azide, diazo, halamine, nitroso, ozonide, and peroxide. For example, diazomethane (CH_2N_2) may decompose explosively when compressed in a ground-glass joint. Compounds containing nitro groups may be highly reactive, especially if other substituents such as halogens are present. Perchlorates, chlorates, nitrates, bromates, chlorites, and iodates, whether organic or inorganic, should be treated with respect, especially at higher temperatures. Acids, bases, and other substances catalyze the explosive polymerization of acrolein and other monomers. Many metal ions can catalyze the violent decomposition of hydrogen peroxide. Any given sample may be just atypical enough (by virtue of impurities and such) to be dangerous.

Not all explosions result from chemical reactions. A dangerous, physically caused explosion can occur if a hot liquid (such as oil) is brought into sudden contact with a lower-boiling-point one (such as water). The instantaneous vaporization of the lower-boiling-point substance can be hazardous to personnel and destructive to equipment.

There are three types of fast chemical decomposition which fit under the general title of explosion. *Deflagration* is a soft explosion when pressures are relatively low. An example is the ignition of a small quantity of a solvent spill. *Explosions* involve pressures of several atmospheres. Ignition of a large solvent spill could be explosive and cause severe injuries and structural damage to the building. *Detonation* is a severe form of explosion when pressures are much higher and are propagated at a high rate (as much as several miles per second). Thus the hazard of detonation is not associated with the total energy released, but rather with the amazingly high rate of the reaction. A high-order explosion of even milligram quantities can drive small fragments of glass or other matter deep into the body.

Explosively unstable materials can be divided into two general categories, those that have a high energy release (called *powerful* groups) and those that are extremely sensitive to initiation (called *sensitive* groups). This does not mean that the sensitive groups are not powerful and the powerful groups are not sensitive, but the behavior is categorized in this way for discussion purposes.

Among the powerful group are nitrates, chlorates, perchlorates, chlorites, and picrates. Trinitrotoluene is one of the most widely used military explosives. Ammonium nitrate has largely displaced dynamite as a commercial explosive. Chlorates and perchlorates are extremely useful in laboratory reactions and in chemical processing. Chlorites supply a ready source of chlorine. Bromates and iodates are less used but are fairly well-known chemical materials. Picrates, combining nitrate and peroxide radicals, are commonly found in initiating devices. Before using any of these materials, check for their reactive hazards.

Among the sensitive groups are acetylides, azides, diazo compounds, nitrosos, and peroxides. The diazos are commonly used in photoengraving work, and azides are used in boiler feedwater to prevent corrosion. As an example, an accident occurred when a laboratory water bath was being cleaned and a deposit at the outlet was being scraped. The materials exploded, causing one serious injury and considerable damage to the equipment. Analysis showed that the material was sodium azide, which is sensitive to friction. Sodium azide is used in vehicle air bags to inflate them quickly. Peroxides are perhaps best known for their formation in ethers. Diethyl ether, isopropyl ether, dioxane, and other such materials form dangerous peroxides upon aging in glass in the presence of air. If one ever sees crystals or any solids showing up on the bottom of a bottle of ether, the bottle should be very carefully disposed of without opening. Contact the Safety Officer immediately. Merely unscrewing the cap may set off an explosion that will cause severe injury. Acetylene combines with some materials (especially metals) used in the laboratory; for example, mercury forms an unstable acetylide which may explode during maintenance work when it is least expected.

A compound is apt to be explosive if its heat of formation is smaller by more than about 100 cal/g than the sum of the heats of formation of its products. In making this calculation, one should assume a reasonable reaction to yield the most exothermic products.

One means of estimating the stability of a material is to calculate an oxygen balance. If the amount of oxygen is approximately equivalent to that which would be needed to form the products of oxidation, then the material should be considered a potentially unstable material. This is of value only as a first estimate and is by no means adequate as a measure of safety of a material. Differential thermal analysis will show exothermic behavior developing in the system. Other tests for thermal stability and mechanical stability are available but are rather complex; hence, reliable results are best obtained by submitting samples to a laboratory that is experienced in this specialty.

Peroxides

Inorganic Peroxides. Peroxides should be handled with care. When mixed with combustible materials, barium, sodium, and potassium peroxides form explosives that ignite easily. Sodium peroxide reacts violently with water, and if mixed with oxidizable materials in the presence of moisture, is liable to inflame. It should not be put into paper or cardboard containers. Barium peroxide is less dangerous than sodium peroxide, but should be protected from moisture, and kept away from organic matter, sulfur, powdered metals, etc. If spilled, solid peroxides should be covered with a sand-soda mixture (9:1) and added slowly to a large volume of sodium sulfite solution with stirring, and then neutralized with sulfuric acid. The sulfate solution should be allowed to settle and then transferred into a disposal container with excess water. The same procedure can be used for general laboratory disposal.

Hydrogen peroxide of high concentration (>30%) will ignite fabric, oil, wood, and some resins with which it comes in contact, and is liable to decompose violently if contaminated with even minute traces of certain catalysts (e.g., iron, copper, or chromium, or other metals or their salts). If handled properly to avoid contamination, this reagent has a long shelf life and is not explosive. With concentrated sulfuric acid, hydrogen peroxide forms monopersulfuric acid (Caro's acid), which can explode in the presence of alcohols and aldehydes. With organic acids such as acetic acid, hydrogen peroxide will form a per acid, especially if catalyzed by strong mineral acids. The per acids are good oxidizing agents and are capable of oxidizing a wide spectrum of organic compounds. Hydrogen peroxide and acetic anhydride (all acid anhydrides and acid chlorides, for that matter) should never be mixed and should never be stored together, since the mixture will lead to the formation of explosive diacetyl peroxide. If spilled, hydrogen peroxide should be considerably diluted with water. It should be stored only in glassware or approved stainless steel containers, and carefully protected against accidental contamination.

Hydrogen peroxide stronger than 3% can cause severe skin burns.

Organic Peroxides. Organic peroxides are explosive and should be handled with precautions. They are some of the most hazardous substances normally handled in laboratories. They should be stored only in the wet condition. Benzoyl peroxide has been known to explode spontaneously. Acetyl peroxide when dry is a very sensitive and violent explosive.

As a class, the organic peroxides are low-power explosives, hazardous because of their extreme sensitivity to shock, sparks, or other forms of accidental ignition. Many peroxides that are handled routinely in laboratories are far more sensitive to shock than most primary explosives (e.g., TNT). Peroxides have a specific half-life, or rate of decomposition, under any given set of conditions. A low rate of decomposition may autoaccelerate and cause a violent explosion, especially in bulk quantities of peroxide. These compounds are sensitive to heat, friction, impact, and light, as well as to strong oxidizing and reducing agents. All organic peroxides are highly flammable, and fires involving bulk quantities of peroxides should be approached with extreme caution. A peroxide present as a contaminant in a reagent or solvent can change the course of a planned reaction.

Precautions for handling peroxides include the following:

1. The quantity of peroxide should be limited to the minimum amount required. Unused peroxides should not be returned to the container.
2. All spills should be cleaned up immediately. Solutions of peroxides can be absorbed on vermiculite.
3. The sensitivity of most peroxides to shock and heat can be reduced by dilution with inert solvents, such as aliphatic hydrocarbons. However, toluene is known to induce the decomposition of diacyl peroxides.
4. Solutions of peroxides in volatile solvents should not be used under conditions in which the solvent might be vaporized, because this will increase the peroxide concentration in the solution.
5. Metal spatulas should not be used to handle peroxides, because contamination by metals can lead to explosive decomposition. Ceramic or wooden spatulas may be used.
6. Smoking, open flames, and other sources of heat should not be permitted near peroxides.
7. Friction, grinding, and all forms of impact should be avoided near peroxides (especially solid ones). Glass containers that have screw-cap lids or glass stoppers should not be used. Polyethylene bottles that have screw-cap lids may be used.
8. To minimize the rate of decomposition, peroxides should be stored at the lowest possible temperature consistent with their solubility or freezing point. Liquid or solutions of peroxides should not be stored at or lower than the temperature at which the peroxide freezes or precipitates, because peroxides in these forms are extremely sensitive to shock and heat.

Peroxide-Forming Compounds. Some organic compounds form dangerous hydroperoxides and polymeric peroxides when allowed to stand in air and light, the reaction being further catalyzed by impurities such as acetaldehyde. The reaction is a peroxidation of a carbon atom adjacent to an oxygen atom.

Types of compounds known to form peroxides include the following:

1. Aldehydes.
2. Ethers, especially cyclic ethers and those containing primary and secondary alcohol groups.
3. Compounds containing benzylic hydrogen atoms. Such compounds are especially susceptible to peroxide formation if the hydrogens are on tertiary carbon atoms [e.g., cumene (isopropyl benzene)].
4. Compounds containing the allylic ($CH_2=CHCH_2R$) structure, including most alkenes.
5. Vinyl and vinylidene compounds (e.g., vinyl acetate and vinylidene chloride).

Specific chemicals that can form dangerous concentrations of peroxides on long exposure to air are the following:

Cellosolves
 ethyleneglycol monomethyl (or ethyl) ether
 ethyleneglycol dimethyl (or ethyl) ether

Cyclohexene

Cyclooctene

Decalin (decahydronaphthalene)

Dibutyl ether

Diethyl ether

Diisopropyl ether

p-Dioxane

Tetrahydrofuran (THF)

Tetralin (tetrahydronaphthalene)

Some of these can be purchased with an antioxidant stabilizer which retards the rate of peroxide formation. These inhibited reagents are recommended if the antioxidant does not interfere with the reagent's use. For example, the butylated hydroxytoluene (BHT) used to inhibit THF is not suitable for liquid chromatography when sample detection is performed in the low ultraviolet, where BHT absorbs.

Ethyl ether tends to form explosive peroxides which may produce unexpected explosions when it is distilled, evaporated, or, in extreme cases, merely heated. The formation of the peroxides is accelerated by sunlight. Stocks of ether should therefore be kept in brown glass bottles or otherwise protected from sunlight, and should be regarded with suspicion if they have been in stock a long time. Explosions have occurred when caps or stoppers were turned.

Diisopropyl ether is extremely hazardous when stored and will form peroxides even in the dark and in the presence of inhibitors. The presence of two tertiary carbon atoms renders it readily peroxidizable to a dihydroperoxide, which then polymerizes to a polymeric peroxide.

Ethers must never be distilled unless they are known to be free of peroxides. Explosions have occurred during heating or refluxing because of the presence of peroxides. Containers of diethyl or diisopropyl ether should be dated when they are opened. If they are still in the laboratory after 1 month, the ether should be tested for peroxides before use.

The National Safety Council (Data Sheet 655) recommends a maximum concentration of 80 ppm. If peroxides are determined to be above this level, the material should be decontaminated or destroyed. As a general rule, a material suspected of containing peroxides should be disposed of. Store peroxide formers away from heat and light in closed vessels, preferably in the container furnished by the supplier.

Testing for and Removing Peroxides from Organic Solvents. Peroxide-containing solvents ordinarily are not dangerous unless they have spontaneously evaporated or have been distilled or evaporated, thereby concentrating the peroxides in the residue to dangerous levels. Ethers which are to be used for extractions and subsequently evaporated to dryness should therefore always be tested for peroxides and the peroxides, if found, destroyed before use.

Dipstick-type test kits are available commercially for colorimetric determination of peroxide levels in solvents. Other acceptable colorimetric tests for peroxides in ether (e.g., Jorissen reagent) are available.

A solution of potassium iodide in glacial acetic acid (0.1 g in 1 mL) will give a brown or yellow color due to iodine in the presence of hydroperoxides. Prepare the solution at the time of the test and add an equal volume of the sample to be tested. For water-miscible ethers (dioxane, tetrahydrofuran, cellosolves) a water-moistened KI-starch paper will be given the characteristic violet color when a drop of a peroxide-containing ether is tested. These tests, however, do not reveal the presence of polymeric peroxides, which are best tested for with titanium sulfate in 50% sulfuric acid. The strong acid decomposes the polymeric peroxides to hydrogen peroxide, which in turn gives the yellow or orange $TiO_2(SO_4)_2$. (The reagent is prepared by dissolving a small amount of TiO_2 in hot concentrated H_2SO_4 and adding this to an equal volume of water.)

If a test for peroxides in ether is positive, the contaminated liquid can be filtered through a column of basic-grade chromatographic aluminum oxide until the test is negative. In this procedure the peroxides are not destroyed but are concentrated and absorbed by the alumina, a fact that should be remembered when using ether for column chromatography. The contaminated alumina should be handled as a flammable solid. The peroxides in the alumina can then be neutralized by use of the acidic ferrous sulfate solution described below. The National Safety Council recommends, in Data Sheet 655, that peroxides can be removed from water-insoluble solvents by mixing with the following solution: 60 g of ferrous sulfate, 6 mL of concentrated sulfuric acid, and 110 mL of water. Remember that water is added by this method.

Disposal of Peroxides. Pure peroxides should never be disposed of directly. They must be diluted before disposal.

Small quantities (25 g or less) of peroxides are generally disposed of by dilution with water to a concentration of 2% or less and then transfer of the solution to a polyethylene bottle containing an aqueous solution of a reducing agent, such as ferrous sulfate or sodium bisulfite. The material can then be handled like any other hazardous waste chemical; however, it must not be mixed with other chemicals for disposal. Spilled peroxides should be absorbed on vermiculite as quickly as possible. The vermiculite-peroxide mixture can be burned directly or may be stirred with a suitable solvent to form a slurry that can be treated with an acidic ferrous sulfate solution as described above. Organic peroxides should never be flushed down the drain.

Large quantities (more than 25 g) of peroxides require special handling. Each case should be considered separately, and handling, storage, and disposal procedures should be determined by the physical and chemical properties of the particular peroxide.

Per Salts. Per salts have similar risks to those of peroxides, the perborates being probably the least dangerous. Persulfates, when moist, are particularly reactive with metals and may produce ignition when mixed with them. Ammonium perchlorate, perbromate, and periodate are explosive, and the periodate is very sensitive to friction.

Oxidizing Agents. In addition to their corrosive properties, powerful oxidizing agents such as perchloric and chromic acids (sometimes used as a glassware-cleaning solution) present fire and explosion hazards on contact with organic compounds and other oxidizable substances. The hazards associated with the use of perchloric acid are especially severe; it should be handled only after thorough familiarization with recommended procedures. Strong oxidizing agents should be stored and used in glass or other inert containers (preferably unbreakable), and corks or rubber stoppers should not be used. Reaction vessels containing significant quantities of these reagents should be heated by using fiberglass mantles or sand baths rather than oil baths.

Other strong oxidizing agents include nitric acid, chlorates, permanganates, peroxides, and persulfates.

Detonators. Detonators are a subset of explosives which are extremely shock-sensitive, decompose with great bricance (shattering power), and do not need oxygen from the air to sustain an ignition, since oxygen is provided as an inherent part of the molecule. This is in contrast to explosives such as TNT, which is relatively shock-insensitive (requiring a primary detonator and booster to set it off) and explosive vapors which require admixture with oxygen of the air and a spark to set them off.

Acetyl nitrate is a powerful nitrating agent which is formed when nitric acid is combined with acetic anhydride. It is known to detonate spontaneously or on slight provocation. For this reason, nitric acid must *never* be stored next to acetic anhydride (or acetic acid), since accidental breakage of the containers will unite the two.

Ammoniacal silver nitrate ($AgNO_3$ + NH_4OH) solutions, when allowed to stand, invariably form silver fulminate, a shock-sensitive explosive. When being used as a diagnostic agent or in a staining procedure, this solution should always be made up immediately before use and discarded immediately thereafter. The correct mode of addition of reagents is critical: Silver nitrate should be dissolved in ammonium hydroxide and then the sodium hydroxide added; ammonium hydroxide must never be added to dissolve precipitated silver hydroxide (from $AgNO_3$ and $NaOH$).

Precipitation of AgCl in ethanolic solutions using ethanolic silver nitrate has also led to the formation of silver fulminate. For this reason the precipitates should be discarded as soon as the test result is known.

Perchloric acid (60–70% aqueous $HClO_4$) is not treated with the respect it is due, particularly in biochemical laboratories, where it is used routinely for digestions. Before any method involving its use is undertaken, the literature should be consulted to ensure that a safe technique is used. A perchloric acid disaster caused the death of 17 people and the destruction of 16 buildings in Los Angeles in 1947.

Perchloric acid in contact with easily oxidizable organic material is sensitive to heat and impact. It must be stored under fireproof conditions away from oxidizable material, including wood cabinets and shelves and all organic chemicals. When handling perchloric acid, rubber gloves and a face shield or goggles should always be worn, and any spillage should be cleaned up immediately (use a dripping, water-soaked sponge, which must then be thoroughly rinsed out with water). Beaker tongs, rather than rubber gloves, should be used when handling fuming $HClO_4$. Discolored perchloric acid (which means it is contaminated) must *not* be used but should be disposed of by the Safety Officer. Seventy percent $HClO_4$ can be boiled safely at approximately 200°C, but contact of the boiling undiluted acid or the hot vapor with organic matter, or even easily oxidized inorganic matter (such as compounds of trivalent antimony), will lead to serious explosions.

Perchloric acid is volatile, and when large amounts of digestions are carried out, the vapors should be aspirated into a glass bottle containing a large amount of cold water and must not be allowed to go up into the fume hood ducts, where the anhydrous acid will condense to create a future explosion hazard. Perchloric acid fumes, reacting with litharge-glycerine cements in ventilating systems, have caused violent explosions during maintenance of the ventilating system components. Frequent (weekly) washing out of the hood and ventilator ducts with water is necessary to avoid danger of spontaneous combustion or explosion if this acid is in common use and the fumes cannot be collected as described above.

Perchloric acid should be disposed of by slow mixing into cold water until the concentration is less than 5%, and then neutralizing with aqueous sodium hydroxide. The neutral solution may then be washed down a laboratory sink drain with a large excess of water. This procedure is described in detail in *Prudent Practices for Disposal of Chemicals from Laboratories* (1983, pp. 83–84).

Perchlorates should not be used unless absolutely necessary. They should never be allowed to contact acids, particularly the organic acids, since perchloric acid will then be formed in nearly anhydrous form, which is even more hazardous than the aqueous solution of the acid. Perchlorates {e.g., magnesium perchlorate [$Mg(ClO_4)_2$], marketed as Anhydrone} should not be used as drying agents if there is a possibility of contact with organic compounds or a strong dehydrating acid (e.g., in a drying train that has a bubble counter containing sulfuric acid). Safer drying agents should be used.

Picric acid (2,4,6-trinitrophenol), either in powder form or in solution, should not be stored in ground-glass-stoppered reagent bottles, since some material invariably becomes lodged between the ground surfaces, and the grinding action of removing the stopper can set off a detonation. Picric acid is in the same class as TNT as an explosive and is more sensitive to compression.

Other Potentially Explosive Chemicals. Following is a list of a number of chemicals which can be explosive, or which, while not themselves explosive, may produce explosions or violent reactions under certain conditions. This should not be regarded as a complete list; experience is constantly adding to the list of chemicals which, either by themselves or in mixtures, may present unexpected risks.

Acetylene forms highly dangerous explosive compounds with copper and silver, and should not be brought into contact with these metals or their salts. Acetylene cylinders or bottles must always be stored in an upright position. At pressures of 2 atm or more, acetylene (C_2H_2) subjected to an electrical discharge or high temperature decomposes with explosive violence.

Acetylenic compounds are explosive in mixtures of 2.5–80% with air. Dry acetylides detonate upon receiving the slightest shock.

Ammonia (NH_3) reacts with iodine to give nitrogen triiodide, which is explosive, and with hypochlorites to give chlorine. Mixtures of NH_3 and organic halides sometimes react violently when heated under pressure.

Dry benzoyl peroxide ($C_6H_5CO_2)_2$ is easily ignited and sensitive to shock. It decomposes spontaneously at temperatures above 50°C. It is reported to be desensitized by addition of 20% water.

Chlorine (Cl_2) may react violently with hydrogen (H_2) or with hydrocarbons when exposed to sunlight.

Chromium trioxide-pyridine complex ($CrO_3 \cdot C_5H_5N$) may explode if the CrO_3 concentration is too high. The complex should be prepared by addition of CrO_3 to excess C_5H_5N.

Diazomethane (CH_2N_2) and related compounds should be treated with extreme caution. They are very toxic, and the pure gases and liquids explode readily. Solutions in ether are safer from this standpoint.

Dimethyl sulfoxide [$(CH_3)_2SO$] decomposes violently on contact with a wide variety of active halogen compounds. Explosions from contact with active metal hydrides have been reported.

Dry ice should not be kept in a container that is not designed to withstand pressure. Containers of other substances stored over dry ice for extended periods generally absorb carbon dioxide (CO_2) unless they have been sealed with care. When such containers are removed from storage and allowed to come rapidly to room temperature, the CO_2 may develop sufficient pressure to burst the container with explosive violence. Upon removal of such containers from storage, the stopper should be loosened or the container itself should be wrapped in towels and kept behind a shield. Dry ice can produce serious burns (this is also true for all types of cooling baths).

Ethylene oxide (C_2H_4O) has been known to explode when heated in a closed vessel. Experiments using ethylene oxide under pressure should be carried out behind suitable barricades.

Finely divided materials. Many metals which are safe in massive form are dangerously reactive when in a finely divided state, so that they constitute a fire and explosion hazard. Zirconium, uranium, and to a lesser extent aluminum and magnesium, together with a few of their compounds, fall into this category.

In general, the finely divided powders (i.e., those with a particle size of less than 100 µm) present the greatest hazard. They sometimes ignite spontaneously or even explode violently when merely exposed to air. It must never be forgotten that this is normal behavior for fine uranium, uranium hydride, uranium carbide, and zirconium powders. The only safe way to handle powders of <100 µm in size is in a glovebox filled with an inert gas. Machine-produced dust is unlikely to be an explosion hazard, but fine dust can easily be ignited.

The following conditions encourage fire or explosion:

1. The presence of oxygen-containing materials, even water. The danger of self-ignition with zirconium and magnesium is enhanced by the presence of moisture.

2. Impact or friction. There is considerable danger in storing pyrophoric metal powders in breakable containers. Serious explosions have resulted from dropping glass bottles containing such material. Mechanical process involving small particles may be dangerous (e.g., grinding of metal powders). Also, the sparks produced in machining, etc., could ignite the fine dust.

3. Heat.

Halogenated compounds. Chloroform ($CHCl_3$), carbon tetrachloride (CCl_4), and other halogenated solvents should not be dried with sodium, potassium, or other active metals; violent explosions are usually the result of such attempts. Many halogenated compounds are toxic.

Liquid nitrogen-cooled traps open to the atmosphere rapidly condense liquid air. Then, when the coolant is removed, an explosive pressure buildup occurs, usually with enough force to shatter glass equipment. Hence, only sealed or evacuated equipment should be so cooled.

Lithium aluminum hydride (LiAlH$_4$) reacts with CO$_2$ to form explosive products. It should not be used to dry methyl ethers or tetrahydrofuran; fires from this are very common. Carbon dioxide or bicarbonate extinguishers should not be used against LiAlH$_4$ fires, which should be smothered with sand or some other inert substance.

Inorganic nitrates give inflammable mixtures with carbonaceous matter and other easily oxidizable substances, but the mixtures are normally much less sensitive than those formed by chlorates.

Nitric acid-based solutions. Alcohol-nitric acid can be extremely dangerous. Methanol should always be used in place of ethanol, since nitric acid-based solutions of ethyl alcohol are very unstable and may decompose violently.

Particular care should be taken during the preparation of these solutions, as an exothermic reaction occurs upon mixing which can raise the temperature dangerously high. Mix the reagents slowly.

The same general precautions should be taken with such solutions as with perchloric acid-based solutions: Wear gloves and face shields, keep the solutions cool, and use the smallest possible quantities consistent with performing the job satisfactorily.

Nitrogen trichloride, which is an extremely sensitive explosive, may be formed by the action of chlorine on ammonia or ammonium salts in solution.

Oxygen tanks. Serious explosions have resulted from contact between oil and high-pressure oxygen. Oil should not be used on connections to an O$_2$ cylinder. Liquid oxygen can cause an explosion in contact with easily oxidizable substances.

Ozone (O$_3$) is a highly reactive oxidizer and toxic gas. It should be kept away from all oxidizable substances.

Palladium or platinum on carbon, platinum oxide, Raney nickel, and other catalysts may cause an explosion if additional catalyst is added to a flask in which hydrogen is present. Any recovered catalyst is usually saturated with hydrogen and highly reactive and thus will inflame spontaneously upon exposure to air. Filter cake should not be allowed to become dry, particularly in large-scale reactions. The funnel containing the still-moist catalyst filter cake should be put into a water bath immediately after completion of the filtration.

Perchloric acid electropolishing solutions. Mixtures of perchloric acid, acetic anhydride (or acetic acid), and water can be explosive or spontaneously flammable in certain composition ranges. Generally, typical electropolishing solutions based on these reagents are spontaneously flammable. Should the solution catch fire, the rate of burning increases as it warms up, and for large volumes of liquid can terminate in an explosion. Never use perchloric acid solutions to electropolish alloys containing bismuth.

In the case of mounted specimens, or specimens with an organic stopping-off compound over parts of them, be careful in the choice of organic medium. Polyethylene- and polystyrene-based plastics or rubber-based materials are much safer than Bakelite, Lucite, or cellulose-based materials. Keep the electropolishing bath as cool as possible during use, particularly during the initial mixing. The temperature should not be allowed to exceed 35°C. Heat is generated during the electropolishing, and cooling and stirring may be necessary to keep the temperature down. Use as small a volume of solution as possible. Always use gloves and a face shield when working with these solutions.

Permanganates are explosive when treated with sulfuric acid. When both compounds are used in an absorption train, an empty trap should be placed between them.

Phosphorus (P) (red and white) forms explosive mixtures with oxidizing agents. White P should be stored under water because it is spontaneously flammable in air. The reaction of P with aqueous hydroxides gives phosphine, which may ignite spontaneously in air or explode.

Phosphorus trichloride (PCl$_3$) reacts with water to form phosphorous acid, which decomposes upon heating to form phosphine, which may ignite spontaneously or explode. Care should be taken in opening containers of PCl$_3$, and samples that have been exposed to moisture should not be heated without adequate shielding to protect the operator.

Potassium (K) is in general more reactive than sodium. The reaction with water produces hydrogen. It ignites quickly upon exposure to humid air and should therefore be handled under the surface of a hydrocarbon solvent such as mineral oil or toluene. Neither carbon dioxide nor bicarbonate nor carbon tetrachloride fire extinguishers should be used on alkali metal fires.

Residues from vacuum distillations (e.g., ethyl palmitate) have been known to explode when the still was vented to the air before the residue was cool. Such explosions can be avoided by venting the still pot with nitrogen, by cooling it before venting, or by restoring the pressure slowly.

Silver salts, when mixed with ammonia—as, for example, in the process for silvering mirrors or as used for silvering specimens prior to electrolytic nickel plating—may produce an explosive compound, silver acetylide. These processes must be carried out with great care, using an approved formula. Solutions must not be kept after use. Pour the solution into a large tank of water and neutralize with an acid.

Sodium (Na) should be stored in a closed container under kerosene, toluene, or mineral oil. Scraps of Na or K should be destroyed by reaction with *n*-butyl alcohol. Contact with water should be avoided, because Na reacts violently with water to form H_2, with evolution of sufficient heat to cause ignition.

Trichloroethylene (Cl_2CCHCl) reacts under a variety of conditions with potassium hydroxide or sodium hydroxide to form dichloroacetylene, which ignites spontaneously in air and detonates readily even at dry-ice temperatures. The compound itself is highly toxic, and suitable precautions should be taken when it is used as a degreasing solvent.

Handling Explosive Materials. Explosive chemicals decompose under conditions of mechanical shock, elevated temperature, or chemical action with forces that release large volumes of gases, heat, toxic vapors, or combinations thereof. Various state and federal regulations cover the transportation, storage, and use of explosives. These regulations should be consulted before explosives and related dangerous materials are used in the laboratory.

Explosive materials should be brought into the laboratory only as required and then in the smallest quantities adequate for the experiment being conducted. Explosives should be segregated from other materials that could create a serious hazard to life or property should an accident occur.

The handling of highly energetic substances without injury demands attention to the most minute detail. The unusual nature of work involving such substances requires special safety measures and handling techniques that must be thoroughly understood and followed by all persons involved. The practices listed below are a guide for use in any laboratory operation that might involve explosive materials.

Safety measures for unstable materials are very similar to those used in handling reactive materials or hazardous reactions. A basic safeguard is to limit quantities of materials so that if a material is initiated, it will not involve a substantial amount of energy release. Another safeguard is, of course, to understand the hazards of the materials, to find out if the materials being used have any stability problems, and, if so, how sensitive they are and what measures need to be taken to prevent their being initiated. Some are shock-sensitive, some are heat-sensitive, some are initiated only by very substantial mechanical or thermal action. If materials are unstable, then minimal protection such as benchtop shields must be used even for handling small quantities. If extremely sensitive materials are used or if quantities must be increased, a suitably engineered barricading must be provided to handle these materials. Good eye, face, hand, and body protection are required (see *Fundamentals of Laboratory Safety: Physical Hazards in the Academic Laboratory*, Chap. 10). Again, the steps of preplanning, procedure development, and emergency preparedness should not be overlooked.

Personal Protective Apparel. Personal protective apparel includes the following:

1. Safety glasses that have a cup-type side shield made of a light, clear plastic material affixed to the frame should be worn by all laboratory personnel.

2. A face shield that has a "snap-on" throat protector in place should be worn at all times when the worker is in a hazardous, exposed position (e.g., when operating or manipulating synthesis systems, when bench shields are moved aside, or when handling or transporting such products).

3. Gloves should be worn whenever it is necessary to reach behind a shielded area while a hazardous experiment is in progress or when handling adducts or gaseous reactants. [Heavy-duty electrical linesman's gloves afford good protection against 2-g-quantity detonations in glass provided the detonation is 3 in. (7.5 cm) away; however, such a detonation in contact with a gloved hand would cause severe injury and probable loss of fingers.] Specially designed protective gloves (Kevlar or woven stainless steel) should be utilized for some manipulations. These gloves can better protect the hands from some of the force of an explosion should one occur.

4. Laboratory coats should be worn at all times while in explosives laboratories. They should be of a slow-burning material and fitted with quick-release cloth buttons. (These coats help reduce minor injuries from flying glass and reduce the possibility of injury from an explosive flash.)

Protective Devices. Barriers such as shields, barricades, and guards should be used to protect personnel and equipment from injury or damage. The barrier should completely surround the hazardous area. On benches and in hoods, a 0.25-in.-thick acrylic sliding shield effectively protects a worker from glass fragments resulting from a laboratory-scale detonation. The shield should be closed whenever hazardous reactions are in progress or whenever hazardous materials are being temporarily stored. Such shielding is not effective against metal shrapnel.

Dry boxes should be fitted with safety glass windows overlaid with 0.25-in.-thick acrylic. This protection is adequate against an internal 5-g-quantity detonation. The problem of hand protection, however, still remains, although electric linesman's gloves over the rubber dry box gloves offer some additional protection. Other safety devices, as required, should be used in conjunction with the gloves (e.g., tongs and labjack turners).

Armored hoods or barricades made with extra-thick (1-in.) polyvinyl butyral resin shielding and heavy metal walls give complete protection against detonations not in excess of the acceptable 20-g limit. These hoods are designed for use with 100 g of material, but an arbitrary 20-g limit is usually set because of the noise level in the event of a detonation. Such hoods should be equipped with mechanical hands that enable the operator to manipulate equipment and handle adduct containers remotely. A sign, such as the following, should be posted:

CAUTION: NO ONE MAY ENTER AN ARMORED HOOD FOR ANY REASON DURING THE COURSE OF A HAZARDOUS OPERATION.

Miscellaneous protective devices such as both long- and short-handled tongs for holding or manipulating hazardous items at a safe distance, and remote-control equipment (e.g., mechanical arms, stopcock turners, labjack turners, and remote cable controllers) should be available as required to prevent exposure of any part of the body to injury.

Reaction Quantities. In conventional explosives laboratories, no more than 0.5 g of product should be prepared in a single run. During the actual reaction period, no more than 2 g of reactants should be present in the reaction vessel. This means that the diluent, the substrate, and the energetic reactant must all be considered when determining the total explosive power of the reaction mixture. Special reviews should be established to examine operational and safety problems involved in scaling up a reaction in which an explosive substance is used.

Incompatible Chemicals

When transporting, storing, using, or disposing of any substance, utmost care must be exercised to ensure that the substance cannot accidentally come in contact with another with which it is incompatible. Such contact could result in a serious explosion or the formation of substances that are highly toxic or flammable or both.

Table 1-2 is a guide to avoiding accidents involving incompatible substances.

Corrosives—Including Acids and Alkalies

Corrosivity, for the purpose of this discussion, means corrosion to metals and other materials. The term *corrosive* is applied not only to metal corrosion but also to the severe effects on the skin which one might experience from a drop of strong acid. The corrosive effects of chemicals on body tissues are discussed in Chapter 2. Some day we may have a better term for destruction of living tissue which will enable us to discriminate between these two meanings. Corrosive materials are typified by acids, alkalis, salts, and many of the gases (some of which fall in the acid or alkali category). Some of the more corrosive acids are sulfuric, nitric, hydrochloric, and hydrofluoric acids. Corrosive alkalis include sodium hydroxide, soda ash, ammonia, etc. An infinite array of salts are corrosive, the most common of which is sodium chloride. The corrosive effects of seawater are well known, and the prevention of such corrosion is the concern in many marine operations.

Conditions affect the rates of corrosion; for example, an elevated temperature would probably increase the rate of corrosion considerably. Concentration of the corrosive material is of great importance. For example, concentrated sulfuric acid, 93% or 98%, is relatively noncorrosive and can be shipped in heavy steel tanks or drums. However, the addition of enough water to reduce the concentration to about 65% will create a much more severe corrosive condition. Sulfuric acid concentrations of less than 65% or so are very corrosive to many ordinary metals. The presence of air, water, or other impurities also has effects.

There are several recognized patterns of corrosion. *Intergranular corrosion* occurs when the corrosive material actually penetrates the grain structure of the metal and removes materials from between the grains, thus weakening its structure and causing failure. A more obvious type of corrosion is *pitting,* in which the surface is actually destroyed. The material is dissolved from either a limited or a widespread area, and the damage can be easily seen upon inspection. A third type of effect is called *stress corrosion.* This type of corrosion takes place particularly when metals are under physical stress, such as would be induced by internal pressure or tension.

Detection or identification of corrosive effects is sometimes possible by visual means. This involves a knowledgeable person inspecting the surfaces carefully to determine whether there is any evidence of cracks, pits, or general roughness from corrosive effects. Ultrasonic instruments have been successfully applied to this type of work; however, absence of any irregularities that would be responsive to ultrasonic testing does not necessarily mean that the vessel is free

TABLE 1-2. Some Incompatible Chemicals

Chemical	Is Incompatible with
Acetic acid	Chromic acid, nitric acid, hydroxyl compounds, ethylene glycol, perchloric acid, peroxides, permanganates
Acetone	Concentrated nitric and sulfuric acid mixtures
Acetylene	Chlorine, bromine, copper, fluorine, silver, mercury
Alkali and alkaline earth metals (such as powdered aluminum or magnesium, calcium, lithium, sodium, potassium)	Water, carbon tetrachloride or other chlorinated hydrocarbons, carbon dioxide, halogens
Ammonia (anhydrous)	Mercury (in manometers, for example), chlorine, calcium hypochlorite, iodine, bromine, hydrofluoric acid (anhydrous)
Ammonium nitrate	Acids, powdered metals, flammable liquids, chlorates, nitrites, sulfur, finely divided organic or combustible materials
Aniline	Nitric acid, hydrogen peroxide
Arsenical materials	Any reducing agent
Azides	Acids
Bromine	See Chlorine
Calcium oxide	Water
Carbon (activated)	Calcium hypochlorite, all oxidizing agents
Carbon tetrachloride	Sodium
Chlorates	Ammonium salts, acids, powdered metals, sulfur, finely divided organic or combustible materials
Chlorine	Ammonia, acetylene, butadiene, butane, methane, propane (or other petroleum gases), hydrogen, sodium carbide, benzene, finely divided metals, turpentine
Chlorine dioxide	Ammonia, methane, phosphine, hydrogen sulfide
Chromic acid and chromium trioxide	Acetic acid, naphthalene, camphor, glycerol, alcohol, flammable liquids in general
Copper	Acetylene, hydrogen peroxide
Cumene hydroperoxide	Acids (organic or inorganic)
Cyanides	Acids
Flammable liquids	Ammonium nitrate, chromic acid, hydrogen peroxide, nitric acid, sodium peroxide, halogens
Fluorine	Everything
Hydrocarbons (such as butane, propane, benzene)	Fluorine, chlorine, bromine, chromic acid, sodium peroxide
Hydrocyanic acid	Nitric acid, alkali
Hydrofluoric acid (anhydrous)	Ammonia (aqueous or anhydrous)
Hydrogen peroxide	Copper, chromium, iron, most metals or their salts, alcohols, acetone, organic materials, aniline, nitromethane, combustible materials
Hydrogen sulfide	Fuming nitric acid, oxidizing gases
Hypochlorites	Acids, activated carbon
Iodine	Acetylene, ammonia (aqueous or anhydrous), hydrogen
Mercury	Acetylene, fulminic acid, ammonia
Nitrates	Sulfuric acid
Nitric acid (concentrated)	Acetic acid, aniline, chromic acid, hydrocyanic acid, hydrogen sulfide, flammable liquids, flammable gases, copper, brass, any heavy metals
Nitrites	Acids

TABLE 1-2. Some Incompatible Chemicals—*continued*

Chemical	Is Incompatible with
Nitroparaffins	Inorganic bases, amines
Oxalic acid	Silver, mercury
Oxygen	Oils, grease, hydrogen, flammable liquids, solids, or gases
Perchloric acid	Acetic anhydride, bismuth and its alloys, alcohol, paper, wood, grease, oils
Peroxides, organic	Acids (organic or mineral), avoid friction, store cold
Phosphorus (white)	Air, oxygen, alkalis, reducing agents
Potassium	Carbon tetrachloride, carbon dioxide, water
Potassium chlorate	Sulfuric and other acids
Potassium perchlorate (see also Chlorates)	Sulfuric and other acids
Potassium permanganate	Glycerol, ethylene glycol, benzaldehyde, sulfuric acid
Selenides	Reducing agents
Silver	Acetylene, oxalic acid, tartartic acid, ammonium compounds, fulminic acid
Sodium	Carbon tetrachloride, carbon dioxide, water
Sodium nitrite	Ammonium nitrate and other ammonium salts
Sodium peroxide	Ethyl or methyl alcohol, glacial acetic acid, acetic anhydride, benzaldehyde, carbon disulfide, glycerin, ethylene glycol, ethyl acetate, methyl acetate, furfural
Sulfides	Acids
Sulfuric acid	Potassium chlorate, potassium perchlorate, potassium permanganate (similar compounds of light metals, such as sodium, lithium)
Tellurides	Reducing agents

Source: NIOSH, *Safety in the School Science Laboratory, Instructors Resource Guide*, 1979, p. 6-30.

from corrosive effects. The ultrasonic instrument is basically a location device. Similarly, radiography may identify particular problems, but again, the freedom of any obvious imperfections on radiographic plates does not necessarily mean that the vessel has no defects.

Corrosion can be prevented by a number of means, one of the most important being selection of the proper material. This should be done by someone knowledgeable not only about the properties of materials but also about the system in which the materials are used and possible corrosive effects on various materials under the planned conditions of operation.

Plastic, rubber, and glass coatings are sometimes used to reduce or eliminate corrosion. Glass-lined equipment is commonly used to protect against chemical attack; as long as the coating is perfect, there is no corrosion of the construction metal. However, if there is any penetration of the coating, corrosive action will start under the coating at the point of the defect. When a vessel is built of metal-clad, the interior layer may be a modest thickness of a corrosion-resistant material, while the actual strength of the vessel is contributed by ordinary steel which comprises most of the thickness of the construction plate.

Inhibitors are sometimes used to reduce the corrosive character of materials; for example, when steel is pickled in muriatic acid, an inhibitor is added to control attack on the base metal and still permit the removal of scale and undesired materials by the pickling agent.

Exothermic and Other Reactions

When the term *routine chemical reaction* is used, it usually implies a safe reaction—safe because the reaction rate is relatively slow or can be easily controlled. Exothermic reactions involving highly reactive chemicals can differ from routine mainly in the rate at which they progress. Reaction rates almost always increase rapidly as the temperature increases. Typically, the reaction rate doubles for approximately every 10°C increase in temperature. The curve in Fig. 1-1 showing a normal reaction is typical of the exotherm potential. It

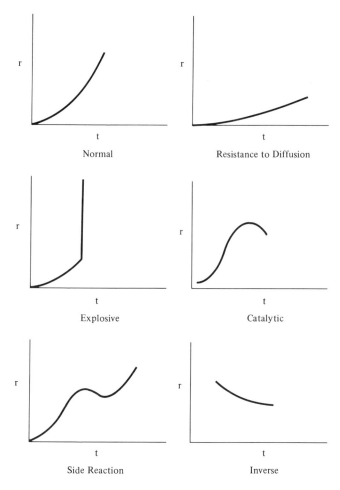

FIGURE 1-1. Characteristics of reactions. r = Rate and t = temperature. (From NIOSH, *Safety in the School Science Laboratory, Instructors Resource Guide,* 1979, p. II-33.)

should be obvious that if one is operating on the part of the curve with moderate slope that a substantial increase in temperature would not lead to a dangerous condition, since the rate of reaction would go up only moderately. On the other hand, if the reaction is being carried out on the part of the curve where the slope is steep, a modest increase in temperature would cause a very great increase in reaction rate, and could conceivably develop pressures that would have explosive consequences. In this case, if the heat evolved in a reaction is not dissipated, the reaction rate can increase until an explosion results. This factor must be considered, particularly when scaling up experiments, so that sufficient cooling and surface for heat exchange can be provided.

Some typical exothermic reactions are combustion, oxidation, nitration, halogenation, polymerization, and hydrogenation. Combustion, if it is complete, is relatively easy to control. Furnaces and other heating equipment, on the other hand, must have control devices in order to protect against explosion due to the presence of unburned fuel after a flame-out. Oxidation is always an exothermic operation and is very difficult to control if both reactants are present in substantial quantities. Usually one component is added gradually so that the rate of addition acts as a control on the reaction rate.

Nitration is also an exothermic procedure, which has the added problem of generating a product which is itself a hazardous material.

Halogenation is actually an oxidation utilizing one of the halogens as the oxidizing agent. Some vigorous halogenations are well known, for example, the explosion produced by exposing a mixture of chlorine and ethylene to sunlight.

Hydrogenation is mildly exothermic and rather easily controlled; a drop in pressure or a decrease in hydrogen content in the gas phase will effectively slow down or stop a hydrogenation reaction. Hydrogen, on the other hand, is frequently used under high-pressure conditions, and a release of this gas can be very destructive if it is ignited.

Catalytic reactions. Two types of catalysts are in common use. Some are unaffected by the reaction or are self-regenerating and thus continue to promote the reaction throughout the operation. Other catalysts are actually used up in the reaction or are killed by by-products such as water which may be formed in the reaction. The first type of catalyst retains its potential for promoting reaction as an exothermic reaction approaches runaway conditions. The catalyst that is destroyed during reaction has little of this potential, since a reaction which causes depletion of the catalyst is self-controlling or self-quenching. An example of this is included in Fig. 1-1 as "Catalytic." Merely handling some catalysts may be hazardous.

Phase separations have many potential problems, one of which is foaming. When gases are evolved from a liquid, there is often a tendency for the liquid to be carried over in the form of a foam. Upon completely filling the vessel, the foam may clog lines, plug relief devices, or foul control sensors. Solids separating from liquid may cause the same problems, and in addition may collect in such a way that mixing is seriously impeded and the reaction does not go as planned. There are cases where unstable materials have separated out as a coating on the sides of the vessel and then decomposed violently when heated by steam in the vessel shell.

Side reactions are so varied that it would be impossible to generalize their effects on the safety of a reaction. Each reaction must be analyzed, and all available information must be supplied to foresee and take care of any side reactions that may develop. Another problem with side reactions is that of disposing of undesired materials that may be formed.

Operations Involving Hazardous Reactions

Controls for potentially hazardous reactions should include process reviews both before and during the course of carrying out experimental work. Process review information is then used in the development of standard operating procedures for the guidance of those performing the reaction. Standard operating procedures are especially useful in training operators and assuring that they will perform all steps exactly as directed. Instrumentation and automation of the operation must be carefully thought out, and the best principles of feedback and fail-safe design must be incorporated. Interlocks may be used on some equipment to prevent incorrect order of addition or incorrect sequence of operating controls. Relief devices must be provided whenever pressure is involved. As mentioned previously, it is often useful to provide dumping or quenching facilities to stop a potential runaway reaction. Hazardous operations are often placed in barricades in order to prevent people from being exposed to the effects of accidental release of energy, and in extreme cases it is customary to limit the number of people that will be associated with the operation or exposed at any given time.

Heating methods must be carefully thought out. The most common are heating tapes and mantles, and sand, water, steam, and silicone oil baths. Hair dryers (heat guns) can be used for certain operations; however, their use should be prohibited when a flammable vapor potential is present. Heater controls can also contribute to the hazard. Sparks from a switch can ignite vapors. All controls for heating and stirring equipment should be operable from outside a shielded area.

When potentially explosive materials are being handled, the area should be posted with a warning sign. For example, in areas where explosive or flammable vapors may accumulate, the following sign may be posted:

WARNING: VACATE THE AREA AT THE FIRST SIGN OR ODOR. STAY OUT UNTIL THE VENTILATION SYSTEM HAS CLEARED THE AIR.

When condensing explosive gases, the temperature of the bath and the effect on the reactant gas of the condensing material selected must be determined experimentally. Very small quantities should be used because detonations may occur. In all cases, a shielded Dewar flask should be used when condensing reactants. Maximum quantity limits should be observed. Heating baths of flammable materials should be prohibited.

When mixing is essential or cooling is required for control of the reaction, one must be concerned with the loss of utilities such as electricity, cooling water, etc. When a reaction is underway, such a loss of utilities can cause a reaction to proceed in an unpredictable and sometimes catastrophic fashion. The same problem must be considered when hoods are used to remove reactive vapors.

Rupture disks or spring-loaded relief valves, commonly installed to protect vessels from overpressure due to reactivity, must be properly designed. If a vessel is designed for 2000 psi and a reaction is being carried out at 200 psi, it might seem logical that a relief device rated at 2000 psi would give adequate protection. However, if the reaction is being carried out on the lower part of the curve and would develop the 2000 psi only under runaway conditions characteristic of the steep part of the curve, the pressures would be building up so fast that the relief device might not be able to handle the release when the relief pressure is obtained. It is therefore highly desirable that any relief device be rated at a pressure not too far above the operating conditions, so that a modest override in temperature and pressure would cause the relief to function and protect the vessel from the escalating increase in reaction rates or conditions. The relief lines connected to the rupture disks and relief valves must also be properly designed. [*Fundamentals of Laboratory Safety: Physical Hazards in the Academic Laboratory,* Chap. 8, contains detailed information on the proper design of pressure systems.]

One of the cheapest and best control measures is an operating procedure that is well thought out and has been drafted by experienced people, taking into account all the foreseeable parameters and variations. Thus, if a well-drafted procedure is carefully followed, it is very likely that the reaction or operation can be carried to a successful conclusion. If departures are made, the consequences are the responsibility of the person who decided to modify the procedure.

SUGGESTED READINGS

National Research Council, *Prudent Practices for Disposal of Chemicals from Laboratories,* National Academy Press, Washington, D.C., 1983.

National Safety Council, Data Sheet 655: Recognition and Handling of Peroxidizable Compounds, Chicago, 1976.

2
TOXIC HAZARDS

INTRODUCTION

People are continually exposed to toxic agents at work, in school, and at play, in the food we eat, the water we drink, the air we breathe, and the things we touch. Toxic materials have many routes of entry into the body. Fortunately, the great majority of these substances are metabolized and cleared from the body without any significant or adverse effect. If the rate of absorption of a toxic substance exceeds its rate of elimination, the accumulation of the substance or its altered form becomes critical, resulting in effects recognized as abnormal or toxic. Reversible and irreversible injury or illness may result. The accumulation may be rapid, in which case the toxic effect is called *acute*. If the accumulation of a toxic material occurs over a long term, the effect is classed as *chronic* toxicity.

Students in the laboratory are usually subject to the potential for acute toxicity. Faculty, however, are more likely to experience chronic toxic effects. Persons in a typical workplace might experience either.

Toxicology can be simply defined as the study of poisons. This overly simple definition can be misleading, however, because most everything can, in some situations, be a poison, particularly if given to a biological organism in a large enough quantity or administered under abnormal circumstances such as during a disease. This concept was recognized by Paracelsus, who, in 1537, wrote "All things are poisons and nothing is without poison; only the dose makes a thing a poison."

Accordingly, even salt and water can be poisons if an organism ingests too much. Likewise, there are safe limits to the use of carcinogens and poisons. For the maintenance of normal biological function it is essential that a well-defined amount of substances be received, an amount called an *optimal dose*. The concept of optimal dosage applies to vitamins, minerals, and nutrients. In the case of vitamins and minerals, if the dose is too small, malnutrition results; if it is too large, toxic manifestations result.

A modern and explicit definition of toxicology may be given as follows: *Toxicology* is an interdisciplinary science dealing predominantly with chemical insults to humans and their dependent environment.

The science of toxicology has evolved into several subdisciplines based on emphasis. These include:

Clinical toxicology
Forensic toxicology
Environmental toxicology
Occupational toxicology
Veterinary toxicology
Analytical toxicology

Of primary interest in this handbook is occupational toxicology, which is defined as the branch of toxicology that determines the limits of safety or chemical agents in the laboratory and workplace.

Many of the chemicals encountered in the laboratory are known to be toxic or corrosive or both. New and untested substances that may be hazardous are also frequently encountered. Thus, it is essential that all laboratory workers understand the types of toxicity, know the routes of exposure, and recognize the major classes of toxic and corrosive chemicals.

BASIC RULES OF TOXICOLOGY

Toxicity has been defined by the Manufacturing Chemists Association as "the capacity of a substance to cause injury or harm to living tissue" (*Guide for Safety in the Chemical Laboratory,* New York, 1972, p. 144). Toxic substances interfere with cell function. Toxicity may be local or systemic. *Local toxicity* affects only the area exposed. *Systemic toxicity* affects the body after absorption into the bloodstream. The systems that may be affected and example chemicals that may cause that effect include:

Respiratory—HCN gas

Nervous system—the nerve gases, insecticides

Gastrointestinal (digestive) tract—cyanides

Kidney and genital urinary tract—mercury

Liver—fatty degeneration from carbon tetrachloride

Circulatory—the cyanogenic aromatic nitro and amino compounds

Muscular—chemicals that poison the oxidative enzyme system (cyanide)

Endocrine system—thyroid inhibitors (thiouracil)

Duration of Exposure

The toxicity of many chemicals depends on the length of time over which exposure has occurred. There are several reasons for this dependency. First, some chemicals are not readily eliminated from the body, so continued exposure to low doses (each too small to produce an effect) may lead to accumulation of the chemical in the body at levels high enough to produce adverse results. For example, cadmium is strongly retained in the body and tends to accumulate in the kidney. When levels become high enough (usually after many years), kidney dysfunction begins to occur.

A second reason why toxic effects may depend on duration of exposure is related to the ability of cells to repair themselves. When an injury to a cell cannot be quickly reversed by repair processes, there is a tendency for the injury to accumulate in the cell as a function of increasing exposure duration. Thus, a dose of a chemical that causes a small, but irreversible, injury may have no immediately apparent effect, but a clear adverse response may develop with continued exposure.

Finally, some adverse effects simply require an extended period of time to develop, even though they may be the result of exposure months or years earlier. Lead exposure, for example, may impair the development of the nervous system in young children, but this effect requires an extended period of exposure and does not become apparent for several years. Similarly, the development of tumors following exposure to a carcinogen may take months or years to occur.

Acute and Chronic Toxicity

The toxicity of a material results from its ability to damage or interfere with the metabolism of living tissue. An acutely toxic substance can cause damage as the result of a single or short-duration exposure. A chronically toxic substance causes damage after repeated or long-duration exposure or becomes evident only after a long latency period. Hydrogen cyanide, hydrogen sulfide, and nitrogen dioxide are examples of acute poisons. Chronic poisons include all carcinogens and many metals and their compounds (such as mercury and lead and their derivatives). Chronic toxins are particularly insidious because of their long latency periods. The cumulative effects of low exposures may not be apparent for many years. Some chemicals—e.g., vinyl chloride—can be either acutely or chronically toxic, depending on the degree of exposure. All new and untested chemicals should be regarded as toxic until proven otherwise.

An example of both acute and chronic effects is exposure to benzene. Inhaling a high concentration of benzene within a short period of time may produce anesthesia as the hydrocarbon enters the blood and begins to affect the nervous system. On the other hand, inhaling a small concentration of benzene over a substantial period of time, six months or so, results in anemia produced by changes in the bone marrow. The marrow, particularly in the long bones of the leg, manufactures the red blood cells which are supplied to the blood to replenish or replace those that are worn out or destroyed. When the bone marrow is affected by benzene, it is no longer able to perform this function. The red blood cells in the blood are not replaced effectively, resulting in anemia. Over an even longer period, possibly 20 years, either acute or chronic exposure to benzene may cause leukemia and lymphomas.

Expressions of Toxicity and Dose Response

The toxicity of chemical agents is often expressed in terms of the effect of the dosage on an animal species. The most significant of these expressions is the LD_{50} value, which is the administered dose at which 50% of all tested animals die. The LD_{50} is usually expressed in terms of milligrams of the chemical per kilogram of body weight.

For example, take a man who weighs 70 kg and sodium cyanide with an LD_{50} of 6 mg/kg. A dose of 420 mg (6 mg $\times$ 70 kg) would probably be fatal. Any substance with an LD_{50} of 50 mg/kg or less is defined as a poison. See Appendix A for further details on a chemical toxicity rating scale.

The LD_{50} is useful for comparing the toxicity of chemicals with each other. The LD_{50} differs with the route of administration and the species of animal used. Caution must be used in attempting to extrapolate the data to humans; it can be only a guideline. The rat is commonly used because of its great metabolic similarity to man.

For toxic materials in the air, an LC_{50} is used. That is, the concentration (C) of material in the air resulting in 50% fatality is reported as parts per million (ppm) for the time period.

The most important dose-response relationship for occupational toxicology is the *threshold limit value,* or TLV. The TLV is the concentration of chemical agent in the air to which it is believed that nearly all workers may be repeatedly exposed, 7–8 hours a day, 40 hours a week, without adverse effect.

In some cases, the Occupational Safety and Health Administration (OSHA) has used the TLV to establish occupational limits. For an OSHA occupational exposure, a time-weighted average (TWA) is determined. The TWA is the sum of the products of the toxicant concentration and the duration of exposure taken over the total exposure time and is represented by the following mathematical expression:

$$TWA = \sum_{i=1}^{i=n} \frac{t_i C_i}{t_{total}}$$

$$= \frac{t_1 C_1 + t_2 C_2 + \cdots + t_n C_n}{t_1 + t_2 + \cdots + t_n}$$

where

t = exposure time in minutes
C = concentration in ppm or mg/m³
n = total number of exposures
i = exposure number

Routes of Entry

For a chemical to exert a toxic effect on an organism, it must first gain access to the cells and tissues of that organism. In humans, the routes by which toxic chemicals enter the body are inhalation, dermal absorption, ingestion, and subcutaneous injection. The absorptive surfaces of the tissues involved in these routes of exposure (lungs, skin, gastrointestinal tract, and vascular system) differ from each other with respect to the rate with which chemicals move across them. A summary of the factors that influence absorption of chemicals through these routes of exposure is presented next. The pathways taken by chemicals in the body after entrance by the primary dermal and respiratory routes are illustrated in Figs. 2–1 and 2–2.

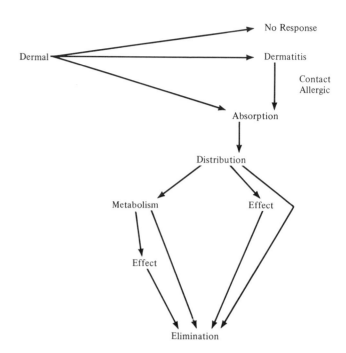

FIGURE 2-1. Chemical pathways via the dermal route. (From NIOSH, *Manual of Safety and Health Hazards in the School Science Laboratory,* 1980, p. A-192.)

Inhalation. Inhalation is the most common route of entry. Inhalation of toxic vapors, mists, gases, or dusts can produce poisoning by absorption through the mucous membranes of the mouth, throat, and lungs and can seriously damage these tissues by local action. Inhaled gases or vapors may pass rapidly into the capillaries of the lungs and be carried into the circulatory system and cause systemic effects within the body. Material that is not absorbed can cause irritation of the lung membrane. Absorption in the lung is usually high, because the surface area is large and blood vessels are close to the exposed surface area. This absorption can be extremely rapid. The rate will vary with the concentration of the toxic substance and its solubility in tissue fluids. If the gas has a low solubility (e.g., ethylene) the rate of absorption is limited by the rate of blood flow through the lung, whereas the absorption of readily soluble gases (e.g., chloroform) is limited only by the rate and depth of respiration. This means that a person who is active will absorb a soluble gas much more rapidly than when he or she is at rest.

Chemicals may also be inhaled in solid or liquid form as dusts or aerosols. Liquid aerosols, if lipid-soluble, will readily cross the cell membrane by passive diffusion. The absorption of solid particulate matter is highly dependent on the size and chemical nature of the particles. The rate of absorption of particulates from the alveoli is determined by the compound's solubility in lung fluids, with poorly soluble compounds being absorbed at a slower rate than readily soluble compounds. Certain small, insoluble particles may remain in the alveoli indefinitely. Larger particles (2–5 μm) are deposited in the tracheobronchiolar regions (air passages) of the lungs, where they are cleared by coughing and sneezing, or they are swallowed and deposited in the gastrointestinal tract. Particles of 5 μm or larger are usually deposited in the nasopharyngeal region (nasal passages above the level of the mouth), where they are subsequently expelled or swallowed.

The degree of injury resulting from exposure to toxic vapors, mists, gases, and dusts depends on the toxicity of the material and its solubility in tissue fluids, as well as on its concentration and the duration of exposure. Chemical activity and the time of response after exposure are not necessarily a measure of the degree of toxicity. Several chemicals (e.g., mercury and its derivatives) and some common solvents (e.g., carbon tetrachloride and benzene) are cumulative poisons that can produce body damage through exposure to small concentrations over a long period of time.

The American Conference of Governmental Industrial Hygienists (ACGIH) produces annual lists of threshold limit values (TLVs) and short-term exposure limits (STELs) for common chemicals used in laboratories. These values are guides, not legal standards. Definitions are given in Appendix B.

Most of the 1968 TLVs were adopted by OSHA in 1972 as legal permissible exposure levels (PELs). The basis for selection of the TLVs appears to be more secure than the justification for the STELs. The TLVs provide a useful estimate of how much ventilation may be needed in laboratories where the occupants typically spend most of their working time.

However, because of the many factors influencing toxicity, each situation should be evaluated individually and the TLVs used as guidelines rather than as fine lines between safe and dangerous concentrations.

The best way to avoid exposure to toxic vapors, mists, gases, and dusts is to prevent the escape of such materials into the working atmosphere and to ensure adequate ventilation by the use of exhaust hoods and other local ventilation (see *Fundamentals of Laboratory Safety: Physical Hazards in the Academic Laboratory,* Chap. 9). Chemicals of unknown toxicity should not be smelled.

Contact with Skin or Eyes. Some materials are absorbed through the skin. Absorption through the epidermal layer of the skin is hindered by the densely packed layer of horny, keratinized epidermal cells. Absorption of chemicals occurs much more readily through scratched or broken skin. There are significant differences in skin structure from one region of the body to another (palms of hands versus facial skin), and these differences further influence dermal absorption.

Absorption of chemicals through the skin is roughly proportional to their lipid solubility, and can be enhanced by mixing the chemical in an oily vehicle and rubbing the resulting preparation into the skin. Some lipid-soluble compounds can be absorbed through the skin in quantities sufficient to produce systemic effects. For example, carbon tetrachloride can be absorbed through the skin in amounts large enough to produce liver injury.

Nitrogen compounds (e.g., aniline, nitrobenzene, and the amines) are also easily absorbed. If material containing these compounds is spilled on the skin, they are absorbed into the body and produce toxic effects on the blood or other organs.

Contact with the skin is a frequent mode of chemical injury. A common result of skin contact is a localized irritation, but an appreciable number of materials are absorbed through the skin with sufficient rapidity to produce systemic poisoning. The main portals of entry for chemicals through the skin are the hair follicles, sebaceous glands, sweat glands, and cuts or abrasions of the outer layers of the skin. The follicles and glands are abundantly supplied with blood vessels, which facilitates the absorption of chemicals into the body.

Contact of chemicals with the eyes is of particular concern because these organs are so sensitive to irri-

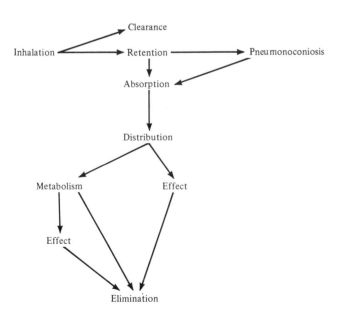

FIGURE 2–2. Chemical pathways via the inhalation route. (From NIOSH, *Manual of Safety and Health Hazards in the School Science Laboratory,* 1980, p. A-10.)

tants. Few substances are innocuous in contact with the eyes; most are painful and irritating, and a considerable number are capable of causing burns and loss of vision. Alkaline materials, phenols, and strong acids are particularly corrosive and can cause permanent loss of vision. Also, eyes are very vascular and provide for rapid absorption of many chemicals.

Skin and eye contact with chemicals should be avoided by use of appropriate protective equipment (See *Fundamentals of Laboratory Safety: Physical Hazards in the Academic Laboratory,* Chap. 10). All persons in the laboratory should wear safety glasses. Face shields, safety goggles, shields, and similar devices provide better protection for the eyes. Protection against skin contact may be obtained by use of gloves, laboratory coats, tongs, and other protective devices. Spills should be cleaned up promptly.

In the event of skin contact, the affected areas should be flushed with water, and medical attention should be sought if symptoms persist; in the event of eye contact, the eye(s) should be flushed with water for 15 min, and medical attention should be sought whether symptoms persist or not.

Another effect that does not involve actual entry into the body but that does produce effects on the body is termed "skin irritation." When some materials come in contact with the skin, an irritation occurs, a reaction develops, and the skin becomes reddened, swollen, or blistered, or may show other effects. Industrial dermatitis is a major problem in the laboratory.

An extension of skin irritation is tissue destruction. This is often termed a corrosive effect. A material such as concentrated sulfuric acid will actually destroy skin and penetrate until it is sufficiently diluted with body fluids that its action is arrested. An alkali such as sodium hydroxide, in contact with tissues, causes saponification and other reactions with the skin that continue to deeper and deeper layers, with dilution producing very little arresting effect. This is particularly serious in the matter of eye contact. A little bit of alkaline material in the eye can cause very severe damage, often including loss of vision.

Ingestion. Ingestion simply means taking materials into the body through the mouth. Many of the chemicals used in the laboratory are extremely dangerous if they enter the mouth and are swallowed. The oral route is significant if people are permitted to eat, drink, smoke, or apply cosmetics in working areas subject to contamination. This is particularly important in the case of laboratories. Laboratory personnel, students, and teachers should never eat, drink, or smoke in the laboratory. Eating without adequate washing can also result in toxic chemical ingestion.

To prevent entry of toxic chemicals into the mouth, laboratory workers should wash their hands before eating, smoking, or applying cosmetics; immediately after use of any toxic substance; and before leaving the laboratory. Food and drink should not be stored or consumed in areas where chemicals are being used, nor should cigarettes, cigars, or pipes be used in such areas. Chemicals should not be tasted; and pipetting and siphoning of liquids should never be done by mouth.

The relative acute toxicity of a chemical can be evaluated by looking up its oral LD_{50} and particular toxic effects. Many chemicals may damage the tissues of the mouth, nose, throat, lungs, and gastrointestinal tract and produce systemic poisoning if absorbed through the tissues.

Ingestion brings chemicals into contact with the tissues of the gastrointestinal tract. The normal function of the gastrointestinal tract is absorption of foods and fluids that are ingested. The gastrointestinal tract is also effective in absorbing toxic chemicals contained in the food or water. The degree of absorption generally depends on whether the chemical is hydrophilic (easily soluble in water) or lipophilic (easily soluble in organic solvents or fats). Lipophilic compounds (e.g., organic solvents) are usually well absorbed, since these chemicals can easily diffuse across the membranes of the cells that line the gastrointestinal tract. Hydrophilic compounds (e.g., metal ions) cannot cross the cell lining in this way, and must be "carried" across by transport systems in the cells. The extent of the transport depends on how efficient the transport system is and how closely the chemical resembles the normal compound for which the transport system is intended.

If a chemical is a weak organic acid or base, it will tend to be absorbed by diffusion in the part of the gastrointestinal tract in which it exists in its most lipid-soluble (least ionized or polar) form. Since gastric juice in the stomach is acidic and the intestinal contents are nearly neutral, the polarity of a chemical can differ markedly in these two areas of the gastrointestinal tract. A weak organic acid is in its least polar form while in the stomach and, therefore, tends to be absorbed through the stomach. A weak organic base is in its least polar form while in the intestine and, therefore, tends to be absorbed through the intestine.

Another important determinant of absorption from the gastrointestinal tract is the interaction of the chemical with gastric or intestinal contents. Many chemicals tend to bind to food, and so a chemical ingested in food is often not absorbed as efficiently as when it is ingested in water.

Additionally, some chemicals may not be stable in the strongly acidic environment of the stomach and others may be altered by digestive enzymes or intestinal flora (bacteria that reside in the intestines) to yield different chemicals with altered toxicological properties. For example, intestinal flora can reduce aromatic nitro groups to aromatic amines, which may be carcinogenic.

Injection. Exposure to toxic chemicals by injection can occur inadvertently through mechanical injury from glass or metal contaminated with chemicals or when chemicals are handled in syringes. Subcutaneous injection can be achieved if one simply contaminates a cut area with a chemical. Careless or improper handling of a hypodermic syringe containing a toxic material can result in injection of the material beneath the skin.

Other Routes. In toxicity assessment, the route of exposure employed in experimental animal studies is normally chosen to be the same as the anticipated route of exposure of humans to the specific chemical. However, toxicologists frequently administer chemicals to laboratory animals by injection (parenterally), the most common routes being subcutaneous (s.c.), intraperitoneal (i.p.), intramuscular (i.m.), and intravenous (i.v.). These routes are employed because they are often more convenient and yield more reproducible results than oral, dermal, or inhalation routes of exposure. However, results from studies of this sort must be interpreted with caution, since parenteral administration bypasses the normal absorptive processes, and a parenteral dose may be more toxic than the same dose given by ingestion, inhalation, or dermal application.

TYPES OF TOXIC CHEMICALS

The physiological action of toxic materials varies greatly. Most of the major classes of toxins are discussed below.

Carcinogens

Carcinogens, materials that cause cancer, are a special problem, not yet very well defined or controlled. Known or potential carcinogens can be safely handled in the laboratory if proper precautions are taken and exposures are minimized. Before using a known or suspected carcinogen, other options should be investigated (e.g., substitutions). Handling materials with good ventilation, use of personal protective equipment, and good personal hygiene will go a long way toward minimizing the possible effect of handling carcinogenic materials (see Chap. 3), but one should always consider the possibility of accidental exposure.

Poisons

Some compounds are especially toxic. Appendix A includes a chemical toxicity rating scale that categorizes toxic hazards on the basis of LD_{50} and LC_{50}. Several classes of compounds should be considered potentially dangerous. These are listed in Table 2-1.

Certain functional groups of organic molecules should be watched for particularly toxic properties. These groups are called *toxophores*. If a molecule contains a toxophore group and an autotoxic group, a poisonous compound is produced. Sample toxophores and autotoxics are shown in Table 2-2.

TABLE 2-1. Dangerous Classes of Chemicals

Acetylides
Aldehydes
Alkaloids
Antimony compounds
Arsenic compounds
Asbestos fibers
Beryllium compounds
Cadmium compounds
Carbonyls (particularly Fe, Co, N)
Chlorates
Chromium compounds
Copper compounds
Cyanogen
Fumigants
Glucosides
Halides
Halogenated hydrocarbons
Halogens
Heavy metals and their dusts, vapors
Isocyanic acid and its esters
Lead compounds
Mercury compounds
Metal cyanides (soluble)
Metal fumes
Nitrates
Nitrides
Nitriles
Oxalic acid
Oxides of nitrogen
Pesticides
Phenolics
Phosphorous, white
Selenium compounds
Silver compounds
Sulfur (thio-) compounds
Thallium compounds
Zinc compounds

Source: NIOSH, *Hazards in the Chemistry Laboratory, Student Manual,* 1978, p. 4-11.

26 MANAGING CHEMICAL HAZARDS IN THE LABORATORY

TABLE 2-2. Predicting What Chemicals May Be Poisonous

Molecules that contain one or more of the following *toxophores*:

$>C=O$	Aldehydes, ketones
$>S$	Sulfide, thiols, thioketones
$>C=C<$	Alkenes
$O=N=O$	Nitro compounds
$-C\equiv N$	Nitriles
$-As-$	Arsines

in combination with one or more of the following *autotoxics*:

OH	Alcohol
Cl	Chloro-
NH_2	Amine
⌬	Phenyl
HO-⌬	Phenolic
CH_3	Methyl
CH_3CH_2	Ethyl
⌬⌬	Polynuclear rings

Source: NIOSH, *Hazards in the Chemistry Laboratory, Student Manual,* 1978, p. 4-10.

Some common gases and vapors are extremely hazardous:

Carbon monoxide has no smell or taste. By the time the first symptoms of headache and dizziness occur, irreversible damage has probably been done. Carbon monoxide has been cited as the cause of more accidents than all other toxic gases combined.

Mercury is an insidious poison because most people deal too casually with the substance. Precautions must be taken in the laboratory to clean up mercury spills thoroughly.

Hydrogen sulfide has been used for years in freshman laboratories with little or no concern for its toxic nature. In analytical laboratories, it is often handled as though it were not toxic at all. Yet hydrogen sulfide is 20 times more poisonous than hydrogen cyanide. It anesthetizes the odor-detecting mechanism of the body, making it more insidious.

Benzene and *carbon tetrachloride* are other examples of the old saying that "Familiarity breeds contempt." The past popularity of these chemicals as cleaning fluids was appalling considering the hazards involved. The "Use with adequate ventilation" warning should always be read as "Use only outdoors or in a laboratory fume hood."

The TLVs of these chemicals are listed in Table 2-3.

Irritants

Irritants can affect the skin and mucous membranes (fluid-secreting tissues of the eyes, nose, mouth, throat, and lungs). Irritants can be either primary irritants or delayed irritants. Contact dermatitis is a common symptom of exposure to an irritant. Individuals may have differing sensitivities to some materials. Congenital sensitivity is inborn. Acquired sensitivity comes from repeated contact with a material.

There are many chemicals in the laboratory that are primary irritants (the corrosive chemicals, acids and alkalies, reactive gases, etc.). A material such as chlorine, hydrogen chloride, or ammonia, when breathed, causes severe discomfort and produces a serious irritation of the lung surface. Because of this immediate distress there is a strong tendency to get out of the affected area. Table 2-4 lists some common irritants.

Phosgene and nitrogen dioxide are the prominent examples of another class termed delayed irritant materials. One can be exposed to a substantial amount of either of these materials and experience only mild immediate discomfort. The irritation of the lung tissues is appreciable, however, and may be so severe that after several hours a substantial amount of fluid will have collected in the lungs. A man who has been exposed to one of these materials may finish his work day, go home, and then find several hours later that he has difficulty breathing and fluid is filling his lungs. Only if he relates the condition to the work that he was doing and the exposure that he had is he likely to make the proper connection and obtain the proper treatment.

TABLE 2-3. Insidiously Poisonous Gases and Vapors Commonly Found in Analytical Laboratories

Chemical	TLV
Carbon monoxide	100 ppm
Benzene	25 ppm
Carbon tetrachloride	10 ppm
Hydrogen sulfide	10 ppm
Mercury	0.1 mg/m^3

Source: NIOSH, *Hazards in the Chemistry Laboratory, Student Manual,* 1978, p. 4-13.

TABLE 2-4. Examples of Irritants

Liquids

Acids
  Acetic acid
  Chloroacetic acid
  Cresylic acid
  Formic acid
  Hydrochloric acid
  Hydrofluoric acid
  Nitric acid
  Phosphoric acid
  Sulfuric acid

Organic solvents
  Most alcohols
  Carbon disulfide
  Chlorinated hydrocarbon solvents
  Coal tar solvents
  Most liquid esters and ketones
  Petroleum solvents
  Turpentine and terpines

Solids

Alkalies
  Alkaline sulfides
  Ammonium carbonate
  Barium hydroxide
  Barium carbonate
  Calcium cyanamide
  Lime (hydrated and dehydrated)
  Potassium carbonate
  Sodium carbonate
  Sodium hydroxide
  Sodium silicate
  Trisodium phosphate

Elements and salts
  Antimony and its salts
  Arsenic and its salts
  Chromium and the alkaline chromates
  Copper cyanide
  Copper sulfate
  Mercuric salts
  Phosphorus
  Potassium
  Silver nitrate
  Sodium
  Zinc chloride

Gases

  Acetic acid
  Acetic anhydride
  Acrolein
  Ammonia
  Arsenic trichloride
  Bromine
  Chlorine
  Chloropicrin
  Dichlorethyl sulfide
  Dichloromethyl ether
  Dimethyl sulfate
  Ethyl chlorosulfonate
  Formaldehyde
  Hydrochloric acid
  Hydrofluoric acid
  Iodine
  Methyl chlorosulfonate
  Nitrogen dioxide
  Ozone
  Phosgene
  Phosphorus pentachloride
  Phosphorus trichloride
  Sulfur dioxide
  Sulfur monochloride
  Sulfuric acid
  Sulfuryl chloride
  Thionyl chloride
  Xylyl bromide

A bothersome problem that occurs in the laboratory or in industry is a phenomenon called *sensitization*. A worker using epoxy resin compounds as adhesives or in a foaming-type operation may have daily contact with toluene diisocyanate and may be able to use this material for many years without any perceptible effect. However, she may then suddenly find that she is developing an allergic reaction, as evidenced by swelling of the face and eyes, serious interference with breathing, and great distress in general. Such a person has become sensitized to toluene diisocyanate and will never be able to be anywhere near this material for the rest of her life. Sensitization can cause severe dislocation in a person's work. There have been some attempts at desensitizing by exposure to tiny amounts, but this is rarely successful and should not be expected to resolve the problem after a person has developed sensitivity. Some materials, called *sensitizers,* can increase a person's sensitivity to irritants or allergens.

Corrosive Chemicals. The major classes of corrosive chemicals are strong acids and bases, dehydrating agents, and oxidizing agents. Some chemicals, such as sulfuric acid, belong to more than one class. Inhalation of vapors or mists of these substances can cause severe bronchial irritation. These chemicals erode the skin and the respiratory epithelium and are particularly damaging to the eyes.

Strong Acids. All concentrated strong acids can damage the skin and eyes. Exposed areas should be flushed promptly with water. Nitric, chromic, and hydrofluoric acids are especially damaging because of the types of burns they inflict. Hydrofluoric acid, which produces slow-healing, painful burns, should be used only after thorough familiarization with recommended handling procedures.

Strong Bases. The common strong bases are potassium hydroxide, sodium hydroxide, and ammonia. Ammonia is a severe bronchial irritant and should always be used in a well-ventilated area. The metal hydroxides are extremely damaging to the eyes. Should exposure occur, the affected areas should be washed at once with copious quantities of water, and an ophthalmologist should evaluate the need for further treatment.

Dehydrating Agents. The strong dehydrating agents include concentrated sulfuric acid, sodium hydroxide, phosphorus pentoxide, and calcium oxide. Because of their affinity for water, these substances cause severe burns upon contact with the skin. Affected areas should be washed promptly with large volumes of water.

Asphyxiants

Certain materials are asphyxiants. Light hydrocarbons such as methane and ethane, nitrogen, and other inert gases are called simple asphyxiants. They simply reduce the amount of oxygen in the air breathed to the extent that the body, particularly the brain, is starved and the person may collapse and die. Removal to fresh air reverses the effect but must be done before breathing stops. If resuscitation is not started within minutes of the end of breathing, brain damage will occur. Carbon monoxide has the same end effect, although its method of operation is different. It binds to hemoglobin and thus interferes with the capacity of the blood to carry oxygen to the tissues, so that the tissues become oxygen-starved. The brain is the tissue most easily affected, and collapse usually comes because of its failure to function. Cyanides fall into this category, since they also interfere with the ability of the body to use oxygen.

Anesthetics

Anesthetics include ether and other substances that are commonly used in surgery. Alcohol is also an anesthetic, as are many of the hydrocarbons in gasoline. Most of these materials cause an initial hilarity or stimulation followed by sleepiness, collapse, and, upon further depression of the nervous system, death.

Embryotoxins

Embryotoxins are substances that act during pregnancy to cause adverse effects on the fetus. These effects include embryolethality (death of the fertilized egg, the embryo, or the fetus), malformations (teratologic effects), retarded growth, and postnatal functional deficits.

A few substances have been demonstrated to be embryotoxic in humans. These include organic mercury compounds and lead compounds. Glycol ethers are a large class of compounds that are suspected of being embryotoxic in humans. Maternal alcoholism is probably the leading known cause of embryotoxic effects in humans, but exposure to ethanol encountered in laboratories is unlikely to be embryotoxic. Many substances, some common (e.g., sodium chloride), have been shown to be embryotoxic to animals at some exposure level, but usually a considerably higher level than is met in the course of normal laboratory work. Some substances, however, do require special controls because of their embryotoxic properties. One example is formamide; women of child-bearing potential should handle this substance only in a hood, and should take precautions to avoid skin contact with the liquid because of the ease with which it passes through the skin.

Because the period of greatest susceptibility to embryotoxins is the first 8–12 weeks of pregnancy, which includes a period when a woman may not know she is pregnant, women of child-bearing potential should take care to avoid skin contact with all chemicals. The following procedures are recommended to be followed routinely by women of child-bearing potential in working with chemicals requiring special control because of embryotoxic properties.

1. Each use should be reviewed for particular hazards by the research supervisor, who will decide whether special procedures are warranted or whether warning signs should be posted. Consultation with appropriate safety personnel may be desirable. In cases of continued use of a known embryotoxin, the operation should be reviewed annually or whenever a change in procedures is made.

2. Embryotoxins requiring special control should be stored in an adequately ventilated area. The container should be labeled in a clear manner, such as:

EMBRYOTOXIN: READ SPECIFIC PROCEDURES FOR USE.

Breakable storage containers should be kept in an impermeable, unbreakable, secondary container that has sufficient capacity to retain the material if the primary container should accidentally break.

3. Women of child-bearing potential should take adequate precautions to guard against spills and splashes. Use only impermeable containers and work in adequately ventilated areas. Appropriate safety apparel, especially gloves, should be worn. Check all hoods, gloveboxes, and other protective systems to be sure that they are operating at required efficiency before work is started.

4. Supervisors should be notified of all incidents of exposure or spills of embryotoxins requiring special control. A qualified physician should be consulted about any exposures of women of child-bearing potential above the acceptable level—that is, any skin contact or any inhalation.

Essentially, a woman of child-bearing potential should treat embryotoxins as if they were carcinogens.

Allergens

A wide variety of substances can produce skin and lung hypersensitivity. Examples include such common substances as diazomethane, chromium, nickel, bichromates, formaldehyde, isocyanates, and certain phenols. Because of their variety and because of the varying responses of individuals, suitable gloves should be used whenever hand contact with products of unknown activity is probable.

Particulates

Particulates in the air can have several physiological effects. Extremely fine particles are breathed in and out again without coming to rest or having any effect whatever. Coarse particles are usually caught in the nose and throat and are expectorated from the body. Particles in the size range of 1 to 5 µm, however, have a tendency to be deposited on the surface of the lung and remain there due to the moisture and other physical conditions. Some materials are quite harmless and simply remain on the surface of the lung as inert particles, causing no particular problem. Other materials, particularly silica, are irritants to the lung tissue, so a process of encapsulation takes place. Tissue is generated around the particle, rendering it no longer irritating, but, on the other hand, building scar tissue, which reduces by a finite amount the area of lung surface that is useful to the person. Over a period of time, when many millions of such particles have been inhaled and encapsulated, lung capacity is decreased and the condition known as silicosis develops. This process is not reversible. The only way that silicosis can be prevented is by preventing the breathing of the dust into the lungs.

UNDERSTANDING TOXIC EFFECTS

When reading about chemicals' toxic hazards, it is helpful to have some understanding of how those toxic effects are measured. Exposure of an organism to a chemical often results in multiple effects. For example, long-term exposure to dioxin results in hepatotoxicity (liver toxicity), genotoxicity (chromosomal damage), teratogenicity (structural/functional abnormality), fetotoxicity (injury to developing fetuses), and carcinogenicity (growth of malignant tumors). Effects that are measured by the toxicologist as an index of a chemical's toxicity are called *endpoints*. The criteria for identifying the endpoint most appropriate for use in toxicity assessment (e.g., determination of a TLV) include dose sensitivity, the severity of the effect, and whether the effect is reversible or irreversible. See Fig. 2–3.

The most appropriate endpoint for use in the toxicity assessment process is usually the one in which a measurable change can first be detected in response to increasing doses. For example, pyridine is toxic to the central nervous system (CNS), the liver, and the kidney. However, CNS toxicity can be demonstrated at much lower doses than adverse kidney and liver effects. In studying pyridine, then, CNS effects are appropriate as the most sensitive endpoint.

The selection of a toxicological endpoint is sometimes based on the extent of damage to a particular organ following exposure. A toxic chemical may produce harmful effects in a number of organs, but the severity of the response may be quite different. For example, carbon tetrachloride exposure may result in mild damage to the kidney, but severe damage and loss of function in the liver. In studying carbon tetrachloride, then, effects on the liver are the most appropriate endpoint.

Sometimes a low dose of a chemical may produce an effect that is not in itself clearly adverse. For example, a low dose of acrylamide may cause slowed axonal transport in nerve cells without measurably affecting the ability of the cells to carry nerve impulses. To distinguish between detectable effects that are ad-

CHEMICAL HAZARDS IN THE LABORATORY

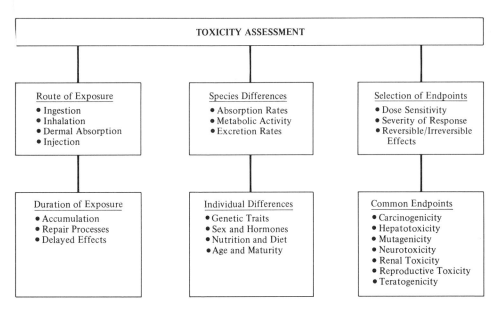

FIGURE 2-3. Factors considered during toxicity assessments. (From Government Institutes, Inc., *Toxicology Handbook,* Rockville, Md., 1986, p. 4-2.)

verse and those that are not, the term LOEL (lowest-observed-effect level) is used, as distinct from LOAEL (lowest-observed-*adverse*-effect level). Similarly, NOEL (no-observed-effect level) implies no detectable effect of any sort, whereas NOAEL (no-observed-*adverse*-effect level) may include some effect that is judged not to be adverse. See Fig. 2–4. The decision as to whether an effect is adverse or not must be made on the basis of whether the change is an early indication of a more serious consequence or is not of significant concern.

Some toxic effects are reversible and others are not. In a tissue that has a strong ability to regenerate (e.g., the liver), most injuries are reversible; whereas injury to the CNS is largely irreversible, since specialized cells of the CNS cannot divide or be replaced. Carcinogenic effects of chemicals are also irreversible. Irreversible effects are often chosen as toxicological endpoints, because these effects are likely to produce serious consequences following chronic (long-term, low-level) exposure to a chemical.

Table 2–5 lists endpoints that are commonly used to assess the toxic effects of a chemical, along with the experimental means of measuring such effects. More detailed descriptions of these endpoints are given next.

Cancer

Cancer is a complex group of diseases whose causes are not yet fully understood, but there is ample evidence that some chemicals can cause or promote certain types of tumors in animals or humans. The carcinogenic potential of a chemical can be measured with lifetime animal bioassays, short-term carcinogenicity tests (with bacterial or cultured mammalian cells), or limited in vivo bioassays. Each of these methods is associated with certain advantages and disadvantages, as discussed below.

Standard lifetime animal bioassays are long-term experiments conducted to measure the effect of a chemical on frequency of tumor occurrence. Typically, large groups of animals (at least 50 per sex per dose) are exposed to the chemical for their lifetime, and the number and types of tumors occurring in exposed animals are compared to control animals.

These studies are considered to be the most predictive of carcinogenicity screening tests. However, substantial controversy exists over certain standard practices used in the bioassays. For example, to compensate for the relative insensitivity of these studies, the maximum tolerated dose (MTD) is frequently used to maximize the likelihood of detecting carcinogenicity. The use of MTD is controversial, because high doses of a chemical may produce physiological conditions that affect the induction and development of tumors. Normal detoxification and repair mechanisms may be overwhelmed by the use of the MTD, or different absorption, distribution, metabolism, or excretion may result from the use of the MTD. These events might result in a response at the MTD that may not be indicative of effects at lower exposure levels.

A similar controversy exists over the use of strains of test animals that are very susceptible to carcinogens. The purpose of using these animals is to increase the ability to detect carcinogenic potential in chemicals. However, these sensitive strains may have very high spontaneous tumor frequencies, and the meaning and validity of a positive test result in such a sensitive strain is not entirely clear. Because of these uncertainties, it is very desirable to perform lifetime carcinogenicity bioassays using two or more different animal species.

There are two major types of short-term carcinogenicity tests used to indicate carcinogenic potential: mutagenicity tests and transformation tests using cultured mammalian cells. Mutagenicity experiments are often used to evaluate the potential for inducing tumors because of basic similarities in the postulated molecular mechanisms of chemical carcinogenesis and mutagenesis. Some mutagenicity tests, especially the Ames test, have been extensively validated and shown to correlate very well with known carcinogens, but there is still a significant frequency of false-positive or false-negative results (approximately 10% each for the Ames test). Positive results in mutagenicity tests support other experimental findings of carcinogenic potential and are generally considered to provide suggestive evidence of carcinogenic hazard. They do not constitute definite proof of a chemical's carcinogenicity in humans, nor do negative results rule out the possibility of carcinogenic potential.

A major disadvantage of mutagenicity tests using bacterial test systems is the basic biological differences between bacterial cells and human cells, making extrapolations to human health effects somewhat tenuous. Testing in mammalian cells provides a stronger basis for extrapolating to human health effects, but the test methods are not as well developed or validated as those using bacteria. The primary short-term test of this sort is based on mammalian cell transformation. Transformation occurs when cultured cells develop uncontrolled growth, an event that is analogous to the formation of a tumor in an organism. A number of transformation tests using mammalian cells have been developed in recent years and are in widespread use.

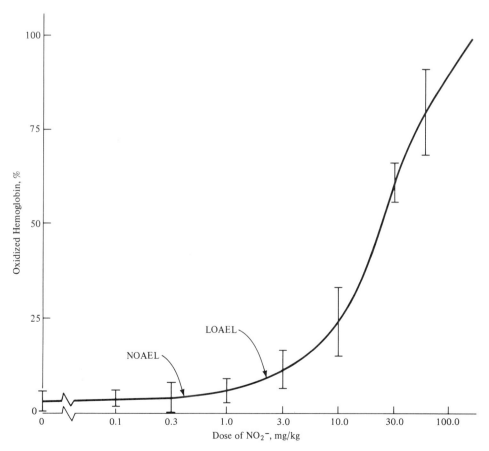

FIGURE 2-4. Dose-response curve for nitrite-induced hemoglobin oxidation demonstrating the NOAEL (no-observed-effect level) and LOAEL (lowest-observed-*adverse*-effect level). (From Government Institutes, Inc., *Toxicology Handbook*, Rockville, Md., 1986, p. 3-7.)

TABLE 2-5. Measurement of Common Toxicological Endpoints

Toxicological Endpoint	Parameters Measured
Behavioral toxicity	Motor function (motor activity, coordination strength), sensory function (vision, audition), integrative systems (learning and memory)
Carcinogenicity	Tumor frequency in tissues, detected by gross observation or histological examination
Hematologic toxicity	Hematocrit, hemoglobin levels, changes in cellular components (erythrocytes, leucocytes, platelets), plasma components, and foreign substances
Hepatoxicity	Gross and microscopic examination, organ weight, liver function (bile formation, lipid metabolism, protein metabolism, carbohydrate metabolism, metabolism of foreign compounds, serum enzyme activities)
Inhalation toxicity	Gross anatomy, microscopic and ultrastructural anatomy, changes in function
Mutagenicity	Chromosome alterations, bacterial mutations, DNA damage
Neurotoxicity	Gross observation, clinical evaluation, neurological exams, behavioral tests, neurohistopathological tests, neurochemical tests
Renal toxicity	Urinalysis, function tests (clearance, glomerular filtration rate), gross and microscopic examinations, organ weight
Reproductive toxicity	Fertility, litter size and survival, gestation survival, postnatal body weight
Teratogenicity	Gross abnormalities, skeletal and visceral malformations, microscopic abnormalities, functional/behavioral deviations

Source: Government Institutes, Inc., *Toxicology Handbook,* Rockville, Md., 1986, p. 4-9.

The cells are treated with the chemical in question and the transformation frequency is measured. Cell transformation is usually detected by observing changes in the cultured cells, and is confirmed by injecting the transformed cells into animals, where they become malignant tumors. A major disadvantage is that the carcinogenic potential of a chemical may depend on its metabolism in the living organism, with one or more metabolic products being more carcinogenic than the original chemical. In a case such as this, a transformation test might yield negative results, while positive results would be obtained in a lifetime animal bioassay.

Limited in vivo bioassays can provide evidence of the tumorogenic potential of a chemical without the great time investment and expense required for a lifetime bioassay. These tests generally yield results in 30 weeks or less and use mice or rats as test animals. Examples of limited in vivo bioassays include skin tumor formation in mice, breast cancer induction in rats, or altered liver foci (an early step in liver tumor formation) in mice or rats. While limited in vivo bioassays are not an adequate substitute for lifetime bioassays, they are more useful as predictors of carcinogenicity than short-term tests, and positive results in well-designed and executed limited in vivo bioassays are additional supportive evidence of carcinogenic hazard.

It is not uncommon that a chemical will yield different results in different tests of carcinogenic potential (positive in some tests, negative in others). In such cases, the probability that the chemical is a human carcinogen is determined by a "weight-of-evidence" approach. In general, positive results in animal systems or positive findings in epidemiological studies are required before a chemical is classified as a probable human carcinogen. This may be supported by positive results in bacterial or cell system tests, but positive results in these systems are insufficient alone. The result is that it may take a long time to classify a chemical as a probable human carcinogen.

Nervous System Damage

The nervous system is of special toxicological concern, because chemical-induced injury to nerve cells is often irreversible and may lead to adverse health effects. There are many means of measuring nervous system functions, including tests of reflexes, coordination, conditioned responses in animals, and intelligence (IQ) tests in humans. In addition, there are sophisticated means of analyzing the status of the individual nerves by measuring the rate at which they transmit nerve impulses or the rate at which they synthesize and transport cellular materials.

The Environmental Protection Agency (EPA) has published guidelines that focus on delayed neurotoxicity as an endpoint. Delayed neurotoxicity is a syndrome in which damage to the peripheral nervous system and some portions of the CNS may result in paralysis. The domestic hen is typically the species chosen for evaluation of delayed neurotoxicity. In the acute delayed neurotoxicity test, a single dose of the test material is administered orally to groups of adult domestic hens. The hens are observed daily for at least 21 days for behavioral abnormalities, ataxia (inability to coordinate muscles), and paralysis. Selected sections of nervous tissues are also examined histopathologically. In tests for subchronic delayed neurotoxicity, groups of hens are administered the test substance orally for 90 days, followed by an observation period of 7 days. As in the acute studies, the hens are observed daily for behavioral abnormalities, ataxia, and paralysis, and selected sections of nerve tissue are examined histopathologically.

Liver Damage

Liver damage is a frequent response to exposures to toxic chemicals. Since the liver is such a vital organ, a variety of procedures have been developed over the years to assess extent of liver damage. Because the liver has considerable reserve capacity, tests that measure its ability to perform its functions may not reveal an effect until the liver is already extensively damaged. A more sensitive test involves measurement of liver enzymes in blood serum. This test is based on the observation that when liver cells are damaged, some of the active enzymes within the cells escape into the blood. This increase in liver enzymes can be measured simply by collecting a sample of blood and measuring enzyme activity. A disadvantage of this test is that liver enzymes do not endure very long in the blood, and so only an ongoing injury can be detected. Finally, evidence of liver damage may be detected both during and well after a chemical-induced injury by microscopic examination of the liver for signs of abnormality. Therefore, microscopic examination for histological changes is another excellent endpoint.

Kidney Damage

Damage to the kidneys is another common and serious consequence of exposure to a toxic chemical. As with the liver, techniques to assess kidney injury may include functional tests in the intact animal, along with direct histological examination of kidney tissue. Urinalysis offers another convenient and sensitive means of detecting kidney damage. For example, detection of substances not normally present in urine (proteins, cells or cell fragments, glucose) is strong evidence that the kidney has been injured.

Reproductive System Effects

Fertility and reproductive toxicity studies are usually performed in rats or mice at dose levels that produce no overt toxicity in the exposed adults. In a typical study, the male parent is exposed to a chemical for 60 to 80 days and the female for 14 days prior to mating. The percentage of females that become pregnant is determined. The number of stillborn and live offspring, and their weight, growth, survival, and general condition during the first 3 weeks of life are also recorded.

The perinatal (during late pregnancy) and lactational (during nursing) toxicities of chemicals may be measured in a similar fashion. Pregnant female rats are exposed to the chemical from the fifteenth day of gestation to the time of weaning. Parameters measured may include all of those above, as well as analysis of milk for presence of the chemical.

Teratology is defined as the study of functional or physical defects induced during development of an animal from the time of conception to birth. Teratogenic studies are usually performed in rats and/or rabbits with doses of the test chemical that produce no maternal toxicity. Teratogens are most effective when administered during the period of organogenesis, so pregnant females are usually exposed on days 6 to 15 of gestation. Prior to delivery, some females are sacrificed and examined for the number of fetal implantations in the uterus. Dead and living fetuses are counted, weighed, and examined for gross malformations. These fetuses are examined microscopically for more subtle effects, and some are cleared of soft tissue and examined for skeletal abnormalities. Since teratogens can produce functional as well as morphologic changes, offspring of other females are sometimes monitored after delivery for changes in behavior or development.

Mutagenesis

Mutagenesis is the induction of changes in genetic material that are transmitted during cell division. If mutations are present in the genetic material of eggs or sperm, the fertilized ovum may not be viable. A mutation may also result in congenital abnormalities or death of a fetus at a later developmental period. There are a number of powerful tests for mutagenic potential of chemicals, most involving bacteria or other cells in culture. For example, the Ames test measures the frequency of a certain type of mutational event in the bacterial species *Salmonella typhimurium*. Other valuable tests examine Chinese hamster ovary (CHO) cells for alterations in chromosome structure, or determine whether unscheduled DNA synthesis (a strong indicator of damage to the DNA) is occurring in other cultured mammal cells.

HOW TOXIC IS "TOXIC"?

Dose Response and Threshold

The dose-response relationship is the most fundamental concept in toxicology. The dose-response curve describes the relationship that exists between degree of exposure to a chemical (dose) and the magnitude of the effect (response) in the exposed organism. By definition, no response is seen in the absence of a chemical. As the amount of chemical exposure increases, the response becomes apparent and increases. Depending on the mechanism by which the chemical acts, the curve may rise with or without a threshold (Fig. 2–5). In both cases, the response normally reaches a maximum, after which the dose-response curve becomes flat.

For some chemicals there may be a certain minimum level of exposure below which no significant or observable effects occur. This is due to one or both of two reasons: (1) there is considerable "reserve capacity" in many tissues, such that limited damage (e.g., a 5% loss of hemoglobin) does not cause any decrease in function; and (2) most cells have at least a limited capacity to repair or compensate for cellular damage. Of course, both the reserve and the repair capacity of a tissue can be overwhelmed by too much of a harmful chemical, and it is at that point that the organism as a whole begins to suffer adverse effects.

One example of this is shown in Fig. 2–6. Nitrite (NO_2^-) produces injury to red blood cells by oxidizing the iron atom in hemoglobin from the ferrous (Fe^{2+}) to the ferric (Fe^{3+}) form. Within the red blood cell is a

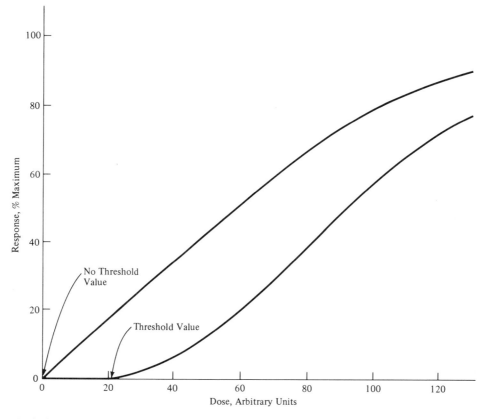

FIGURE 2–5. Hypothetical dose-response curves demonstrating the presence and absence of a threshold value. (From Government Institutes, Inc., *Toxicology Handbook,* Rockville, Md. 1986, p. 3-2.)

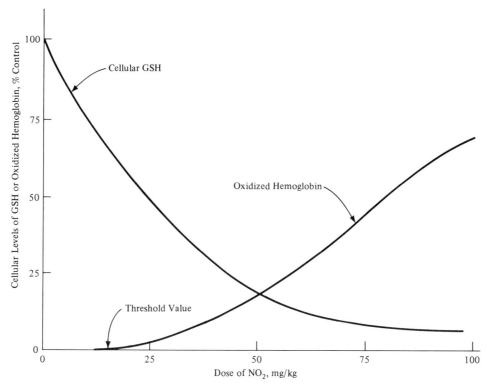

FIGURE 2-6. Cellular defense mechanisms as a basis for the threshold phenomenon. (From Government Institutes, Inc., *Toxicology Handbook*, Rockville, Md., 1986, p. 3-3.)

system designed to protect the cell against this kind of damage. It involves providing a "sacrificial" chemical, glutathione (GSH), that is oxidized by the nitrite instead of the iron. Only when the amount of nitrite exceeds the ability of the cell to supply glutathione is the iron of hemoglobin oxidized. And even then there is usually a considerable excess of oxygen-carrying capacity in normal blood, so that when some hemoglobin becomes oxidized (and hence nonfunctional), the amount of oxygen supplied to tissues is still in excess of the tissues' needs. Only when a large amount of hemoglobin has been oxidized will there be an adverse effect in tissues consuming oxygen (Fig. 2–7).

Some chemicals produce adverse effects that are characterized by a dose-response curve with no threshold (see Fig. 2–5), which means that there may be no dose that is without risk of causing adverse effects. This is usually because the cells that are affected have little or no "defense" against the chemical, and little or no ability to repair or compensate for damage that is done. For example, recent research suggests that there may be no threshold for the effects of lead on the nervous systems of infants and children. Chemicals that produce cancer (carcinogens) are also considered to belong to this group.

Variables to the Dose-Response Relationship

Table 2–6 lists some factors that influence the significance of the exposure and resulting response. These variables affect all areas of toxicology.

Practicality of Using Available Information

Data from human exposure are often not in the literature; judgment must often be based on information derived from laboratory animal evaluation. To extrapolate to human situations is extremely difficult and subject to human judgment.

Experimental data can be obtained from the following types of tests to determine the general range of toxicity:

Acute and chronic oral
Acute and chronic dermal
Acute and chronic inhalation
Hypersensitivity
Eye irritation

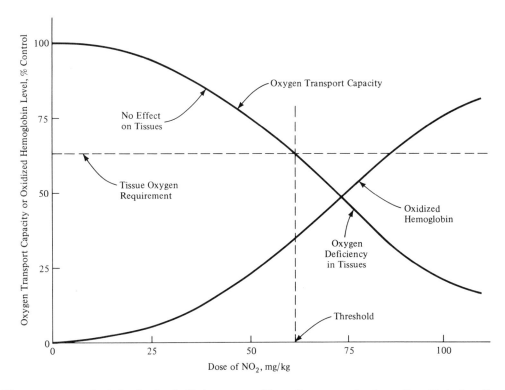

FIGURE 2-7. **Tissue reserve as a basis for the threshold phenomenon.** (From Government Institutes, Inc., *Toxicology Handbook,* Rockville, Md., 1986, p. 3-4.)

Judgments of hazard are then based on:

Human exposure information

Basic toxicologic tests

General toxicologic guidelines

Carcinogenicity, mutagenicity, teratogenicity, and reproduction information

Basic physiochemical information

A semiquantitative basis for toxicity ratings can be summarized by assigning a scale rating to toxicity. Hodge and Sterner proposed the scale described in Appendix A.

The Chemical Entries section contains a list of common laboratory reagents with LD_{50}, LC_{50}, TLV, and TWA data. This list is not exhaustive, but includes chemicals found in most laboratories.

Extrapolating from Animal Experiments to Humans

When evaluating toxicity data from animal studies, the doses administered to the animals must be converted to an equivalent dose in humans. When an animal is exposed to a chemical in a laboratory study involving oral exposure, it is customary to describe the amount of chemical the animal ingests in units of milligrams of chemical per kilogram of body weight (mg/kg). If humans and animals were equally sensitive to the chemical on this basis (weight equivalence), then a dose without effect in the animal would also be without effect in humans. However, a considerable body of laboratory data in animals and clinical data in humans indicates that doses expressed in these units are not toxicologically equivalent in animals of different sizes, and that a dose that produces no effect in mice might indeed produce an effect in humans.

Correcting for Toxicokinetic Differences. The principal reason that data from animals are not directly applicable to humans is that there are toxicokinetic differences (differences in ingestion, inhalation, absorption, or metabolism) among species. Whenever there are good quantitative data that characterize these differences, it is possible to correct for them. For example, if an oral dose of 100 mg/kg is 30% absorbed in rats, this corresponds to an absorbed dose of 30 mg/kg. If an oral dose of 100 mg/kg of the same chemical is 60% absorbed in humans, this corresponds to an absorbed dose of 60 mg/kg. Clearly, the same oral dose (100 mg/kg) would be more effective in humans than in rats. This is simply corrected for, as follows:

$$\frac{D_A A_A}{A_H} = D_H$$

where

D_A = dose in animal (mg/kg)
D_H = dose in humans (mg/kg)
A_A = absorption in animal (%)
A_H = absoprtion in humans (%)

For the above example the calculation would be as follows:

$$(100 \text{ mg/kg})\left(\frac{30\%}{60\%}\right) = 50 \text{ mg/kg}$$

where

100 mg/kg = dose in rat
30% = absorption in rat
60% = absorption in humans
50 mg/kg = equivalent dose in humans

Similar corrections may be employed to adjust for differences in absorption across skin or lung and differences in rates of metabolism. Unfortunately, reliable quantitative toxicokinetic data in animals and/or humans are frequently lacking, and objective corrections for differences in absorption or metabolism are not possible in all cases.

The Surface-Area Concept. Studies of effect levels of chemicals indicate that a better correlation among different species exists between toxicity and dose when doses are expressed in units of milligrams of chemical per unit surface area (mg/m²). The theoretical basis for this correlation is not obvious, but a large number of physiological parameters, including surface area, are approximately proportional to the two-thirds power of body weight. These parameters include metabolic rates, oxygen consumption, blood volume, kidney function, thyroid function, brain weight, liver weight, cardiac output, blood pressure, and extracellular water volume. Since many of these parameters are related to the absorption, distribution, excretion, metabolism, and mechanisms of toxicity of chemicals, surface area is used as a convenient physical parameter that is proportional to the physiological rates and functions that are directly affected.

The use of surface-area equivalence in dose conversions yields calculated doses that are lower than if weight equivalence is used. Comparative data are presented in Table 2–7, which shows the doses in several species calculated to be equivalent to a dose of 2 mg in a 20-g mouse (100 mg/kg). Using weight equivalence, the same dose in a 70-kg human is 7000 mg (100 mg/kg), but if surface-area equivalence is used, the dose is only 776 mg (11 mg/kg).

Conversion of doses (expressed in units of milligrams per kilogram) in one species to surface-area-equivalent doses (also expressed as milligrams per kilogram) in another species may be accomplished by simply multiplying by a conversion factor, as follows:

$$D_H = (D_A)\left(\frac{W_A}{W_H}\right)^{1/3}$$

where

D_H = human equivalent dose (mg/kg)
D_A = animal dose (mg/kg)
W_H = human body weight (kg)
W_A = animal body weight (kg)

Table 2–8 lists the factors for converting doses in animals to doses in humans employing the weights given in Table 2–7. The term $(W_A/W_H)^{1/3}$ to estimate body-surface area ratios is employed because weight can be measured easily although surface area cannot. If actual surface areas were employed, the result would differ slightly. A sample calculation using this equation is presented next.

TABLE 2–6. Factors Influencing Toxicology

Composition of the sample
Manufacturing or formulation process
Vehicle, adjuvants, binding agents, coatings
Concentration and volume
Site and rate of administration
Diurnal and seasonal effects
Species and genetic studies
Immunological status
Sex and hormone status
Nutritional and dietary factors
Age, maturity, and weight
Emotion, stress, and behavior
Preexistent factors
Unusual temperature or pressure
Radiation effects
Environmental factors
Molecular size and solubility
Previous exposure to chemical agents
Synergistic and antagonistic effects
Metabolic pathway effects

Source: NIOSH, *Hazards in the Chemistry Laboratory, Student Manual,* 1978, p. A-15.

TABLE 2-7. Comparison of Dose Conversions Using Surface Area and Weight Equivalence

Species	Weight, g	Surface area, cm²	Calculated dose, mg[a]	
			Weight equivalence	Surface-area equivalence
Mouse	20	46	2	2
Rat	200	325	20	14
Guinea pig	400	564	40	24
Rabbit	1,500	1,272	150	55
Cat	2,000	1,381	200	59
Monkey	4,000	2,975	400	128
Dog	12,000	5,766	1,200	248
Human	70,000	18,000	7,000	776

[a]Based on a dose of 2 mg in a 20-g mouse (100 mg/kg).
Source: Government Institutes, Inc., *Toxicology Handbook,* Rockville, Md., 1986, p. 6-5.

Sample Calculation. To convert a dose of 100 mg/kg in the mouse to an equivalent dose in humans, simply multiply by the conversion factor for the mouse:

$$(100 \text{ mg/kg})(0.066) = 6.6 \text{ mg/kg}$$

where

- 100 mg/kg = dose in mouse
- 0.066 = dose conversion factor (assuming surface area equivalence)
- 6.6 mg/kg = calculated equivalent dose in humans

Differences in Species' Sensitivity

It is generally true that if a chemical is found to be toxic in one species of organism (e.g., rat), it will also be toxic in similar organisms (e.g., other mammals, including humans). However, there are often significant differences as much as 100-fold, in the sensitivity of different species to a chemical. Sometimes there are also qualitative differences in the types of effects that occur. The reason for these differences among species is usually related to differences in the absorption or metabolism of the chemical or to differences in anatomic function. These differences are not accounted for in scaling corrections and assumptions.

The rate of absorption of chemicals across the skin, lungs, or gastrointestinal tract is determined primarily by the properties of the cells at the surfaces of these tissues, and there are some significant differences in these cells among species. For example, the skins of the rat and rabbit are more permeable, the skin of the cat is less permeable, and the skin of the pig, guinea pig, and monkey are similar in permeability characteristics to those observed in humans. Additionally, physical and chemical conditions that influence gastrointestinal absorption may also differ among species. For example, gastrointestinal transit time, surface area-to-volume ratio, and pH in various parts of the gastrointestinal tract often differ among species. Finally, the bacterial populations in the gastrointestinal tract vary among species. Some bacteria may convert one chemical into another one that is more or less absorbable and thus alter the apparent toxicity of the chemical, or they may convert a nontoxic chemical into a toxic one.

Metabolism is the name applied to any chemical reaction that a chemical may undergo while in the body. The liver and kidney are especially active in these reactions, but metabolism of a chemical may occur in any tissue. Nearly all chemicals are modified by one or more reactions, but the nature and extent of these reactions vary widely among different organisms. The rate of metabolism of chemicals is often the limiting step in detoxification and/or excretion of chemicals, so differences in metabolic activity can markedly influence how long a toxic chemical endures in the body. In addition, metabolism of a chemical may sometimes generate a more toxic chemical. For example, pyridine is extensively methylated in cats, gerbils, guinea pigs, and hamsters, while it is poorly methylated in mice, rabbits, rats, and humans. Since methylation (addition of a methyl group, $-CH_3$) may increase the toxicity of pyridine, the effects produced by equal doses in these two groups of animals may be more adverse in the animals that methylate pyridine efficiently.

Differences Between Individuals

Just as there are significant differences among species with respect to toxicity of some chemicals, so there may also be significant differences among subgroups of a population (as a function of sex, race, or age) and among individuals in a population. The principal factors underlying these variations in sensitivity are outlined next.

The genetic makeup of an individual is expressed by the presence or absence of key enzymes in cells, and differences in these enzymes underlie much of the variation in susceptibility of individual members of a particular strain or population. For example, the variation in the susceptibility of some rabbits to the toxic effects of atropine is explained by the presence of an enzyme, atropine esterase, in the blood of the resistant animal.

It should be recognized that laboratory animals used in most experimental studies of chemical toxicity are highly inbred in order to achieve a uniform genetic composition, so variations in chemical sensitivity between individual animals of the same species are usually small. Conversely, humans are genetically highly heterogeneous, and variations between individuals, even of the same age, sex, and race, can be significant. For example, two subgroups have been identified in the human population with respect to the ability to acetylate certain chemicals: slow and rapid acetylators. Acetylation (addition of the acetyl group, $-CH_3CO^-$) is significant, because it can lead to detoxification of some chemicals. Slow acetylators have less N-acetyltransferase in their livers than rapid acetylators, and this is the enzyme that catalyzes the acetylation process. Therefore, slow acetylators are more likely to develop toxic effects to certain chemicals.

In the future, tests for sensitivity to chemical exposure may be used to determine who can work at particular jobs where such exposure may occur.

Differences in toxicity between sexes have been demonstrated in studies on the effects of chloroform, benzene, and some organophosphate insecticides. For example, female mice show little response to chloroform exposures that are lethal for males. The difference has been shown to be under direct endocrine (hormonal) control. As another example, female rats and rabbits are more susceptible to the toxic effects of parathion and benzene, respectively, than are male rats and rabbits. These sex-related effects become reversed after castration and administration of hormones. Pregnancy, with its increased hormonal activity, has been shown to markedly increase the susceptibility of mice to some types of pesticides, and similar effects have been reported for a lactating animal exposed to heavy metals. Hyperthyroidism (excessive secretion of thyroid hormone) and hyperinsulinism (excessive secretion of insulin) may also alter the susceptibility of animals, including humans, to toxic chemicals.

Humans are able to achieve adjustments in the absorption and metabolism of foods and minerals to compensate for fluctuations in dietary intake levels. These metabolic adaptations frequently influence the absorption and/or metabolism of toxic chemicals as well. For example, long-term ingestion of a diet low in essential minerals (iron, calcium, zinc) leads the body to increase absorption and retention of these minerals. However, along with this adaptation to retain essential minerals, the absorption of toxic metals (cadmium, barium) also increases. Generally, low-calorie or protein-deficient diets result in increased sensitivity to a number of toxic chemicals.

Some chemicals are more toxic to one age group than another (usually being more toxic to infants and children than adults). In some cases, this is only because infants and children drink and eat proportionately larger amounts than do adults, and thereby ingest proportionately larger doses. However, infants and children may be inherently more sensitive to chemicals for reasons related to the development process. For example, lead ingestion has much more severe effects on the nervous system of infants and children than it does on adults. Additionally, the ability of the young to metabolize and detoxify chemicals is usually less than for adults. Elderly humans may be more sensitive to some chemicals, since the detoxifying capacity of liver and the excretory capacity of kidney tend to decrease in old age.

The range of human sensitivities creates an important source of uncertainty of unquantified magnitude in assessing human risks. A safety factor of 10 is generally employed to account for variations between individuals and to provide a basic margin of safety. Additional factors are added (usually powers of 10) to account for use of data from animals, or for use of limited or poor-quality data, or data from short-term studies.

TABLE 2-8. Dose Conversion Factors for Calculating Human Equivalent Doses

Species	Weight, kg	$(W_A/W_H)^{1/3}$
Mouse	0.02	0.066
Rat	0.2	0.142
Guinea pig	0.4	0.179
Rabbit	1.5	0.278
Cat	2.0	0.306
Monkey	4.0	0.385
Dog	12.0	0.555
Human	70.0	1.000

Source: Government Institutes, Inc., *Toxicology Handbook,* Rockville, Md., 1986, p. 6-7.

Interactions and Synergisms

When considering possible toxicity hazards while planning an experiment, it is important to recognize that the combination of the toxic effects of two substances may be significantly greater than the toxic effect of either substance alone. Animals are usually exposed to only one test chemical in carefully controlled settings. Humans, however, are exposed to a wide variety of other chemicals and environmental conditions.

Because most chemical reactions are likely to contain mixtures of substances whose combined toxicities have never been evaluated, it is prudent to assume that mixtures of different substances (i.e., chemical reaction mixtures) will be more toxic than the most toxic ingredient in the mixture. The risk from multiple exposures may be even greater than the sum of the risks from exposure to individual toxic chemicals. Most such synergisms have not been identified, let alone quantified.

Furthermore, chemical reactions involving two or more substances may form reaction products that are significantly more toxic than the starting reactants. This possibility of generating toxic reaction products may not be anticipated by the laboratory worker in cases where the reactants are mixed unintentionally. For example, successive treatments of a surface or an object with aqueous ammonia and then with sodium hypochlorite or other positive-halogen reagents may generate hydrazine, a substance that poses both acute and chronic toxicity hazards. Similarly, inadvertent mixing of formaldehyde (a common tissue fixative and a suspected human carcinogen) and hydrogen chloride could result in the generation of bis(chloromethyl)ether, a potent human carcinogen.

If the effects of the different hazards are to be considered as additive, then the following equation would apply for gases:

$$\frac{1}{(LD_{50})_1} + \frac{1}{(LD_{50})_2} + \cdots + \frac{1}{(LD_{50})_n} = \frac{1}{(LD_{50})_{mixture}}$$

If C_n is the observed atmospheric concentration of a substance with a TLV of T_n, then the following rule should apply:

$$\frac{C_1}{T_1} + \frac{C_2}{T_2} + \cdots + \frac{C_n}{T_n} < 1$$

Approaches in Dealing with Uncertainty

When there are insufficient data to permit clear decisions in assessment of a chemical in terms of human toxicity, two strategies are available. The first is to assume worst-case values; this strategy almost certainly ensures that calculated values will not be too high, but may yield values that are lower than really necessary. The EPA's calculation of cancer risk estimates is a prominent example of the application of this strategy.

A second approach to dealing with uncertainty is to employ uncertainty factors. These factors are intended to provide a sufficient "margin of error" to account for uncertainties arising from all of the possible sources described previously. The following list provides information on uncertainty factors.

GUIDELINES FOR SELECTION OF UNCERTAINTY FACTORS

1. Use a 10-fold factor when extrapolating from valid experimental results of studies on prolonged ingestion by humans. This 10-fold factor protects the sensitive members of the human population estimated from data gathered on average healthy individuals.

2. Use a 100-fold factor when extrapolating from valid results of long-term feeding studies on experimental animals when results of studies of human ingestion are not available or scanty (e.g., acute exposure only). This represents an additional 10-fold uncertainty factor in extrapolating data from the average animal to the average human.

3. Use a 1000-fold factor when extrapolating from short-term study results from experimental animals when no useful long-term or acute human data are available. This represents an additional 10-fold uncertainty factor in extrapolating from short-term to chronic exposures.

4. Use an additional uncertainty factor of between 1 and 10, depending on the sensitivity of the adverse effect, when deriving an ADI (acceptable daily intake) from a LOAEL (lowest-observed-*adverse*-effects level). This uncertainty factor drops the LOAEL into the range of a NOAEL (no-observed-*adverse*-effects level).

SUGGESTED READINGS

Other pertinent information may be obtained from the following sources.

American Conference of Governmental Industrial Hygienists, *Threshold Limit Values for Chemical Substances and Physical Agents in the Work Environment and Biological Exposure Indices with Intended Changes,* Cincinnati, Ohio. Use the latest edition.

Burgess, W., *Recognition of Health Hazards in Industry,* Wiley, New York, 1981.

Chemical Rubber Company, *Handbook of Chemistry and Physics,* Cleveland, Ohio. Use the latest edition.

Government Industries, Inc., *EPA Toxicology Handbook,* Rockville, Md., 1988.

International Agency for Research on Cancer, *IARC Monographs on the Evaluation of the Carcinogenic Risk of Chemicals to Man,* World Health Organization, Geneva, 1972–present (multivolume work).

Manufacturing Chemists Association, *Guide for Safety in the Chemical Laboratory,* Van Nostrand Reinhold, New York, 1972.

Merck and Company, *The Merck Index: An Encyclopedia of Chemicals and Drugs,* Rahway, N.J. Use the latest edition.

Ottoboni, M., *The Dose Makes the Poison,* Vincente Books, Berkeley, Calif., 1984.

Patty, F. A., *Industrial Hygiene and Toxicology,* Wiley, New York, 1962.

Proctor, N. and J. Hughes, *Chemical Hazards of the Workplace,* Lippincott, New York, 1988.

Sax, N. (ed.), *Condensed Chemical Dictionary,* Van Nostrand Reinhold, New York, 1987.

Sax, N., *Dangerous Properties of Industrial Materials,* Van Nostrand Reinhold, New York, 1988.

3
PROCEDURES FOR HANDLING TOXIC MATERIALS

INTRODUCTION

Controlling chemical hazards in the laboratory requires knowledge of the material and its hazards; a means to detect the presence and measure the amount of the material; procedures for protecting personnel from the hazard; the use of protective equipment, whether personal protection or built-in protection; and preparation and preplanning for emergencies.

As noted in Chap. 2, toxic substances can enter the body in three ways: by ingestion, by absorption through the skin, and by inhalation. The dangers of ingestion from food can be eliminated by thorough washing of the hands before eating. Direct contact with substances that can be absorbed through the skin should be avoided. Any splashes on unprotected skin should be washed off immediately with soap and water. The most widespread danger is the inhalation of harmful vapors, gases, and dusts. Aerosols should be treated with care. Materials likely to give rise to this hazard should be handled in a fume hood. Where such materials can occur in the laboratory air, TLVs should be adhered to.

Remember that, even for substances of no known significant hazard, it is prudent to observe good laboratory practice, minimizing exposure by working in an exhaust hood and wearing eye and hand protection and a laboratory coat or apron.

GENERAL CONTROL MEASURES

General control measures in the laboratory begin with knowing the material, whether or not it is toxic, and what the literature says about its effects. If this information is not readily available, the manufacturer should be asked to submit information about its safety and use in the form of a Material Safety Data Sheet (MSDS).

Other control measures have the goal of reducing exposure to a chemical. It is always a good idea to minimize the amount of a chemical that is handled in the laboratory at any time. In this way the amount that enters the atmosphere or can be spilled will be reduced, and the chances of exposure will be reduced. Maintain an effective barrier between the worker and the potentially hazardous environment in which toxic (and corrosive, flammable, or explosive) chemicals are present. Minimize contact through equipment and apparatus design. Use totally enclosed systems where possible. Adequate ventilation reduces the level of airborne materials such as hazardous gases, vapors, and dusts. This can be achieved in most laboratories by handling all toxic materials in the hood.

Protective clothing (safety glasses, gloves, and a lab coat) should be worn routinely when handling toxic chemicals. Other types of protective equipment should be used when engineering design and other methods of control are inadequate or unavailable. Aprons, face shields, and respirators are commonly used in the laboratory to minimize exposure.

Unnecessary contact with chemicals should be avoided unless it is well known that the material is completely harmless. If contact is unavoidable, the next precaution is to use excellent personal hygiene by washing the hands and all exposed skin frequently and taking a shower or bath no less frequently than at the end of each workday. Contaminated clothing should be removed following any severe exposure, and fresh clothing should be worn daily whenever there is substantial exposure.

Other Safety Precautions

1. Maintain a clean, orderly, and uncrowded workplace. Insist on meticulous housekeeping in the laboratory.

2. Perform routine preventive maintenance and cleaning to reduce contamination of the workplace to an absolute minimum. Clean up spills (e.g., mercury) at once.
3. Place trays, pans, or other containers under reactions carried out in glass and under chemicals in storage to retain contents in case of spill or breakage.
4. Provide the means to transport chemicals safely: All hazardous chemicals, especially corrosive and toxic liquids, and solvents should be carried in safety carriers fabricated from rubber, metal, or plastic to protect the glass container and provide containment in case of breakage.
5. Carry out reactions and transfers of highly hazardous chemicals behind barricades with slave-type manipulators (e.g., preparation and use of hydrazoic acid).
6. When methods are available, analyze the atmosphere in the work environment for the presence of hazardous airborne contaminants and compliance with OSHA regulations.
7. Develop and enforce safety rules and good practice techniques.
8. Train students and other laboratory personnel in the following:
   (a) Techniques in the assembly of equipment to reduce chemical contact
   (b) Techniques in the transfer of chemicals from container to container, weighing, pipetting, volumetric procedures, filtration, drying, grinding, and mixing
   (c) Proper handling of compressed gases
   (d) Preparation for the unexpected
   (e) Recognition of hazardous conditions, and how to apply corrective measures

Potential Problem Areas

Prepare and be vigilant for these problems with toxic chemicals:

1. Careless technique—the first and most important hazard may be the individual worker. Eternal vigilance is the cost of an injury-free career.
2. Unknown toxicity—Assume that a chemical is toxic until proven otherwise. Nitrogen-containing compounds are usually physiologically active.
3. Contamination of equipment and work area—the most serious source of chemical contact, especially chronic exposure.
4. Airborne hazards—failure to employ sufficient ventilation, inadequate hood design, or abuse of hood facilities.
5. Spills caused by upsetting containers (beakers, bottles, flasks, etc.) or broken containers account for a significant percentage of accidental release of chemicals.
6. Careless handling of compressed gases—leaking fittings, improper use of accessories such as gauges and valves.
7. Storage of chemicals (see Chap. 4). The following are particular problems:
   (a) Failure to label or use of incorrect labels leads to disaster sooner or later.
   (b) Incompatible chemicals stored together, i.e., concentrated ammonium hydroxide and concentrated nitric acid, methanol and perchloric acid, etc.
   (c) Failure to provide a tray, pan, or other container to retain contents of container in case of breakage.
   (d) Unventilated cabinets, closets, etc.
   (e) Chemicals should be dated upon receipt, upon opening, and when made up into solutions or mixtures.
   (f) Failure to discard chemicals after the safe storage period has expired (e.g., decomposition of cellulose nitrate).
   (g) Excessive temperature in storage.
8. Disposal of waste chemicals—for toxic materials, incineration, and lab packs are usually the best options. Sometimes neutralization of a chemical's toxic characteristics is a possibility.
9. Explosions and implosions—scattering of the hazardous chemical in addition to flying debris (glass hazard).
10. Fire can generate toxic vapors and disperse toxic materials.
11. Highly hazardous materials
    (a) Carcinogens—usually involve long induction periods.
    (b) Acute toxicity hazards—ricin from castor beans, the acetylcholine esterase inhibitors (e.g., many insecticides), Kreb's cycle inhibitors (e.g., sodium fluoroacetate, curare, etc.), cyanides, mercury.
12. Chronic toxicity hazards (beryllium, asbestos, lead, mercury, cadmium, chromium, etc.)—

avoid accumulating small doses over long periods.
13. The mercury problem—many laboratories are contaminated by broken thermometers, manometers, etc. (See *Fundamentals of Laboratory Safety: Physical Hazards in the Academic Laboratory,* Chap. 12.)
14. Reaction of nitric acid with reducing agents to produce toxic nitrogen oxides.
15. Atmospheric oxidation may produce toxic products: Paraphenylenediamine produces a skin sensitizer, probably the aldoxime.
16. Phosgene from chlorinated hydrocarbons on hot metal surfaces or flames.
17. Dust control—usually a minor problem, but poor planning, poor housekeeping, or carelessness can generate problems. Welding, brazing, grinding, sand blasting, and even just handling some materials may disperse toxic dusts.

SPECIFIC PROCEDURES FOR TOXIC CHEMICALS

The high toxicity of some chemicals demands special handling techniques. Some specific procedures are given below.

1. *Carcinogens.* OSHA has published detailed procedures (29 CFR 1910, parts 1000–1500) for working with substances that have been classified as carcinogens. Anyone contemplating work with materials on this list should consult the regulations regarding the necessary approvals, training, working conditions, monitoring, recordkeeping, and medical surveillance. In addition, if a worker anticipates that an OSHA-regulated carcinogen might be a product or an impurity, the regulations should be consulted. For example, N-phenyl-1,4-benzenediamine (simidine) can contain substantial benzidine impurity.

    See Appendix D for a list of carcinogens and suspected carcinogens. In addition, the International Agency for Research on Cancer (IARC) has published a list of chemicals considered to pose a carcinogenic risk, and the Environmental Protection Agency (EPA) has published a list of substances subject to labeling requirements under the Toxic Substances Act of 1976.

2. *High chronic toxicity.* Procedure A (described later in this chapter) should be followed in laboratory operations involving those substances believed to be moderately to highly toxic, even when used in small amounts. A substance that has caused cancer in humans or has shown high carcinogenic potency in test animals (but for which a regulatory standard has not been issued by OSHA) will generally require the use of Procedure A. However, before choosing Procedure A, other factors, such as the physical form and volatility of the substance, the kind and duration of exposure, and the amount to be used, should also be considered.

    A substance is deemed to have moderate to high carcinogenic potency in test animals if it causes statistically significant tumor incidence (a) after inhalation exposure of 6–7 hr per day, 5 days per week, for a significant portion of a lifetime to dosages of less than 10 mg/m^3; or (b) after repeated skin application of less than 300 mg/kg of body weight per week; or (c) after oral dosages of less than 50 mg/kg of body weight per day.

3. *High acute toxicity or moderate chronic toxicity.* Procedure B should be followed in laboratory operations, employing substances for which infrequent, small quantities do not constitute a significant carcinogenic hazard but which can be dangerous to those exposed to high concentrations or repeated small doses. A substance that is not known to cause cancer in humans but that has shown statistically significant, though low, carcinogenic potency in animals, generally should be handled according to Procedure B.

Procedure B: For Substances of Moderate Chronic or High Acute Toxicity

Before beginning a laboratory operation, each worker is strongly advised to consult MSDSs, the Chemical Entries section in this volume, or one of the standard compilations that list toxic properties of known substances and learn what is known about the substances to be used. The precautions and procedures described below (termed Procedure B) should be followed if any of the substances to be used in significant quantities is known to be moderately or highly toxic. (If any of the substances being used is known to be highly toxic, it is desirable that two people be present in the area at all times.) These procedures should also be followed if the toxicological properties of any of the substances being used or prepared are unknown. If any of the substances to be used or prepared are known to have high chronic toxicity (e.g., compounds of heavy metals and strong carcinogens), then the precautions and procedures described below should be supplemented with additional precautions (termed Procedure A) to aid in containing and ultimately destroying substances that have high chronic toxicity.

The overall objective of Procedure B is to minimize exposure of the laboratory worker to toxic substances, by any route of exposure, by taking all reasonable precautions. Thus, the general precautions outlined in other chapters should normally be followed whenever a toxic substance is being transferred from one container to another or is being subjected to some chemical or physical manipulation. The following three precautions should always be followed:

1. Protect the hands and forearms by wearing either gloves and a laboratory coat or suitable long gloves (gauntlets) to avoid contact of toxic material with the skin.
2. Procedures involving volatile substances and those involving solid or liquid toxic substances that may result in the generation of aerosols should be conducted in a hood or other suitable containment device (see *Fundamentals of Laboratory Safety: Physical Hazards in the Academic Laboratory*, Chap. 9). (The hood should have been evaluated previously to establish that it is providing adequate ventilation and has an average face velocity of not less than 150 linear feet per minute.
3. After working with toxic materials, wash the hands and arms immediately. Never eat, drink, smoke, chew gum, apply cosmetics, take medicine, or store food in areas where toxic substances are being used.

These standard precautions will provide laboratory workers with good protection from most toxic substances. In addition, records that include amounts of material used and names of workers involved should be kept as part of the laboratory notebook record of the experiment. To minimize hazards from accidental breakage of apparatus or spillage of toxic substances in the hood, containers of such substances should be stored in pans or trays made of polyethylene or other chemically resistant materials, and apparatus should be mounted above trays of the same type of material. Alternatively, the working surface of the hood can be fitted with a removable liner of adsorbent, plastic-backed paper. Such procedures will greatly simplify subsequent cleanup and disposal. Vapors that are discharged from the apparatus should be trapped or condensed to avoid adding substantial amounts of toxic vapor to the hood exhaust air. Areas where toxic substances are being used and stored should have restricted access, and specific warning signs should be posted if a special toxicity hazard exists.

The general waste disposal procedures described in this volume should be followed. However, certain additional precautions should be observed when waste materials are known to contain substances of moderate to high toxicity.

Volatile toxic substances should never be disposed of by evaporation in the hood. If practical, waste materials and waste solvents containing toxic substances should be decontaminated chemically by some procedure that can reasonably be expected to convert essentially all of the toxic substance to nontoxic substances. If chemical decontamination is not feasible, waste materials and solvents containing toxic substances should be stored in closed, impervious containers so that personnel handling the containers will not be exposed to their contents. In general, liquid residues should be contained in glass or polyethylene bottles half-filled with vermiculite. In some institutions, nonreactive liquids are incinerated in a solvent burner. In such cases it may be appropriate to add used motor oil to reduce flammability; vermiculite or other solids should not be added. All containers of toxic wastes should be suitably labeled to indicate the contents (chemicals and approximate amounts) and the type of toxicity hazard that contact may pose. For example, containers of wastes from experiments involving appreciable amounts of weak or moderate carcinogens should carry the warning: CANCER-SUSPECT AGENT. All wastes and residues that have not been chemically decontaminated in the exhaust hood where the experiment was carried out should be disposed of in a safe manner that ensures that personnel are not exposed to the material (see Chap. 5).

The laboratory worker should be prepared for possible accidents or spills involving toxic substances. If a toxic substance contacts the skin, the area should be washed well with water or a safety shower should be used. If there is a major spill outside the hood, the room or appropriate area should be evacuated and necessary measures to prevent exposure of other workers should be taken. Spills should be cleaned up by personnel wearing suitable personal protective apparel. If a spill of a toxicologically significant quantity of toxic material occurs outside the hood, a supplied-air, full-face respirator should be worn. Contaminated clothing and shoes should be thoroughly decontaminated or incinerated.

Procedure A: For Substances of High Chronic Toxicity

All the procedures and precautions described in Procedure B should be followed when working with substances known to have high chronic toxicity. In addition, when such substances are to be used in quantities in excess of a few milligrams to a few grams (depending on the hazard posed by the particular substance), the additional precautions described below should be used. Each laboratory worker's plans for experimental work and for disposing of waste materials should be approved by the laboratory supervisor. Consultation with the departmental safety coordinator may be appropriate to ensure that the toxic material is effectively contained during the experiment and that waste materials are disposed of in a safe manner. Substances in this high-chronic-toxicity category include certain heavy metal compounds (e.g., dimethylmercury and nickel carbonyl) and compounds normally classified as strong carcinogens. Examples of compounds frequently considered to be strong carcinogens include the following: benzo[a]pyrene; 3-methylcholanthrene; 7,12-dimethylbenz[a]anthracene; dimethylcarbamoyl chloride; hexamethylphosphoramide; 2-nitronaphthalene; propane sultone; many N-nitrosamines; many N-nitrosamides; bis(chloromethyl)ether; aflatoxin B_1; and 2-acetylaminofluorene.

An accurate record of the amounts of such substances being stored and of the amounts used, dates of use, and names of users should be maintained. It may be appropriate to keep such records as part of the record of experimental work in the laboratory workers' research notebooks, but it must be understood that the research supervisor is responsible for ensuring that accurate records are kept. Any volatile substances exhibiting high chronic toxicity should be stored in a ventilated storage area in a secondary tray or container with sufficient capacity to contain the materials should the primary container accidentally break. All containers of substances in this category should have labels that identify the contents and include a warning such as the following:

WARNING! HIGH CHRONIC TOXICITY
or
CANCER-SUSPECT AGENT

Storage areas for substances in this category should have limited access, and special signs should be posted if a special toxicity hazard exists. Any area used for storing substances of high chronic toxicity should be maintained under negative pressure with respect to surrounding areas.

All experiments with and transfers of such substances or mixtures containing such substances should be done in a controlled area. (*Note:* A controlled area is a laboratory, a portion of a laboratory, or a facility such as an exhaust hood or a glove box that is designated for the use of highly toxic substances. Its use need not be restricted to the handling of toxic substances if all personnel who have access to the controlled area are aware of the nature of the substances being used and the precautions that are necessary.) When a negative-pressure glove box in which work is done through attached gloves is used, the ventilation rate in the glove box should be at least two volume changes per hour, the pressure should be at least 0.5 in. of water lower than that of the external environment, and the exit gases should be passed through a trap or HEPA filter. Positive-pressure glove boxes are normally used to provide an inert anhydrous atmosphere. If these glove boxes are used with highly toxic compounds, then the box should be thoroughly checked for leaks before each use and the exit gases should be passed through a suitable trap or filter. Laboratory vacuum pumps used with substances having high chronic toxicity should be protected by high-efficiency scrubbers or HEPA filters and vented into an exhaust hood. Motor-driven vacuum pumps are recommended because they are easy to decontaminate. (*Note:* Decontamination of a vacuum pump should be carried out in an exhaust hood.) Controlled areas should be clearly marked with a conspicuous sign such as the following:

WARNING: TOXIC SUBSTANCE IN USE
or
CANCER-SUSPECT AGENT: AUTHORIZED PERSONNEL ONLY

Only authorized and instructed personnel should be allowed to work in or have access to controlled areas.

Proper gloves (see *Fundamentals of Laboratory Safety: Physical Hazards in the Academic Laboratory,* Chap. 10) should be worn when transferring or otherwise handling substances or solutions of substances having high chronic toxicity. In some cases, the laboratory worker or the research supervisor may deem it advisable to use other protective apparel, such as an apron of reduced permeability covered by a disposable coat. Precautions such as these might be taken, for example, when handling large amounts of certain heavy metals and their derivatives or compounds known to be potent carcinogens. Surfaces on which high-chronic-toxicity substances are handled should be protected from contamination by using chemically resistant trays or pans that can be decontaminated after the experiment or by using dry, absorbent, plastic-backed paper that can be disposed of after use.

Upon leaving a controlled area, laboratory workers should remove any protective apparel that has been used and thoroughly wash hands, forearms, face, and neck. If disposable apparel or absorbent paper liners have been used, these items should be placed in a closed and impervious container that should then be labeled in some manner such as the following:

CAUTION: CONTENTS CONTAMINATED WITH SUBSTANCES OF HIGH CHRONIC TOXICITY

Nondisposable protective apparel should be thoroughly washed, and containers of disposable apparel and paper liners should be incinerated.

Wastes and other contaminated materials from an experiment involving substances of high chronic toxicity should be collected together with the washings from flasks and such and either decontaminated chemically at the work area or placed in closed, suitably labeled containers for disposal. If chemical decontamination is to be used, a method should be chosen that can reasonably be expected to convert essentially all of the toxic materials into nontoxic materials. For example, residues and wastes from experiments in which β-propiolactone has been used can be treated with sodium hydroxide (as described in *Prudent Practices for the Disposal of Chemicals from Laboratories*, pp. 67–68) to produce β-hydroxypropionic acid, a relatively nontoxic material that may be flushed down the laboratory drain. Consult the pertinent sections in Chap. 5 before disposing of any chemical down the drain.

In the event that chemical decontamination is not feasible, wastes and residues should be placed in an impervious container that is closed and labeled in some manner such as the following:

CAUTION: COMPOUNDS OF HIGH CHRONIC TOXICITY
or
CAUTION: CANCER-SUSPECT AGENT

Transfer of contaminated wastes from the controlled area should be done under the supervision of authorized personnel and in such a manner as to prevent spill or loss. (Chapter 5 describes the packaging and handling of chemical wastes to be disposed of by incineration or labpack.) In general, liquid wastes containing such compounds should be placed in glass or (usually preferable) polyethylene bottles half-filled with vermiculite, and these should be transported in plastic or metal pails of sufficient capacity to contain the material in case of accidental breakage of the primary container.

Normal laboratory work should not be resumed in a space that has been used as a controlled area until it has been adequately decontaminated. Work surfaces should be thoroughly washed and rinsed. If experiments have involved the use of finely divided solid materials, dry sweeping should not be done. In such cases, surfaces should be cleaned by wet mopping or by use of a vacuum cleaner equipped with a HEPA filter. All equipment (e.g., glassware, vacuum traps, and containers) that is known or suspected to have been in contact with substances of high chronic toxicity should be washed and rinsed before they are removed from the controlled area.

In the event of continued experimentation with a substance of high chronic toxicity (i.e., if a worker regularly uses toxicologically significant quantities of such a substance three times a week), a qualified physician should be consulted to determine if it is advisable to establish a regular schedule of medical surveillance or biological monitoring.

A Specific Example for Polycyclic Aromatic Hydrocarbons

This information has been taken from the chemical entry for benzo[a]pyrene (3,4-benzpyrene). Although the information given below applies specifically to benzo[a]pyrene (a potent carcinogen), the general precautions and procedures are also applicable to other carcinogenic polycyclic aromatic hydrocarbons.

All work with benzo[a]pyrene in quantities in excess of a few milligrams or capable of resulting in the formation of aerosols should be carried out in a well-ventilated hood or in a glove box equipped with a HEPA filter. All work should be carried out in apparatus that is contained in or mounted above unbreakable pans that will contain any spill, except that, for very small amounts, a disposable mat may be adequate to contain possible spills. All containers should bear a label such as CANCER-SUSPECT AGENT. All personnel who handle benzo[a]pyrene should wear plastic or latex gloves and a fully buttoned laboratory coat.

Storage and use—All bottles of benzo[a]pyrene should be stored and transported in unbreakable outer containers.

Cleanup of spills and waste disposal—Disposal of benzo[a]pyrene is best carried out by oxidation; this can be accomplished by high-temperature incineration or by the use of strong oxidants such as chromic acid cleaning solution. The latter procedure is especially applicable to small residues on glassware and such. For incineration of liquid wastes, solutions should be neutralized if necessary, filtered to remove solids, and put in a polyethylene container for transport. All equipment should be thoroughly rinsed with solvent to decontaminate it, and this solvent should be added to the wastes to be incinerated. Great care should be exercised to prevent contamination of the outside of the solvent container.

Solid reaction wastes should be incinerated or decomposed by other oxidation procedures. Alternatively, solid reaction wastes can be extracted with solvent that is added to other liquid waste for incineration. Any contaminated rags, paper, and such should be incinerated. Contaminated solid materials should be enclosed in sealed plastic bags that are labeled CANCER-SUSPECT AGENT and with the name and amount of the carcinogen. The bags should be stored in a well-ventilated area until they are incinerated.

Animal Experiments with Substances of High Chronic Toxicity

The use in experimental animals of substances that have shown high chronic toxicity can present a special exposure hazard, particularly because of the possibility of the formation of aerosols or dusts that contain the toxic substance. Such dusts and aerosols can be dispersed throughout the laboratory or animal quarters through animal food, urine, or feces. Accordingly, procedures should be devised that reduce the formation of such aerosols and dusts to the lowest possible level. All procedures should be designed to minimize the possible exposure of personnel (see *Fundamentals of Laboratory Safety: Physical Hazards in the Academic Laboratory,* Chaps. 9 and 10).

Administration of substances by injection or gavage is preferable. However, if substances are to be administered in the diet, a relatively closed caging system—either one in which the cages are under negative pressure or one in which there is a horizontal laminar airflow directed toward HEPA filters—should be used. Procedures such as the use of a vacuum cleaner equipped with a HEPA filter or wetting the bedding to reduce dusts should be used for the removal of contaminated bedding or cage matting. All toxic substance-containing diets should be mixed within closed containers and within a hood.

Workers carrying out such operations should wear plastic or rubber gloves and a fully buttoned laboratory coat or equivalent clothing at all times. If exposure to aerosols cannot be controlled in other ways, a respirator should be used.

When large-scale studies are being carried out with highly toxic substances, special facilities or rooms with restricted access are preferable. If the caging system for test animals does not adequately protect personnel, the use of a jumpsuit or similar clothing and shoe and head coverings should be considered.

Other Specific Hazards

Beryllium. Beryllium is exceptionally poisonous—about 50 times more toxic than mercury. Stringent precautions are therefore needed in handling it, similar to those necessary with toxic radioactive materials. Before using or even ordering beryllium or any beryllium compounds, the safety officer or committee should be consulted.

Carbonyls. Carbon monoxide forms highly toxic carbonyls with metals. Nickel carbonyl vapor is exceedingly poisonous. Cases of lung cancer have occurred as a result of long exposure. Nickel carbonyl may also explode when heated to 60°C.

Chromium Solutions. Solutions used in chromium plating and chromic acid can cause chrome ulceration of the skin and nasal mucous membranes. Some chromium salts are carcinogens. Chromium may also produce sensitization dermatitis. Protective clothing should therefore be worn when working with these substances, and operations should be adequately ventilated. Chromium solutions must be disposed of properly. See Chromium compounds in the Chemical Entries section.

Corrosives. Many of the general recommendations for safe practices in the laboratory are especially applicable to the handling of corrosive substances. It is also important that attention be given to the use of protective apparel and safety equipment. In addition, the storage, disposal, and cleanup of corrosive substances require special care. Bottles of corrosive liquids should be stored in acid containers or in polyethylene or lead trays or containers large enough to contain the contents of the bottles; most major suppliers will provide acids in plastic-coated glass bottles, which are much less likely to break than ordinary bottles. To ensure that mutually reactive chemicals cannot accidentally contact one another, such substances should not be stored in the same trays unless they are in unbreakable, corrosion-resistant secondary containers. In disposing of corrosive substances, care must be taken not to mix them with other potentially reactive wastes. In most cases, spills involving these substances should be contained, carefully diluted with water, and then neutralized.

Cyanides. Cyanide poisoning can occur by adsorption through the skin or the digestive or respiratory systems. If cyanide is taken through the mouth, the effects can be instantly fatal.

Cyanide stocks should be kept in a closed vessel clearly marked "POISON." Containers that have been used for cyanide should be destroyed. When cyanide solutions are spilled, the area should be evacuated and the spill treated (by a responsible officer wearing self-contained breathing apparatus) with an excess of bleaching powder or sodium hypochlorite solution.

Cyanide solutions should not be left open to the atmosphere, as they can become acidic via the absorption of CO_2 and release HCN to the atmosphere.

Cyanide solutions that are no longer needed can be rendered harmless by treating them with an excess amount of sodium hypochlorite in a fume hood. The solution will then be safe to be disposed of. An alternative method is to add excess sulfur to an alkaline slurry of the cyanide. When the mixture is heated, a thiocyanate (e.g., NaCNS) is formed, which can be safely disposed of.

Epoxy Resins. Epoxy resins are liable to cause a severe form of dermatitis with consequent inflammation, swelling, and/or pustules. Symptoms are usually localized on the hands and arms, but sometimes they affect the whole body. They are most likely to occur in people who have sensitive skins, but anyone can develop this sensitivity. The hardener is more dangerous than the resin itself.

The following rules are recommended by the manufacturers:

1. Always work in a well-ventilated room.
2. Keep strictly to the mixing percentages indicated by the suppliers, as the resin/hardener reaction gives an immediate increase in the temperature, and if a certain critical mass is reached, instantaneous boiling or an explosion is possible.
3. The utmost personal cleanliness is essential. Before working with the resins, wash the hands carefully, then smear the hands and the fingertips with a barrier cream. Hands must also be washed well as soon as the work is finished and before eating, smoking, drinking, etc. Special resin removal creams are also available. Particular care should be taken to avoid scratching the skin, rubbing the eyes, and any other action likely to spread contamination.
4. In the event that unprotected skin is splashed, it should be washed with soapy water or with a solution containing 3% acetic acid. Such a solution should be permanently available in workshops where such resins are used.
5. If the skin is affected, a doctor should be consulted.
6. Special care should be observed when machining a finished article to remove excess adhesive. This process can release surplus hardener in the form of fine dust, which is particularly dangerous.
7. Persons who find themselves allergic to this material should not, of course, continue to work with it.

Gases and Vapors. TLVs for many compounds can be found in the Chemical Entries section. They refer to conditions in which people are exposed to chemicals throughout the working day. Where this applies, regular checks must be made of the concentrations in the air.

When two or more hazardous substances are present, their combined effect, rather than that of either individually, should be given primary consideration: The effects of the different hazards can sometimes be additive. If C_n is the observed atmospheric concentration of a substance with a TLV of T_n, then the following rule should apply:

$$\frac{C_1}{T_1} + \frac{C_2}{T_2} + \cdots + \frac{C_n}{T_n} < 1$$

Heavy Metals. Many toxic heavy metals can be inhaled as vapor or dust. The appropriate TLVs should be strictly adhered to. Particular compounds of these metals can have quite different toxicities, certain substances having an affinity for particular organs.

Many heavy metals should be machined or used at elevated temperatures only in a well-ventilated room. The use of a protective mask will reduce the quantity inhaled. It is also important to wash thoroughly after handling these materials.

Particular care should be taken when using molten tin baths.

Welding Vapors and Ozone. The melting of the flux around arc-welding electrodes generates vapors that, over a long period of time, can have a toxic effect on the digestive tract. Ozone is also formed when air is ionized, as in an electric arc. Nitrous fumes are produced under certain circumstances. Exposure to any of these materials can lead to acute lung conditions. In addition, welding of painted metal parts is also dangerous because of the possibility of lead in the paint. All arc welding should therefore be performed in the open air or in a well-ventilated shop. Welding in confined spaces is particularly dangerous.

Other Rules for Handling Toxic Materials

In addition to those substances, procedures, and precautions discussed in this section, certain state and federal regulatory agencies have listed substances whose use and disposal is regulated. These lists of regulated substances are usually accompanied by specific requirements for use and disposal. The lists of substances and the requirements change frequently. Because compliance with these regulations is required by law, every departmental safety coordinator and all research supervisors should know the current lists of regulated substances and the requirements for their use and disposal.

QUALITY CONTROL—EVALUATION OF THE EFFICIENCY OF THE EXPOSURE CONTROL PROGRAM

Determining how well a laboratory's toxic hazards have been controlled is best done by monitoring concentrations of the toxic chemicals in the laboratory environment or in the worker:

1. Personnel monitoring of airborne hazards by means of sampling instruments attached to the clothing of employees. Samples of breathing air should be collected continuously during the work assignment.
2. Biological monitoring—laboratory workers are the most reliable samplers of their ambient environment. Analysis of their blood, breath, or urine will disclose the presence of foreign chemicals absorbed from contact in the workplace. By combining personnel monitoring with biological monitoring, absorbed doses and their physiological effects can be determined and appropriate exposure control procedures developed to maintain severity within tolerable limits.

MATERIAL SAFETY DATA SHEETS

The information provided on Material Safety Data Sheets (MSDS) can be disappointing. Manufacturers may make a minimal attempt to comply with the letter of the law. Misspellings and inaccurate information on some MSDSs indicate that the individuals filling out the forms are less than familiar with the subject. Despite their drawbacks, however, the MSDS still represents a simple method of obtaining valuable information about materials that are used daily in a laboratory. The data sheets must be provided by the manufacturer upon request.

MSDSs are very useful in a production environment where only a few different materials are used. In a laboratory, however, where hundreds of different chemicals may be kept on hand, maintaining an up-to-date, organized file or notebook of MSDSs can be a big, ongoing project.

SUGGESTED READING

National Research Council, *Prudent Practices for Disposal of Chemicals from Laboratories.* National Academy Press, Washington, D.C., 1983.

4
CHEMICAL LABELING AND STORAGE

INTRODUCTION

Mishandling of chemical reagents has long been a problem in science laboratories of all types. The literature describes many instances of explosions, fires, poisonings, burns, and other bodily injuries caused by improper or careless handling of chemical reagents.

Misuse of chemical reagents does not necessarily involve just problems occurring while the reagents are actually being used. Misuse can also consist of improper chemical reagent storage, improper recordkeeping, improper labeling, or improper purchasing and procurement programs. The proper handling of chemical reagents is a total program in itself, comprised of procurement, recordkeeping, storage, education, usage, and disposal.

LABELING

Proper labeling is one of the fundamental aspects of safe and effective laboratory operation. Materials that are made in the laboratory, in the course of experiments, require labeling just as much as purchased chemicals that reside on the storeroom shelf.

Labeling has two principal functions. First is adequate identification—what the material is and where it came from. Second is precautionary information about safe handling of any chemical posing a significant hazard.

Purchased reagents always have either the chemical or common name of the material, the name of the manufacturer, and a lot number. If the chemical is flammable, a flash point may be shown on the label, accompanied by a precaution regarding fire hazard. If the material is corrosive, toxic, reactive, or unstable, other precautionary information is usually shown. Absence of precautions on the label does not necessarily mean, however, that there is no hazard, since not all manufacturers are diligent in presenting such information.

Solutions, mixtures, reaction products, and other materials generated in the school laboratory should be labeled unless their existence is momentary. Laboratory glassware or bottles containing such materials must be labeled for identification purposes with the name of the materials, the name of the person who made it, and the date it was made. If there is any material hazard, an indication of the type of hazard should at least be indicated on the label. If lids are labeled, this must be in addition to the body label.

Absence of labels leads to errors and can be very dangerous. It is not unusual to walk into a science laboratory and see reagent bottles without identifying labels. Labels sometimes become illegible from contact with chemicals during storage. No such containers should be allowed in the laboratory. Proper handling and disposal of a chemical requires knowing what it is. If the contents of a container cannot be identified, they must be analyzed to determine identity before disposal (see Chap. 5).

There are several federal government requirements for labeling chemical containers. The Federal Hazardous Substances Act, now administered by the Consumer Product Safety Commission, requires precautionary labeling on all flammable, corrosive, reactive, toxic, or radioactive substances intended for nonindustrial use. Unlabeled reagents violate the Occupational Safety and Health Act. The U.S. Department of Transportation requires certain shipping labels on packages of hazardous materials carried interstate. The Occupational Safety and Health Act has a general-duty clause requiring an employer to provide a safe place of employment. Unlabeled or inadequately labeled reagents or other materials can be construed to be a violation of this act.

In 1944 the Manufacturing Chemists Association, realizing the crucial importance of precautionary labeling, formed its Labels and Precautionary Information Committee and subsequently published six editions of *Guide to Precautionary Labeling of Haz-*

ardous Chemicals (see Manufacturing Chemists Association, 1945). The work of this committee and its publications have been the basis of much current legislation and practice.

Liability suits often are based on questions of adequacy of the label of a material involved in an accident. Good labeling practice will help protect the teacher from such litigation.

Information on Labels

Labels should serve the following functions:

1. Show clearly what the material is
2. Show the manufacturer's name and address
3. Indicate the age of the material
4. Indicate possible hazards of the contents and suggest handling precautions

Material hazards fall into two classes: physical hazards and health hazards. Physical hazards are immediately evident and include fire, chemical burns, reactivity, and explosion potential. Health hazards are more subtle, and sometimes their effects are not evident for a period of time. They result from inhalation, skin absorption, and ingestion of toxic substances.

The Manufacturing Chemists Association in the *Guide for Safety in the Chemical Laboratory* has adopted the philosophy that "Chemicals in any form can be safely stored, handled, or used if the physical, chemical, and hazardous properties are fully understood and the necessary precautions, including the use of proper safeguards and personal protective equipment, are observed." The label should convey the information necessary to promote safe handling.

The NIOSH criteria document, *An Identification System for Occupationally Hazardous Materials, A Recommended Standard,* recommends that the label contain the following:

1. The trade name or chemical name of the product
2. A hazard symbol consisting of three rectangles containing terse indications of relative health hazard, fire hazard, and reactivity hazard
3. Appropriate statements on the nature of the hazard
4. Appropriate action statements
5. Emergency action and first-aid statements
6. Cleanup and disposal statements where appropriate

The American National Standards Institute (ANSI), in Standard Z129.1, requires, at a minimum, the following label components:

1. Identification of the container's contents
2. A signal word and summary description of any hazards
3. Precautionary information—what to do to minimize the hazard or prevent an accident from happening
4. First aid in case of exposure
5. Spillage and cleanup procedures
6. If appropriate, special instructions to physicians

At times it may not seem practical to put all the desired information on the label, especially if the container is small. In this case the instructor must use judgment in selecting the information that should appear on the label. The most essential item is the chemical or product name. A word giving the principal hazard characteristic may be next in importance.

A Model Labeling System

The following is a suggested labeling system that provides the types of information to be included in a label and a general format for the label.

A LABELING SYSTEM

Substance name/formula

Supplier

Date received

Date opened

Waste disposal procedure (code)

TLV (ACGIH) ppm (mg/m^3)

NFPA hazard signals

Specific gravity

Vapor density

Flash point

Ignition temperature

Flammable or explosive limits in air

Boiling point

Melting point

Solubility in H_2O, g/100 g

Emergency treatment

BLANK SAMPLE

Chemical name						
Formula		Flash point				
Kinds of hazards	Health	0	1	2	3	4
	fire	0	1	2	3	4
	reactivity	0	1	2	3	4
	specific					
Emergency treatment						
Supplier						
Date received _____ Prepared _____ Purchased _____						
Usage						
Other information						

Appendix E contains a list of standard chemical label statements describing health hazards, fire hazards, reactivity hazards, first-aid actions, and disposal features.

Interpreting Labels

Chemical labels are generally self-explanatory. However, existing regulations do not require that all chemical labels contain complete information. Commonly an explanation of the hazard presented by a particular compound is handled by a phrase such as "avoid skin contact," "avoid inhalation," or "keep away from fire, sparks, and open flame." While these statements may apprise the user of the general hazard involved, they do not necessarily indicate the degree of hazard.

Laboratory workers become inured to the phrases "avoid skin contact" and " avoid inhalation of vapor." They are rarely appreciated and routinely ignored. After spending time working in science laboratories, students may acquire a similar lack of concern for the health hazards presented by chemical reagents. Assigned research into the actual health effects of chemical reagents used in laboratory work should be an important part in any laboratory project.

Labels used on mixtures of chemical reagents or solvent mixtures can also be inadequate. Generally, only the primary components are listed. Minor chemical compounds that may actually be more toxic and present a greater danger to the user may be ignored. Instructors who are involved in purchasing chemicals for science laboratory use should make a practice of requesting the material safety data sheet (MSDS) from the supplier or manufacturer of the chemical. The MSDS will give relevant information about the physical and toxicological properties of the chemical. Frequently the MSDS will contain information about minor components or impurities that is lacking on the label.

Many chemical labels are now using the National Fire Protection Association (NFPA) diamond symbol as a shorthand method of informing the user of the dangers presented by the particular chemical. Figure 4–1 shows a typical diagram. This diagram is based on a hazard identification system proposed in the NFPA's *Recommended System for the Identification of the Fire Hazards of Materials* (NFPA 704-1975). See Appendix B. This system is designed primarily to aid in understanding the hazards encountered during a fire. However, it is also useful as a general laboratory warning system.

The NFPA diagram is divided into four segments. The top segment indicates the flammability hazard; the left, the health hazard; the right, the reactivity hazard. The bottom segment is used to identify any special characteristics of which the user should be aware. For example, a W with a line through the middle indicates unusual reactivity with water. Oxidizing chemicals are identified by an OXY in the bottom segment, and radiation hazards are identified with the radiation symbol.

The degree of each hazard is indicated by a number in the appropriate segment. The number scale ranges from 0 to 4, with materials designated as 0 presenting little or no hazard in that particular category. Materials with a 4 designation are extreme hazards in the category so designated.

As an example, for the health hazard category (the left segment), the numbers have the following meanings:

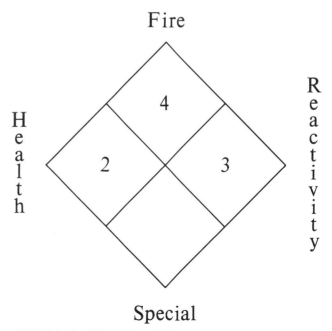

FIGURE 4–1. NFPA Hazard Diagram. (From NIOSH, *Safety in the School Science Laboratory, Instructor's Resource Guide*, 1979, Exhibit 8-1, p. 8-17.)

A 4 indicates that a few whiffs of the gas or vapor or skin contact with a small amount could cause death. Special protective clothing designed specifically to protect against this particular hazard should be worn by persons entering an area where exposure is likely.

A 3 indicates that the material is extremely hazardous to health, but areas may be entered with extreme care. Full protective clothing and self-contained breathing equipment must be worn in areas where exposure is likely.

A 2 indicates that the material is hazardous to health, but that areas may be entered freely with a self-contained breathing apparatus.

A 1 indicates that the material is only slightly hazardous to health.

A 0 indicates that the material presents no greater health hazard under fire conditions than ordinary combustible materials.

The health hazard designations refer only to the immediate, acute effects of exposure to the chemical. Chronic, long-term health effects are not taken into account.

Examples

Nitric acid

The top number represents flammability. Zero indicates that nitric acid is nonflammable.

The left number indicates health hazard. The fumes are toxic and the liquid causes severe tissue burns, which equates to a number 3 rating.

The right number indicates reactivity. Nitric acid has a zero rating because of its lack of reaction with water. This is somewhat misleading, however, since nitric is a strong oxidizer and is very reactive with many other chemicals.

The oxy indicates that nitric acid is an oxidizer.

Sodium

The top number represents flammability. Sodium may ignite spontaneously upon exposure to air.

The left number indicates health hazard. Sodium causes severe skin and eye burns, which equates to a number 3 rating.

The right number indicates reactivity. The 2 means that sodium may react violently with water.

The ⩘ indicates that sodium reacts dangerously with water.

All chemicals and solvents, especially those with hazard ratings of 2 or higher in any category, should be carefully controlled, well labeled, properly stored, and discretely used. Remember, they can be handled safely if their hazardous properties are understood and precautions are heeded.

ORDERING AND RECEIVING CHEMICALS

Purchasing Philosophies: Cost Versus Safety

Procurement programs generally have one fundamental rule: Get the largest quantity for the lowest possible price. Only occasionally are other factors considered important enough to override the lowest-possible-price rule. The purchase of chemical reagents is one area where the cost factor is frequently outweighed by safety and health considerations.

Chemical reagents can be purchased at a much lower cost per unit size in bulk quantities than in smaller quantities. Purchasing chemical reagents in bulk quantities may be a very poor practice, however, if the rate of usage of those particular reagents does not result in their consumption in a reasonable amount of time.

From a safety standpoint, chemical reagents should be purchased in the smallest quantities possible consistent with the manner in which they are used. This holds for both flammable and nonflammable chemical reagents. For school science laboratories, a semester's supply of chemical reagents is satisfactory for most situations.

If even a fraction of a purchased chemical becomes a waste material after sitting unused for several months, the price advantage in large-quantity purchasing is gone. Disposing of a chemical can be considerably more costly than the original purchase.

Chemical Grades

Several grades of reagent chemicals are usually purchased for laboratory use:

Technical grade: Labels show name, formula, lot number, hazards, and sometimes the manufacturer. This is a commercial grade, of reasonable purity, usually used in manufacturing. Sometimes an assay is given. This grade is most economical.

Practical, purified: Labels carry name, formula, lot number, manufacturer, and sometimes an index of purity such as a boiling range, melting range, or assay. Hazard notice is usually supplied. "Purified" is used when there are no official standards. This grade is adequate for many syntheses.

NF, USP, FCC: These letters stand for National Formulary, United States Pharmacopia or Food Chemicals Codex. Their presence on a label indicates that the products are of sufficient quality to pass these standards.

Reagent grade: Characterized by having a purity adequate for analytical use. There are two types of reagent labels: Lot Analyzed Reagents give actual amounts of impurities and Maximum Impurities Reagents provide the upper acceptable limits of impurities.

ACS certified: Meets standards for purity set by the American Chemical Society. These chemicals generally will have a lot analysis on the label.

Primary standards: Reagent-grade chemicals with lot assays in addition to amounts of impurities.

There are many other grades, most of which are intended for specific applications such as chromatography, spectrophotometry, electrophoresis, etc.

Ordering and Procuring of Chemicals

The achievement of safe handling, use, and disposal of hazardous substances begins with the persons who requisition such substances and those who approve their purchase orders. These people must be aware of the potential hazards of the substances being ordered, know whether or not adequate facilities and trained personnel are available to handle them, and be able to ensure that a safe disposal route exists.

Before a new substance that is known or suspected to be hazardous is received, information concerning its proper handling, including proper disposal, should be given to all those who will be involved with it. If the distribution system involves receiving room or storeroom personnel, they should be advised that the substance has been ordered. It is the responsibility of the laboratory supervisor to ensure that the facilities are adequate and that those who will handle any material have received proper training and education to do so safely.

Receiving Ordered Chemicals

The U.S. Department of Transportation (DOT) requires that shippers furnish and attach department-prescribed labels on all shipments of hazardous substances (see Appendix C). These labels indicate the nature of the hazards of substances shipped and thus provide some indication to receiving room, storeroom, and stockroom personnel of the type of hazard received, but they should not be relied upon after the container has been opened.

Because storage in laboratories is usually restricted to small containers, it is sometimes preferable to order in small-container lots to avoid the hazards associated with repackaging. Some chemical suppliers ship solvents in small metal containers to avoid the hazard of breakage.

All substances should be received at a central location for distribution to storerooms, stockrooms, and laboratories. Central receiving is also helpful in monitoring substances that may eventually enter the waste disposal system. An inventory of substances kept in storerooms and stockrooms can serve to alert those responsible for disposal as to what they may expect regarding the quantity and nature of substances they may be required to handle.

No container of chemical or cylinder of compressed gas should be accepted that does not have an identifying label. For chemicals, it is desirable that this label should have, at a minimum, the information required by ANSI Z129.1, which was described earlier in this chapter.

Receiving room, storeroom, and stockroom personnel should be knowledgeable about or trained in the handling of hazardous substances. Such training should include the physical handling of containers of chemicals so that they are not dropped, bumped, or subjected to crushing by being piled one on top of another. Information should be provided about environmental and hazard-initiating exposures that must be avoided. Some of the more common items with which receiving room, storeroom, and stockroom personnel need to be familiar include the following:

1. The use of proper material handling equipment, protective apparel, and safety equipment.
2. Emergency procedures, including the cleanup of spills and the disposal of broken containers.
3. The dangers of contacting chemicals by skin absorption, inhalation, or ingestion.
4. The meanings of the various DOT labels on shipping packages. See Appendix C.

5. The proper methods of material handling and storage, especially the incompatibility of some common substances, the dangers associated with alphabetical storage, and the sensitivity of some substances to heat, moisture, light, and other storage hazards.
6. The special requirements of heat-sensitive materials, including those shipped refrigerated or packed in dry ice.
7. The problems associated with compressed gases, including unique situations such as the construction of an acetylene cylinder.
8. The hazards associated with flammable liquids, especially the danger of their vapors catching fire some distance from the container, and explosives, and of toxic gases and vapors, and oxygen displacement.
9. Substances that react with water, giving rise to hazardous conditions (e.g., alkali metals, burning magnesium, metal hydrides, acid chlorides, phosphides, and carbides).
10. Federal and state regulations governing controlled substances such as radioactive materials, drugs, ethyl alcohol, explosives, and needles and syringes.
11. Chemicals that have offensive smells.
12. Packages that exhibit evidence that the inside container has broken and leaked its contents.

RECORDKEEPING

Inventory Control

A science laboratory should have readily available inventory records of the chemical reagents in stock (as well as existing instrumentation and miscellaneous items such as glassware, fire extinguishers, and personal protective devices). Inventory records are essential for school laboratories because of the constant turnover of students and the frequent turnover of instructors.

There is no legal requirement that the school or teacher keep records of this nature. However, these records can be extremely advantageous to the laboratory instructor and the school administration.

Adequate written records allow inventory searches to be made rapidly. As the hazards presented by certain chemicals become recognized, it may become necessary to remove the chemical compounds from school shelves. A case in point is the list of proscribed carcinogens developed by OSHA. Many school science laboratories contained one or more of these compounds, and the instructors were, and in many cases still are, unaware of the potential hazards. As more compounds are placed on the proscribed list, more compounds will have to be removed from the shelves of school laboratories.

A uniform, orderly system of recording purchase date, receiving dates, quantities received, quantities used, and disposal date is necessary if the school and the instructor are to keep up with the ever-increasing flow of information on hazardous chemicals.

Dated Receiving System

All chemical reagents, upon receipt in the laboratory, should be immediately marked with the receiving date. The reason for dating the reagent is the time sensitivity of certain chemical reagents. Many chemical reagents are oxidized by atmospheric oxygen over a period of time. The effect of this can be minimal; i.e., the reagent will not perform properly in a chemical test and a new batch of reagent will have to be acquired or prepared. On the other hand, some materials are time-sensitive in a dangerous fashion. Ethyl ether and dioxane, for example, form explosive peroxides after sitting for varying periods of time. These chemical compounds should be used before peroxides can form. An inventory control system allows for routine removal and disposal of time-sensitive chemical compounds. Manufacturers or suppliers can provide specific information relating to these problems.

Stored chemicals should be examined periodically (at least annually). At this time, those that have been kept beyond their appropriate shelf life or have deteriorated, have questionable labels, are leaking, have corroded caps, or have developed any other problem should be disposed of in a safe manner. A first-in, first-out system of stock keeping should be used.

Recordkeeping Detail

The amount of time that an instructor can devote to maintaining chemical usage and inventory records is limited. However, maintaining a complete records system may require considerable time. There should be a record for each chemical reagent listing the date received, quantities withdrawn, when withdrawn and by whom, the date disposed of, and the disposal method. Inventory records can be designed to contain even more information. They may list toxicity data, safety data, and special precautions to take in handling the chemical. Compiling all this information might be prohibitive for the average science instructor, but it might serve as a useful project for one or more students. If students are involved in the recordkeeping system, much of the burden can be removed from the instructor.

CHEMICAL STORAGE

The storage of chemical reagents is a key element in a program of safe and proper chemical usage. Proper storage is frequently overlooked, not only in school science laboratories but in all science laboratories. There are four functions of chemical storage:

1. To minimize exposure to personnel using the room
2. To provide security against unauthorized removal of chemicals by students or others
3. To protect the outside environment by restricting emissions from stored chemicals
4. To protect the reagents from fire

An ideal chemical storage system will fulfill all four functions and require a minimal expenditure of funds.

Satisfactory storage for chemical reagents in school science laboratories can be achieved with a specially designed storage room or with commercial, nonflammable storage cabinets for flammables.

General Rules for Chemical Storage

1. Reagents should be stored in an uncluttered manner on shelves with solid back and sides so that bottles cannot be pushed off the shelf.
2. Shelved chemicals can walk, creep, and even tip over. Such chemicals can be prevented from falling by using shelves with antiroll lips on the front edge, placing retaining shock cords or similar restraining devices across the open face of the shelf, or by raising the forward face of the shelf about ¼ in.
3. Secure shelf assemblies firmly to walls. Avoid island shelf assemblies.
4. Ideally, shelving assemblies should be made of wood. Metal corrodes. Avoid adjustable metal shelf supports and clips. Fixed, wooden supports are better.
5. Avoid storing chemicals on the floor (even temporarily).
6. Do not store chemicals on the top shelf.
7. Do not store chemicals above eye level.
8. Store large bottles on lower shelves.
9. Never store chemicals by strict alphabetical order only. Incompatible reagents (see the list of incompatible chemicals in Chap. 1) should not be stored near each other. Alkalies should not be stored near acids. Oxidants should be separated from reducing agents.
10. Control the environment. Storage should be away from heat and strong direct sunlight. Some chemicals require refrigeration.
11. Provide resistant trays, pans, or other containers for storage of liquids (especially corrosives) to retain contents in case of breakage or spillage.
12. Flammable reagents should not be stored in ordinary refrigerators. A spark from the motor or light switch can ignite such a substance.
13. Reagent bottles with attached delivery tubes must be secured so they cannot tip over.
14. Never return a reagent to the storage bottle.
15. Label all chemicals as received. The label should clearly state any hazards or unusual properties.
16. Date all chemicals when received and when opened. All reagents, solutions, and mixtures also should be dated.
17. Dispose of samples and chemical supplies when the expiration date has passed [e.g., isopropyl ether (peroxide hazard)].
18. Store materials that evaporate or produce airborne contaminants in ventilated cabinets.
19. Store carcinogens and poisons under lock and key in a dedicated poisons cabinet.
20. Provide containers that are not attacked by the contents—solvents in plastic bottles, etc.
21. Store acids in a dedicated acid cabinet. Store nitric acid in that same cabinet *only* if it can be isolated from other acids. Store both inorganic and some organic acids in the acid cabinet.
22. Store flammable substances in a dedicated flammables cabinet.

Be sure to follow local fire codes when storing flammable chemicals in separate cabinets.

Chemical Storerooms and Stockrooms

There is a range of possibilities for storing chemical substances: central storerooms, stockrooms, laboratory storage cabinets, and others. The arrangements made will depend on the size of the organization, the quantities handled, and the nature of the problems.

Often, the provision of adequate storage space is given little consideration in the design of laboratory buildings. Lack of sufficient storage space can create hazards due to overcrowding, storage of incompatible chemicals together, and poor housekeeping. Adequate, properly designed, well-ventilated storage facilities should be provided to ensure personnel safety and property protection.

In many instances, chemicals are delivered after receipt in the institution directly to the individual who initiated the order. If the facilities of the laboratory are appropriate for the kinds and quantities of materials used, this system may be eminently satisfactory.

However, experience has shown that it is usually necessary to maintain a reserve of supplies in excess of the amounts that can be kept safely in the laboratory. If the quantities are large or the volumes of the individual containers are such that repackaging is necessary, then a safe place is needed to store these containers and to perform these functions. Depending on needs, this could be a stockroom for the laboratory or a central storeroom for the particular organization.

Central Storerooms. Usually at least one chemical storage room under the supervision of a qualified person is essential for each school. Inside storage rooms allow the use of an inventory control system and provide fire protection and security. There are three critical elements that must be dealt with when designing and constructing inside storage facilities: fire protection, chemical exposure protection, and security. Facilities should include the following features to assure safety:

1. Proper materials of construction to contain spills and fires
2. A fire extinguishing system that includes sprinklers and fire extinguishers of approved type, including sand and soda positioned near an escape route
3. Control of the types of chemicals stored
4. Spill-control and cleanup materials
5. Master-control shutoff valves for gas, water, and security
6. Approved eye/face wash
7. Shower
8. Smoke detector
9. Forced ventilation from floor to ceiling with exhaust above roof level
10. Lip-edged shelving secured to wall, with top shelf below eye level
11. Safety cabinets for specific groups of compatible substances
12. A communication system to the main office or emergency center
13. Controlled access, using locked doors, limited key distribution, and responsible officers to minimize chemical exposures and unauthorized use of materials

Fundamentals of Laboratory Safety: Physical Hazards in the Academic Laboratory, Chap. 12, contains more information on the design requirements of a chemical storeroom.

Stockrooms. Stockrooms are similar to central storerooms except that the quantities of materials involved are usually much smaller (the materials stored will be the inventory for a particular laboratory or group of laboratories), and such rooms are usually within or close to the areas served.

Stockrooms should not be used as preparation areas because of the possibility of an accident contaminating a large quantity of materials. Preparation and repackaging should be performed in a separate area.

Stockrooms should be conveniently located and open during normal working hours so that laboratory workers need not store excessive quantities of chemicals in their laboratories. However, this does not imply that all laboratory workers should have unlimited access to the chemicals in the stockroom. Procedures must be established for the operation of any stockroom that places responsibility for its safety and inventory control in the hands of one person. If it is not feasible to have a full-time stockroom clerk, then one person who is readily available should be assigned that responsibility.

Stockrooms should be well ventilated. If opened containers are returned to the stockroom, then extra local exhaust ventilation and the use of spill trays is necessary.

Suggested Chemical Storage Patterns. The alphabetical method of storing chemicals presents hazards because chemicals that react violently with each other may be stored in close proximity. The J. T. Baker Chemical Company has devised a simple color-coding scheme to address this problem. The code includes both solid and striped colors, which are used to designate specific hazards as follows:

COLOR CODE SYSTEM FOR CHEMICAL STORAGE

Red — Flammability hazard: Store in a flammable chemical storage area.

Red stripe — Flammability hazard: Do not store in the same area as other flammable substances.

Yellow — Reactivity hazard: Store separately from other chemicals.

Yellow stripe — Reactivity hazard: Do not store with other yellow-coded chemicals; store separately.

White — Contact hazard: Store separately, in a corrosion-proof location.

White stripe — Contact hazard: Not compatible with chemicals in solid white category.

Blue — Health hazard: Store in a secure poison area.

Orange — Not suitably characterized by any of the foregoing categories.

For safe handling purposes, the chemical container should be marked with the appropriate color tape. This simplifies identification of hazardous chemicals in use in the laboratory.

These classes then require separate storage cabinets, shelves, or areas for:

Orange—standard chemical storage cabinet

Red—flammables storage cabinet

Red stripe—another flammables cabinet

Yellow—a separate standard chemical cabinet

Yellow stripe—another chemical cabinet

White—corrosion-resistant cabinet

White stripe—another corrosion-resistant cabinet

Blue—locked poisons cabinet

Of course, if the laboratory does not have any chemicals that fall into a particular category, there will be no need for a cabinet for this group. The size of the chemical stock will be the major factor in how many storage cabinets are needed.

Once all chemicals are sorted according to their color codes, sorting into organic and inorganic classes within a color should be done. A further separation into related and compatible families will define the storage of a chemical to a particular shelf. Appendix G contains a list of chemicals commonly found in school science laboratories along with their hazards, color coding, and suggested shelf storage position based on this scheme.

Flammable Liquids. Centralized storage of bulk quantities of flammable liquids provides the best method of controlling the associated fire hazard.

Because the most effective way to minimize the impact of a hazard is to isolate it, a storage and dispensing room for flammable liquids is best located in a special building separated from the main building. If this is not feasible and the room must be located in a main building, the preferred location is a "cutoff" area at grade level (the building's ground level) that has at least one exterior wall. (Note: *Cutoff* is the fire-protection term for "separated from other areas by fire-rated construction.") In any case, storage rooms for flammable liquids should not be placed on the roof or located below grade level, on an upper floor, or in the center of the building. All these locations are undesirable because they are less accessible for fire fighting and potentially dangerous to the safety of the personnel in the building.

The walls, ceilings, and floors of an inside storage room for flammable liquids should be constructed of materials having at least a 2-hr fire resistance, and there should be self-closing Class B fire doors [see OSHA Standard 29 CFR 1910.106(d) (4) (i) or NFPA Standard No. 30.4310]. All storage rooms should have adequate mechanical ventilation controlled by a switch outside the door and explosion-proof lighting and switches. Other potential sources of ignition, such as burning tobacco and lighted matches, should be forbidden.

Fundamentals of Laboratory Safety: Physical Hazards in the Academic Laboratory, Chap. 12, provides more information on flammables storage.

See Table 4–1 for a list of acceptable flammable-liquid storage container sizes.

Toxic Substances. Toxic substances should be segregated from other substances and stored in a well-defined or identified area that is cool, well ventilated, and away from light, heat, acids, oxidizing agents, and moisture. Highly toxic materials should be stored in a lockable poisons cabinet.

The storage of unopened containers of toxic substances normally presents no unusual requirements.

TABLE 4-1. Container Size Limitations for Flammable and Combustible Liquids

	Flammable Liquids						Combustible Liquids			
	Class IA		Class IB		Class IC		Class II		Class IIIA	
Type of Container	Liters	Gallons	Liters	Gallons	Liters	Gallons	Liters	Gallons	Liters	Gallons
Glass	0.5	0.12	1	0.25	4	1	4	1	4[a]	1[a]
Metal (other than DOT drums)	4	1	20	5	20	5	20	5	20	5
Safety cans	7.5	2	20	5	20	5	20	5	20	5
Metal drums (DOT specifications)[b]	225	60	225	60	225	60	225	60	225	60
Approved portable tanks[c]	2500	660	2500	660	2500	660	250	660	2500	660

[a] OSHA Limitation: NFPA No. 30 and 45 allow 20 L (5 gal).
[b] Maximum size permitted in a laboratory for Class I materials is 20 L (5 gal); drum size is permitted only in an inside storage room (OSHA 1910.106 and NFPA No. 30).
[c] Permitted only outside buildings.
Source: From National Research Council, *Prudent Practices for Handling Hazardous Chemicals in Laboratories,* National Academy Press, Washington, D.C., 1981, p. 228.

However, because containers occasionally develop leaks or are broken, storerooms should be equipped with exhaust hoods or equivalent local ventilation devices in which containers of toxic substances can be handled.

Opened containers of toxic substances should be closed with tape or other sealant before being returned to the storeroom and should not be returned unless some type of local exhaust ventilation is available. (See Chap. 3 for specific storage and recordkeeping procedures for compounds of known high chronic toxicity.)

Compressed Gases. Cylinders of compressed gases should be stored in well-ventilated, dry areas. Where practicable, storage rooms should be of fire-resistant construction and above ground. Cylinders may be stored out of doors, but some protection must be provided to prevent corrosion of the cylinder bottoms, and air circulation must not be restricted. All storage and use of compressed gases should be in compliance with OSHA regulations.

Compressed gas cylinders should not be stored near sources of ignition nor where they might be exposed to corrosive chemicals or vapors. They should not be stored where heavy objects might strike or fall on them, such as near elevators, service corridors, or unprotected platform edges.

The cylinder storage area should be posted with the names of the gases stored. Where gases of different types are stored at the same location, the cylinders should be grouped by type of gas (e.g., flammable, toxic, or corrosive). If possible, however, flammable gases should be stored separately from other gases, and provision should be made to protect them from fire. Full and empty cylinders should be stored in separate portions of the storage area, and the layout should be arranged so that older stock can be used first with minimum handling of other cylinders.

Cylinders and valves are usually equipped with various safety devices, including a fusible metal plug that melts at 70–95°C. Although most cylinders are designed for safe use up to a temperature of 50°C, they should not be placed where they can become overheated (e.g., near radiators, steam pipes, or boilers). Cylinder caps to protect the container withdrawal valve should be in place at all times during storage and movement to and from storage.

Cylinders should be stored in an upright position where they are unlikely to be knocked over, or they should be secured in an upright or horizontal position.

Acetylene cylinders should always be stored valve end up to minimize the possibility of solvent discharge. Oxygen should be stored in an area that is at least 20 ft away from any flammable or combustible materials (especially oil and grease) or separated from them by a noncombustible barrier at least 5 ft high and having a fire-resistance rating of at least ½ hr.

Cylinders are sometimes painted by the vendor to aid in the recognition of their contents and make separation of them during handling easier. However, this color coding is not a reliable method for identification of their contents; the stenciled or printed name on the cylinder is the only accepted method. If it is suspected that a stored cylinder is leaking, the procedures described in *Fundamentals of Laboratory Safety: Physical Hazards in the Academic Laboratory,* Chap. 8, should be followed.

Transport of Chemicals from Stockrooms to Laboratories

Moving reagents creates the greatest potential for an accident. The method of transport of chemicals be-

CHEMICAL LABELING AND STORAGE   61

tween stockrooms and laboratories must reflect the potential danger posed by the specific substance.

Solid and Liquid Chemicals. Transport reagents in safety containers. Reagents in vacuum containers should be moved only after being shielded in a wooden box or metal screen. Resistant, plastic containers of appropriate capacity should be used for corrosive liquid chemicals. Jars of solids can be transported in plastic boxes. Move glass bottles and carboys in special carriers. Beakers or flasks should be handled by grasping the body, not the lip. When chemicals are hand-carried, they should be placed in an outside container or acid-carrying bucket to protect against breakage and spillage. When they are transported on a wheeled cart, the cart should be stable under the load and have wheels large enough to negotiate uneven surfaces (such as expansion joints and floor drain depressions) without tipping or stopping suddenly.

To avoid exposure to persons on passenger elevators, if possible, chemicals should be transported on freight-only elevators.

Provisions for the safe transport of small quantities of flammable liquids also include:

1. The use of rugged, pressure-resistant, non-venting containers
2. Storage during transport in a well-ventilated vehicle
3. Elimination of potential ignition sources

Cylinders of Compressed Gases. The cylinders that contain compressed gases are primarily shipping containers and should not be subjected to rough handling or abuse. Such misuse can seriously weaken the cylinder and render it unfit for further use or transform it into a rocket having sufficient thrust to drive it through masonry walls. To protect the valve during transportation, the cover cap should be left screwed on hand-tight until the cylinder is in place and ready for actual use. Cylinders should never be rolled or dragged. The preferred transport, even for short distances, is by suitable hand truck with the cylinder strapped in place. Only one cylinder should be handled at a time.

Chemical Storage in Laboratories

The amounts of toxic, flammable, unstable, or highly reactive materials that should be permitted in laboratories are an important concern. To arbitrarily restrict quantities may interfere with laboratory operations, but, conversely, unrestricted quantities can result in the undesirable accumulation of such materials in the laboratory. It is necessary to balance the needs of laboratory workers and established requirements for safety. Decisions in this area will be affected by the level of competence of the workers, the level of safety features designed into the facility, the location of the laboratory, the nature of the chemical operations, and the accessibility of the stockroom. It is also necessary to comply with any local statutory restrictions or insurance requirements on allowable quantities.

General Considerations. Quantities of chemicals should not be stored in labs with only one exit. All laboratories should have two exits (one may be an emergency exit), so that a fire at one exit will not block occupants' escape; doors that open outward are desirable.

Every chemical in the laboratory should have a definite storage place and should be returned to that location after each use. The storage of chemicals on bench tops is undesirable; in such locations, they are unprotected from potential exposure to fire and are also more readily knocked over. Storage in hoods is also inadvisable, but placing a ventilated storage cabinet under the hood is a good idea. Because of its proximity to the hood, this encourages the safe practice of making transfers of hazardous materials to and from the hood.

Storage trays or secondary containers (vessels that fit around the primary container) should be used to minimize the distribution of material should a container break or leak.

Just as for storerooms and stockrooms, care should be taken in the laboratory to avoid exposure of chemicals to heat or direct sunlight and to observe precautions regarding the proximity of incompatible substances (see the list of incompatible chemicals in Chap. 1).

Laboratory refrigerators should be used for the storage of chemicals only; food must not be placed in them. All containers placed in the refrigerator should be properly labeled (identification of contents and owner, date of acquisition or preparation, and nature of any potential hazard) and, if necessary, should be sealed to prevent escape of any corrosive vapors. Flammable liquids should not be stored in laboratory refrigerators unless the unit is an approved, explosion-proof, or laboratory-safe type.

The chemicals stored in the laboratory should be inventoried periodically, and unneeded items should be returned to the stockroom or storeroom. At the same time, containers that have illegible labels and chemicals that appear to have deteriorated should be properly disposed of.

Upon termination, transfer, graduation, or other departure of any laboratory personnel, those personnel and the laboratory supervisor should arrange for the removal of safe storage of all hazardous materials those persons have on hand.

Flammable Liquids. OSHA regulations for the laboratory storage of flammable and combustible liquids are not based on fire prevention and protection principles but rather address the types and sizes of containers allowable (and would permit the storage of 60-gal metal drums in laboratories of colleges and universities). The NFPA standard (No. 45), on the other hand, has a quantity limit per 100 ft^2 that depends on the construction and fire protection afforded in the laboratory and restricts instructional laboratories to half the quantities for industrial or graduate student laboratories. A second NFPA standard (No. 30) addresses the amounts that may be stored outside of an approved flammable-liquid storage room or cabinet, but does not consider fire protection features available.

Whenever feasible, quantities of flammable liquids greater than 1 L should be stored in metal containers. Portable approved safety cans are one of the safest methods of storing flammable liquids. These cans are available in a variety of sizes and materials. They have spring-loaded spout covers that can open to relieve internal pressure when subjected to a fire and will prevent leakage if tipped over. Some are equipped with a flame arrester in the spout that will prevent flame propagation into the can. If possible, flammable liquids received in large containers should be repackaged into safety cans for distribution to laboratories. Such cans must be properly labeled to identify their contents.

Small quantities of flammable liquids should be stored in approved, ventilated, flammables storage cabinets. These are usually made of 18-gauge steel and have riveted and spot-welded seams. They have double-wall construction with a 1.5-in. air space between the inner and the outer walls. The door is 2 in. above the bottom of the cabinet, and the cabinet is liquid-tight to this point. It is provided with vapor-venting provisions and can be equipped with a sprinkler system. (Materials that react with water should not be stored in sprinkler-equipped cabinets.) Some models have doors that close automatically in the event of fire.

If, for reasons of cost or space limitations, storage cabinets must be constructed of wood, they should be built according to the Los Angeles Fire Department specifications (see also NFPA Standard No. 30). The hazard of storage of flammable materials in wooden cabinets in existing laboratories can be decreased by the use of intumescent fire-retardant coating or other means that provide effective fire insulation. (Note: Upon heating, intumescent materials expand from a thin, paintlike coating to a thick, puffy coating that insulates or excludes oxygen and protects the subsurface from ignition.)

Other considerations in the storage of flammable liquids in the laboratory include ensuring that aisles and exits are not blocked in the event of fire; that accidental contact with strong oxidizing agents such as chromic acid, permanganates, chlorates, perchlorates, and peroxides is not possible; and that sources of ignition are excluded.

See Table 4–1 on p. 60 for a list of acceptable flammable liquid storage container sizes.

Toxic Substances. Chemicals known to be highly toxic, including those classified as carcinogens, should be stored in ventilated storage areas in unbreakable, chemically resistant secondary containers (vessels that fit around the primary container).

Only minimum working quantities of toxic materials should be present in the work area. Storage vessels containing such substances should carry a label such as the following:

CAUTION: HIGH CHRONIC TOXICITY
or
CANCER-SUSPECT AGENT

Storage areas for substances that have high acute or chronic toxicity should exhibit a sign warning of the hazard, have limited access (i.e., be lockable), and be adequately ventilated. An inventory of these toxic materials should be maintained. For those chronically toxic materials designated as regulated carcinogens, this inventory is required by federal and state regulations. Adequate ventilation is of particular concern for hazardous materials that have a high vapor pressure (such as bromine, mercury, and mercaptans). See Chap. 3 for more information on handling toxic substances.

Compressed Gases. Cylinders of compressed gases should be securely strapped or chained to a wall or bench top to prevent their being knocked over accidentally. When they are not in use, it is good practice to keep them capped. Care should be taken to keep them away from sources of heat or ignition. *Fundamentals of Laboratory Safety: Physical Hazards in the Academic Laboratory,* Chap. 8, discusses in detail the storage, handling, and hazards of compressed gases.

Radioactive Materials. The possession and use of radioactive material is regulated by law in all states. Educational institutions must take special care with radioactive materials. Radioactive materials include some of the most hazardous substances on earth. There are two key elements in protecting users and the general public from the health hazards of radiation: security and shielding.

Security is important because radioactive materials must be kept out of the wrong hands. Radioactive materials must always be stored in locked cabinets, drawers, or boxes.

The hazards of exposure to radiation are well documented; it is not necessary to repeat them here. Each state as well as the federal government has regulations governing the acquisition, possession, and use of radioactive materials. Each school must comply with the regulations of its own particular state.

Fundamentals of Laboratory Safety: Physical Hazards in the Academic Laboratory, Chap. 11, discusses in detail the storage, handling, and hazards of radioactive materials.

POOR STORAGE PRACTICES

Reagents in Hoods

In most school science laboratories, fume hoods are used for storing chemical reagents at least part of the time. This is a practice that should be discouraged. Many school fume hood systems are not designed to operate 24 hr a day, 7 days a week. If hoods are going to be used for storage, they must operate continuously, or volatile, flammable chemicals cannot be stored in them.

Other problems that occur when hoods are used for storage include the loss of labels due to the corrosive environment frequently found in hoods, the loss of space needed for laboratory work because it has been devoted to storing reagents, and possible hazards from storing or using incompatible chemicals (see the list of incompatible chemicals in Chap. 1). Hoods also do not provide any security for chemical reagents.

Flammable Reagents in Standard Refrigerators

Refrigerators are commonly found in biological and chemical science laboratories. They are used to store biological materials, volatile chemical reagents, and heat-sensitive chemical reagents.

Standard household or commercial refrigerators should not be used in school science laboratories. Flammable liquids should never be stored in standard refrigerators. The light switches, the lights themselves, and the thermostat can serve as ignition sources in standard refrigerators.

Refrigerators used in laboratories should be of explosion-proof or explosion-safe design. Explosion-proof refrigerators can be used in dangerous atmospheres. All possible sources of ignition, both inside and out, are sealed. Explosion-safe refrigerators can be used to store flammable materials, but cannot be used in hazardous atmospheres, because the external circuitry may cause sparks.

Food in Refrigerators

Food for human consumption should never be stored in refrigerators used for the storage of chemical reagents or biological materials. The chances of contaminating the food are too great to allow this practice. With all the hazards existing in a school science laboratory, there is no point in adding the danger of ingestion of toxic substances.

STORAGE AND HANDLING OF SPECIFIC MATERIALS

The storage in laboratories of dangerously reactive chemicals in excessive quantities or under unsuitable conditions may involve hazards comparable to those arising from incorrect storage of explosives. The dangers of handling and storing reagents are not always readily apparent to the observer and may result from insufficient care on the part of a previous user. Therefore the following guidelines are of significant importance to yourself as well as other persons within the laboratory. They may also be used as a guide in choosing suitable conditions for other dangerous substances not mentioned.

More information on handling specific materials can be found in the Chemical Entries section.

Oxidizing Materials

Chlorates, bromates, iodates, perchlorates, permanganates, persulfates, and chromium trioxide may be kept in laboratories in glass bottles containing not more than 500 g.

Perchloric acid, concentrated hydrogen peroxide, and metallic peroxides: Quantities not greater than 100 g may be kept in laboratories. Larger quantities must be kept in a dangerous chemicals storage outside the laboratory.

Store perchloric acid away from all inflammable materials; a mixture of perchloric acid and sawdust, for example, has more explosive power than TNT.

Spontaneously Flammable or Easily Oxidizable Substances

White phosphorus must be stored under water in glass (not metal) containers in a dangerous chemicals store. The container must be kept at temperatures above freezing, and the water levels must be checked at intervals.

Red phosphorus, sodium sulfide (anhydrous), metallic alkyls, and pyrophoric metals must be stored in a dangerous chemicals store.

Peroxide Formers

Some organic compounds (e.g., cyclohexene, cyclooctene, decalin, dibutyl ether, *p*-dioxane, diethyl ether, diisopropyl ether, tetrahydrofuran, tetralin, and cellosolves) can form dangerous concentrations of peroxides upon long exposure to air. These materials should be stored away from heat and light in closed vessels, preferably in the container furnished by the supplier. More information on peroxide-forming compounds can be found in Chap. 1.

Corrosive Chemicals

The storage of corrosive chemicals, even in an approved dangerous chemicals store, presents another hazard that is not always appreciated. Such storage can produce an atmosphere in which ordinary labels disintegrate so that doubts as to the contents of various bottles arise.

To prevent this, labels should be given a protective coating. One method is to cover the label with a silicone resin varnish. The area of the varnish should be extended beyond the edges of the labels on to the glass for at least ½ in.

Toxic and Corrosive Materials

Bromine may be stored in the original bottle as supplied, standing on a porcelain tray in a fume hood (temporarily), ventilated cabinet, or in a dangerous chemicals store.

Sulfur dioxide must be kept in a fume hood or ventilated cabinet. The normal precautions must be taken when this is supplied in a cylinder.

Fuming nitric acid may be kept in quantities not greater than 1000 mL in a bottle standing on a porcelain tray in a fume hood (temporarily), ventilated cabinet, or in a dangerous chemicals store.

Concentrated nitric, sulfuric, hydrochloric or acetic acids, mixed acids, oleum, concentrated ammonia, and potassium hydroxide or sodium hydroxide solutions may be kept as bench reagents in bottles of not greater than 500-mL capacity standing on porcelain trays in a fume hood. Small stocks should be assigned to a reserved area in the laboratory where the bottles can be kept on a tile surface or in an acid-resisting tray. All large quantities must be kept in a special acids store outside the laboratory.

Dimethyl sulfate, sulfur trioxide, chlorosulfonic acid, titanium tetrachloride, stannic chloride, and hydrofluoric acid may only be kept in a dangerous chemicals store outside the laboratory.

Acetyl chloride and acetic anhydride may be kept in quantities not exceeding 100 g in a fume hood with the bottles standing on a porcelain tray. All other stocks must be kept in a special store for acids or dangerous chemicals outside the laboratory.

Materials That React Dangerously with Water

Some chemicals react with water to evolve heat and flammable or explosive gases. For example, potassium and sodium metals and many metal hydrides react on contact with water to produce hydrogen, and these reactions evolve sufficient heat to ignite the hydrogen with explosive violence. Certain polymerization catalysts, such as aluminum alkyls, react and burn violently on contact with water. Bulk quantities of aluminum dust may heat spontaneously when damp. The hazard increases with smaller particle size.

Magnesium powder will react with water and acids to release hydrogen.

Potassium, sodium, magnesium powder, aluminum powder, and other water-reactive materials may be kept in laboratories in amounts not exceeding 500 g in glass bottles with sealed caps. Larger amounts must be stored under conditions similar to those applying to explosives.

Storage facilities for water-sensitive chemicals should be constructed to prevent their accidental contact with water. This is best accomplished by eliminating all sources of water in the storage area; for example, areas where large quantities of water-sensitive chemicals are stored should not have automatic sprinkler systems. Storage facilities for such chemicals should be of fire-resistant construction, and other combustible materials should not be stored in the same area.

Finely Divided Materials

The storage of very fine powders (<100 μm) should be avoided if possible, but if it is essential, only small amounts should be stored in metal containers under an atmosphere of argon or helium. Courser material may be stored in air in air-tight steel drums. In both cases, any material not actually in use in the laboratory should be stored in a dangerous chemicals store.

As an example, magnesium powders form explosive mixtures with air, which may be ignited by a spark.

Pyrophoric substances should never be stored near sources of heat or near flammable gases and vapors. Unwanted dust should not be allowed to accumulate.

Chapter 1 describes the proper handling of finely divided materials.

SUGGESTED READINGS

The American National Standards Institute, *Precautionary Labeling Hazardous Industrial Chemicals,* Standard Z129.1. Use the most current version.

Compressed Gas Association, *Safe Handling of Compressed Gases in Containers* and *Characteristics and Safe Handling of Medical Gases,* New York, 1965, pamphlets P1 and P2.

Flinn Scientific Inc., *The Flinn Chemical Catalog Reference Manual,* Batavia, Ill. Use the latest edition.

Gatson, P. J., *Handling, and Disposal of Dangerous Chemicals,* Northern Publishers, Aberdeen, Scotland, 1970.

Los Angeles Fire Department, *Hazardous Materials Storage Cabinets,* Standard No. 40, Los Angeles, Calif., January 1, 1960.

Manufacturing Chemists Association, *Guide for Safety in the Chemical Laboratory,* Van Nostrand Reinhold, New York, 1972.

Manufacturing Chemists Association, *Guide to Precautionary Labeling of Hazardous Chemicals,* Vols. 1–6, New York, 1945.

National Fire Protection Association, *Flammable and Combustible Liquids Code,* Standard No. 30; *Fire Hazard Properties of Flammable Liquids, Gases, and Volatile Solids,* Standard No. 325M; *Recommended System for the Identification of Fire Hazards of Materials,* Standard No. 704; *Fire Protection for Laboratories Using Chemicals,* Standard No. 45; *Hazardous Chemical Data,* Standard No. 49; *Manual of Hazardous Chemical Reactions,* Standard No. 491M; *Flash Point Index of Trade Name Liquids,* Standard No. SPR-51, Boston. Use the most current version.

NIOSH, *An Identification System for Occupationally Hazardous Materials, A Recommended Standard,* U.S. Government Printing Office, Washington, D.C., 1974.

5
LABORATORY WASTES

INTRODUCTION

One of the most important aspects of laboratory operation is disposing of waste chemicals. Because the disposal of hazardous chemicals is strictly regulated by state and federal laws, wastes can be difficult and costly to get rid of. Disposal problems can be minimized by good laboratory management.

The disposal of chemicals from instructional laboratories is a special problem, because students in such laboratories are inexperienced, the quantities of wastes may be relatively large, and the facilities may not be optimum. Teachers have a legal and moral obligation to dispose of laboratory waste in ways that have the least risk to health and environment. Formal instruction in the disposal of waste chemicals should be a fundamental part of every experimental lesson. The disposal of waste or outdated chemicals can be a valuable educational experience for instructors and students alike.

LEGAL ASPECTS

The disposal of toxic and hazardous laboratory wastes is restricted by law. Before an instructor tells students to dump hazardous material into the sink and flush it down the drain or to pour hazardous or toxic material onto the ground, he or she must check to see if such action will violate federal, state, or municipal law.

The federal government's responsibility for control of chemicals in the environment is managed by the Environmental Protection Agency (EPA). State and local regulations also limit what can be discharged into the sanitary sewer system or put in landfills. Recent community "right-to-know" laws place strict recordkeeping and reporting requirements on some chemical users. Contact local environmental offices for more information.

The many laws controlling chemical wastes were written primarily to cover large-volume industrial disposal. Laboratories generate only about 1% of the total hazardous waste produced in the United States. However, these laws do apply to the disposal of laboratory wastes.

RCRA

The Resources Conservation and Recovery Act (RCRA) of 1976, which is intended to provide cradle-to-grave control of hazardous waste, establishes a national hazardous waste management program. RCRA is administered by the EPA. The EPA is authorized to establish a system that will track hazardous chemical wastes from the time they are generated, through their storage, transportation, and treatment, to their ultimate disposal. The act requires that each step be documented, and regulations have been promulgated aimed at implementing this law.

The act is amended from time to time, and thus it is possible that any written information about waste disposal (such as this book) may be out of date. To get current information, call the EPA, which has a national 800 number for RCRA and Superfund information (1-800-424-9346). The nearest EPA regional office should also be helpful (see *Fundamentals of Laboratory Safety: Physical Hazards in the Academic Laboratory*, App. E). The EPA will be able to discuss problems with specific chemicals that you may have and provide information on suitable waste disposal facilities and contractors.

Definitions. Under RCRA, a "waste" is defined as something either named as a waste substance or something cast aside because it is worthless and has no further utility. Once a chemical is categorized as a waste, then it is regulated by RCRA and legal disposal can be difficult. Thus it is best to avoid quickly labeling chemicals as wastes. Most of the chemicals that were discarded in the past were not worthless, even though they were cast aside. Attempt to reuse chemicals in some way. A used solvent can be burned as fuel or redistilled to be used again. RCRA was designed to encourage reuse of materials.

Under RCRA, a "solid waste" can have any physical state. It can be a liquid, gas, mist, dust, or any other physical material.

The EPA defines waste as hazardous if it meets criteria of toxicity, ignitability, corrosiveness, reactivity, radioactivity, infectivity, phytotoxicity, mutagenicity, or if it is identified by name as acutely hazardous. Each of these terms is uniquely defined. For example, ignitable waste is defined as (1) any liquid that has a flash point of less than 60°C, (2) any substance that can cause a fire by reaction or self-ignition, (3) any ignitable compressed gas, or (4) any oxidizer. See Title 40 of the Code of Federal Regulations, Parts 261 and 262 (40 CFR 261 and 40 CFR 262), or equivalent state or local regulations for detailed definitions.

The EPA considers some hazardous waste chemicals so dangerous to the environment that they are placed in a special category called "acutely hazardous solid waste." Table 5-1 lists some of these materials.

Table 5-1. Examples of Acutely Hazardous Solid Wastes

Acrolein
Allyl alcohol
Aluminum phosphide
Barium cyanide
Calcium cyanide
Carbon disulfide
Chloroacetaldehyde
Dieldrin
2,4-Dinitrophenol
Ethyleneimine
Fluorine
Heptachlor
Hydrocyanic acid
Mercury fulminate
Methyl hydrazine
Nitrogen dioxide
Nitrogen tetraoxide
Osmium tetraoxide
Parathion
Phenyl mercaptan
Phosgene
Phosphorus trihydride
Potassium cyanide
1,2-Propanediol
Sodium azide
Sodium cyanide
Strontium sulfide
Thallium(III) oxide
Vanadium pentoxide
Zinc phosphide

Source: Young, J. A., "Academic Laboratory Waste," *Journal of Chemical Education* 60(6), 1983:491.

RCRA regulates only hazardous solid wastes, as defined in the act and in the regulations. As you can see, these definitions of hazardous, solid, and waste are not intuitive.

State Laws Governing Waste Disposal

Waste disposal regulations are in a dynamic state and, additionally, regulations differ in different parts of the country. Therefore, waste generators should obtain detailed and up-to-date information from any city, county, or state environmental offices that may have jurisdiction. Local regulations may be more strict than those specified by RCRA.

Many states have laws that regulate the discharge of toxic and hazardous materials into sewerage systems and receiving waters and the atmosphere. It is impossible to review here the regulations of all states. Disposal procedures should be confirmed with all regulating agencies at least annually.

TYPES OF WASTE

A laboratory generates wastes of many sorts. The identity and character of this waste depends on the kind of laboratory and the materials with which it is engaged. The types of waste are categorized as nonhazardous, hazardous, and acutely hazardous. Nonhazardous waste includes ordinary trash, glass, biological waste (which is not contaminated by hazardous chemicals or infectious organisms), and some chemicals (see Table 5-2). The hazardous materials are divided into the following classes:

Reactive

Toxic

Ignitable

Corrosive

Infectious

Phytotoxic

Mutagenic

Radioactive

The EPA also has specifically named some chemicals as hazardous. See Fig. 5-1 for a decision guide on determining if a material is a hazardous waste.

Laboratory workers should never begin to use a chemical without knowing how they are going to dispose of it. This requires knowledge of the material's chemical properties and the relevant regulations. Re-

Table 5-2. Low-Toxic-Hazard Cations and Anions Safest for Drain Disposal

Cations	Anions
Aluminum	Bisulfite, HSO_3^-
Ammonium	Borate, BO_3^{3-}
Bismuth	$B_4O_7^{2-}$
Calcium	Bromide, Br^-
Cerium	Carbonate, CO_3^{2-}
Cesium	Chloride, Cl^-
Copper[a]	Cyanate, OCN^-
Gold	Hydroxide, OH^-
Hydrogen	Iodide, I^-
Iron[a]	Nitrate, NO_3^-
Lanthanides	Oxide, O^{2-}
Lithium	Phosphate, PO_4^{3-}
Magnesium	Sulfate, SO_4^{2-}
Molybdenum(VI)	Sulfite, SO_3^{2-}
Niobium(V)	Thiocyanate, SCN^-
Palladium	
Potassium	
Rubidium	
Scandium	
Sodium	
Strontium	
Tantalum	
Tin	
Titanium	
Yttrium	
Zinc[a]	
Zirconium	

[a] Large amounts should not be put into sanitary sewer systems.
Source: From National Research Council, *Prudent Practices for Disposal of Chemicals from Laboratories*, National Academy Press, Washington, D.C., 1983, pp. 78-79.

member that the chemical form may well change during an experiment.

Disposal of radioactive waste is strictly regulated. See *Fundamentals of Laboratory Safety: Physical Hazards in the Academic Laboratory*, Chap. 11, for information on handling radioactive waste.

Nonhazardous Waste

The disposal options for nonhazardous waste are relatively open. These materials are subject to the least regulation. Small amounts of soluble nonhazardous chemical waste are generally washed down the drain. Insoluble nonhazardous chemical wastes are more difficult to dispose of, since most chemicals are now banned from landfill operations. The sections on sewer system and landfill disposal describe the kinds of nonhazardous chemicals that are suitable for these methods.

Ordinary Trash. The ordinary trash generated in the laboratory includes paper from all sources, cotton packing material, and other combustible materials. Rubber, plastics, waxes, and such laboratory scrap do not create much of a problem, and in many laboratories are disposed of in open waste cans, preferably of metal construction. Once in a while, some combination of materials or events causes a fire in this type of trash. Combustible trash should be removed from the building promptly after collection and put into closed metal containers for disposal.

Glass. Broken glass has been a particular disposal problem. When one puts glass into a container with ordinary trash, there is a strong possibility that the person emptying containers will be cut or punctured by the glass. Sometimes, rather than pick up the container and dumping it, someone will reach into it for a handful of trash; if the handful includes some broken glass, a severe cut to the hand may result. Consequently, it is good policy to have a separate container for broken glass and to arrange for special handling in the emptying of this container.

Nonhazardous Biological Wastes. Biological wastes are not broadly regulated. Disposal requirements are primarily defined by any contamination, either chemical or microbiological, of the material. Nonhazardous biological wastes generally constitute little problem if a satisfactory incinerator is available. Laboratories that have numerous animal carcasses to dispose of will have their own incinerator and will burn the carcasses as they are discarded. Materials discarded in cleaning cages warrant consideration also. Biological wastes become hazardous solid wastes if they are contaminated by hazardous chemicals or pathogenic organisms.

Hazardous Waste

Hazardous chemical wastes include very toxic substances (e.g., carcinogens, mutagens, and nerve gases) and reactive substances (e.g., flammables, explosives, and corrosives). The laboratory worker has the responsibility of ensuring that proper arrangements are made for disposal of these materials. Wherever possible, chemical reaction in the laboratory to produce less hazard substances should be undertaken. In the case of those chemicals regulated as carcinogens, EPA disposal rules must be followed.

A spill of one of these substances can be an especially serious hazard. Personnel working with such substances should have contingency plans, equipment, and materials available for coping with potential accidents.

LABORATORY WASTES 69

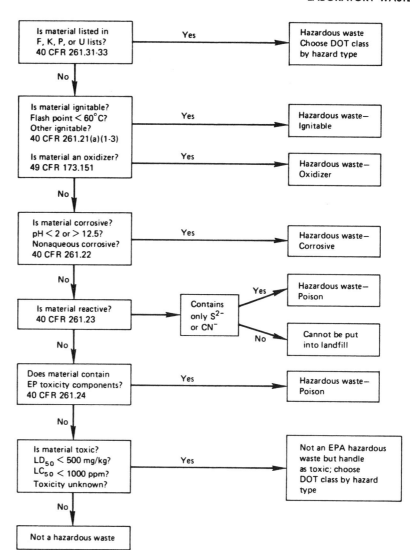

FIGURE 5-1. Flowchart for the Identification of Hazardous Wastes. (From National Research Council, *Prudent Practices for Disposal of Hazardous Chemicals from Laboratories*, National Academy Press, Washington, D.C., 1983, p. 133.)

Toxic Waste. A material is considered by the EPA to be toxic if it has toxic components listed in 40 CFR 261. For other materials, a good rule of thumb is that a material should be considered toxic if it has an LD_{50} less than 500 mg/kg or an LC_{50} less than 1000 ppm (see Chap. 1).

Ignitable Waste. An ignitable hazardous waste is any liquid that has a flash point of less than 60°C, any substance that can cause a fire by reaction or self-ignition, any ignitable compressed gas, or any moderate or strong oxidizer. It includes substances that ignite spontaneously in air under ordinary conditions, and substances that burn vigorously and persistently once they are ignited. Acetone, isobutyl nitrile, 80% ethanol (20% water), ethyl dichlorosilane, cyclohexane, methyl ethyl ketone, potassium permanganate, sodium hypochlorite, and toluene are examples of ignitable hazardous solid wastes.

Corrosive Materials. A corrosive hazardous waste has a pH of less than 2 or more than 12.5 or corrodes SAE 1020 steel at 55°C at a rate greater than 6.35 mm per year, no matter what its pH might be. Examples of corrosive hazardous wastes include bromine, 30% hydrogen peroxide, moderate to high concentrations of aqueous solutions of common mineral acids, alkali metal hydroxides, many primary amines, and saturated and some subsaturated solutions of alkaline earth hydroxides.

Corrosive materials are generally water-soluble. Putting them into glass jugs is about the only way of collecting and discarding them, and they must be handled in a bucket that will minimize spillage in case of accident.

Reactive Waste. A material is reactive and thus a hazardous solid waste if it is explosive, detonatable, pyrophoric, reacts with water, forms a potential explosive with water, generates toxic fumes when in contact with water, or contains cyanide or sulfide and will generate toxic fumes if exposed to aqueous solutions that are mildly acidic or basic. All of the explosives forbidden to be transported by the Department of Transportation are also included in the reactive classification. See Table 5-3 for examples of reactive hazardous solid wastes.

Pyrophoric Materials. Materials that ignite upon exposure to air are termed *pyrophoric*. Diborane is one of these, and the aluminum alkyls constitute another class of materials that have this characteristic to varying degrees. The best means of disposal for small amounts of these materials is to take the containers, usually pressure cylinders of some sort, to a safe location and vent them gradually so that the contents ignite and burn as they leave the vessel. Be sure to consult local ordinances on burning. A short length of metal tubing will protect the valve from overheating. Care must be taken not to release too much material at once, and that no flame impinging upon a cylinder causes overheating and possible failure of the vessel. Alkyls may be rendered nonpyrophoric by diluting them to about 10% in a dry hydrocarbon. They can then be burned safely, but should not be mixed with other flammable liquids for disposal. The combustion products from an alkyl fire are usually not too highly toxic, but it is advisable to avoid the smoke.

GENERAL CONSIDERATIONS FOR DISPOSAL

Every organization should have a system for the disposal of chemical wastes that is safe and environmentally acceptable. This system should be under the direction of a safety coordinator or department specifically charged with that responsibility. Written plans for handling emergencies such as spills should be readily available. To serve the range of laboratory requirements (solid versus liquid, toxic versus nontoxic wastes), the disposal system should be complete and permit no loopholes. Storage of wastes at the institution should be minimal. A recommended maximum for temporary accumulation of nonacute hazardous waste is 55 gal.

The plan for safe disposal of the substances used is as much a part of the plan for the experiment as is the acquisition of materials and the experimental procedures. If an experiment involves new types of disposal problems, the laboratory worker should discuss the disposal plan with the laboratory supervisor and, if necessary, with the safety coordinator for the organization.

All persons using chemicals in the laboratory should be generally aware of the toxic properties of the substance used, including consideration of the toxicity of possible reaction products. If the toxic properties of possible products are not known, the products should be treated with respect and the disposal method should take account of the uncertain hazards. Some products may be disposed of as an integral part of the experiment (e.g., by using a scrubber for a gaseous product).

All laboratory workers are responsible for proper disposal of the substances they use. Arrangements for disposal may vary from laboratory to laboratory, depending on the facilities and the types of substances used, but the basic principle is that substances must be disposed of in ways that avoid harm to people and the environment. Wastes should be transported in a way that is safe and acceptable to the people involved in disposal operations. It is also important to consider the future fate of the waste substance.

Table 5-3. Examples of Reactive Hazardous Solid Wastes

Reactive elements:	Alkali metals
	Alkaline earth elements
Water-reactive compounds:	Lithium aluminum hydride
	Calcium phosphide
	Magnesium nitride
	Calcium carbide
Peroxides:	Acetyl peroxide
	Benzoyl peroxide
Peroxide formers:	Diethyl ether
	Dioxane
	Tetrahydrofuran
Pyrophorics:	Finely divided metals
	Iron
	Lead
	Zinc
	Alkyl zinc
	Alkyl aluminum compounds
Explosives:	Picric acid
	Urea nitrate
	Ammonium nitrate
Ignitable solids:	Nitrocellulose
	Sodium amide
Polymerizable, exothermic:	Styrene monomer
	Acetaldehyde

Source: Young, J. A., "Academic Laboratory Waste," *Journal of Chemical Education* 60(6), 1983:491.

Material Safety Data Sheets (MSDSs) provided by most chemical suppliers give details for safe handling, including waste disposal. These should be made readily available to students and laboratory personnel.

Always try to reduce the magnitude of future disposal problems by planning present procedures to reduce the use of hazardous materials or their quantities.

The First Step

The first step in waste chemical disposal is to prepare a complete list of all the chemicals to be disposed. This list consists of your inventory of unwanted chemicals on hand and any chemical products that will be produced in laboratory work, lecture demonstrations, and so on. For the unwanted chemicals on hand, this list should include:

1. The name of each chemical (or mixture of chemicals)
2. Quantity (usually in grams)
3. Date
4. A description of each container (e.g., 1-L screw-capped glass bottle)
5. A description of the label on the container (enough information to identify the container)

For waste chemicals that are produced from time to time in laboratory or other activities, the list should show for each chemical the period of time (e.g., months) in which the waste chemical will be produced, the estimated quantities that will be produced in that period, a description of the container used for each chemical, and a description of the label.

Next, classify each chemical on the list as a nonhazardous waste chemical or as a hazardous waste chemical.

In many cases individual chemicals need not be kept separate. For example, all chlorinated hydrocarbons can be put into one container labeled "*xx* mL chlorinated hydrocarbons," plus the date and the bottler's initials. Consolidation of wastes is discussed in more detail below.

Disposal Choices

School laboratories use several means of handling unwanted chemicals from the laboratory. Some are

1. Recovery, recycling, and reuse
2. Return of unused chemicals to the supplier or transfer to a laboratory that can use them
3. Incineration
4. Commercial contractor
5. Sanitary sewer disposal
6. In-lab destruction of hazardous properties of the waste (neutralization, dilution, etc.)
7. Landfill disposal

For each chemical there is a particular choice that will be best.

Recycling

Because of the constraints being placed on chemical waste disposal, the cost of acceptable disposal methods is increasing dramatically. This process will result in increased emphasis on recovery and recycling of chemicals that were formerly discarded. Metals may be recovered by extraction or precipitation procedures. Solvents may be recovered by distillation. The heat content of a flammable material may be recovered by burning as fuel.

Ideally, all laboratory wastes would be recycled so that they could be reused in some fashion. But in most instances this is impractical. Large generators of solvent wastes usually recycle them. Smaller generators can contract with a firm that will recover solvents. All organizations should look into the possibility of establishing procedures for this purpose, depending on the types of materials used and their suitability for recovery operations.

Return to Supplier or Give to Another Organization

Chemical suppliers, in some situations, will accept the return of unopened materials that were purchased from them. Contact the supplier about specific chemicals to be returned in order to determine the feasibility and the necessary packaging/labeling procedure. Shipping of chemicals is regulated by the DOT.

Also consider contacting nearby school systems or other laboratories that may want unused reagents you wish to discard. However, the safety issues connected with this practice are very complicated. State and/or local laws may prohibit it.

Incineration and Solvent Burners

Incineration is the most environmentally acceptable method of chemical waste disposal. Combustion of organic materials with excess oxygen at high temperatures for sufficient time results in degradation to elemental constituents or by-products that pose less risk to the environment. In addition to heat, the principal products of incineration are carbon dioxide, water, and oxides of sulfur and nitrogen; depending on what is burned, other volatile materials may also be formed. Nonvolatile products include fly ash and solid residues.

Incinerator technology is highly developed; a wide variety of sizes and types of equipment that can be used to handle solids, liquids, and gases is available. Incinerator regulations require high-destruction combustion efficiencies, set particulate limits, and require secondary-treatment devices such as afterburners, scrubbers, electrostatic precipitators, and filters. Incinerators are complex pieces of equipment that require competent operating personnel as well as operating permits. Some incinerators can be equipped to burn discarded solvents for the heat value.

Organizations that do not have access to an incinerator can arrange for the collection and incineration of chemical wastes by contract with a supplier of such services. However, some commercial incinerator operators may not be interested in the small amounts of material generated by school laboratories. The transportation of hazardous chemicals to an incinerator must be in accordance with the DOT regulations.

One of the most common uses of incinerators for laboratory wastes is in disposal of biological materials (e.g., animal carcasses, tissues, animal wastes, and cultures). This is a clean and efficient process that leaves little residue. The heat of the incinerator sterilizes these materials and thus minimizes the transmission of pathologic agents.

Wastes that contain mercury, thallium, gallium, osmium, selenium, or arsenic cannot legally be incinerated, because the combustion may produce toxic fumes of these elements.

Disposal to the Sewer System

Many water-soluble chemicals can safely be flushed down the drain into sewers that go to treatment plants. However, care must be taken to avoid using this disposal system for materials that can create problems in the sewer system or constitute a violation of regulations. In the United States, local municipal codes and regulations govern discharge into public sanitary and storm drain systems and usually have limitations regarding the introduction of chemical wastes into these systems. The laboratory supervisor should know the local regulations and communicate this information to laboratory workers so that they can conform to the regulations.

In general, the following rules regarding disposal into a sewer system should be followed:

1. Never put chemicals into storm sewer systems. These systems usually run directly into lakes and streams without treatment. Dispose only into sanitary sewer systems, typically the laboratory drain.

2. Only water-soluble substances should be disposed of in the laboratory sink. Water-soluble is defined as materials that will form a 3% aqueous solution (0.1 g in 3 mL of water). Organic materials usually allowable include short-chain alcohols, aldehydes, amides, amines, carboxylic acids, esters, ethers, and ketones. This rule excludes hydrocarbons, halogenated hydrocarbons, nitro compounds, mercaptans, and most oxygenated compounds that contain more than five carbon atoms. Even the most benign material should not be disposed of more than a few hundred grams or milliliters at a time. Always flush with at least 100-fold excess water to create high dilution. Substances that boil below 50°C, even though they are water-soluble (e.g., diethyl ether), should not be put down the drain. Solutions of flammable solvents must be sufficiently dilute that they do not pose a fire hazard. See Table 5–2 for a list of cations and anions that may be disposed of down the drain. To be sure, check with the EPA and local environmental offices.

3. Strong acids and bases should be diluted or neutralized to the pH 3–11 range before they are poured into the sewer system. Acids and alkalis should not be poured into the sewer drain at a rate exceeding the equivalent of 50 mL of concentrated substance per minute.

4. Mixtures of materials can be put down the drain if the water-soluble component is greater than 98% (e.g., ethanol with hydrocarbon denaturant).

5. Highly toxic, malodorous (e.g., mercaptans), flammable (e.g., gasoline, kerosene), explosive (e.g., azides, peroxides), or lachrymatory chemicals should not be disposed of down the drain. Laboratory drains are generally interconnected; a substance that goes down one sink may well come up as a vapor in another or mix with an incompatible chemical that was poured down another drain. Beware of disposing of incompatible materials without dilution. The sulfide poured into one drain may contact the acid poured into another, with unpleasant consequences for everyone in the building. Some sim-

ple reactions can even cause explosions (e.g., ammonia plus iodine, silver nitrate plus ethanol, or picric acid plus lead salts).

6. Viscous materials or water-soluble polymers that might clog the plumbing should not be allowed into the sewer system.
7. Small amounts of some heavy-metal compounds may generally be disposed of in the sink, but larger amounts may pose a hazard for the sewer system or water supply. See Table 5–2.

Regulations do not exempt material that might be discharged because of an accidental spill. Therefore, it is necessary to provide for control of spilled substances by assuring that they do not go down the drain if they are not allowed.

Conversion of Hazardous Materials to a Less Hazardous Form

If practical, very hazardous substances should be converted to less hazardous substances or rendered innocuous in the laboratory rather than being disposed. For example, acids and alkalis should be neutralized; oxidizing agents should be reduced; strong carcinogens should be oxidized in solution in the laboratory before disposal; highly reactive substances, such as metallic sodium and peroxides, should be converted to less reactive substances; cyanides should be treated with bleaching powder or with excess sulfur. Reaction mixtures may be made less hazardous by dilution, cooling, or the slow addition of a neutralizing agent. For water-miscible materials, pouring the reaction mixture onto a bed of ice can often be a way to cool and dilute it simultaneously.

An extensive set of procedures for converting hazardous wastes into less or nonhazardous compounds is found in the National Research Councils *Prudent Practices for Disposal of Chemicals from Laboratories* (1983). Since some of the procedures can be dangerous, suitable precautions should always be taken. It is recommended that these procedures only be performed by, or under the supervision of, a trained professional who understands the chemistry involved. However, it would be good training for students to be involved in the destruction of chemicals as part of their laboratory experience.

It is possible that in the process of neutralizing a hazardous solid waste, you may become, in the eyes of the EPA, a treater of hazardous solid waste. You would then require a permit for this operation. However, using a material that would otherwise be a hazardous solid waste in an educational laboratory experiment that neutralized the properties of this material may be perfectly acceptable.

Consolidation of Hazardous Materials Before Disposal

Generally, the larger the volume or mass of waste, the more expensive it is to dispose of it. However, under RCRA, the volume of the container is important. A partially full 1-L bottle of chloroform is considered to be 1 L of hazardous solid waste. It is better to add other chlorinated hydrocarbons to the bottle until it is full and then dispose of it. The full bottle is also 1 L of hazardous waste.

Similarly, it is better to concentrate an aqueous solution of chromic acid by evaporation to a smaller volume. A hazardous material in solution can sometimes be precipitated to form a solid and a nonhazardous solvent. The decreased volumes will be much less costly to dispose of.

Disposal of Unlabeled Chemicals

A laboratory chemical in a container with a missing label first must be characterized as to general type (e.g., aromatic hydrocarbon, chlorinated solvent, mineral acid, soluble metal salt) before it can be disposed of safely. This can be a difficult or expensive process. Obviously, it is much better to avoid having unlabeled containers.

For do-it-yourself types, *Prudent Practices for Disposal of Chemicals from Laboratories* contains a chapter on identification of unlabeled materials. A more complete set of procedures for identification of mystery materials is detailed in Shriner et al.'s *The Systematic Identification of Organic Compounds* (1980).

A more expensive option is to ship the material to a landfill operator who will analyze the substance for a fee. Such operators will supply information on the correct procedures to follow.

Landfills

Landfill or burial has been a common method for disposal of chemical wastes; both public and private landfills were important outlets for such wastes. This disposal procedure, however, has led to a dispersal of wastes in the environment and is often only a postponement of the ultimate problem. Virtually all chemicals are now prohibited from landfill operations. Before sending wastes to a landfill, federal, state, and local ordinances should be checked. All stages of the disposal should be in compliance with applicable regulations. Great care should be taken to ensure that potentially hazardous chemicals are treated or disposed of in a manner that avoids personal or environmental risks either currently or in the future.

Nonhazardous wastes that are not suitable for drain disposal may be acceptable for disposal at a nearby sanitary landfill. Table 5-4 lists wastes that are not hazardous by EPA criteria and thus are suitable for disposal in sanitary landfills. To be sure, check local regulations. Materials sent to a sanitary landfill should be marked "Nonhazardous Waste" to assure handlers of their safety.

Hazardous solid wastes cannot be disposed of in sanitary landfills. They must go to secure landfills approved by the EPA and the state. Ignitable, explosive, or reactive (except for cyanides and sulfides) wastes cannot even be placed in secure landfills. There are not many secure landfills in the country, and wastes may have to be shipped a considerable distance, but for many chemicals a secure landfill will be the best disposal choice.

Many hazardous materials can be rendered nonhazardous by oxidation, reduction, complexation, or other process and then sent to a sanitary landfill for disposal.

Table 5-4. Typical Nonhazardous Laboratory Wastes

Organic chemicals
  Sugars and sugar alcohols
  Starch
  Naturally occurring α-amino acids and salts
  Citric acid and its Ba, Ca, K, Mg, Na, NH_4, and Sr salts
  Lactic acid and its Ba, Ca, K, Mg, Na, NH_4, and Sr salts

Inorganic chemicals
  Sulfates: Ba, Ca, K, Mg, Na, NH_4, Sr
  Phosphates: Ca, K, Mg, Na, NH_4, Sr
  Carbonates: Ba, Ca, K, Mg, Na, NH_4, Sr
  Oxides: Al, B, Ca, Co, Cu, Fe, Mg, Mn, Si, Sr, Ti, Zn
  Chlorides: K, Mg, Na
  Fluorides: Ca
  Borates: Ca, K, Mg, Na

Laboratory materials not contaminated with hazardous chemicals
  Chromatographic adsorbent
  Glassware
  Filter paper
  Filter aids
  Rubber and plastic protective clothing

Source: From National Research Council, *Prudent Practices for Disposal of Hazardous Chemicals from Laboratories*, National Academy Press, Washington, D.C., 1983, p. 128.

Contractor Disposal

Some commercial waste disposal contractors will perform some or all of the operations necessary to dispose of laboratory wastes. Because the procedures are elaborate and troublesome, these services can be expensive. For the same reason, smaller institutions, which have fewer personnel resources, must rely on contractors to a greater extent than larger schools. Commercial contractors are commonly used for lab pack disposals. Organizations that contract for waste disposal should ensure that the contractor is performing in a safe, legal, and responsible manner.

In the simplest but most costly arrangement, you simply present your list of chemicals to the contractor and ensure that he has the necessary permits to do the work. The contractor packs the wastes, labels the containers, prepares the necessary forms, and distributes them to the appropriate offices, transports the wastes to an approved disposal site, and then certifies that they have been properly disposed of.

It is important that substances that must by RCRA regulations be disposed of separately be provided to the disposal contractor correctly. If they are mixed, the contractor must separate the components—and charge the waste generator for the cost of separation.

It is less costly for the institution to do some of the work required for disposal and let the contractor handle the remainder.

SPECIFIC PROCEDURES

Disposal of Liquid Chemical Wastes

Each organization should have a procedure for collecting liquid chemical wastes from the laboratories and arranging for their disposal by the institution. Suitable containers should be provided, and laboratory workers should understand what may, or may not, be placed in these containers and which materials require special labeling.

Waste solvents that are free of solids and corrosive or reactive substances may be collected in a common bottle or can, which is taken away when full. If this system is used, it is essential to consider exactly what mixtures will go into the can and whether the substances involved are compatible (this may include waste from the neighboring laboratory, should its waste solvent bottle be full). Segregation into two or three types of waste is often useful (e.g., chlorinated solvents, hydrocarbons), as is the use of completely separate bottles for waste that poses special difficulties. In particular, because chlorinated solvents form hydrogen chloride on combustion, they often must be segregated from materials destined for incineration, as their burning will violate local air pollution ordinances. Generally speaking, separated and well-defined waste is easier to dispose of, and if an outside contractor is used, is also less expensive. All wastes that pose hazards should be so labeled.

Some solvents (such as ethers and secondary alcohols) form explosive peroxides upon standing. Some chemicals can react explosively (e.g., acetone plus chloroform in the presence of a base). Others, such as with acid–base interactions, can generate sufficient heat to vaporize or ignite flammable materials such as carbon disulfide. The addition of hot materials can cause the buildup of pressure in a tightly closed solvent container, with the potential for compressive ignition. The acid formed when halogenated solvents are left moist can corrode cans, as can any dissolved corrosive in a discarded mixture.

When large quantities of a solvent are involved, consideration should be given to recycling rather than disposal. This operation also involves some potential hazard and expense, but these limitations may be less severe than those for disposal, especially as disposal costs are increasing.

Lab Packs

Small quantities of hazardous waste are most conveniently disposed of by placement in a lab pack and burial in a secure landfill. A lab pack is a steel drum that is filled with small containers of chemicals cushioned and separated by vermiculite. A commercial contractor is commonly used to help with the process of lab pack disposal.

Typically, a lab pack is a liquid-tight, 55-gal steel drum (maximum 110 gal) with a securable, tight lid. The drum must meet DOT specifications. The chemicals must be sealed into small, leak-proof containers that will not react with their contents. The small containers are surrounded by an absorbent that will not react with the contents of the pack. There must be enough absorbent to protect each container from hitting another and to completely absorb all of the liquid contained in the pack. When the pack is closed up, the absorbent must fill all spaces in the pack. Vermiculite and fuller's earth are the most common absorbents, but others are available. Once filled, a list of contents is attached to the outside.

There are many types of hazardous waste that cannot be put into lab packs:

Incompatible wastes in the same lab pack

Ignitable wastes

Reactive wastes that can explode or release toxic gases, vapors, or fumes when at standard pressure and temperature; when mixed with or exposed to water; when subjected to a strong initiating force; or when heated under confinement (cyanide- and sulfide-bearing reactive wastes are permitted in lab packs).

DOT-forbidden materials

Explosives, either Class A or Class B.

Other requirements may be placed on the waste generator by the shipper or land-disposal facility.

The requirement that incompatible wastes not be packed together is usually avoided by using several lab packs for different materials (e.g., one drum contains chlorides and another contains ammonium salts). Segregation by DOT classes may be suitable for large waste generators.

Materials that cannot be put into a lab pack can be chemically altered to lessen their hazardous properties or to make them nonhazardous (in which case a lab pack is not needed).

Once a lab pack has been sealed and a list of contents has been attached, there is a host of paperwork to be completed. DOT requirements as well as those of the landfill must be complied with. It may be wise to have a commercial contractor handle the rest of the disposal process.

Disposal of Hazardous Biological Wastes

The most appropriate method of disposing of animal tissues, excretion products, and other biological materials is by incineration. This is true even if the materials are contaminated by hazardous materials.

Before incineration, it is important to consider the chemical properties and thermal stability, as well as the quantity, of hazardous substances present in the items to be disposed of. For example, if a toxic contaminant is thermally stable, incineration in a conventional facility may lead to contamination of the immediate environment of the incinerator. It is also important to know whether any product is formed during combustion that is thermally stable and likely to have significant toxicity. In rare cases, the combustion of animal tissues containing toxic substances may result in the formation of products that are more toxic than the material being disposed of. For example, recent work has indicated that the incineration of polychlorinated biphenyls at 500–600°C may lead to the formation of highly toxic chlorinated benzofurans as a side product.

Thus, alternative methods of disposal should be used for animal remains and products containing a thermally stable toxic or if it is anticipated that the combustion process will result in the formation of thermally stable toxins. One alternative is combustion in a special incinerator capable of very high temperatures (1000°C) and having sufficient dwell time for the vaporized material to ensure its degradation. Another, but less acceptable, alternative is burial of animal remains and products in an EPA-approved burial site.

During the disposal process, care must be taken to package the animals and animal products in a way that minimizes potential exposure of all personnel to any toxic substances. Wastes should be sealed in polyethylene bags. To protect from tearing, the bags should be placed in cardboard boxes or in heavy paper bags. If the bags are not incinerated daily, they should be stored at 4°C or less. Adequate protective apparel and respirators should be available if needed.

STORAGE OF WASTES

Temporary storage of wastes before final disposal is a standard operating procedure for laboratories. The amount of material in storage should be limited, and a maximum amount of 55 gal should be adhered to. Incompatible chemicals should be segregated.

The storage facility should be separate from the laboratory and protected from the elements. It should have limited access and be posted with warning signs. Ideally, the on-site storage containers are also used to ship the wastes. If so, they must be DOT approved.

A waste storage area should be capable of containing a leak from the storage containers. This means that the floor must be impervious to all the stored materials and that the edge around the floor must be high enough so that it will contain the complete contents of the largest container. If the stored chemicals are flammable, the area should have some kind of sprinkler system and fire extinguishers. See *Fundamentals of Laboratory Safety: Physical Hazards in the Academic Laboratory*, Chap. 12, for more details on flammables storage.

A running inventory of wastes and contents of waste containers in storage should be kept at the storage site. Establish a log-in/log-out procedure for the storage area. When someone brings waste to storage, it is recorded. When waste is removed, it is recorded in the same book. This helps with the necessary paperwork, usually assures more accurate information, and is safer. It is not good practice to store containers whose contents are not known. The waste storage area should be inspected on a regular basis.

In most laboratories there are chemicals that have been stored away because no one wants to go to the trouble of disposing of them. This is poor laboratory practice. No type of laboratory waste should be stored for more than a short time before disposal. Flammables, reactive chemicals, and toxins can rapidly accumulate in a fume hood, waste container, or storage cabinet. Waste containers in school laboratories often become dumping grounds for a variety of unwanted substances. Do not allow this to happen.

TRANSPORTATION OF CHEMICAL WASTES

Adherence to the DOT regulations regarding the transportation of hazardous substances must be ensured. Laboratory workers should know that no quantity of chemically hazardous material may be transported on any public conveyance, by the U.S. Postal Service, or by a private conveyance over public highways without proper packaging classification, labeling, and documentation.

MINIMIZATION OF WASTE

Once the complexities of laboratory waste disposal are understood, the need to minimize these wastes is obvious. The first approach to waste minimization is to limit the amount of chemicals purchased. Do not buy more of a chemical than you will use in a short period, possibly a quarter or a semester. It may be initially less costly per unit to buy a larger amount, but if even a small amount of the order is unused and becomes a waste material, the cost advantage will have been lost in the cost and effort required to dispose of the material. Storage of excessive quantities of chemicals is also an unnecessary safety hazard. Purchase only what will foreseeably be used in the near future.

Experiments can be redesigned to use smaller quantities of chemicals and to produce less or fewer hazardous wastes.

The Government Institute's *Waste Minimization Manual* (1987) is a useful tool for setting up an institutional system for minimizing chemical wastes.

RECORDKEEPING

It is important to keep records of waste chemicals on hand and of those that have been disposed. These records should contain the name of the person responsible, the chemical identity or description of the waste, the amount, the date it was designated as a waste, the date of disposal, and the location of the disposal site. All such records should be retained permanently.

For chemical mixtures, a description of the waste means components and percentages. For lab packs, description means a complete contents list.

SUGGESTED READINGS

American Chemical Society, *Safety in Academic Chemistry Laboratories*, Washington, D.C., 1985.

Code of Federal Regulations, Title 40, parts 261 and 262. [CFR titles (chapters) are available from the U.S. Government Printing Office, Washington, D.C.; each title is revised annually.]

Gerlovich, J. A. (ed.), *Secondary School Science Safety*, Flinn Scientific, Inc., Batavia, Ill., 1988.

Government Institutes, *Waste Minimization Manual*, Rockville, Md., 1987.

National Research Council, *Prudent Practices for Disposal of Chemicals from Laboratories*, National Academy Press, Washington, D.C., 1983.

Shriner, R. L., Fuson, R. C., Curtin, D. Y., and Morril, T. C., *The Systematic Identification of Organic Compounds*, 6th ed., Wiley, New York, 1980.

Young, J. A., "Academic Laboratory Waste," *Journal of Chemical Education* 60(6), 1983:490–492.

Part 2
CHEMICAL ENTRIES

ACETALDEHYDE

CAS #75-07-0   mf: C_2H_4O
DOT #1089   mw: 44.06

STRUCTURE
CH_3CHO

NFPA LABEL

SYNONYMS
Acetic aldehyde
Ethanal
Ethyl aldehyde

PROPERTIES   Colorless, fuming liquid or gas; pungent, fruity odor; miscible in water, alcohol, and ether; melting point −123.5°C, boiling point 20.8°C; density 0.7827 at 20°C/20°C; flash point −37.8°C (closed cup); lower explosive limit 4.0%, upper explosive limit 57%; autoignition temperature 175°C; vapor density 1.52

OSHA PEL   TWA 200 ppm; IDLH 10,000 ppm

ACGIH TLV   TWA 100 ppm; STEL 150 ppm

DOT Classification   Flammable liquid

TOXIC EFFECTS   Eye, skin, and respiratory irritant. Can produce severe eye burns. Prolonged inhalation may have narcotic effect. A teratogen and mutagen.

The vapor is irritating to the eyes, nose, and throat. If inhaled, it will cause nausea, vomiting, headache, difficult breathing, or loss of consciousness.

The liquid will burn the skin and eyes. It is harmful if swallowed.

Routes   Inhalation, ingestion

Symptoms   Eye, nose, throat irritation; conjunctivitis; cough; central nervous system depressant/depression; eye, skin burns; dermatitis; delayed pulmonary edema

Target Organs   Respiratory system, skin, kidneys

HAZARDOUS REACTIONS

Acid anhydrides: violent reaction
Acids (traces): polymerizes violently
Air: mixtures of 30–60% of the vapor in air ignite above 100°C
Alcohols: violent reaction
Alkalies (strong): violent reaction
Amines: violent reaction
Ammonia: violent reaction
Cobalt chloride: reacts to form sensitive, explosive product
Halogens: violent reaction
Hydrogen cyanide: violent reaction
Hydrogen sulfide: violent reaction
Isocyanates: violent reaction
Ketones: violent reaction
Mercury(II) chlorate: reacts to form sensitive, explosive product
Mercury(II) perchlorate: reacts to form sensitive, explosive product
Metals (traces): polymerizes violently
Oxidizers (strong): incompatible
Oxygen: reaction may lead to detonation
Phenols: violent reaction
Phosphorus: violent reaction

When heated to decomposition, it may produce acrid smoke and toxic fumes.

FIRST AID   In case of contact with the eyes, immediately wash the eyes with large amounts of water, occasionally lifting the lower and upper lids. Get medical attention immediately. Contact lenses should not be worn when working with this chemical.

In case of contact with the skin, immediately flush the contaminated skin with water. If this chemical penetrates the clothing, immediately remove the clothing and flush the skin with water. If irritation persists after washing, get medical attention.

If a person breathes in large amounts of this chemical, move the victim to fresh air at once. If breathing has stopped, perform artificial respiration. If breathing is difficult, give oxygen. Keep the affected person warm and at rest. Get medical attention as soon as possible.

If this chemical has been swallowed, get medical attention immediately.

FIRE HAZARD   Extremely flammable. Vapor may explode if ignited in an enclosed area.

HANDLING   Wear appropriate clothing and equipment to prevent repeated or prolonged skin contact. Wear eye protection to prevent any possibility of eye contact. Personnel should wash promptly when skin becomes wet. Immediately remove any clothing that becomes wet to avoid flammability hazard. The following equipment should be available: an eyewash station.

SPILL CLEANUP   Avoid contact with liquid and vapor. Wear self-contained breathing apparatus. Equipment should not have exposed rubber parts.

ACETIC ACID

CAS #64-19-7 mf: $C_2H_4O_2$
DOT #2789/2790 mw: 58.08

STRUCTURE NFPA LABEL
CH_3COOH

2 / 2 / 1

SYNONYMS
Acetic acid, glacial   Glacial acetic acid
Ethanoic acid   Methanecarboxylic acid
Ethylic acid   Vinegar acid

PROPERTIES   Clear liquid with a strong, vinegar-like odor; miscible with water, alcohol, and ether; boiling point 118.1°C, melting point 16.7°C; density 1.049 at 20°/4°C; flash point 42.8°C; lower explosive limit 5.4%, upper explosive limit 16.0% at 100°C; autoignition temperature 465°C; vapor pressure 11.4 mm at 20°C; vapor density 2.07

OSHA PEL   10 ppm; IDLH 1000 ppm

ACGIH TLV   TWA 10 ppm; STEL 15 ppm

DOT Classification   Label: Corrosive, flammable liquid

TOXIC EFFECTS   A poison. The vapor is irritating to the nose and throat. If inhaled it will cause coughing, nausea, vomiting, or difficult breathing. Repeated inhalation of vapor may cause chronic inflammation of upper respiratory tract and chronic bronchitis. The liquid or solid will burn skin or eyes and is harmful if swallowed.

Routes   Inhalation

Symptoms   Conjunctivitis lacrimation; irritation of nose, throat; pharyngeal edema, chronic bronchitis; burns eyes, skin; skin sensitization; dental erosion; black skin, hyperkeratosis

Target Organs   Respiratory system, skin, eyes, teeth

HAZARDOUS REACTIONS
Acetaldehyde: potentially violent reaction
Acetic anhydride: potentially violent reaction
Acids: incompatible
Alcohol: incompatible
Amines: incompatible
2-Amino-ethanol: incompatible
Ammonia: incompatible
Ammonium nitrate: incompatible
5-Azidotetrazole: potentially explosive reaction
Bases: incompatible
Bromine pentafluoride: potentially explosive reaction
Caustics (strong): incompatible
Chlorosulfonic acid: incompatible
Chromic acid: incompatible
Chromium trioxide: potentially explosive reaction
Chlorine trifluoride: incompatible
Ethylene imine: incompatible
Ethylenediamine: incompatible
Hydrogen cyanide: incompatible
Hydrogen peroxide: potentially explosive reaction
Hydrogen sulfide: incompatible
Ketones: incompatible
Nitric acid + acetone: incompatible
Nitric acid: incompatible
Oleum: incompatible
Oxidizers (strong): incompatible
Ozone + diallyl methyl carbinol: incompatible
$P(OCN)_3$: incompatible
Perchloric acid: incompatible

Permanganates: incompatible
Phenols: incompatible
Phosphorus trichloride: potentially explosive reaction
Potassium hydroxide: incompatible
Potassium permanganate: potentially explosive reaction
Potassium-*tert*-butoxide: ignites on contact
Sodium hydroxide: incompatible
Sodium peroxide: potentially explosive reaction
n-Xylene: incompatible

When heated to decomposition, it may produce irritating fumes.

FIRST AID   In case of contact with the eyes, immediately wash the eyes with large amounts of water, occasionally lifting the lower and upper lids. Get medical attention immediately. Contact lenses should not be worn when working with this chemical.

In case of contact with the skin, immediately flush the contaminated skin with water. If this chemical penetrates the clothing, immediately remove the clothing and flush the skin with water. Get medical attention promptly.

If a person breathes in large amounts of this chemical, move the victim to fresh air at once. If breathing has stopped, perform artificial respiration. Keep the affected person warm and at rest. Get medical attention as soon as possible.

If the chemical has been swallowed, get medical attention immediately. Move the person to fresh air. If breathing has stopped, give artificial respiration. If breathing is difficult, give oxygen.

FIRE HAZARD   Combustible. The vapor is explosive if ignited in an enclosed area.

Use CO_2, dry chemical, alcohol foam, foam, or mist extinguishers to fight a fire.

SPILL CLEANUP   Avoid contact with liquid or vapor. Wear goggles, self-contained breathing apparatus, and rubber overclothing.

HANDLING AND STORAGE   Wear appropriate equipment to prevent any possibility of skin contact with liquids containing >50% of the acid, and repeated or prolonged skin contact with liquids containing 10–49%. Wear goggles to prevent any possibility of eye contact. Workers should wash immediately when skin is contaminated with liquids containing >50% acid, and promptly when skin is contaminated with 10–49%. Remove clothing contaminated by >50% solutions immediately, and remove clothing contaminated by 10–49% solutions promptly.

Wear appropriate equipment to prevent any possibility of skin contact with liquids containing >50% of the acid, and repeated or prolonged skin contact with liquids containing 10–49%. Wash immediately when skin is contaminated with liquids containing >50% acid. Remove clothing that has been contaminated with liquids containing >50% acid, and promptly remove clothing contaminated with liquids containing 10–49% acid.

Wear eye protection to prevent any possibility of eye contact.

Provide an eyewash station if liquids containing >5% acid are present. Provide a quick-drench shower if liquids containing >50% acid are present.

ACETIC ANHYDRIDE

CAS #108-24-7　　　　　　　　　　mf: $C_4H_6O_3$
DOT #1715　　　　　　　　　　　　mw: 102.10

STRUCTURE　　　　　　　NFPA LABEL
$(CH_3CO)_2O$

SYNONYMS
Acetic acid, anhydride　　　Acetyl ether
Acetic oxide　　　　　　　　Acetyl oxide
Acetyl anhydride　　　　　　Ethanoic anhydrate

PROPERTIES   Colorless, very mobile, strongly refractive liquid; very strong acetic odor; somewhat soluble in cold water, decomposes in hot water and hot alcohol, miscible in alcohol and ether; melting point −73.1°C, boiling point 140°C; density 1.082 at 20°C/4°C; flash point 53.9°C (closed cup); lower explosive limit 2.9%, upper explosive limit 10.3%; autoignition temperature 390°C; vapor pressure 10 mm at 36.0°C; vapor density 3.52

OSHA PEL   TWA 5 ppm

ACGIH TLV   ceiling 5 ppm

IMO Classification   Corrosive material; Label: Corrosive, flammable liquid

TOXIC EFFECTS   Moderately toxic by inhalation, ingestion, and skin contact. Corrosive to body tissues. Especially hazardous to the eyes; a delayed action will cause burns. The vapor will burn the eyes and is irritating to the nose and throat. The liquid will burn skin and eyes and is harmful if swallowed.

Routes   Inhalation, ingestion, contact

Symptoms   Conjunctivitis, lacrimation, corneal edema, opacity, photophobia, nose/nasal pharyngeal irritation, cough, dyspnea, bronchitis, skin burns, vesiculation, dermatitis

Target Organs   Respiratory system, eyes, skin

HAZARDOUS REACTIONS
Acetic acid + water: potentially explosive reaction
Alcohols: incompatible
Amines: incompatible
2-Aminoethanol aniline: incompatible
Ammonium nitrate + hexamethylenetetraminium acetate + nitric acid: reaction forms the military explosives RDX and HMX
Barium peroxide: potentially explosive reaction
Boric acid: potentially explosive reaction
N-*tert*-butylphthalimic acid + tetrafluoroboric acid: violent reaction
Caustics (strong): incompatible
Chlorosulfonic acid: incompatible
Chromic acid: violent reaction
Chromium trioxide: potentially explosive reaction
Chromium trioxide + acetic acid: incompatible
1,3-Diphenyltriazene: potentially explosive reaction
Ethanol + sodium hydrogen sulfate: reacts to form explosive products
Ethylenediamine: incompatible
Ethyleneimine: incompatible
Glycerol: incompatible
Glycerol + phosphoryl chloride: violent reaction
Hydrochloric acid + water: potentially explosive reaction
Hydrogen fluoride: incompatible
Hydrogen peroxide: reacts to form explosive products
Hypochlorous acid: potentially explosive reaction
Metal nitrates (e.g., copper or sodium nitrates): violent reaction
N_2O_2: incompatible
Nitric acid: potentially explosive reaction
Oleum: incompatible
Oxidizers (strong): incompatible
Oxidizing materials: vigorous reaction
Perchloric acid + water: potentially explosive reaction
Permanganates: incompatible
Peroxyacetic acid: potentially explosive reaction
Potassium permanganate: potentially explosive reaction
Sodium hydroxide: incompatible
Sodium peroxide: incompatible
Steam: violent reaction forms acetic acid
Sulfuric acid: incompatible
Tetrafluoroboric acid: potentially explosive reaction
4-Toluenesulfonic acid + water: potentially explosive reaction
Water: violent reaction forms acetic acid

When heated to decomposition, it may produce toxic fumes.

FIRST AID   In case of contact with the eyes, immediately wash the eyes with large amounts of water, occasionally lifting the lower and upper lids. Get medical attention immediately. Contact lenses should not be worn when working with this chemical.

In case of contact with the skin, immediately flush the contaminated skin with water. If this chemical penetrates the clothing, immediately remove the clothing and flush the skin with water. Get medical attention promptly.

If a person breathes in large amounts of this chemical, move the victim to fresh air at once. If breathing has stopped, perform artificial respiration. Keep the affected person warm and at rest. Get medical attention as soon as possible.

If this chemical has been swallowed, get medical attention immediately.

FIRE HAZARD   Flammable liquid. Vapor may explode if ignited in an enclosed area.

To extinguish fires, use CO_2, dry chemical, water mist, or alcohol foam.

HANDLING   Wear appropriate clothing and equipment to prevent a reasonable probability of skin contact. Wash immediately when skin becomes contaminated. Immediately remove any nonimpervious clothing that becomes contaminated.

Wear eye protection to prevent any possibility of eye contact.

The following equipment should be available: an eyewash station and a quick-drench shower.

SPILL CLEANUP   Avoid contact with liquid and vapor. Wear goggles, self-contained breathing apparatus, and rubber overclothing (including gloves).

ACETONE

CAS #67-64-1   mf: CH_3COCH_3
DOT #1090   mw: 58.08

STRUCTURE   NFPA LABEL
CH_3COCH_3

3 / 1 / 0

SYNONYMS

Dimethylformaldehyde
Dimethylketal
Dimethyl ketone
Ketone propane
β-Ketopropane
Methyl ketone
Propanone
2-Propanone
Pyroacetic acid
Pyroacetic ether

PROPERTIES   Colorless liquid with characteristic odor; miscible with water, alcohol, and ether; melting point −94.6°C, boiling point 56°C; density 0.7972; flash point −17.8°C; lower explosive limit 2.6%, upper explosive limit 12.8%; autoignition temperature 465°C; vapor pressure 400 mm at 39.5°C; vapor density 2.00

Highly flammable; avoid breathing vapor, prevent contact with eyes.

OSHA PEL   TWA 1000 ppm, 2400 mg/m^3; IDLH 20,000 ppm

ACGIH TLV   TWA 750 ppm; STEL 1000 ppm

NIOSH REL   TWA 250 ppm; 590 mg/m^3

Reported in EPA TSCA Inventory and Community Right to Know List

DOT Classification   Flammable liquid; Label: Flammable liquid

TOXIC EFFECTS   Moderately toxic. Inhalation of vapor may cause dizziness, narcosis, and coma. The liquid irritates the eyes and may cause severe damage, but it is not irritating to the skin. If swallowed, it may cause gastric irritation, narcosis, and coma. Narcotic in high concentrations. The vapor is irritating to the eyes, nose, and throat. If inhaled, it may cause difficult breathing or loss of consciousness.

Routes   Inhalation, ingestion, contact

Symptoms   Irritated eyes, nose, throat; headache, dizziness; dermatitis

Target Organs   Respiratory system, skin

HAZARDOUS REACTIONS

Acids: incompatible
Bromine + air: vigorous oxidation reaction
Bromine trifluoride: potentially explosive reaction
Bromine: violent reaction
Bromoform: violent reaction
Bromoform + bases: violent reaction
Carbon (active) + air: vigorous oxidation reaction
Chloroform: incompatible
Chloroform + alkalies: violent reaction
Chromium oxide: incompatible
Chromium trioxide: ignites on contact
Chromyl chloride: potentially explosive reaction
Dioxygen difluoride + carbon dioxide: ignites on contact
Fluorine oxide + air: vigorous oxidation reaction
Fluoroform + bases: violent reaction
Hydrogen peroxide: reacts to form explosive peroxide products
Hydrogen peroxide + air: vigorous oxidation reaction
2-Methyl-1,3-butadiene: reacts to form explosive peroxide products
NaOBr: incompatible
Nitric acid: incompatible
Nitric acid + acetic acid: incompatible
Nitric acid + air: vigorous oxidation reaction
Nitric acid + sulfuric acid: potentially explosive reaction
Nitrosyl chloride + air: vigorous oxidation reaction
Nitrosyl chloride + platinum: potentially explosive reaction
Nitrosyl perchlorate: potentially explosive reaction
Nitryl perchlorate: incompatible
Nitryl perchlorate + air: vigorous oxidation reaction
NOCl: incompatible
Oxidizing materials: can react vigorously
Permonosulfuric acid: incompatible
Peroxomonosulfuric acid: reacts to form explosive peroxide products
Peroxomonosulfuric acid + air: vigorous oxidation reaction
Potassium-*tert*-butoxide: ignites on contact
Sulfur dichloride: violent reaction
Sulfuric acid: incompatible
Sulfuric acid + potassium dichromate: incompatible
thio-Diglycol + hydrogen peroxide: incompatible
Thiotrithiazyl perchlorate: potentially explosive reaction
Trichloromelamine: incompatible
2,4,6-Trichloro-1,3,5-triazine + water: potentially explosive reaction

FIRST AID   In case of contact with the eyes, immediately wash the eyes with large amounts of water, occasionally lifting the lower and upper lids. Get medical attention immediately. Contact lenses should not be worn when working with this chemical.

In case of contact with the skin, immediately wash the contaminated skin with soap and water. If this chemical penetrates through the clothing, immediately remove the clothing and wash the skin with soap and water. Get medical attention promptly.

If a person breathes in large amounts of this chemical, move the victim to fresh air at once. If breathing has stopped, perform artificial respiration. Keep the affected person warm and at rest. Get medical attention as soon as possible.

If this chemical has been swallowed, get medical attention immediately.

FIRE HAZARD   Very dangerous. Flammable. Vapor may explode if ignited in an enclosed area.

Use water spray, dry chemical, or carbon dioxide extinguishers to fight a fire.

HANDLING   Wear appropriate clothing and equipment to prevent repeated or prolonged skin contact. Wash promptly when skin becomes wet. Immediately remove any clothing that becomes wet to avoid flammability hazard.

Wear eye protection to prevent a reasonable probability of eye contact.

SPILL CLEANUP   Shut off possible sources of ignition. Wear face shield and gloves. Mop up with plenty of water and run to waste, diluting greatly with running water. Ventilate area well to evaporate remaining liquid and dispel vapor.

ACETONITRILE

CAS #75-05-8 mf: C_2H_3N
DOT #1648 mw: 41.06

NFPA LABEL

SYNONYMS
Cyanomethane   Methanecarbonitrile
Ethanenitrile   Methyl cyanide
Ethyl nitrile

PROPERTIES   Colorless liquid with an aromatic odor; miscible in water, alcohol, and ether; melting point −45°C, boiling point 81.1°C; density 0.7868 at 20°C; lower explosive limit 4.4%, upper explosive limit 16%; autoignition temperature 523.9°C; vapor pressure 100 mm at 27°C; vapor density 1.42

OSHA PEL   TWA 40 ppm; IDLH 4000 ppm

ACGIH TLV   TWA 40 ppm; STEL 60 ppm (skin)

NIOSH REL   TWA 20 ppm (10 h TWA), 34 mg/m³

Community Right to Know List

DOT Classification   Flammable liquid; Label: Flammable liquid and poison

TOXIC EFFECTS   Poisonous by ingestion. Moderately toxic by other routes. The vapor is irritating to the eyes, nose, and throat. If inhaled, it will cause difficult breathing. The liquid is irritating to skin and eyes. A teratogen.

Routes   Inhalation, ingestion, absorption, contact

Symptoms   Asphyxiation; nausea, vomiting; chest pain; weakness; stupor, convulsions; eye irritation

Target Organs   Kidneys, liver, cardiovascular system, central nervous system, lungs, skin, eyes

HAZARDOUS REACTIONS
Acids: reacts to form toxic and flammable vapors
Chlorosulfonic acid: incompatible
Dinitrogen tetraoxide: incompatible
Indium: incompatible
Lanthanide perchlorates: potentially explosive reaction
N-fluoro compounds (i.e., perfluorourea): incompatible
Nitrating agents: incompatible
Nitric acid: incompatible
Nitrogen-fluorine compounds: potentially explosive reaction
Oleum: incompatible
Oxidizers (strong): incompatible
Perchlorates: incompatible
Steam: reacts to form toxic and flammable vapors
Sulfur trioxide: incompatible
Sulfuric acid: exothermic reaction at 53°C

When heated to decomposition, it may emit highly toxic fumes of CN^- and NO_x.

FIRST AID   In case of contact with the eyes, immediately wash the eyes with large amounts of water, occasionally lifting the lower and upper lids. Get medical attention immediately. Contact lenses should not be worn when working with this chemical.

In case of contact with the skin, immediately flush the contaminated skin with water. If this chemical penetrates the clothing, immediately remove the clothing and flush the skin with water. Get medical attention promptly.

If a person breathes in large amounts of this chemical, move the victim to fresh air at once. If breathing has stopped, perform artificial respiration. Keep the affected person warm and at rest. Get medical attention as soon as possible.

If this chemical has been swallowed, get medical attention immediately.

FIRE HAZARD   Dangerous. Flammable liquid. Vapor may explode if ignited in an enclosed area. Do not expose to heat, flame, or oxidizers.

Use foam, CO_2, or dry chemical extinguishers to fight a fire.

HANDLING   Wear appropriate clothing and equipment to prevent repeated or prolonged skin contact. Wash immediately when skin becomes contaminated. Immediately remove any clothing that becomes wet to avoid flammability hazard.

Wear eye protection to prevent a reasonable probability of eye contact.

The following equipment should be available: a quick-drench shower.

SPILL CLEANUP   Wear self-contained breathing apparatus and goggles.

ACETYL PEROXIDE

CAS #110-22-5   mf: $C_4H_6O_4$
mw: 118.04

STRUCTURE

$$CH_3-\overset{\overset{O}{\|}}{C}-O-O-\overset{\overset{O}{\|}}{C}-CH_3$$

SYNONYM
Diacetyl peroxide

PROPERTIES   Solid or colorless crystals or liquid; slightly soluble in water, soluble in alcohol and hot ether; melting point 30°C, boiling point 63°C at 21 mm Hg; density 1.18. Explosive.

DOT Classification   Forbidden

TOXIC EFFECTS   A severe eye irritant. Irritating to the eyes, skin, and mucous membranes by ingestion and inhalation. Application of two drops of a 30% solution (in dimethyl phthalate) has caused very severe corneal damage to rabbits.

HAZARDOUS REACTIONS   Severe explosion hazard when shocked or exposed to heat. It may explode spontaneously in storage and should be used as soon as prepared.

Acetyl peroxide is a powerful oxidizing agent and can cause ignition of organic materials on contact. There are reports of detonation of the pure material; the 25% solution also has explosive potential, and inadvertent partial evaporation of even weak solutions can create explosive solutions or shock-sensitive crystalline material.

Acids: reaction may produce toxic fumes
Acid fumes: reaction may produce toxic fumes
Organic materials: may ignite on contact
Reducing materials: vigorous reaction
Steam: exothermic reaction
Water: exothermic reaction

FIRE HAZARD   A dangerous fire hazard due to spontaneous chemical reaction.

Use CO_2 or dry chemical extinguishers to fight a fire.

HANDLING   Must be kept below 27°C and not warmed over 30°C. Do not add to hot materials. Do not add accelerator to this material. Store in original container with vented cap. Avoid bodily contact. This material is nearly always stored and handled as a 25% solution in an inert solvent.

A face shield and rubber gloves should be worn when handling acetyl peroxide, and a safety shield or hood door should be in front of apparatus containing it.

ACETYLENE

CAS #74-86-2   mf: C_2H_2
DOT #1001   mw: 26.04

STRUCTURE   NFPA LABEL
HC≡CH

SYNONYMS
Acetylen   Ethine
Narcylen   Ethyne

PROPERTIES   Colorless gas, garliclike odor; soluble in water, very soluble in alcohol, almost miscible in ether; melting point −81.8°C, boiling point −84.0°C (sublimes); density 1.173 g/L at 0°C, density (liquid) 0.613 at −80°C, density (solid) 0.730 at −85°C; flash point −17.8°C (closed cup); lower explosive limit 2.5%, upper explosive limit 82%; autoignition temperature 305°C; vapor pressure 40 atm at 16.8°C; vapor density 0.91

OSHA PEL   Ceiling 2500 ppm

ACGIH TLV   Simple asphyxiant

88 ACETYLENE

NIOSH REL Ceiling 2500 ppm

DOT Classification Forbidden; Flammable gas; Label: Flammable gas

TOXIC EFFECTS Slightly toxic by inhalation. Narcotic at concentrations greater than 40%.

Symptoms Headache, difficult breathing, loss of consciousness

HAZARDOUS REACTIONS When ignited, it burns with an intensely hot flame. At high pressures and moderate temperatures, and in the absence of air, acetylene has been known to decompose explosively.

Brass: incompatible
Cesium hydride: incompatible
Cobalt powder: incompatible
Copper carbide: incompatible
Copper: reacts to form the explosive copper acetylide
Copper salts: incompatible
Halides + UV: potentially explosive reaction
Halogens: incompatible
Mercury: incompatible
Mercury salts: incompatible
Nitric acid: incompatible
Oxidizing materials: vigorous reaction
Oxygen: potentially powerful explosion
Potassium: incompatible
Potassium (molten): ignites and then explodes
Rubidium hydride: incompatible
Silver: incompatible
Silver salts: incompatible
Sodium hydride: incompatible
Trifluoromethyl hypofluorite: vigorous reaction

The most thermodynamically unstable common substance. It has a very wide explosive range. In addition to being explosive and shock-sensitive, acetylene can deflagrate in the absence of air to give (presumably) oligomers or polymers. Its stability is markedly enhanced by the presence of small amounts of other compounds, such as methane, and, in fact, acetylene from cylinders is rather safe to handle, because it is dissolved in acetone. For some uses, such as the preparation of acetylides, it is necessary to scrub the gas, thus removing the acetone. Such purified acetylene is incomparably more dangerous than acetylene straight from the tank. Among other peculiarities of pure acetylene is the fact that its stability seems to be related to the diameter of the pipe used to transport it; it is actually less stable in wide-bore piping. The handling and use of acetylene under pressure is extremely hazardous. In the absence of compelling reasons to the contrary, all reactions and operations involving acetylene should be run in a pressure laboratory that has the necessary facilities, as well as expertise and experience, for its safe handling.

FIRE HAZARD Very dangerous. A fire and explosion hazard when exposed to heat, flame, or oxidizers.

Use CO_2, water spray, or dry chemical extinguishers to fight a fire.

HANDLING AND SPILL CLEANUP Under no circumstances should acetylene be used under pressure in unbarricaded equipment. In any case, the lowest pressure necessary for the desired work should always be used.

A pressure of 15 lb_f/in.^{2} (103 kPa) is generally accepted as presenting the maximum allowable hazard for supply lines and regulator systems. However, even below this pressure, a serious hazard still exists, particularly in closed systems containing more than 1 L of gaseous acetylene.

Only Underwriters Laboratory-approved equipment should be used in acetylene service. Acetylene forms shock-sensitive and explosive compounds, including copper and silver acetylides. Alloys of these metals, including solders, should not be used for acetylene service unless they have been specifically approved for this purpose. If it is known or suspected that acetylides have been formed, a supervisor or safety officer should be consulted regarding safe methods of disposal.

Contaminated piping should not be used in acetylene service. Acetylene reacts with oxidizing agents such as chlorine or oxygen with explosive violence. Explosive decomposition is known to be initiated by a variety of conditions, particularly elevated temperatures. When an acetylene cylinder is connected to a pressure reactor, a valving system should be used to prevent flashback into the supply system.

Only pressure regulators approved for acetylene service should be used. These are fitted with a flame arrestor. All repairs or modifications of acetylene regulators should be done by qualified personnel.

Used gauges may be contaminated and should be put in acetylene service only after thorough reconditioning and inspection by qualified personnel. Only gauges that have Bourdon tubes constructed of stainless steel or an alloy containing less than 60% copper should be used for acetylene service. Ordinary gauges usually contain brass and bronze parts that can lead to acetylide formation.

Cylinders of acetylene must be protected from mechanical shock. The acetylene is in solution, under pressure, in a porous, acetone-impregnated, monolithic filler and is safe to handle only in this state. Such cylinders must always be positioned vertically and have the valve end up when gas is withdrawn from them.

The purification of acetylene at atmospheric pressure and room temperature is most efficiently done by scrubbing the gas through concentrated sulfuric acid and caustic traps. Activated alumina (F.1 grade), an all-purpose solid absorbent, is recommended when purification over a solid is desired. Activated carbon should be avoided because the heat of absorption may be sufficient to trigger thermal decomposition of the acetylene.

If gas is drawn from an acetylene cylinder too rapidly, there is a risk of its being diluted with acetone vapor. The hourly rate of withdrawal should not normally be greater than 20% of the content of the cylinder. Follow the supplier's recommendations in this regard.

ACROLEIN

CAS #107-02-8   mf: C_3H_4O
DOT #1092   mw: 56.07

STRUCTURE
$$H_2C=CHCH\!\!=\!\!O$$

NFPA LABEL

SYNONYMS
Acquinite
Acraldehyde
Acryaldehyde
Acrylic aldehyde
Allyl aldehyde
Aqualine
Biocide
Crolean
Ethylene aldehyde
Magnacide H
2-Propenal
Prop-2-en-1-al
Propylene aldehyde
Slimicide

PROPERTIES   Colorless to yellowish liquid with a pungent and intensely irritating odor, threshold = 0.3–0.4 ppm; soluble 20.8% by weight in water (with which it forms an azeotrope boiling at 52.4°C containing 97.4% acrolein); melting point −87°C, boiling point 52.7°C; density 0.8427 at 20°C/20C°C; flash point >-26.1°C; lower explosive limit 2.8%, upper explosive limit 31%; autoignition temperature unstable (235°C); vapor pressure 214 mm Hg at 20°C; vapor density 1.94

OSHA PEL   TWA 0.1 ppm; IDLH 5 ppm

ACGIH TLV   TWA 0.1 ppm; STEL 0.3 ppm

EPA Extremely Hazardous Substances List, Community Right to Know List, and Reported in EPA TSCA Inventory

DOT Classification   Flammable liquid; Label: Flammable liquid and poison

TOXIC EFFECTS   A poison by inhalation and most other routes. A carcinogen. Severe eye and skin irritant. May cause serious burns.

A 1-min exposure to 1 ppm acrolein causes slight nasal irritation. In 5 min, moderate nasal irritation and almost intolerable eye irritation develop. At an acrolein level of 5 ppm, the latter effects were seen in 1 min. Inhalation of air containing 10 ppm acrolein may be fatal in a few minutes. Inhalation sufficient to cause intense lacrymation and nasal irritation may lead to slowly developing pulmonary edema in the course of 24 h. Liquid acrolein in the eye or on the skin can produce serious injury, and acrolein vapors can also cause damage to the eyes. Chronic exposures of 3.7 ppm produce toxic effects in monkeys and dogs.

Routes   Inhalation, ingestion, contact

Symptoms   Irritation of the eyes, skin, mucous membrane; abnormal pulmonary function; delayed pulmonary edema, chronic respiratory disease; lacrimation; delayed hypersensitivity with multiple organ involvement

Target Organs   Heart, eyes, skin, respiratory system

HAZARDOUS REACTIONS   Vapor may explode if ignited in an enclosed area. An explosion hazard.

Acids (strong): violent polymerization reaction on contact
Alkalies: incompatible
Amines: incompatible
Ammonia: incompatible
Bases (strong): violent polymerization reaction on contact
Carbon dioxide: violent polymerization reaction on contact
Dimethylamine: violent polymerization reaction on contact
Light + heat: incompatible
Metal salts: incompatible
Nitrous fumes: violent polymerization reaction on contact
Oxidizing materials: vigorous reaction
Sulfur dioxide: violent polymerization reaction on contact

Thiourea: violent polymerization reaction on contact

Weak acid conditions: violent polymerization reaction on contact

When heated to decomposition, it may produce highly toxic fumes.

FIRST AID   In case of contact with the eyes, immediately wash the eyes with large amounts of water, occasionally lifting the lower and upper lids. Get medical attention immediately. Contact lenses should not be worn when working with this chemical.

In case of contact with the skin, immediately flush the contaminated skin with water. If this chemical penetrates the clothing, immediately remove the clothing and flush the skin with water. Get medical attention promptly.

If a person breathes in large amounts of this chemical, move the exposed person to fresh air at once. If breathing has stopped, perform artificial respiration. Keep the affected person warm and at rest. Get medical attention as soon as possible.

If this chemical has been swallowed, get medical attention immediately.

FIRE HAZARD   Dangerous fire hazard when exposed to heat, flame, or oxidizers.

Use CO_2, dry chemical, or alcohol foam extinguishers to fight a fire.

HANDLING   Although the irritating odor of acrolein provides a useful warning, its high toxicity makes it advisable to carry out laboratory operations using it in a hood. Because of its high volatility and flammability, acrolein should not be handled near open flames.

SPILL CLEANUP   Wear goggles, self-contained breathing apparatus, and impervious protective clothing.

ACRYLAMIDE

CAS #79-06-1   mf: C_3H_5NO
                          mw: 71.09

STRUCTURE
$CH_2CHCONH_2$

SYNONYMS
Acrylamide monomer   Propenamide
Acrylic amide   2-Propenamide
Ethylenecarboxamide

PROPERTIES   Colorless crystalline solid; soluble in water, alcohol, and ether; melting point 84.5°C, boiling point 125°C at 25 mm Hg; density 1.122 at 30°C; vapor pressure 1.6 mm Hg at 84.5°C; vapor density 2.45

OSHA PEL   TWA 0.3 mg/m³

ACGIH TLV   TWA 0.03 mg/m³; suspected carcinogen

NIOSH REL   0.3 mg/m³, 10-h TWA

DOT Classification   IMO: Poison B; Label: St. Andrew's Cross

TOXIC EFFECTS   Poisonous by ingestion, skin contact, and other routes. A carcinogen. A skin and eye irritant.

Human intoxication is mainly through the skin, next by inhalation, and last by ingestion. Onset of symptoms from chronic exposure can be delayed for months or years. It seems to be a central nervous system toxin.

Target Organs   Central nervous system, peripheral nervous system, skin, eyes

Routes   Inhalation, absorption, ingestion, contact

Symptoms   Ataxia; numbness of the limbs, paresthesia; muscle weakness; absent deep tendon reflex; hand sweating; fatigue; lethargy; irritates eyes, skin; erythema and peeling of the palms.

HAZARDOUS REACTIONS   Polymerizes violently at its melting point.

Strong oxidizers: incompatible

When heated to decomposition, it may produce acrid fumes and NO_x.

FIRST AID   In case of contact with the eyes, immediately wash the eyes with large amounts of water, occasionally lifting the lower and upper lids. Get medical attention immediately. Contact lenses should not be worn when working with this chemical.

In case of contact with the skin, immediately flush the contaminated skin with water. If this chemical penetrates the clothing, immediately remove the clothing and flush the skin with water. Get medical attention promptly.

If a person breathes in large amounts of this chemical, move the exposed person to fresh air at once. If breathing has stopped, perform artificial respiration. Keep the affected person warm and at rest. Get medical attention as soon as possible.

If this chemical has been swallowed, get medical attention immediately.

HANDLING   Wear appropriate clothing and equipment to prevent repeated or prolonged skin contact. Wash immediately when skin becomes contaminated. Immediately remove any nonimpervious clothing that becomes contaminated. Work clothing should be changed daily if it is reasonably probable that the clothing is contaminated.

Wear eye protection to reduce probability of eye contact.

The following equipment should be available: a quick-drench shower.

ACRYLONITRILE (INHIBITED)

CAS #107-13-1   mf: C_3H_3N
DOT #1093   mw: 53.07

STRUCTURE   NFPA LABEL
$CH_2=CHC\equiv N$

SYNONYMS
AN   Propenenitrile
Acrylonitrile monomer   2-Propenenitrile
Carbacryl   Ventox
Cyanoethylene   VCN
Fumigrain   Vinyl cyanide

PROPERTIES   Colorless, mobile liquid; mild odor; water soluble; melting point $-82°C$, boiling point $77.3°C$; density 0.806 at $20°C/4°C$; flash point $-1°C$ (TCC), flash point of 5% aqueous solution $<10°C$; lower explosive limit 3.1%, upper explosive limit 17%; autoignition temperature $481°C$; vapor pressure 100 mm at $22.8°C$; vapor density 1.83

Community Right to Know List and EPA Extremely Hazardous Substances List

OSHA PEL   TWA 2 ppm; 10 ppm ceiling

ACGIH TLV   TWA 2 ppm (skin); suspected carcinogen

NIOSH REL   1 ppm 8-h TWA; 10 ppm 15-min ceiling

DOT Classification   Flammable liquid and poison

TOXIC EFFECTS   Poisonous by inhalation, ingestion, skin contact, and other routes. A teratogen. May be a carcinogen. Human systemic irritant, somnolence, general anesthesia, cyanosis, and diarrhea by inhalation and skin contact.

The action on the body resembles hydrocyanic acid. Depending on the amount and rapidity of absorption into the body, acrylonitrile can produce nausea, vomiting, headache, sneezing, weakness, light-headedness, asphyxia, and even death by inhalation, skin contact, or inadvertent ingestion.

These toxic effects may be partially due to conversion of acrylonitrile to cyanide in the body. By inhibiting the respiratory enzymes of tissue, it renders the tissue cells incapable of oxygen absorption.

The liquid can irritate eye and skin, and blistering has resulted after prolonged, apparently harmless contact with previously contaminated clothing. OSHA reduced the workplace exposure limit to 2 ppm with a ceiling value of 10 ppm for a single daily 15-min excursion following reports of cancer induction in animals and concern regarding its possible carcinogenicity to humans. Exposures at or lower than these levels should afford protection against that health risk.

Routes   Inhalation, absorption, ingestion, contact

Symptoms   Asphyxia; irritation of the eyes; headache; sneezing; vomiting; weakness, light-headedness; skin vesiculation, scaling dermatitis; carcinogenic

Target Organs   Cardiovascular system, liver, kidneys, central nervous system, skin; brain tumor, lung and bowel cancer

HAZARDOUS REACTIONS   Unstable and easily oxidized

Acids (strong): violent reaction
Amines: violent reaction
Ammonia: incompatible
Azoisobutyronitrile: violent reaction
Bases (strong): violent reaction
Benzyltrimethylammonium hydroxide + pyrrole: potentially explosive reaction
Bromine: violent reaction
di-tert-Butylperoxide: violent reaction
Copper: incompatible
Copper alloys: incompatible
Dibenzoyl peroxide: violent reaction
Nitric acid: violent reaction
Oxidizing materials: vigorous reaction
Silver nitrate: explosive polymerization may occur on storage
Sulfuric acid: violent reaction
Tetrahydrocarbazole + benzyltrimethylammonium hydroxide: potentially explosive reaction

When heated to decomposition, it may produce toxic fumes of NO_x and CN^-.

FIRST AID   In case of contact with the eyes, immediately wash the eyes with large amounts of water, occasionally lifting the lower and upper lids. Get medical attention immediately. Contact lenses should not be worn when working with this chemical.

In case of contact with the skin, immediately wash the contaminated skin with water. If this chemical penetrates the clothing, immediately remove the clothing and wash the skin with water. If symptoms occur after washing, get medical attention immediately.

If a person breathes in large amounts of this chemical, move the victim to fresh air at once. If breathing has stopped, perform artificial respiration. Keep the affected person warm and at rest. Get medical attention as soon as possible.

If this chemical has been swallowed, get medical attention immediately.

FIRE HAZARD   Highly flammable liquid; do not expose to heat, flame, or oxidizing agents. Vapor may explode if ignited in an enclosed area.

Use CO_2, dry chemical, or alcohol foam extinguishers to fight a fire.

HANDLING   Acrylonitrile is regulated as a human carcinogen by OSHA. The PEL is 2 ppm as an 8-h time-weighted average or 10 ppm as averaged over any 15-min period. Dermal or eye contact with liquid acrylonitrile is also prohibited. Where feasible, worker exposure must be controlled by engineering methods or work practices. Laboratory hoods that have been demonstrated to provide sufficient protection should be used, and closed systems are recommended for laboratory operations. Use of gloves and goggles when handling liquid acrylonitrile is also recommended. OSHA regulations also require that exposure monitoring be conducted for all acrylonitrile operations to determine the airborne exposure levels for workers. In situations where the 15-min or 8-h exposure limits are exceeded and engineering or administrative controls are not feasible, respiratory protection must be employed based on the expected exposure level. In cases of unknown concentration or fire fighting, supplied-air or self-contained breathing apparatus with a full facepiece operated in the positive-pressure mode is required. There are other detailed requirements in the OSHA standard related to housekeeping, waste disposal, hygiene facilities, employee training, and medical monitoring. Managers and laboratory supervisors should review these requirements before starting work with acrylonitrile.

Wear appropriate clothing and equipment to prevent repeated or prolonged skin contact. Personnel should wash immediately when the skin becomes wet. Clothing soaked by a spill should be removed immediately to avoid a flammability hazard.

Wear eye protection to prevent any possibility of eye contact.

The following equipment should be available: a quick-drench shower.

SPILL CLEANUP   Avoid contact with liquid and vapor. Wear goggles, self-contained breathing apparatus, and rubber overclothing (including gloves).

ALUMINUM

CAS #7429-90-5   mf: Al
DOT #1309; 1383; 1396   mw: 26.98

SYNONYMS
Alumina fiber
Aluminum flake
Aluminum, dehydrated
Aluminum, metallic powder
Aluminum powder
Aluminum powder, uncoated, nonpyrophoric
Emanay atomized aluminum powder
Metana aluminum paste
Noral ink grade aluminum
Pap-1

PROPERTIES   A silvery, ductile metal; soluble in HCl, H_2SO_4, and alkalies; melting point 660°C, boiling point 2450°C; density 2.702; vapor pressure 1 mm at 1284°C

ACGIH TLV   (Metal and oxide) TWA 10 mg/m³ (dust); (pyro powders and welding fumes) TWA 5 mg/m³; (soluble salts and alkyls) TWA 2 mg/m³

Community Right to Know List (fume or dust) and Reported in the EPA TSCA Inventory

DOT Classification   Label: Flammable solid; Label: Spontaneously combustible (pyrophoric)

IMO Classification   Flammable solid; Label: Dangerous when wet (nonpyrophoric)

TOXIC EFFECTS   Inhalation of finely divided powder has been reported as a cause of pulmonary fibrosis. Aluminum in aerosols has been implicated in Alzheimer's disease. It is a reactive metal, and the greatest hazards are with chemical reactions. As with other metals, the powder and dust are the most dangerous forms.

HAZARDOUS REACTIONS   Dust is moderately explosive by heat, flame, or chemical reaction with powerful oxidizers.

Powdered aluminum undergoes the following dangerous interactions:

Ammonium nitrate: mixtures are powerful explosives
Ammonium nitrate + calcium nitrate + formamide + water: mixtures are powerful explosives
Ammonium peroxodisulfate: forms sensitive explosive mixtures
Ammonium peroxodisulfate + water: mixture is explosive
Antimony: reacts violently on heating
Antimony trichloride vapor: ignites on heating
Nitrates (aqueous solutions): forms sensitive explosive mixtures
Arsenic: reacts violently on heating
Arsenic trichloride vapor: ignites on contact
Barium nitrate + potassium nitrate + sulfur + vegetable adhesives + water: explosive reaction after a delay period
Barium peroxide: ignites on contact
Bromates: forms sensitive explosive mixtures
Bromine liquid: foil reacts vigorously
Bromine pentafluoride: violent reaction or ignition
Bromomethane: violent or explosive reactions have occurred in industry
Bromotrifluoromethane: violent or explosive reactions have occurred in industry
Carbon dioxide: burns when heated
Carbon disulfide vapor: ignites spontaneously
Carbon tetrachloride (during ball milling operations): potentially explosive reaction
Chlorates: forms sensitive explosive mixtures
Chlorine fluoride: violent reaction or ignition
Chlorine gas: ignites spontaneously
Chlorine liquid: forms sensitive explosive mixtures
Chloroform amidinium nitrate: potentially explosive reaction
Copper oxide (worked while hot with an iron or steel tool): violent or explosive "thermite" reaction when heated
1,2-Dichloroethane: violent or explosive reactions have occurred in industry
Dichloromethane: violent or explosive reactions have occurred in industry
1,2-Dichloropropane: violent or explosive reactions have occurred in industry
1,2-Difluorotetrafluoroethane: violent or explosive reactions have occurred in industry
Dinitrogen tetroxide: forms sensitive explosive mixtures
Iron powder + water: exothermic reaction releases explosive hydrogen gas
Fluorotrichloroethane: violent or explosive reactions have occurred in industry
Halocarbons: violent or explosive reactions have occurred in industry
Halogens: forms sensitive explosive mixtures
Hexachloroethane + alcohol: violent or explosive reactions have occurred in industry
Hydrochloric acid: violent reaction
Hydrofluoric acid: violent reaction
Hydrogen chloride gas: violent reaction
Interhalogens: violent reaction or ignition
Iodates: forms sensitive explosive mixtures
Iodine + water: violent reaction
Iodine chloride: violent reaction or ignition
Iodine heptafluoride: violent reaction or ignition
Iodine pentafluoride: violent reaction or ignition
Potassium chlorate: forms sensitive explosive mixtures
Potassium perchlorate + barium nitrate + potassium nitrate + water: explosive reaction after a delay period
Potassium perchlorate: forms sensitive explosive mixtures
Metal oxides: violent or explosive "thermite" reaction when heated
Sodium chlorate: forms sensitive explosive mixtures
Sodium nitrate: forms sensitive explosive mixtures
Nitrates: violent or explosive "thermite" reaction when heated
Nitryl fluoride: forms sensitive explosive mixtures
Oxidants: forms sensitive explosive mixtures
Oxosalts: violent or explosive "thermite" reaction when heated
Oxygen + water: mixtures ignite and react violently
Oxygen: ignites on contact
Perchlorate salts: forms sensitive explosive mixtures
Peroxides: forms sensitive explosive mixtures
Phosphorus pentachloride vapor: ignites on contact
Picric acid + water: mixture ignites after a delay period
Polytetrafluoroethylene (PTFE) powder: forms sensitive explosive mixtures
Polytrifluoroethylene oils and greases: violent or explosive reactions have occurred in industry
Red phosphorus: forms sensitive explosive mixtures
Selenium chloride: ignites on contact
Silver chloride powder: mixtures are powerful explosives
Sodium acetylide: potentially violent reaction
Sodium peroxide: mixtures may ignite or react violently
Sodium sulfate: explosive reaction above 800°C
Sulfates: violent or explosive "thermite" reaction when heated
Sulfides: violent or explosive "thermite" reaction when heated

Sulfur + heat: violent reaction
Sulfur dichloride vapor: ignites on contact
Tetrachloroethylene: violent or explosive reactions have occurred in industry
Tetrafluoromethane: violent or explosive reactions have occurred in industry
Tetranitromethane: forms sensitive explosive mixtures
1,1,1-Trichloroethane: violent or explosive reactions have occurred in industry
Trichloroethylene: violent or explosive reactions have occurred in industry
1,1,2-Trichlorotrifluoroethane: violent or explosive reactions have occurred in industry
Trichlorotrifluoroethane-dichlorobenzene: violent or explosive reactions have occurred in industry
Zinc peroxide: forms sensitive explosive mixtures

Bulk aluminum may undergo the following dangerous interactions:

Alcohols: exothermic reaction releases explosive hydrogen gas
Arsenic trioxide + sodium arsenate + sodium hydroxide: reaction forms toxic arsine gas
Butanol: exothermic reaction releases explosive hydrogen gas
Chlorine trifluoride: violent reaction
Diborane: reaction forms pyrophoric product
Formic acid: incandescent reaction
Mercury(II) salts + moisture: vigorous amalgamation reaction
Metal oxides (molten): explosive reaction
Methanol + carbon tetrachloride: vigorous dissolution reaction
Methanol: exothermic reaction releases explosive hydrogen gas
Niobium oxide + sulfur: ignition on contact
Nitrates: explosive reaction
Oxosalts: explosive reaction
Palladium: potentially violent alloy formation at melting point of Al, 600°C
Platinum: potentially violent alloy formation at melting point of Al, 600°C
2-Propanol: exothermic reaction releases explosive hydrogen gas
Silicon steels (molten): violent reaction
Sodium carbonate: explosive reaction
Sodium diuranate: violent exothermic reaction above 600°C
Sodium hydroxide: exothermic reaction releases explosive hydrogen gas
Sulfates: explosive reaction
Sulfides: explosive reaction

FIRE HAZARD   Dust is moderately flammable by heat, flame, or chemical reaction with powerful oxidizers.

Use special mixtures of dry chemical extinguishers to fight a fire.

ALUMINUM SULFATE (2:3)

CAS #10043-01-3   mf: $Al_2(SO_4)_3$
DOT #1760; 9078   mw: 342.14

NFPA LABEL

SYNONYMS
Alum
Aluminum trisulfate
Cake alum
Dialuminum sulphate
Dialuminum trisulfate
Sulfuric acid, aluminum salt (3:2)

PROPERTIES   White powder; solubility in water 36.4% at 20°C; decomposes at 770°C; density 2.71

ACGIH TLV   TWA 2 mg(Al)/m³

DOT Classification   ORM-E; Label: None, solid; ORM-B; Label: None, solution

TOXIC EFFECTS   Moderately toxic by ingestion and some routes. Hydrolyzes to form sulfuric acid, which irritates tissue, especially lungs.

HAZARDOUS REACTIONS   When heated to decomposition, it may produce toxic fumes of SO_x.

AMMONIA

CAS #7664-41-7   mf: NH_3
DOT #1005   mw: 17.04

NFPA LABEL

SYNONYMS

Ammonia anhydrous   Anhydrous ammonia
Ammonia gas   Spirit of Hartshorn

PROPERTIES   Colorless gas with an extremely pungent odor; liquefied by compression; very soluble in water, moderately soluble in alcohol; melting point −77.7°C, boiling point −33.35°C; density 0.771 g/L at 0°C; lower explosive limit 16%, upper explosive limit 25%; autoignition temperature 651.1°C; vapor pressure 10 atm at 25.7°C; vapor density 0.6

EPA Extremely Hazardous Substances List and Community Right to Know List

OSHA PEL   TWA 50 ppm

ACGIH TLV   TWA 25 ppm; STEL 35 ppm

NIOSH REL   ceiling 50 ppm

DOT Classification   Nonflammable gas; Label: Nonflammable gas

TOXIC EFFECTS   A poison by inhalation, ingestion, and possibly other routes. An eye, mucous membrane, and systemic irritant by inhalation.

Target Organs   Respiratory system, eyes

Routes   Inhalation, ingestion, contact

Symptoms   Irritation of the eye, nose, and throat; dyspnea; bronchospasm; chest pain; pulmonary edema; pink frothy sputum; skin burns; vesiculation

HAZARDOUS REACTIONS

Acetaldehyde: incompatible
Acrolein: incompatible
Air + hydrocarbons: forms sensitive explosive mixture
Ammonium peroxodisulfate: potentially violent or explosive reaction on contact
Antimony: incompatible
Bleaches: incompatible
Boron halides: potentially violent or explosive reaction on contact
Boron triiodide: incompatible
Boron: incompatible
Bromine pentafluoride: potentially violent or explosive reaction on contact
Bromine: reacts to form explosive product
Calcium + heat: incandescent reaction
Calcium chloride: reacts to form potentially explosive product
Chlorine azide: reacts to form explosive product
Chlorine oxide: incompatible
Chlorine trifluoride: potentially violent or explosive reaction on contact
Chlorine: forms sensitive explosive mixture
Chlorites: incompatible
1-Chloro-2,4-dinitrobenzene: forms sensitive explosive mixture
Chloroformamidinium nitrate: potentially violent or explosive reaction on contact
2- or 4-Chloronitrobenzene (above 160°C/30 bar): forms sensitive explosive mixture
Chlorosilane: incompatible
Chromium trioxide: potentially violent or explosive reaction on contact
Chromyl chloride: potentially violent or explosive reaction on contact
Dichlorine oxide: potentially violent or explosive reaction on contact
1,2-Dichloroethane + liquid ammonia: potentially violent or explosive reaction on contact
Dinitrogen tetraoxide: potentially violent or explosive reaction on contact
Ethanol + silver nitrate: forms sensitive explosive mixture
Ethylene dichloride + liquid ammonia: incompatible
Ethylene oxide: potentially violent or explosive polymerization reaction on contact (polymerization reaction)
Fluorine: potentially violent or explosive reaction on contact
Germanium derivatives: forms sensitive explosive mixture
Gold(III) chloride: reacts to form explosive product
Gold: incompatible
Halogens: incompatible
$HClO_3$: incompatible
Heavy metals and their compounds: react to form explosive product
Hexachloromelamine: incompatible
Hydrazine + alkali metals: incompatible
Hydrobromic acid: incompatible
Hydrogen peroxide: potentially violent or explosive reaction on contact
Hypochlorites: incompatible
Hypochlorous acid: incompatible
Interhalogens: potentially violent or explosive reaction on contact
Iodine: reacts to form explosive product
Iodine + potassium: reacts to form explosive product
Magnesium perchlorate: potentially violent or explosive reaction on contact
Mercury: reacts to form explosive product
Nitric acid: potentially violent or explosive reaction on contact
Nitrogen oxide: potentially violent or explosive reaction on contact
Nitrogen tetroxide: incompatible

96 AMMONIA

Nitrogen trichloride: potentially violent or explosive reaction on contact
Nitrogen trifluoride: incompatible
Nitryl chloride: potentially violent or explosive reaction on contact
Oxidants (strong): potentially violent or explosive reaction on contact
Oxygen (liquid): potentially violent or explosive reaction on contact
Oxygen + platinum: potentially violent or explosive reaction on contact
Oxygen difluoride: incompatible
P_2O_3: incompatible
Pentaborane(9): reacts to form explosive product
Phosphorus pentoxide: incompatible
Picric acid: incompatible
Potassium + arsine: incompatible
Potassium + phosphine: incompatible
Potassium + sodium nitrite: incompatible
Potassium chlorate: potentially violent or explosive reaction on contact
Potassium ferricyanide: incompatible
Potassium mercuric cyanide: incompatible
Potassium thallium amide ammoniate: reacts to form explosive product
Silver azide: reacts to form explosive silver nitride
Silver chloride: reacts to form explosive silver nitride
Silver nitrate: reacts to form explosive silver nitride
Silver oxide: reacts to form explosive silver nitride
Silver: incompatible
Sodium + carbon monoxide: incompatible
Stibine: forms sensitive explosive mixture
Sulfinyl chloride: potentially violent or explosive reaction on contact
Sulfur oxychloride: incompatible
Sulfur dichloride: incompatible
Sulfur: incompatible
Tellurium halides: reacts to form explosive product
Tellurium hydropentachloride: incompatible
Tellurium tetrabromide: reacts to form explosive product
Tellurium tetrachloride: reacts to form explosive product
Tetramethylammonium amide: potentially violent or explosive reaction on contact
Thiocarbonyl azide thiocyanate: potentially violent or explosive reaction on contact
Thiotriazyl chloride: potentially violent or explosive reaction on contact
Thiotrithiazylchloride: incompatible in contact
Trichloromelamine: incompatible in contact
Trioxygen difluoride: potentially violent or explosive reaction on contact

May produce toxic fumes of NH_3 and NO_x when exposed to heat.

FIRST AID   In case of contact with the eyes, immediately wash the eyes with large amounts of water, occasionally lifting the lower and upper lids. Get medical attention immediately. Contact lenses should not be worn when working with this chemical.

In case of contact with the skin, immediately flush the contaminated skin with water. If this chemical penetrates the clothing, immediately remove the clothing and flush the skin with water. Get medical attention promptly.

If a person breathes in large amounts of this chemical, move the victim to fresh air at once. If breathing has stopped, perform artificial respiration. Keep the affected person warm and at rest. Get medical attention as soon as possible.

If this chemical has been swallowed, get medical attention immediately.

FIRE HAZARD   Difficult to ignite. Explosion hazard when exposed to flame or in a fire. In a fire, NH_3 + Air can detonate.

Stop flow of gas to fight fire.

HANDLING   A direct flame or steam jet must never be applied against a cylinder of ammonia. If it becomes necessary to increase the pressure in a cylinder to promote more rapid discharge, the cylinder should be moved into a warm room. Extreme care should be exercised to prevent the temperature from rising above 50°C.

Only steel valves and fittings should be used on ammonia containers. Neither copper, silver, nor zinc nor their alloys should be permitted to come into contact with ammonia.

Respiratory protective equipment of a type approved by the U.S. Mine Safety and Health Administration (MSHA) and NIOSH for anhydrous ammonia service should always be readily available in a place where this material is used and so located as to be easily reached in case of need. Proper protection should be afforded the eyes by the use of goggles or large-lens spectacles to eliminate the possibility of liquid ammonia coming in contact with the eyes and causing injury.

Leaks may be detected with the aid of sulfur tapers or sensitive papers. Both of these items and instructions as to their use may be obtained from the cylinder supplier.

Ammonia is also used in laboratories in solution and is usually supplied as a 35% solution in water. If this strong solution is stored in a warm place, a pressure develops in the bottle. The cap must be removed with care; in extreme cases the bottle may crack or explode.

Ammonia should not be dried by calcium chloride because an addition compound forms that is potentially explosive.

Before disposal, pour the strong solution into a large tank of water and neutralize with an acid. Flush away with a large excess of water.

Wear appropriate equipment to prevent any possibility of skin contact with liquids containing >10% of ammonia, and repeated or prolonged skin contact with liquids containing <10%. Wash immediately when skin is contaminated with liquids containing >10% ammonia, and promptly when skin is contaminated with <10%. Remove nonimpervious clothing that has been contaminated with liquids containing >10% ammonia immediately, and promptly remove clothing contaminated with liquids containing <10%.

Wear eye protection to prevent any possibility of eye contact with materials containing >10% of ammonia, and repeated or prolonged contact with any amount of the material.

The following equipment should be available: an eyewash station. A quick-drench shower should also be available when liquids containing >10% of ammonia are present.

AMMONIUM CHLORIDE

CAS #12125-02-9   mf: H_4NCl
DOT #9085   mw: 53.50

STRUCTURE   NFPA LABEL
$(NH_4)Cl$

SYNONYMS
Ammonium muriate
Sal ammonia
Sal ammoniac

PROPERTIES   White crystals; melting point 337.8°C, boiling point 520°C; density 1.520; vapor pressure 1 mm at 160.4°C (sublimes)

Reported in the EPA TSCA Inventory

ACGIH TLV   TWA 10 mg(Al)/m^3; STEL 20 mg/m^3

DOT Classification   Label: ORM-E

TOXIC EFFECTS   Poisonous by some routes. A severe eye irritant.

HAZARDOUS REACTIONS
Ammonia: violent reaction or ignition
Bromine pentafluoride: violent reaction or ignition
Bromine trifluoride: explosive reaction
Hydrogen cyanide: may react to form explosive nitrogen trichloride
Iodine heptafluoride: violent reaction or ignition
Nitrates: violent reaction or ignition
Potassium chlorate: explosive reaction

When heated to decomposition, it may produce very toxic fumes of NO_x, Cl^-, and NH_3.

SPILL CLEANUP   Wear goggles and protective clothing. Self-contained breathing apparatus may be required if large amounts or high concentrations are involved.

AMMONIUM HYDROXIDE

CAS #1336-21-6   mf: H_5NO
DOT #2672   mw: 35.06

STRUCTURE   NFPA LABEL
$(NH_4)OH$

SYNONYMS
Ammonia, aqueous
Ammonia solution
Aqua ammonia

PROPERTIES   Colorless liquid with pungent odor; soluble in water; melting point −77°C; solutions contain less than 44% ammonia

NIOSH REL   ceiling 50 ppm

DOT Classification   Corrosive material; Label: Corrosive

TOXIC EFFECTS   A poison by ingestion and inhalation. A severe eye irritant. A systemic irritant by inhalation. The liquid can cause burns.

98 AMMONIUM NITRATE

HAZARDOUS REACTIONS
Acrolein: incompatible
Acrylic acid: incompatible
Chlorosulfonic acid: incompatible
Dimethyl sulfate: incompatible
Gold + aqua regia: incompatible
Halogens: incompatible
Hydrogen chloride: incompatible
Hydrogen fluoride: incompatible
Nitric acid: incompatible
Nitromethane: incompatible
Oleum: incompatible
β-Propiolactone: incompatible
Propylene oxide: incompatible
Silver nitrate: incompatible
Silver oxide + ethanol: incompatible
Silver oxide: incompatible
Silver permanganate: incompatible
Sulfuric acid: incompatible

When heated to decomposition, it may produce toxic fumes of NH_3 and NO_x.

HANDLING   Use with adequate ventilation.

AMMONIUM NITRATE

CAS #6484-52-2   mf: $HNO_3 \cdot H_3N$
DOT #1942/0222/2426   mw: 80.06

STRUCTURE   NFPA LABEL
$(NH_3)HNO_3$

SYNONYMS
Nitric acid, ammonium salt

PROPERTIES   Colorless crystals; soluble in water; melting point 169.6°C, decomposes above 210°C; density 1.725 at 20°C

Community Right to Know List and Reported in the EPA TSCA Inventory

DOT Classification   Oxidizer; Label: Oxidizer

TOXIC EFFECTS   A powerful oxidizer and an allergen. The dust is irritating to eyes, nose, and throat, and if inhaled may cause coughing or difficult breathing.

HAZARDOUS REACTIONS   A relatively stable explosive that has, however, caused many industrial explosions.

Acetic acid: can ignite when mixed; forms heat- or shock-sensitive explosive mixture
Acetic anhydride + nitric acid: violent or explosive spontaneous reaction
Alkali metals: reacts to form an explosive product
Aluminum powder: forms heat- or shock-sensitive explosive mixture
Aluminum + calcium nitrate + formamide (a blasting explosive): forms heat- or shock-sensitive explosive mixture
Aluminum chloride: forms heat- or shock-sensitive explosive mixture
Ammonia: forms heat- or shock-sensitive explosive mixture
Ammonium chloride: forms heat- or shock-sensitive explosive mixture
Ammonium chloride + water + zinc: violent or explosive spontaneous reaction
Ammonium dichromate: ignites on contact
Ammonium sulfate + potassium: violent or explosive spontaneous reaction
Ammonium sulfate + sodium: mixture can explode
Antimony powder: forms heat- or shock-sensitive explosive mixture
Barium chloride: ignites on contact
Barium nitrate: violent or explosive spontaneous reaction
Bismuth powder: forms heat- or shock-sensitive explosive mixture
Brass powder: forms heat- or shock-sensitive explosive mixture
Cadmium powder: forms heat- or shock-sensitive explosive mixture
Calcium chloride: forms heat- or shock-sensitive explosive mixture
Carbon + heat: incompatible
Charcoal: forms heat- or shock-sensitive explosive mixture
Charcoal + metal oxides (e.g., rust, copper oxide: zinc oxide above 80°C): forms heat- or shock-sensitive explosive mixture
Chloride salts: forms heat- or shock-sensitive explosive mixture
Chromium powder: forms heat- or shock-sensitive explosive mixture
Chromium(VI) salts: ignites on contact
Cobalt powder: forms heat- or shock-sensitive explosive mixture

Copper powder: forms heat- or shock-sensitive explosive mixture
Cyanoguanidine: forms heat- or shock-sensitive explosive mixture
Fertilizers: forms heat- or shock-sensitive explosive mixture
Hydrocarbon oils: forms heat- or shock-sensitive explosive mixture
Iron powder: forms heat- or shock-sensitive explosive mixture
Iron(II) sulfide: violent or explosive spontaneous reaction
Iron(III) chloride: forms heat- or shock-sensitive explosive mixture
Lead powder: forms heat- or shock-sensitive explosive mixture
Magnesium powder: forms heat- or shock-sensitive explosive mixture
Manganese powder: forms heat- or shock-sensitive explosive mixture
Metal powders: forms heat- or shock-sensitive explosive mixture
NaOCl: incompatible
Nickel powder: forms heat- or shock-sensitive explosive mixture
Nonmetals: forms heat- or shock-sensitive explosive mixture
Oils: forms heat- or shock-sensitive explosive mixture
Organic fuels: forms heat- or shock-sensitive explosive mixture
Organic matter: incompatible
Phosphorus: forms heat- or shock-sensitive explosive mixture
Potassium powder: forms heat- or shock-sensitive explosive mixture
Potassium chromate: ignites on contact
Potassium dichromate: ignites on contact
Potassium nitrate: ignites on contact
Potassium permanganate: forms heat- or shock-sensitive explosive mixture
Reducing materials: can react vigorously
Sawdust: violent or explosive spontaneous reaction
Sodium: reacts to form an explosive product
Sodium chloride: ignites on contact
Sodium perchlorate: incompatible
Stainless steel powder: forms heat- or shock-sensitive explosive mixture
Stearates: forms heat- or shock-sensitive explosive mixture
Sugar: forms heat- or shock-sensitive explosive mixture
Sulfur: forms heat- or shock-sensitive explosive mixture
Superphosphate + organic materials (above 90°C): forms heat- or shock-sensitive explosive mixture
Tin powder: forms heat- or shock-sensitive explosive mixture
Titanium powder: forms heat- or shock-sensitive explosive mixture
Trinitroanisole: forms heat- or shock-sensitive explosive mixture
Urea: violent or explosive spontaneous reaction
Water (hot): violent or explosive spontaneous reaction
Wax: forms heat- or shock-sensitive explosive mixture
Zinc powder: forms heat- or shock-sensitive explosive mixture

When heated to decomposition, it may produce toxic fumes of NO_x.

FIRST AID   Move to fresh air if the dust has been inhaled. If in eyes, flush with plenty of water.

FIRE HAZARD   Use water in large amounts to fight fire. Containers may explode in a fire. It is important that containers be kept cool and the fire be extinguished promptly. Ventilate well.

SPILL CLEANUP   Evacuate area of large discharge.

AMMONIUM PERSULFATE

CAS #7727-54-0   mf: $H_8N_2O_8S_2$
DOT #1444   mw: 228.22

STRUCTURE
$H_4NOSO_2OOSO_2ONH_4$

SYNONYMS
Ammonium peroxydisulfate
Ammonium persulfate

PROPERTIES   White crystals; decomposes at 120°C; density 1.982

Reported in the EPA TSCA Inventory

ACGIH TLV   TWA 5 mg $(S_2O_8)/m^3$

DOT Classification   IMO: Oxidizer; Label: Oxidizer

TOXIC EFFECTS   Moderately toxic by ingestion. A poison by some other routes.

HAZARDOUS REACTIONS   A powerful oxidizer, which can react vigorously with reducing agents. Releases oxygen when heated.

Aluminum powder + water: mixture is explosive
Ammonia + silver salts (solution): violent reaction
Iron: violent reaction
Sodium peroxide: mixtures are explosives that are sensitive to friction, heating above 75°C, or contact with CO_2 or water
Sulfuric acid: mixture is a strong oxidizing cleaning solution
Zinc + ammonia: mixture is explosive

When heated to decomposition, it may produce toxic fumes of SO_x, NH_3, and NO_x.

ANILINE

CAS #62-53-3 mf: C_6H_7N
DOT #1547 mw: 93.14

STRUCTURE NFPA LABEL
$C_6H_5NH_2$

SYNONYMS
Aminobenzene Benzenamine
Aminophen Blue oil
Aniline oil Phenylamine

PROPERTIES Colorless, oily liquid with a characteristic odor and burning taste, threshold = 0.5 ppm; soluble in water 3.5 g/100 mL at 20°C, miscible with most organic solvents; melting point −6.3°C, boiling point 184 °C; density 1.0217 at 20 °C; flash point 70°C (closed cup); vapor pressure 0.67 mm Hg at 20°C, 1.0 mm at 30.6°C, 10.0 mm at 68.3°C; vapor density 3.22

EPA Extremely Hazardous Substances List, Community Right to Know List, and Reported in the EPA TSCA Inventory

OSHA PEL TWA 5 ppm (skin); IDLH 100 ppm

ACGIH TLV TWA 2 ppm

DOT Classification Label: Poison B

TOXIC EFFECTS A poison by most routes, including inhalation and ingestion. A skin and severe eye irritant, and a mild sensitizer.

The oral lethal dose of aniline for humans is 15–30 mL. Skin contact is the most common route of entry; the outstanding feature of aniline poisoning in humans is cyanosis due to formation of methemoglobin. The symptoms of severe exposure are cyanosis, headache, weakness, dizziness, nausea, and chills. Onset of symptoms, however, may be delayed up to 4 h.

If aniline is a carcinogen, it is not a potent one in animals.

Target Organs Blood, cardiovascular system, liver, kidneys

Routes Inhalation, absorption, ingestion, contact

Symptoms Headache; weakness, dizziness; cyanosis; ataxia; dyspnea on effort; tachycardia; eye irritation

HAZARDOUS REACTIONS
Acetic anhydride: violent reaction
Acids: violent reaction
Anilinium chloride: forms heat- or shock-sensitive explosive mixture (detonates at 240°C/7.6 bar)
Benzenediazonium-2-carboxylate: spontaneously explosive reaction
Boron trichloride: violent reaction
1-Chloro-2,3-epoxypropane: forms heat- or shock-sensitive explosive mixture
Chlorosulfonic acid: violent reaction
Dibenzoyl peroxide: spontaneously explosive reaction
Diisopropyl peroxydicarbonate: violent reaction
Fluorine: violent reaction
Fluorine nitrate: spontaneously explosive reaction
FO_3Cl: violent reaction
Formaldehyde + perchloric acid: violent reaction
n-Haloimides: violent reaction
Hexachloromelamine: violent reaction
Hydrogen peroxide: forms heat- or shock-sensitive explosive mixture
Nitric acid + dinitrogen tetroxide + sulfuric acid: violent reaction
Nitrobenzene + glycerin: violent reaction
Nitromethane: forms heat- or shock-sensitive explosive mixture
Nitrosyl perchlorate: spontaneously explosive reaction
Oleum: violent reaction
Oxidizers: incompatible
Ozone: reaction forms explosive product
Perchloric acid: reaction forms explosive product
Perchloryl fluoride: reaction forms explosive product
Perchromates: violent reaction
Peroxodisulfuric acid: spontaneously explosive reaction
Peroxomonosulfuric acid: forms heat- or shock-sensitive explosive mixture
Peroxydisulfuric acid: violent reaction
Peroxyformic acid: violent reaction

Potassium peroxide: violent reaction
β-Propiolactone: violent reaction
Nitric acid (red fuming): spontaneously explosive reaction
Silver perchlorate: violent reaction
Sodium peroxide: violent reaction
Sodium peroxide + water: ignites on contact
Acids (strong): incompatible
Sulfuric acid: violent reaction
Tetranitromethane: spontaneously explosive reaction
Trichloromelamine: violent reaction
Trichloronitromethane (145°C): violent reaction

When heated to decomposition, it may produce highly toxic fumes of NO_x.

FIRST AID   In case of contact with the eyes, immediately wash the eyes with large amounts of water, occasionally lifting the lower and upper lids. Get medical attention immediately. Contact lenses should not be worn when working with this chemical.

In case of contact with the skin, promptly wash the contaminated skin with soap and water. If this chemical penetrates through the clothing, promptly remove the clothing and wash the skin with soap and water. Get medical attention promptly.

If a person breathes in large amounts of this chemical, move the victim to fresh air at once. If breathing has stopped, perform artificial respiration. Keep the affected person warm and at rest. Get medical attention as soon as possible.

If this chemical has been swallowed, get medical attention immediately.

FIRE HAZARD   Moderately flammable when exposed to heat or flame.

Use alcohol foam, CO_2, or dry chemical extinguishers to fight a fire.

HANDLING   The OSHA limits include a warning about the potential contribution of skin absorption to the overall exposure.

Because aniline, like many aromatic amines, is a rather toxic substance that readily penetrates the skin, it should be handled carefully. Most laboratory operations should be carried out in a hood, and skin contact should be avoided by appropriate use of protective apparel, e.g., rubber gloves and aprons.

Wear appropriate clothing and equipment to prevent a reasonable probability of skin contact. Wash immediately when skin becomes contaminated. Immediately remove any nonimpervious clothing that becomes contaminated.

Wear eye protection to reduce probability of eye contact.

The following equipment should be available: a quick-drench shower.

ANTIMONY

CAS #7440-36-0   mf: Sb
DOT #2871   mw: 121.75

SYNONYMS
Antimony black
Antimony regulus
Stibium

PROPERTIES   Silvery or gray, lustrous metal; insoluble in water, soluble in hot concentrated sulfuric acid; melting point 630°C, boiling point 1635°C; density 6.684 at 25°C; vapor pressure 1 mm at 886°C

OSHA PEL   TWA 0.5 mg(Sb)/m^3; IDLH 100 ppm

ACGIH TLV   TWA 0.5 mg(Sb)/m^3

NIOSH REL   0.5 mg(Sb)/m^3

Community Right to Know List (antimony and its compounds)

DOT Classification   Poison B; Label: Poison

TOXIC EFFECTS   A poison by some routes.

Target Organs   Respiratory system, cardiovascular system, skin, eyes

Routes   Inhalation, contact

Symptoms   Irritation of the skin, nose, throat, and mouth; cough; dizziness; headache; nausea, vomiting, diarrhea; cramps; insomnia; anorexia; inability to smell; cardiac effects

HAZARDOUS REACTIONS   Electrolysis of acid sulfides and stirred Sb halide yields explosive Sb.

Acids: incompatible
Ammonium nitrate: can react violently
Bromine azide: can react violently
Bromine trifluoride: can react violently
Chlorine trifluoride: can react violently
ClO: can react violently
Halogenated acids: incompatible
Halogens: can react violently
$HClO_3$: can react violently
Nitric acid: can react violently
Oxidants: can react violently

nitrate: can react violently
permanganate: can react violently
Potassium peroxide: can react violently
Sodium nitrate: can react violently

When heated or on contact with acid, it may produce toxic fumes of SbH_3.

FIRST AID   In case of contact with the skin, immediately wash the contaminated skin with soap and water. If this chemical penetrates through the clothing, immediately remove the clothing and wash the skin with soap and water. Get medical attention promptly.

If a person breathes in large amounts of this chemical, move the victim to fresh air at once. If breathing has stopped, perform artificial respiration. Keep the affected person warm and at rest. Get medical attention as soon as possible.

If this chemical has been swallowed, get medical attention immediately.

FIRE HAZARD   The dust and vapor are moderate fire and explosion hazards when exposed to heat or flame.

HANDLING   Wear appropriate clothing and equipment to prevent a reasonable probability of skin contact. Wash promptly when the skin becomes contaminated. Work clothing should be changed daily if it is reasonably probable that the clothing has been contaminated. Promptly remove any nonimpervious clothing that becomes contaminated.

Wear eye protection to reduce probability of eye contact.

ARSENIC

CAS #7440-38-2   mf: As
DOT #1558   mw: 74.92

SYNONYMS
Arsenicals   Grey arsenic
Arsenic black   Metallic arsenic
Colloidal arsenic

PROPERTIES   Silvery to black, brittle, crystalline and amorphous metalloid; insoluble in water, soluble in HNO_3; melting point 814°C at 36 atm, sublimes at 612°C; density of black crystals 5.724 at 14°C, black amorphous 4.7; vapor pressure 1 mm at 372°C (sublimes)

OSHA PEL   TWA 0.01 mg(As)/m³
ACGIH TLV   TWA 0.2 mg(As)/m³
NIOSH REL   ceiling 2 µg(As)/m³
DOT Classification   Poison B; Label: Poison
Community Right to Know List (arsenic and its compounds) and Reported in the EPA TSCA Inventory

TOXIC EFFECTS   Poisonous by some routes. A carcinogen and teratogen.

Target Organs   Liver, kidneys, skin, lungs, lymphatic system

Routes   Inhalation, absorption, contact, ingestion

Symptoms   Ulceration of nose/nasal septum; dermatitis; gastrointestinal disturbances; perineurologic, respiratory irritation; hyperpigmentation of skin; a carcinogen

HAZARDOUS REACTIONS
Acid or acid fumes: may produce highly toxic fumes on contact
Bromates: may ignite on contact
Bromine azide: may ignite on contact
Bromine pentafluoride: may ignite on contact
Bromine trifluoride: may ignite on contact
Chlorates: may ignite on contact
Chlorine monoxide: may ignite on contact
Chlorine trifluoride: may ignite on contact
Chromium trioxide: incompatible
CsC_3BCH: may ignite on contact
Dirubidium acetylide: incompatible
Halogens: incompatible
Hexafluoroisopropylideneamino lithium: incompatible
Iodates: may ignite on contact
Iodine pentafluoride: may ignite on contact
Lithium: may ignite on contact
Nitrogen trichloride: may ignite on contact
NOCl: may ignite on contact
Oxidizing materials: can react vigorously on contact
Palladium: incompatible
Peroxides: may ignite on contact
Platinum: incompatible
Potassium nitrate: may ignite on contact
Potassium permanganate: may ignite on contact
RbC_3BCH: may ignite on contact
Rubidium carbide: may ignite on contact
Silver nitrate: may ignite on contact
Sodium peroxide: incompatible
Zinc: incompatible

When heated, it may produce highly toxic fumes.

FIRST AID   In case of contact with the eyes, immediately wash the eyes with large amounts of water, occasionally lifting the lower and upper lids. Get medical attention immediately. Contact lenses should not be worn when working with this chemical.

In case of contact with the skin, immediately wash the contaminated skin with soap and water. If this chemical penetrates through the clothing, immediately remove the clothing and wash the skin with soap and water. Get medical attention promptly.

If this chemical has been swallowed, get medical attention immediately.

FIRE HAZARD   Flammable and explosive in the form of dust when exposed to heat or flame.

HANDLING   Wear appropriate clothing and equipment to prevent any possibility of skin contact. Wash immediately when skin becomes contaminated, and at the end of each work shift. Immediately remove any clothing that becomes contaminated. Work clothing should be changed daily if there is any possibility that the clothing may be contaminated.

Wear eye protection to prevent any possibility of eye contact.

The following equipment should be available: an eyewash station and a quick-drench shower.

ARSENIC TRIOXIDE

CAS #1327-53-3   mf: As_4O_6
DOT #1561   mw: 395.68

SYNONYMS
Arsenic oxide
Arsenic(III) oxide
Arsenic sesquioxide
Arsenious acid
Arsenious oxide
Arsenious trioxide
Arsenous acid
Arsenous acid anhydride
Arsenous anhydride
Arsenous oxide
Arsenous oxide anhydride
Crude arsenic
Diarsenic trioxide
White arsenic

PROPERTIES   The dimer, claudetite, is colorless, rhombic crystals; soluble in water and alcohol; boiling point 460°C, melting point 278°C. The form with colorless cubic crystals has a melting point of 309°C.

OSHA PEL   TWA 0.01 mg(As)/m³

ACGIH TLV   Suspected carcinogen

NIOSH REL   Ceiling 2 µg(As)/m³/15 min

DOT Classification   Poison B; Label: Poison

TOXIC EFFECTS   Poisonous by ingestion, subcutaneous, intradermal, intravenous, and possibly other routes. A human carcinogen by inhalation.

HAZARDOUS REACTIONS
Rubidium chloride: vigorous reaction
Chlorine trifluoride: vigorous reaction
Fluorine: vigorous reaction
Mercury: vigorous reaction
Oxygen difluoride: vigorous reaction
Sodium chlorate: vigorous reaction

ASBESTOS

CAS #1332-21-4
DOT #2212; 2590

SYNONYMS
Amianthus
Amphibole
Asbestos fiber
Fibrous grunerite
Serpentine

Other forms:
Actinolite
Amosite
Anthophylite
Chrysotile
Crocidolite
Tremolite

PROPERTIES   Fine, slender, flaxy fibers; resists fire and most solvents.

OSHA PEL   TWA 2 million fibers/m³; ceiling 10 million fibers/m³

ACGIH TLV   TWA 2 fibers/cc; human carcinogen

NIOSH REL   100,000 fibers/m³ over 5 µm in length

Community Right to Know List

DOT Classification   ORM-C

TOXIC EFFECTS   A carcinogen. Typically 4 to 7 years of exposure are required before serious lung damage (fibrosis) results.

Routes   Inhalation, ingestion

Symptoms   Dyspnea; interstitial fibrosis; restricted pulmonary function; finger clubbing; cancer

Target Organs   Lung

FIRST AID   In case of contact with the eyes, immediately wash the eyes with large amounts of water, occasionally lifting the lower and upper lids. Get medical attention immediately. Contact lenses should not be worn when working with this chemical.

HANDLING   Wear appropriate clothing and equipment to prevent any possibility of skin contact. Wash at the end of each work shift. Work clothing should be changed daily if it is reasonably probable that the clothing has been contaminated.

Wear eye protection to reduce probability of eye contact.

AZIDES

STRUCTURE
$-N_3$

PROPERTIES   Chemical containing the monovalent $-N_3$ radical and hydrogen or a metal ion.

TOXIC EFFECTS   Many azides are poisonous. Some cause a decrease in blood pressure or inhibit enzyme action; actions are similar to those of nitrates or cyanides.

HAZARDOUS REACTIONS   All azide salts and the acids are unstable, and some may be explosive. One of the most common, lead azide, is a relatively insensitive explosive.

Carbon disulfide: reacts to form explosive salts
Organic azides:
  Metal salts: mixture is a sensitive explosive
  Strong acids (small amounts): mixtures are sensitive explosives

BARIUM

CAS #7440-39-3   mf: Ba
DOT #1399/1400/1854   mw: 137.36

PROPERTIES   Silver-white, slightly lustrous, somewhat malleable metal; melting point 725°C, boiling point 1640°C; density 3.5 at 20°C; vapor pressure 10 mm Hg at 1049°C

OSHA PEL   TWA 0.5 mg/m^3; IDLH 250 mg/m^3

ACGIH TLV   TWA 0.5 mg/m^3

DOT Classification   Some barium compounds are flammable or explosive.

Community Right to Know List (barium and its compounds)

TOXIC EFFECTS   Barium salts are poisons.

Target Organs   Heart, central nervous system, skin, respiratory system, eyes

Routes   Inhalation, ingestion, contact

Symptoms   Upper respiratory and gastrointestinal irritation; muscle spasm; slow pulse, extra systoles; hypokalemia; irritation of the eyes; skin burns

HAZARDOUS REACTIONS
Acids: incompatible
C_2Cl_4: violent or explosive reaction
$C_2H_2FCl_3$: incompatible
C_2HCl_3 + water: incompatible
Carbon tetrachloride: violent or explosive reaction
Fluorotrichloroethane: incompatible
Fluorotrichloromethane: violent or explosive reaction
Trichloroethylene: violent or explosive reaction
1,1,2-Trichlorotrifluoro ethane: incompatible
Water: violent or explosive reaction

FIRST AID   In case of contact with the eyes, immediately wash the eyes with large amounts of water, occasionally lifting the lower and upper lids. Get medical attention immediately. Contact lenses should not be worn when working with this chemical.

In case of contact with the skin, immediately flush the contaminated skin with water. If this chemical penetrates the clothing, immediately remove the clothing and flush the skin with water. Get medical attention promptly.

If a person breathes in large amounts of this chemical, move the exposed person to fresh air at once. If breathing has stopped, perform artificial respiration. Keep the affected person warm and at rest. Get medical attention as soon as possible.

If this chemical has been swallowed, get medical attention immediately.

FIRE HAZARD   Dangerous. The dust may explode when exposed to heat, flame, or chemical reaction with oxidizing gases.

BARIUM CHLORIDE

CAS #10361-37-2   mf: $BaCl_2$
    mw: 208.24

SYNONYMS
Barium dichloride

PROPERTIES   Colorless, flat crystals; melting point (transition to cubic crystals at 925°C), boiling point 1560°C; density 2.856 at 24°C

OSHA PEL   TWA 0.5 mg(Ba)/m^3

ACGIH TLV   TWA 0.5 mg(Ba)/m^3

Reported in the EPA TSCA Inventory, Community Right to Know List (barium and its compounds), and EPA Genetic Toxicology Program

TOXIC EFFECTS   A poison by ingestion and other routes.

HAZARDOUS REACTIONS   When heated to decomposition, it emits toxic fumes of Cl⁻.

BENZALDEHYDE

CAS #100-52-7   mf: C_7H_6O
DOT #1989   mw: 106.13

NFPA LABEL

SYNONYMS
Almond artificial essential oil
Artificial almond oil
Benzenecarbaldehyde
Benzenecarbonal
Benzoic aldehyde

PROPERTIES   Refractive liquid; melting point −26°C, boiling point 179°C; density 1.050 at 15°C/4°C; flash point 64.4°C; autoignition temperature 191.7°C; vapor pressure 1 mm Hg at 26.2°C; vapor density 3.65

DOT Classification   Combustible liquid; Label: None

EPA Genetic Toxicology Program and Reported in the EPA TSCA Inventory

TOXIC EFFECTS   Poisonous by ingestion and some other routes. An allergen. Local contact may cause contact dermatitis. A central nervous system depressant in small doses. Large doses cause convulsions. A skin irritant.

HAZARDOUS REACTIONS   A strong reducing agent.
Peroxyformic acid: violent reaction
Oxidizers: violent reaction

FIRE HAZARD   Flammable when exposed to heat or flame.

Use water, alcohol foam, or dry chemical extinguishers to fight a fire.

BENZENE

CAS #71-43-2   mf: C_6H_6
DOT #1114   mw: 78.12

STRUCTURE   NFPA LABEL

SYNONYMS
Benzol   Motor benzol
Benzole   Nitration benzene
Benzolene   Phene
Carbon oil   Phenyl hydride
Coal naphtha   Pyrobenzol
Cyclohexatriene   Pyrobenzole
Mineral naphtha

PROPERTIES   Colorless liquid; miscible with most organic solvents; melting point 5.51°C, boiling point 80.9°C; density 0.8794 at 20°C; flash point −11°C; lower explosive limit 1.4%, upper explosive limit 8.0%; autoignition temperature 562.2°C; vapor pressure 100 mm Hg at 26.1°C; vapor density (air = 1) 2.77

OSHA PEL   TWA 1 ppm/8 h; peak 5 ppm/15 min

ACGIH TLV   TWA 10 ppm (suspected human carcinogen)

NIOSH REL   ceiling 1 ppm/60 min

DOT Classification   Flammable liquid; Label: Flammable liquid

TOXIC EFFECTS   A poison by inhalation, skin contact, and other routes. Moderately toxic by ingestion. A severe eye and moderate skin irritant. A carcinogen that produces myeloid leukemia and lymphomas by inhalation. A narcotic. Effects are seen at less than 1 ppm. Exposures must be reduced to 0.1 ppm before no effects are observed.

Target Organs   Blood, central nervous system, skin, bone marrow, eyes, respiratory system

Routes   Inhalation, absorption, ingestion, contact

Symptoms   Irritation of the eyes, nose, respiratory system; giddiness; headache; nausea; staggered gait;

tigue; anorexia; lassitude; dermatitis; bone marrow depression; abdominal pain; carcinogen

HAZARDOUS REACTIONS

Arsenic pentafluoride + potassium methoxide: mixture explodes above 30°C
Bromine + iron: incompatible
Bromine pentafluoride: explodes on contact
Bromine trifluoride: vigorous or incandescent reaction
Chlorine + iron: incompatible
Chlorine: vigorous reaction
Chromium trioxide: vigorous reaction
Diborane: explodes on contact
Dioxygen difluoride: ignites on contact
Dioxygenyl tetrafluoroborate: ignites on contact
Hydrogen + Raney nickel: vigorous or incandescent reaction (above 210°C)
Iodine heptafluoride: ignites on contact
Iodine pentafluoride: forms sensitive, explosive mixture
$NClO_4$: vigorous reaction
Nitric acid: forms sensitive, explosive mixture
Nitryl perchlorate: forms sensitive, explosive mixture
Oxidizing materials: vigorous reaction
Oxygen (liquid): forms sensitive, explosive mixture
Ozone: forms sensitive, explosive mixture
Perchlorates: vigorous reaction
Permanganic acid: explodes on contact
Peroxodisulfuric acid: explodes on contact
Peroxomonosulfuric acid: explodes on contact
Potassium dioxide: vigorous reaction
Silver chloride + $FClO_4$: vigorous reaction
Silver perchlorate: forms sensitive, explosive mixture
Silver perchlorate + acetic acid: vigorous reaction
Sodium peroxide: vigorous reaction
Sodium peroxide + water: ignites on contact
Oxidizers (strong): incompatible
Sulfuric acid + permanganates: vigorous reaction
Uranium hexafluoride: vigorous or incandescent reaction

Moderate explosion hazard when exposed to heat or flame. Use with adequate ventilation.

FIRE HAZARD   Dangerous. The liquid is flammable and should not be exposed to heat or flame. The vapors are explosive.

Use foam, CO_2, or dry chemical extinguishers to fight a fire.

HANDLING   Benzene operations in laboratories should be carried out in closed systems or in laboratory hoods that have been shown to have adequate protection factors to prevent significant worker exposure.

When contact with liquid benzene is possible, skin-protection measures should be employed. Before starting work with benzene, the worker should consult the OSHA standard, which requires more stringent precautions than Procedure A.

Wear appropriate clothing and equipment to prevent repeated or prolonged skin contact. Clothing soaked by a spill should be removed immediately to avoid flammability hazard.

Wash promptly with soap when skin becomes contaminated.

Wear eye protection to reduce probability of eye contact.

BENZO(a)PYRENE

CAS #50-32-8   mf: $C_{20}H_{12}$
mw: 252.32

STRUCTURE

SYNONYMS

Benzo(def)chrysene
3,4-Benzopyrene
6,7-Benzopyrene
Benz(a)pyrene
3,4-Benz(a)pyrene
3,4-Benzypyrene
3,4-Benzpyrene

PROPERTIES   A pale yellow crystalline solid; soluble in benzene, toluene, xylene, and acetone; sparingly soluble in ethanol and methanol and almost insoluble (0.005 mg/L at 27°C) in water; melting point 179°C, boiling point 312°C at 10 mm; light sensitive and oxidized by chromic acid and by ozone.

TOXIC EFFECTS   A poison by some routes. A potent carcinogen and teratogen. Both local (skin and subcutaneous tissues) and systemic (lung and liver) carcinogenic effects have been observed. Tumors have resulted from topical application to the skin, ingestion, and parenteral injection. Acute oral toxicity is very low. Repeated contact can cause systemic effects.

HAZARDOUS REACTIONS   When heated to decomposition, it emits acrid smoke and fumes.

HANDLING   Although the information given below applies specifically to benzo[a]pyrene, the general precautions and procedures are also applicable to other carcinogenic polycyclic aromatic hydrocarbons.

No TLV for benzo[a]pyrene has been set. Its carcinogenic potency in animals is high enough to justify the use of Procedure A (see Chap. 1) when handling more than a few milligrams in the laboratory.

All work with benzo[a]pyrene in quantities in excess of a few milligrams or capable of resulting in the formation of aerosols should be carried out in a well-ventilated hood or in a glove box equipped with a HEPA filter. All work should be carried out in apparatus that is contained in or mounted above unbreakable pans that will contain any spill, except that, for very small amounts, a disposable mat may be adequate to contain possible spills. All containers should bear a label such as the following: CANCER-SUSPECT AGENT. All personnel who handle benzo[a]pyrene should wear plastic or latex gloves and a fully buttoned laboratory coat.

Storage and use—All bottles of benzo[a]pyrene should be stored and transported in unbreakable outer containers.

SPILL CLEANUP Cleanup of spills and waste disposal—disposal of benzo[a]pyrene is best carried out by oxidation; this can be accomplished by high-temperature incineration or by the use of strong oxidants such as chromic acid cleaning solution. The latter procedure is especially applicable to small residues on glassware and such. For incineration of liquid wastes, solutions should be neutralized if necessary, filtered to remove solids, and put in a polyethylene container for transport. All equipment should be thoroughly rinsed with solvent to decontaminate it, and this solvent should be added to the wastes to be incinerated. Great care should be exercised to prevent contamination of the outside of the solvent container.

Solid reaction wastes should be incinerated or decomposed by other oxidation procedures. Alternatively, solid reaction wastes can be extracted with solvent that is added to other liquid waste for incineration. Any contaminated rags, paper, and such should be incinerated. Contaminated solid materials should be enclosed in sealed plastic bags that are labeled CANCER-SUSPECT AGENT and with the name and amount of the carcinogen. The bags should be stored in a well-ventilated area until they are incinerated.

BENZOYL PEROXIDE

CAS #94-36-0 mf: $C_{14}H_{10}O_4$
DOT #2085 mw: 242.24

SYNONYMS

Acetoxyl	Garox
Acnegel	Incidol
Aztec BPO	Loroxide
Benoxyl	Lucidol
Benzac	Luperco
Benzaknew	Novadelox
Benzoic acid, peroxide	Oxy-5
Benzoperoxide	Oxy-10
Benzoyl	Oxylite
Benzoyl superoxide	Oxy wash
BZF-60	Panoxyl
Cadet	Persadox
Cadox	Quinolor compound
Clearasil benzoyl peroxide lotion	Sulfoxyl
Cuticura acne cream	Superox
Debroxide	Theraderm
Dibenzoyl peroxide	Topex
Diphenylglyoxal peroxide	Vanoxide
Dry and Clear	Xerac
Epi-clar	
Fostex	

PROPERTIES White, granular, tasteless, odorless powder; insoluble in water, soluble in benzene, acetone, chloroform; decomposes explosively at 103–105°C; autoignition temperature 80°C

OSHA PEL TWA 5 mg/m³; IDLH 7000 mg/m³

ACGIH TLV TWA 5 mg/m³

NIOSH REL 5 mg/m³

Community Right to Know List

DOT Classification Organic peroxide; Label: Organic peroxide

TOXIC EFFECTS A poison by ingestion. An allergen and eye irritant.

Target Organs Skin, respiratory system, eyes

Routes Ingestion, inhalation, contact

Symptoms Irritation of the skin, eyes, mucous membranes; dermatitis; asthmatic effects; testicular atrophy; vasodilation.

HAZARDOUS REACTIONS A powerful oxidizer. Dangerous explosion hazard; may explode spontaneously when heated to above melting point (103°C), or when overheated under confinement. Moderately sensitive to heat, shock, friction, and contact with combustible materials.

Acids (various organic or inorganic): violent reaction on contact
Alcohols: violent reaction on contact
Amines: violent reaction on contact

Aniline: explosive or violent reaction on contact
N-Bromosuccinimide + 4-toluic acid: explosive or violent reaction on contact
Carbon tetrachloride + acetylene: violent reaction on contact
Carbon tetrachloride + ethylene: mixture explodes at elevated temperatures and pressures
Charcoal: violent reaction when heated above 50°C
Combustible substances: incompatible
Dimethyl sulfide: explosive or violent reaction on contact
N, N-dimethylaniline: explosive or violent reaction on contact
Lithium aluminum hydride: explosive at high temperatures
Lithium tetrahydroaluminate: explosive or violent reaction on contact
Metallic naphthenates: violent reaction on contact
Methylmethacrylate: vigorous reaction leading to ignition
Paper: incompatible
Polymerization accelerators: violent reaction on contact
Vinyl acetate + ethyl acetate: vigorous reaction leading to ignition
Wood: incompatible

Decomposition produces dense white smoke of flammable benzoic acid, phenyl benzoate, terphenyls, biphenyls, benzene, and carbon dioxide.

FIRST AID   In case of contact with the eyes, immediately wash the eyes with large amounts of water, occasionally lifting the lower and upper lids. Get medical attention immediately. Contact lenses should not be worn when working with this chemical.

In case of contact with the skin, promptly wash the contaminated skin with soap and water. If this chemical penetrates through the clothing, promptly remove the clothing and wash the skin with soap and water. Get medical attention promptly.

If this chemical has been swallowed, get medical attention immediately.

FIRE HAZARD   It ignites readily and burns rapidly. Moderate fire hazard by spontaneous chemical reaction in contact with reducing agents.

Use water spray or foam extinguishers to fight a fire.

HANDLING   All precautions must be taken to guard against explosion hazards. Keep in a cool place, out of the direct rays of the sun, away from sparks, open flames, and other sources of heat; avoid shock, rough handling, friction from grinding, etc. Isolated storage is required; keep away from possible contact with acids, alcohols, ethers, or other reducing agents or polymerization catalysts such as dimethylaniline. Complete instructions on storage and handling are available from the manufacturer.

Wear appropriate clothing and equipment to prevent repeated or prolonged skin contact. Wash promptly when the skin becomes contaminated. Promptly remove any nonimpervious clothing that becomes contaminated. Work clothing should be changed daily if it is reasonably probable that the clothing has been contaminated.

Wear eye protection to prevent any possibility of eye contact.

BERYLLIUM

CAS #7440-41-7   mf: Be
DOT #1567   mw: 9.01

PROPERTIES   Grayish-white, hard, light metal; melting point 1278°C, boiling point 2970°C; density 1.85

OSHA PEL   TWA 0.002 mg(Be)/m^3, ceiling 0.005 mg/m^3, peak 0.025 mg/m^3/30 min/8 h

ACGIH TLV   TWA 0.002 mg/m^3

NIOSH REL   ceiling not to exceed 0.0005 mg(Be)/m^3/130 min

TOXIC EFFECTS   A deadly poison by intravenous route. A carcinogen and mutagen.

Target Organs   Lung, skin, eyes, mucous membranes

Route   Inhalation

Symptoms   Respiratory symptoms; lung fibrosis; dyspnea; weakness; fatigue; weight loss; carcinogen

HAZARDOUS REACTIONS
Carbon tetrachloride: mixtures with Be powder will flash or spark on impact
Chlorine: incandescent reaction
Fluorine: incandescent reaction
Halocarbons: incompatible
Lithium: hazardous reaction
Phosphorus: hazardous reaction
Trichloroethylene: mixtures with Be powder will flash or spark on impact

When heated to decomposition, very toxic fumes of BeO are emitted.

FIRST AID   In case of contact with the eyes, immediately wash the eyes with large amounts of water, occasionally lifting the lower and upper lids. Get medical attention immediately. Contact lenses should not be worn when working with this chemical.

FIRE HAZARD   A moderate fire hazard in the form of dust or powder, or when exposed to flame or by spontaneous chemical reaction. Slight explosion hazard in the form of powder or dust.

HANDLING   Wear appropriate clothing and equipment to prevent repeated or prolonged skin contact. Promptly remove nonimpervious clothing that becomes contaminated. Work clothing should be changed daily if there is any possibility that the clothing may be contaminated.

Goggles are recommended.

Personnel should wash at the end of each work shift.

BIPHENYL

CAS #92-52-4   mf: $C_{12}H_{10}$
mw: 154.22

SYNONYMS
Bibenzene
1,1'-Biphenyl
Diphenyl
Lemonene
Phenador-x
Phenylbenzene
PHPH
Xenene

PROPERTIES   Colorless to pale yellow solid (white scales) with a very characteristic pleasant odor; melting point 70°C, boiling point 255°C; density 0.991 at 75°C/4°C; flash point 112.8°C (closed cup); lower explosive limit 0.6% at 232°C, upper explosive limit 5.8% at 166.1°C; autoignition temperature 540°C; vapor density 5.31

OSHA PEL   TWA 0.2 ppm; IDLH 300 mg/m^3

ACGIH TLV   TWA 0.2 ppm

Community Right to Know List, Reported in the EPA TSCA Inventory, and EPA Genetic Toxicology Program

TOXIC EFFECTS   Moderately toxic by ingestion. Poisonous by some other routes. A powerful irritant by inhalation in humans. Inhalation of very small amounts sufficient to produce systemic effects.

Routes   Inhalation, absorption, ingestion, contact

Symptoms   Irritation throat, eyes; headache; nausea; fatigue; numbness in limbs; liver damage; flaccid paralysis

Target Organs   Liver, skin, central nervous system, upper respiratory system, eyes

HAZARDOUS REACTIONS
Oxidizing materials: incompatible

When heated to decomposition, it emits acrid smoke and irritating fumes.

FIRST AID   In case of contact with the eyes, immediately wash the eyes with large amounts of water, occasionally lifting the lower and upper lids. Get medical attention immediately. Contact lenses should not be worn when working with this chemical.

In case of contact with the skin, immediately flush the contaminated skin with water. If this chemical penetrates the clothing, immediately remove the clothing and flush the skin with water. Get medical attention promptly.

If a person breathes in large amounts of this chemical, move the victim to fresh air at once. If breathing has stopped, perform artificial respiration. Keep the affected person warm and at rest. Get medical attention as soon as possible.

If this chemical has been swallowed, get medical attention immediately.

FIRE HAZARD   Combustible when exposed to heat or flame.

Use CO_2, dry chemical, water spray, mist, or fog extinguishers to fight a fire.

HANDLING   Wear appropriate clothing and equipment to prevent repeated or prolonged skin contact. Wash promptly when the skin becomes contaminated. Promptly remove any nonimpervious clothing that becomes contaminated. Work clothing should be changed daily if it is reasonably probable that the clothing has been contaminated.

Wear eye protection to prevent any possibility of contact with the molten material, or repeated/prolonged contact with other forms.

BIS(CHLOROMETHYL)ETHER

CAS #542-88-1   mf: $C_2H_4Cl_2O$
DOT #2249   mw: 114.96

STRUCTURE
ClCH$_2$OCH$_2$Cl

SYNONYMS
BCME
Bis-CME
Chloro(chloromethoxy)methane
Chloromethyl ether
sym-Dichlorodimethyl ether
sym-Dichloromethyl ether
Dimethyl-1,1'-dichloroether
Oxybis(chloromethane)

PROPERTIES   Colorless, volatile liquid with a suffocating odor; miscible in all proportions with ethanol, ether, and many organic solvents; decomposes in water to give HCl and formaldehyde; melting point −41.5°C, boiling point 104°C; density 1.315 at 20°C/4°C; flash point <19°C; vapor density 4.0

OSHA PEL   Carcinogen

ACGIH TLV   TWA 0.001 ppm (1 ppb; 5 µg/m^3)

EPA Extremely Hazardous Substances List and Community Right to Know List

TOXIC EFFECTS   A poison by inhalation, ingestion, and skin contact. A carcinogen. Because of the high volatility of bis(chloromethyl)ether, inhalation is the route of exposure that presents the greatest hazard to humans. BCME vapor is severely irritating to the skin and mucous membranes and can cause corneal damage that heals slowly. The substance has caused lung cancer in humans.

It is carcinogenic to mice following inhalation, skin application, or subcutaneous administration. In newborn mice, it is carcinogenic after a single subcutaneous exposure. In the rat, it is carcinogenic by inhalation and subcutaneous administration. It is a lung carcinogen in humans.

HAZARDOUS REACTIONS   When heated to decomposition, it emits highly toxic fumes of Cl$^-$.

FIRE HAZARD   A dangerous fire hazard.

HANDLING   OSHA has classified BCME as a cancer-suspect agent and has stringent regulations for its use if its concentration in a material exceeds 0.1% (see 29 CFR, Part 1910.1008). The regulations, which call for more precautions than Procedure A (see Chap. 2), should be consulted before starting work with BCME.

BORIC ACID

CAS #10043-35-3   mf: BH$_3$O$_3$
                         mw: 61.84

SYNONYMS
Boracic acid   Orthoboric acid
Borofax   Three elephant

PROPERTIES   White crystals or powder; decomposes at 185°C; density 1.435 at 15°C

TOXIC EFFECTS   A poison by ingestion and inhalation. Moderately toxic by skin contact.

Symptoms   Wakefulness; anorexia; nausea and vomiting; diarrhea; abdominal cramps; skin irritation; erythematous lesions on skin and mucous membranes; circulatory collapse; tachycardia; cyanosis; delirium; convulsions and coma. Ingestion of less than 5 g in infants and from 5 to 20 g in adults can cause death. Chronic exposure may result in borism, which has the symptoms of dry skin, eruptions, and gastrointestinal disturbances. Used in the home as a cleaner and insecticide.

HAZARDOUS REACTIONS
(CH$_3$CO)$_2$O: incompatible
Potassium: incompatible

BORON TRICHLORIDE

CAS #10294-34-5   mf: BCl$_3$
DOT #1741   mw: 117.16

SYNONYM
Boron chloride

PROPERTIES   Colorless, fuming liquid; pungent, irritating odor; melting point −107°C, boiling point 12.5°C; density 1.434 at 0°C; vapor pressure 1 atm at 12.7°C; vapor density 4.03

EPA Extremely Hazardous Substances List

DOT Classification   Corrosive material; Label: Corrosive

IMO Classification   Nonflammable gas; Label: Nonflammable gas, corrosive

TOXIC EFFECTS   A poison by inhalation and probably other routes. The vapors are irritating to the eyes and mucous membranes. Generally much less toxic than chlorine.

HAZARDOUS REACTIONS
Aniline: violent reaction
Grease: incompatible
Hexafluorisopropylidene amino lithium: incompatible
Nitrogen dioxide: incompatible
Organic matter: incompatible
Oxygen: incompatible
Phosphine: violent reaction
Water or steam: reaction produces heat + toxic and corrosive fumes

When heated to decomposition, it emits highly toxic fumes of Cl^-.

HANDLING AND SPILL CLEANUP   Although boron trichloride has a very low vapor pressure at normal temperatures, it should be handled with care. The use of goggles and gloves is recommended.

A trap should always be used when transferring the liquid or gas into a solution to prevent impurities from being sucked back into the cylinder, which might create an explosive mixture.

A boron trichloride cylinder contains a large amount of liquid and should never be permitted to become overheated. If the container is completely filled with liquid, an increase in temperature could build up tremendous pressure and cause the container to burst. For this reason, closed containers such as cylinders should be filled only by experts in the field.

Boron trichloride cylinders and valves should be equipped with a safety device containing a fusible metal that melts at approximately 70°C.

BORON TRIFLUORIDE

CAS #7637-07-2   mf: BF_3
DOT #1008   mw: 67.81

SYNONYM
Boron fluoride

PROPERTIES   Colorless gas; pungent, irritating odor; melting point −126.8°C, boiling point −99.9°C; density 2.99 g/L

OSHA PEL   ceiling 1 ppm; IDLH 100 ppm

ACGIH TLV   ceiling 1 ppm

NIOSH REL   No exposure limit, due to the absence of a reliable monitoring method

EPA Extremely Hazardous Substances List

DOT Classification   Nonflammable gas; Label: Nonflammable gas and poison

IMO Classification   Poison A; Label: Poison gas

TOXIC EFFECTS   A poison by inhalation. A strong irritant.

Skin burns produced by high concentrations of boron trifluoride are similar to those caused by hydrogen fluoride, although not as deep.

Target Organs   Respiratory system, kidneys, eyes, skin

Routes   Inhalation, contact

Symptoms   Nasal irritation, epistaxis, burns eyes, skin; in animals, pneumonia, kidney damage, nosebleed

HAZARDOUS REACTIONS
Alkali metals: incompatible
Alkaline earth metals (except magnesium): incompatible
Alkyl nitrates: incompatible
Calcium oxide: incompatible
Water or steam: reaction forms toxic and corrosive fumes of F^-

When heated to decomposition, it will produce toxic and corrosive fumes of F^-.

FIRST AID   In case of contact with the eyes, immediately wash the eyes with large amounts of water, occasionally lifting the lower and upper lids. Get medical attention immediately. Contact lenses should not be worn when working with this chemical.

In case of contact with the skin, immediately flush the contaminated skin with water. If this chemical penetrates the clothing, immediately remove the clothing and flush the skin with water. Get medical attention promptly.

If a person breathes in large amounts of this chemical, move the victim to fresh air at once. If breathing has stopped, perform artificial respiration. Keep the affected person warm and at rest. Get medical attention as soon as possible.

HANDLING PROCEDURES   Boron trifluoride has one characteristic that occasionally proves to be quite disconcerting. In contact with the atmosphere, the gas forms dense white fumes. Even after a cylinder valve has been tightly closed, the fumes will linger around the outlet for as much as half an hour. This frequently causes the user to believe that the valve itself is leaking.

In addition, the gas is inherently difficult to control through valves and piping, and even the best of equipment is apt to show slight signs of leaking, which will produce an abundance of fumes.

It is essential when using boron trifluoride to have a trap in the delivery tube to prevent impurities from being sucked back into the cylinder. Certain chemicals, if drawn back into a boron trifluoride cylinder, can build up tremendous pressure that may cause the cylinder to burst.

Every boron trifluoride valve is equipped with a device consisting of a platinum disk in back of a plug containing a metal that will melt at approximately 70°C. Frequently, a similar safety device is inserted in the base of the cylinder.

BROMINE

CAS #7726-95-6 mf: Br_2
DOT #1744 mw: 159.82

NFPA LABEL

SYNONYM
Bromine solution

PROPERTIES   Rhombic crystals or dark red liquid with an irritating and penetrating odor, threshold = 1.5–3.5 ppm; soluble in alcohol, ether, chloroform, and carbon disulfide; melting point −7.27°C, boiling point 58.73°C; density (liquid) 3.12 at 20°C, 2.928 at 59°C; vapor pressure 175 mm Hg at 21°C; vapor density 3.5

OSHA PEL   TWA 0.1 ppm (0.7 mg/m^3); IDLH 10 ppm

ACGIH TLV   TWA 0.1 ppm; STEL 0.3 ppm

DOT Classification   Corrosive material; Label: Corrosive

IMO Classification   Corrosive material; Label: Corrosive and poison

TOXIC EFFECTS   A poison by ingestion and inhalation. Corrosive. The action of bromine is essentially the same as that of chlorine.

A lethal oral dose for humans is 14 mg/kg of bromine. Inhalation of bromine has caused coughing, nosebleeds, dizziness, and headache, followed after some hours by abdominal pain, diarrhea, and skin rashes. Severe irritation of the respiratory passages and pulmonary edema can also occur. Lacrimation occurs at levels of less than 1 ppm. It is reported that 40–60 ppm are dangerous for short exposures, and 1000 ppm can be fatal. The substance produces irritation and destruction of the skin with blister formation. Severely painful and destructive eye burns may result from contact with either liquid or concentrated vapors of bromine.

Usually, however, the irritant qualities of the chemical force the worker to leave the exposure area before serious poisoning can result.

Target Organs   Respiratory system, eyes, central nervous system

Routes   Inhalation, ingestion, contact

Symptoms   Dizziness; headache; lacrimation; epistaxis; cough; feeling of oppression; pulmonary edema; pneumonia; abdominal pain; diarrhea; measleslike eruptions; severe burns to eyes, skin

HAZARDOUS REACTIONS   A very powerful oxidizer.

Acetaldehyde: incompatible
Acetylides (mono- or dialkali metal): ignition on contact
Acrylonitrile: incompatible
Alcohols: vigorous reaction
Aldehydes: violent reaction
Aluminum: violent reaction
Ammonia: explosive reaction
Antimony: explosive reaction
Azides (metal, particularly silver azide or sodium azide): explosive reaction
Boron: incompatible
Calcium nitride: incompatible
Carbonyl compounds: violent reaction
Carboxylic acids: violent reaction
Cesium carbide: incompatible
Cesium oxide: incompatible
ClF_3C_2: incompatible
Copper acetylides: ignition on contact
Copper hydride: incompatible
CsC_2H: incompatible
Diethyl ether: violent reaction
Diethylzinc: explosive reaction
Dimethylformamide: explosive reaction
Disilane: explosive reaction
Ethanol + phosphorous: vigorous reaction
Ethyl phosphine: incompatible
Ethylene: incompatible

Fluorine: incompatible
Germane: explosive reaction
Germanium: ignition on contact
Hydrogen: explosive reaction
Iron carbide: incompatible
Isobutyrophenone: explosive reaction
Ketones: violent reaction
Lithium: mixture is a shock-sensitive explosive
Li_2Si_2: incompatible
Lithium carbide: incompatible
Mercury: violent reaction
Metals: incompatible with wet bromine
Methanol: vigorous reaction
Mg_2P_2: incompatible
Na_2C_2: incompatible
NaC_2H: incompatible
Nickel carbonyl: incompatible
Nitrogen triiodide: incompatible
Olefins: incompatible
Organics (combustible): incompatible
Oxidizable materials: incompatible
Oxygen difluoride: incompatible
Ozone: incompatible
Phosphine: violent reaction
Phosphorous: incompatible
Potassium: explosive reaction
PO_x: incompatible
Praseodymium: explosive reaction
Rubber (natural): violent reaction
Rubidium carbide: incompatible
Silane and homologs: explosive reaction
Sodium: mixture is a shock-sensitive explosive
Sr_3P: incompatible
Tetrahydrofuran: vigorous reaction
Titanium: violent reaction
Trialkyl boranes: ignition on contact
Trimethylamine: explosive reaction
Uranium dicarbide: incompatible
Water or steam: reaction produces toxic and corrosive fumes of HBr
Zirconium dicarbide: incompatible

When heated, it emits highly toxic fumes.

FIRST AID   In case of contact with the eyes, immediately wash the eyes with large amounts of water, occasionally lifting the lower and upper lids. Get medical attention immediately. Contact lenses should not be worn when working with this chemical.

In case of contact with the skin, immediately wash the contaminated skin with soap and water. If this chemical penetrates through the clothing, immediately remove the clothing and wash the skin with soap and water. Get medical attention promptly.

If a person breathes in large amounts of this chemical, move the victim to fresh air at once. If breathing has stopped, perform artificial respiration. Keep the affected person warm and at rest. Get medical attention as soon as possible.

If this chemical has been swallowed, get medical attention immediately.

FIRE HAZARD   Both the liquid and vapor are flammable by spontaneous chemical reaction with reducing materials.

HANDLING   Splash goggles and rubber gloves should be worn when handling more than a few milliliters of pure liquid bromine. Although the irritating odor provides a warning, it is best to carry out laboratory operations with it in a hood.

Wear appropriate clothing and equipment to prevent any possibility of skin contact. Wash immediately when skin becomes contaminated. A worker whose clothing has been doused with liquid bromine is in severe danger unless the affected clothing is removed immediately.

Wear eye protection to prevent any possibility of eye contact.

The following equipment should be available: an eyewash station and a quick-drench shower.

1-BUTANOL

CAS #71-36-3   mf: $C_4H_{10}O$
DOT #1120   mw: 74.14

STRUCTURE   NFPA LABEL
$CH_3CH_2CH_2CH_2OH$

                    3
                  1   0

SYNONYMS
Butanol   1-Hydroxybutane
n-Butanol   Methylolpropane
Butan-1-ol   Normal primary butyl
Butyl alcohol   alcohol
n-Butyl alcohol   Propylcarbinol
Butyl hydroxide   Propylmethanol
Butyric alcohol

PROPERTIES   Colorless liquid; freezing point −88.9°C, boiling point 117.5°C; density 0.80978 at 20°C/4°C; flash point 35–37.8°C; lower explosive limit 1.4%, upper explosive limit 11.2%; autoignition temperature 365°C; vapor pressure 5.5 mm Hg at 20°C; vapor density 2.55; Underwriters Laboratory Classification 40

OSHA PEL   TWA 100 ppm; IDLH 8000 ppm

ACGIH TLV   ceiling 50 ppm (skin)

Community Right to Know List

DOT Classification   Flammable or combustible liquid; Label: Flammable liquid

TOXIC EFFECTS   A poison by intravenous route. Moderately toxic by skin contact, ingestion, and other routes. A skin and severe eye irritant.

Target Organs   Skin, eyes, respiratory system

Routes   Inhalation, ingestion, contact

Symptoms   Irritation of the eyes, nose, throat; headache; vertigo; drowsiness; corneal inflammation, blurred vision, lacrimation, photophobia; dry, cracked skin

HAZARDOUS REACTIONS
Aluminum: incompatible
Chromium trioxide: incompatible
Oxidizing materials: incompatible

When heated to decomposition, it emits acrid smoke and fumes.

FIRST AID   In case of contact with the eyes, immediately wash the eyes with large amounts of water, occasionally lifting the lower and upper lids. Get medical attention immediately. Contact lenses should not be worn when working with this chemical.

In case of contact with the skin, immediately flush the contaminated skin with water. If this chemical penetrates the clothing, immediately remove the clothing and flush the skin with water. If irritation persists after washing, get medical attention.

If a person breathes in large amounts of this chemical, move the victim to fresh air at once. If breathing has stopped, perform artificial respiration. Keep the affected person warm and at rest. Get medical attention as soon as possible.

If this chemical has been swallowed, get medical attention immediately.

FIRE HAZARD   A dangerous fire hazard when exposed to heat, flame, or oxidizers. The vapors are explosive when exposed to flame.

Use water spray, alcohol foam, CO_2, or dry chemical extinguishers to fight a fire.

HANDLING   Wear appropriate clothing and equipment to prevent repeated or prolonged skin contact. Personnel should wash promptly when skin becomes wet. Clothing soaked by a spill should be removed to avoid flammability hazard.

Wear eye protection to reduce probability of eye contact.

2-BUTANOL

CAS #78-92-2   mf: $C_4H_{10}O$
DOT #1120   mw: 74.14

STRUCTURE   NFPA LABEL
$CH_3CH(OH)CH_2CH_3$

SYNONYMS
Butan-2-ol   Ethylmethyl carbinol
sec-Butanol   2-Hydroxybutane
2-Butyl alcohol   Methylethylcarbinol
Butylene hydrate   S.B.A.

PROPERTIES   Colorless liquid; melting point −89°C, boiling point 99.5°C; density 0.808 at 20°C/4°C; flash point 14°C; lower explosive limit 1.7% at 100°C, upper explosive limit 9.8% at 100°C; autoignition temperature 406.1°C; vapor pressure 10 mm Hg at 20°C; vapor density 2.55

OSHA PEL   TWA 150 ppm; IDLH 10,000 ppm

ACGIH TLV   TWA 100 ppm

Community Right to Know List

DOT Classification   Flammable or combustible liquid; Label: Flammable liquid

TOXIC EFFECTS   Poison by intravenous and intraperitoneal routes. Mildly toxic by ingestion.

Target Organs   Eyes, skin, central nervous system

Routes   Inhalation, ingestion, contact

Symptoms   Eye irritation, narcosis, dry skin

HAZARDOUS REACTIONS   Auto-oxidizes to an explosive peroxide.

Chromium trioxide: ignites on contact
Oxidizing materials: incompatible

When heated to decomposition, it emits acrid smoke and fumes.

FIRST AID   In case of contact with the eyes, immediately wash the eyes with large amounts of water, occasionally lifting the lower and upper lids. Get medical attention immediately. Contact lenses should not be worn when working with this chemical.

In case of contact with the skin, immediately flush the contaminated skin with water. If this chemical penetrates the clothing, immediately remove the clothing and flush the skin with water. If irritation persists after washing, get medical attention.

If a person breathes in large amounts of this chemical, move the victim to fresh air at once. If breathing has stopped, perform artificial respiration. Keep the affected person warm and at rest. Get medical attention as soon as possible.

If this chemical has been swallowed, get medical attention immediately.

FIRE HAZARD   Dangerous fire hazard when exposed to heat or flame.

Use water spray, alcohol foam, CO_2, or dry chemical extinguishers to fight a fire.

HANDLING   Wear appropriate clothing and equipment to prevent repeated or prolonged skin contact. Wash promptly when skin becomes wet. Clothing soaked by a spill should be removed immediately to avoid flammability hazard.

Wear eye protection to reduce probability of eye contact.

tert-BUTANOL

CAS #75-65-0   mf: $C_4H_{10}O$
DOT #1120   mw: 74.14

STRUCTURE   NFPA LABEL
$(CH_3)_3COH$

3 / 1 / 0

SYNONYMS
tert-Butyl alcohol   2-Methyl 2,2-propanol
tert-Butyl hydroxide   Trimethylcarbinol
1,1-Dimethylethanol

PROPERTIES   Colorless liquid or rhombic prisms or plates; melting point 25.3°C, boiling point 82.8°C; flash point 10°C (closed cup); density 0.7887 at 20°C/4°C; autoignition temperature 480°C; lower explosive limit 2.4%, upper explosive limit 8.0%; vapor pressure 400 mm Hg at 24.5°C; vapor density 2.55

OSHA PEL   TWA 100 ppm; IDLH 8000 ppm

ACGIH TLV   TWA 100 ppm; STEL 150 ppm

NIOSH REL   100 ppm (300 mg/m³)

Community Right to Know List, Reported on the EPA TSCA Inventory, and EPA Genetic Toxicology Program

TOXIC EFFECTS   Moderately toxic by ingestion and other routes.

Target Organs   Eyes, skin

Routes   Ingestion, inhalation, contact

Symptoms   Drowsiness; irritation of skin, eyes; headache; dizziness; dry skin

HAZARDOUS REACTIONS
Hydrochloric acid (conc.): incompatible
Hydrogen peroxide: incompatible
Mineral acids (conc.): incompatible
Oxidizing materials: incompatible
Potassium-sodium alloys: ignites on contact

FIRST AID   In case of contact with the eyes, immediately wash the eyes with large amounts of water, occasionally lifting the lower and upper lids. Get medical attention immediately. Contact lenses should not be worn when working with this chemical.

In case of contact with the skin, immediately flush the contaminated skin with water. If this chemical penetrates the clothing, immediately remove the clothing and flush the skin with water. If irritation persists after washing, get medical attention.

If a person breathes in large amounts of this chemical, move the victim to fresh air at once. If breathing has stopped, perform artificial respiration. Keep the affected person warm and at rest. Get medical attention as soon as possible.

If this chemical has been swallowed, get medical attention immediately.

FIRE HAZARD   Dangerous fire hazard when exposed to heat or flame. The vapor is explosive when exposed to flame.

Use alcohol foam, CO_2, or dry chemical extinguishers to fight a fire.

n-BUTYL ACETATE

CAS #123-86-4 mf: $C_6H_{12}O_2$
DOT #1123 mw: 116.18

NFPA LABEL

SYNONYMS
Acetic acid-*n*-butyl ester
Butyl acetate
1-Butyl acetate
Butyl ethanoate

PROPERTIES   Colorless liquid; freezing point −73.5°C, boiling point 126°C; density 0.88 at 20°C/20°C; flash point 22.2°C; lower explosive limit 1.4%, upper explosive limit 7.5%; autoignition temperature 425°C; vapor pressure 15 mm Hg at 25°C; Underwriters Laboratory Classification 50-60

OSHA PEL   TWA 150 ppm; IDLH 10,000 ppm

ACGIH TLV   TWA 150 ppm; STEL 200 ppm

DOT Classification   Flammable liquid; Label: Flammable liquid

TOXIC EFFECTS   Mildly toxic by inhalation and ingestion. Moderately toxic by intraperitoneal route. A skin and severe eye irritant. A mild allergen. A teratogen.

Target Organs   Eyes, skin, respiratory system

Routes   Ingestion, inhalation, contact

Symptoms   Headache; drowsiness; eye irritation and dryness; respiratory system irritation

HAZARDOUS REACTIONS   Explosive when exposed to flame.
Acids: incompatible
Alkalies: incompatible
Nitrates: incompatible
Oxidizers (strong): incompatible
Oxidizing materials: incompatible
Potassium-*tert*-butoxide: ignites on contact

When heated to decomposition, it emits acrid smoke and irritating fumes.

FIRST AID   In case of contact with the eyes, immediately wash the eyes with large amounts of water, occasionally lifting the lower and upper lids. Get medical attention immediately. Contact lenses should not be worn when working with this chemical.

In case of contact with the skin, immediately flush the contaminated skin with water. If this chemical penetrates the clothing, immediately remove the clothing and flush the skin with water. If irritation persists after washing, get medical attention.

If a person breathes in large amounts of this chemical, move the victim to fresh air at once. If breathing has stopped, perform artificial respiration. Keep the affected person warm and at rest. Get medical attention as soon as possible.

If this chemical is swallowed, get medical attention immediately.

FIRE HAZARD   Dangerous fire hazard when exposed to heat or flame.

Use alcohol foam, CO_2, or dry chemical extinguishers to fight a fire.

HANDLING   Wear appropriate clothing and equipment to prevent repeated or prolonged skin contact. Personnel should wash promptly when skin becomes wet. Clothing soaked by a spill should be removed immediately to avoid flammability hazard.

Wear eye protection to reduce probability of eye contact.

n-BUTYL CHLORIDE

CAS #109-69-3 mf: C_4H_9Cl
DOT #1127 mw: 92.58

SYNONYMS
Butyl chloride
1-Chlorobutane
N-Propylcarbinyl chloride

PROPERTIES   Colorless liquid; melting point −123.1°C, boiling point 78°C; density 0.884; flash point −9.4°C; lower explosive limit 1.9%, upper explosive limit 10.1%; autoignition temperature 460°C; vapor density 3.20

DOT Classification   Flammable liquid; Label: Flammable liquid

TOXIC EFFECTS   Moderately toxic by ingestion and possibly other routes. A skin and eye irritant.

HAZARDOUS REACTIONS   Moderately explosive when exposed to flame.

Oxidizing materials: incompatible

When heated to decomposition, it emits highly toxic fumes of phosgene and Cl⁻.

FIRE HAZARD   Dangerous fire hazard when exposed to heat or flame.

Use foam, CO_2, or dry chemical extinguishers to fight a fire.

tert-BUTYL HYDROPEROXIDE

CAS #75-91-2   mf: $C_4H_{10}O_2$
DOT #2093; 2094   mw: 90.12

STRUCTURE
$(CH_3)_3COOH$

NFPA LABEL

SYNONYMS
Cardox TBH
1,1-Dimethylethyl hydroperoxide
2-Hydroperoxy-2-methylpropane

PROPERTIES   Water white liquid; moderately soluble in water, very soluble in organic (ester and alcohols) solvents and alkali metal hydroxide solutions; melting point −8°C, boiling point 35°C; density 0.896 at 20°C/4°C; flash point 27°C (80°C); vapor density 2.07.

DOT Classification: Organic peroxide

TOXIC EFFECTS   A poison by ingestion and inhalation. The vapor is irritating to the eyes, nose, and throat. The liquid is irritating to the skin and eyes and is capable of causing severe eye injury. A skin sensitizer.

Symptoms   Severe depression, incoordination, cyanosis, and death due to respiratory arrest.

HAZARDOUS REACTIONS   Moderately explosive; may explode during distillation.

Acids (trace): violent reaction
Cobalt salts: incompatible
Combustible materials: spontaneous chemical reaction and/or explosion
Copper salts: incompatible
1,2-Dichloroethane: forms an unstable solution
Manganese salts: incompatible
Molecular sieve: concentrated solutions may ignite spontaneously
Reducing agents: incompatible
Transition metal salts: may react vigorously and release oxygen

Spontaneous chemical reaction and/or explosion may occur if mixed with readily oxidizable, organic, or flammable materials.

When heated to decomposition, it emits acrid smoke and fumes.

FIRE HAZARD   A highly flammable liquid. Vapor is explosive if ignited in an enclosed area. Flashback may occur along vapor trail.

Use alcohol foam, CO_2, or dry chemical extinguishers to fight a fire.

SPILL CLEANUP   Avoid contact with liquid and vapor.

n-BUTYRIC ACID

CAS #107-92-6   mf: $C_4H_8O_2$
DOT #2820   mw: 88.12

NFPA LABEL

SYNONYMS
Butanoic acid
Ethylacetic acid
1-Propanecarboxylic acid
Propylformic acid

PROPERTIES   Liquid; melting point −7.9°C, boiling point 163.5°C; density 0.9590 at 20°C/20°C; flash point 71.7°C; lower explosive limit 2.0%, upper explosive limit 10.0%; autoignition temperature 452.2°C; vapor pressure 0.43 mm Hg at 20°C; vapor density 3.04

DOT Classification   Corrosive material; Label: Corrosive

TOXIC EFFECTS   Moderately toxic by ingestion, skin contact, and other routes. A severe skin and eye irritant. Corrosive.

HAZARDOUS REACTIONS
Chromium trioxide: incandescent reaction above 100°C
Oxidizing materials: incompatible

When heated to decomposition, it emits acrid smoke and irritating fumes.

FIRE HAZARD   Flammable when exposed to heat or flame.

Use alcohol foam, CO_2, or dry chemical extinguishers to fight a fire.

CADMIUM

CAS #7440-43-9   mf: Cd
                            mw: 112.40

PROPERTIES   Silver-white malleable metal; melting point 320.9°C, boiling point 767°C; density 8.642; vapor pressure 1 mm Hg at 394°C

OSHA PEL   TWA 0.1 mg(Cd)/m^3; CL 0.6 mg(Cd)/m^3 (fume)

ACGIH TLV   TWA 0.01 mg(Cd)/m^3 (dust); human carcinogen

NIOSH REL   Reduce to lowest feasible level

Community Right to Know List (cadmium and its compounds), Reported in the EPA TSCA Inventory, and EPA Genetic Toxicology Program

TOXIC EFFECTS   A poison by inhalation, ingestion, and most other routes. A carcinogen and teratogen.

Target Organs   Respiratory system, kidneys, prostate, blood

Routes   Ingestion, inhalation

Symptoms   Pulmonary edema, dyspnea, cough, tight chest, substernal pain; headache; chills, muscle aches; nausea, diarrhea; anosmia, emphysema; proteinuria (an excess of protein in the urine); anemia; a carcinogen

HAZARDOUS REACTIONS
Ammonia: violent or explosive reaction
Ammonium nitrate: violent or explosive reaction when heated
Hydrozoic acid: explodes on contact
Metals: violent or explosive reaction
Nitryl fluoride: vigorous reaction when heated
Oxidizing agents: violent or explosive reaction
Selenium: violent or explosive reaction
Sulfur: incompatible
Tellurium: incompatible
Tellurium: violent or explosive reaction
Zinc: violent or explosive reaction

When heated strongly, it emits toxic fumes of Cd.

FIRST AID   In case of contact with the eyes, immediately wash the eyes with large amounts of water, occasionally lifting the lower and upper lids. Get medical attention immediately. Contact lenses should not be worn when working with the dust.

In case of contact with the skin, wash the contaminated skin with soap and water.

If a person breathes in large amounts of the dust or fumes, move the victim to fresh air at once. If breathing has stopped, perform artificial respiration. Keep the affected person warm and at rest. Get medical attention as soon as possible.

If this chemical has been swallowed, get medical attention immediately.

FIRE HAZARD   The dust ignites spontaneously in air, and is flammable and explosive when exposed to heat or flame.

CADMIUM CHLORIDE

CAS #10108-64-2   mf: $CdCl_2$
DOT #2570   mw: 183.30

NFPA LABEL

SYNONYMS
Caddy
Cadmium dichloride

PROPERTIES   Hexagonal, colorless crystals; melting point 568°C, boiling point 960°C; density 4.047 at 25°C; vapor pressure 10 mm Hg at 656°C

OSHA PEL   TWA 0.2 mg(Cd)/m^3; ceiling 0.6 mg(Cd)/m^3 (dust)

ACGIH TLV   TWA 0.01 mg(Cd)/m³ (dust); human carcinogen

NIOSH REL   (Cadmium) Reduce to lowest feasible level

Community Right to Know List

DOT Classification   ORM-E; Label: None

TOXIC EFFECTS   Poisonous by ingestion, inhalation, skin contact, and most other routes. Cadmium compounds are considered to be carcinogens. A carcinogen and teratogen.

HAZARDOUS REACTIONS
Bromine trifluoride: violent reaction
Potassium: violent reaction

When heated to decomposition, it emits toxic fumes of Cd and Cl⁻.

CADMIUM NITRATE TETRAHYDRATE

CAS #10022-68-1   mf: $N_2O_6 \cdot Cd \cdot 4H_2O$
mw: 308.50

SYNONYMS
Nitric acid, cadmium salt, tetrahydrate

OSHA PEL   TWA 0.2 mg(Cd)/m³; ceiling 0.6 mg(Cd)/m³ (dust)

ACGIH TLV   TWA 0.01 mg(Cd)/m³ (dust); human carcinogen

NIOSH REL   (Cadmium) Reduce to lowest feasible level

Community Right to Know List (cadmium and its compounds)

TOXIC EFFECTS   Poisonous by ingestion. Cadmium compounds are considered to be carcinogens. A severe skin and moderate eye irritant.

HAZARDOUS REACTIONS   When heated to decomposition, it emits toxic fumes of Cd and NO_x.

CADMIUM OXIDE

CAS #1306-19-0   mf: CdO
mw: 128.40

PROPERTIES   (1) Amorphous, brown crystals; melting point <1426°C; density 6.95 and (2) cubic, brown crystals; decomposes at 950°C, boiling point 1559°C; density 8.15; vapor pressure 1 mm Hg at 1000°C

OSHA PEL   TWA 0.2 mg(Cd)/m³; ceiling 0.6 mg(Cd)/m³ (dust)

ACGIH TLV   TWA 0.01 mg(Cd)/m³ (dust); human carcinogen

NIOSH REL   (Cadmium) Reduce to lowest feasible level

EPA Extremely Hazardous Substances List and Community Right to Know List (cadmium and its compounds)

TOXIC EFFECTS   Poisonous by ingestion, inhalation, and intraperitoneal routes. Cadmium compounds are considered to be carcinogens.

Symptoms   Change in the sense of smell, change in heart rate, blood pressure increase, an excess of protein in the urine, and other kidney or bladder changes

HAZARDOUS REACTIONS
Magnesium: mixtures explode when heated

When heated to decomposition, it emits toxic fumes of Cd.

CADMIUM NITRATE

CAS #10325-94-7   mf: CdN_2O_6
mw: 236.42

SYNONYMS
Cadmium dinitrate
Nitric acid, cadmium salt

PROPERTIES   White, prismatic needles; hygroscopic; melting point 350°C

OSHA PEL   TWA 0.2 mg(Cd)/m³; ceiling 0.6 mg(Cd)/m³ (dust)

ACGIH TLV   TWA 0.01 mg(Cd)/m³ (dust); human carcinogen

NIOSH REL   (Cadmium) Reduce to lowest feasible level

Community Right to Know List

TOXIC EFFECTS   Poison by ingestion and possibly other routes. Moderately toxic by inhalation. Cadmium compounds are considered to be carcinogens.

HAZARDOUS REACTIONS   When heated to decomposition, it emits toxic fumes of NO_x and Cd.

CALCIUM

CAS #7440-70-2   mf: Ca
DOT #1401; 1855   mw: 40.08

SYNONYMS
Calcicat
Calcium, nonpyrophoric
Calcium, pyrophoric
Calcium, metal
Calcium metal, crystalline

PROPERTIES   Silver-white, soft metal; melting point 842°C, boiling point 1484°C; density 1.54 at 20°C; vapor pressure 10 mm Hg at 983°C

DOT Classification   Flammable solid; Label: Flammable solid and Dangerous when wet

IMO Classification   Flammable solid; Label: Spontaneously combustible

HAZARDOUS REACTIONS   Reacts with moisture or acids to liberate large quantities of hydrogen; can develop explosive pressure in containers.

Air: incompatible
Acids: violent reaction may evolve explosive hydrogen gas
Alkali metal carbonates: potentially explosive reaction
Alkali metal hydroxides: potentially explosive reaction
Asbestos cement: explosive reaction with molten calcium
Chlorine fluorides: hypergolic reaction
Chlorine pentafluoride: hypergolic reaction
Chlorine trifluoride: hypergolic reaction
Chlorine: ignition on contact
Dinitrogen tetraoxide: potentially explosive reaction
Fluorine: ignition on contact
Halogens: ignition on contact
Lead chloride + heat: potentially explosive reaction
Mercury: violent reaction (at 390°C)
Phosphorus(V) oxide + heat: potentially explosive reaction
Silicon: violent reaction (above 1050°C)
Sodium + mixed oxides + heat: violent reaction
Sulfur + heat: potentially explosive reaction
Sulfur + vanadium(V) oxide: ignition on contact
Water: violent reaction may evolve explosive hydrogen gas

FIRE HAZARD   Flammable when heated or in contact with powerful oxidizing agents.
Use special mixtures of dry chemical extinguishers to fight a fire.

CALCIUM ACETATE

CAS #62-54-4   mf: $C_4H_6O_4 \cdot Ca$
   mw: 158.18

SYNONYMS
Brown acetate   Sorbo-calcian
Calcium diacetate   Sorbo-calcion
Gray acetate   Teltozan
Lime acetate   Vinegar salts
Lime pyrolignite

TOXIC EFFECTS   Poisonous by intravenous route

HAZARDOUS REACTIONS   When heated to decomposition, it emits acrid smoke and fumes.

CALCIUM CARBONATE

CAS #1317-65-3   mf: $CO_3 \cdot Ca$
   mw: 100.09

SYNONYMS
Agricultural limestone   Dolomite
Agstone   Limestone
Aragonite   Lithographic stone
Atomit   Marble
Bell mine pulverized limestone   Natural calcium carbonate
Calcite   Portland stone
Carbonic acid, calcium salt (1:1)   Sohnhofen stone
   Vaterite
Chalk

PROPERTIES   White, odorless powder; practically water-insoluble; soluble in dilute acids; aragonite forms orthorhombic crystals with melting point 825°C and decomposes to CaO and CO_2; calcite forms hexagonal-rhombohedral crystals with melting point 1339°C (β) at 102.5 atm; density 2.7–2.95. The major component of the minerals limestone, marble, aragonite, calcite, and vaterite.

ACGIH TLV   TWA 10 mg/m³ (total dust)

TOXIC EFFECTS   A severe eye and a moderate skin irritant

HAZARDOUS REACTIONS
Acids: incompatible

Alum: incompatible
Ammonium salts: incompatible
Fluorine: ignites on contact
Magnesium + hydrogen: incompatible

CALCIUM CHLORIDE

CAS #10043-52-4 mf: $CaCl_2$
  mw: 110.98

SYNONYMS
Calplus
Caltac
Dowflake
Liquidow
Peladow
Snomelt
Superflake anhydrous

PROPERTIES   Cubic, colorless, deliquescent crystals; melting point 772°C, boiling point >1600°C; density 2.512 at 25°C

TOXIC EFFECTS   Moderately toxic by ingestion. Poisonous by intravenous, intramuscular, intraperitoneal, and subcutaneous routes. Tumorigenic and mutagenic data.

HAZARDOUS REACTIONS
Boric oxide + calcium oxide: violent reaction
Bromine trifluoride: violent reaction
Methyl vinyl ether: catalyzes exothermic polymerization of the ether
Water: exothermic reaction
Zinc: reaction releases explosive hydrogen gas

When heated to decomposition, it emits toxic fumes of Cl^-.

CALCIUM HYDROXIDE

CAS #1305-62-0 mf: CaH_2O_2
  mw: 74.10

SYNONYMS
Bell mine
Calcium hydrate
Hydrated lime
Kemikal
Lime water
Slaked lime

PROPERTIES   Rhombic, trigonal, colorless crystals; loses H_2O at 580°C, decomposes before boiling; density 2.343

ACGIH TLV   TWA 5 mg/m³

TOXIC EFFECTS   Mildly toxic by ingestion. A severe eye irritant.

Symptoms   Dermatitis, severe eye irritation; skin, mucous membrane, and respiratory system irritant

HAZARDOUS REACTIONS
Maleic anhydride: violent reaction
Nitroethane: violent reaction
Nitromethane: violent reaction
Nitroparaffins: violent reaction
Nitropropane: violent reaction
Phosphorus: violent reaction
Polychlorinated phenols + potassium nitrate: reaction forms extremely toxic products

CALCIUM HYPOCHLORITE

CAS #7778-54-3 mf: $Cl_2O_2 \cdot Ca$
DOT #2208 mw: 142.98

NFPA LABEL

SYNONYMS
B-K powder
Bleaching powder
Calcium chlorohydrochlorite
Calcium hypochloride
Calcium oxychloride
Caporit
CCH
Chloride of lime
Chlorinated lime
HTH
Hypochlorous acid, calcium salt
Lime chloride
Lo-bax
Losantin
Perchloron
Pittchlor
Pittcide
Pittclor
Sentry

PROPERTIES   White powder; compound contains 39% or less available chlorine

DOT Classification   ORM-C; Label: None

TOXIC EFFECTS   Moderately toxic by ingestion. The solid can cause severe irritation of skin and mucous membranes. The fumes can cause pulmonary edema.

HAZARDOUS REACTIONS   A powerful oxidizer. The bulk material may ignite or explode in storage. Traces of water may initiate the reaction. A rapid exothermic decomposition above 175°C releases oxygen and chlorine. Moderately explosive in its solid form when heated. Dangerous; when heated to decomposition or on contact with acid or acid fumes, it emits highly toxic fumes of HCl and explodes.

Acetic acid + potassium cyanide: explosive reaction
Acetylene: reacts to form explosive products
Acids or acid fumes: explodes on contact
Algacide: ignites on contact
Amines: explosive reaction
Ammonium chloride: explosive reaction
Carbon or charcoal + heat: explosive reaction
Carbon tetrachloride + heat: explosive reaction
Combustible substances: deflagration on contact
N,N-dichloromethylamine + heat: explosive reaction
Diethylene glycol monomethyl ether: ignites on contact
Ethanol: explosive reaction
Glycerol: ignites on contact
Hydroxy compounds: ignites on contact
Iron oxide: explosive reaction
Isobutanethiol: explosive reaction
Methanol: explosive reaction
Nitrogenous bases: reacts to form explosive products
Nitromethane-reducing materials: vigorous reaction
Organic matter: violent reaction above 100°C
Organic sulfur compounds: ignites on contact
Phenol: ignites on contact
1-Propanethiol: explosive reaction
Rust: explosive reaction
Sodium hydrogen sulfate + starch + sodium carbonate: potentially explosive reaction
Sulfur: violent reaction
Turpentine: explosive reaction
Water or steam: releases HCl gas

CALCIUM OXIDE

CAS #1305-78-8   mf: CaO
DOT #1910   mw: 56.08

NFPA LABEL

SYNONYMS

Burnt lime
Calcia
Calx
Lime
Lime, burned
Lime, unslaked
Quicklime
Pebble lime

PROPERTIES   Cubic, colorless crystals; melting point 2580°C, boiling point 2850°C; density 3.37

OSHA PEL   TWA 5 mg/m³; IDLH 250 mg/m³

ACGIH TLV   TWA 2 mg/m³

DOT Classification   ORM-B; Label: None

TOXIC EFFECTS   A caustic and irritating material. A powerful caustic to living tissue.

Routes   Ingestion, inhalation, skin contact

Symptoms   Irritated eyes and upper respiratory tract; ulcers, perforated nasal septum; pneumonia; dermatitis, burns

Target Organs   respiratory system, skin, eyes

HAZARDOUS REACTIONS

Boric oxide + calcium chloride: violent reaction
Bromine trifluoride: violent reaction
Chlorine trifluoride: violent reaction
Ethanol: mixtures may ignite if heated and can cause an air-vapor explosion
Fluorine: violent reaction
Hydrogen fluoride liquid: incandescent reaction
Interhalogens: violent reaction
Phosphorus pentoxide + heat: violent reaction
Water: the powdered oxide may react explosively

FIRST AID   In case of contact with the eyes, immediately wash the eyes with large amounts of water, occasionally lifting the lower and upper lids. Get medical attention immediately. Contact lenses should not be worn when working with this chemical.

In case of contact with the skin, immediately flush the contaminated skin with water. If this chemical penetrates the clothing, immediately remove the clothing and flush the skin with water. Get medical attention promptly.

If a person breathes in large amounts of this chemical, move the victim to fresh air at once. If breathing has stopped, perform artificial respiration. Keep the affected person warm and at rest. Get medical attention as soon as possible.

If this chemical has been swallowed, get medical attention immediately.

HANDLING   Wear appropriate clothing and equipment to prevent a reasonable probability of skin contact. Wash promptly when skin becomes contaminated and at the end of each work shift. Promptly remove any nonimpervious clothing that becomes contaminated. Work clothing should be changed daily if it is reasonably probable that the clothing has been contaminated.

Wear eye protection to prevent any possibility of eye contact.

The following equipment should be available: an eyewash station and a quick-drench shower.

CALCIUM SULFATE DIHYDRATE

CAS #10101-41-4 mf: $CaSO_4 \cdot 2H_2O$
mw: 172.18

SYNONYMS
Alabaster
Annaline
Gypsum
Gypsum stone
Land plaster
Light spar
Magnesia white
Mineral white
Native calcium sulfate
Precipitated calcium sulfate
Satinite
Satin spar
Sulfuric acid, calcium-(2+) salt, dihydrate
Terra alba

PROPERTIES   Colorless crystals; loses $1.5(H_2O)$ at 128°C; loses $2(H_2O)$ at 163°C; density 2.32

ACGIH TLV   TWA 10 mg/m³ as a nuisance particulate as long as toxic impurities are not present (e.g., containing <1% quartz)

TOXIC EFFECTS   An experimental carcinogen. Human systemic effects by ingestion: fibrosing alveolitis (growth of fibrous tissue in the lung); unspecified respiratory system effects and unspecified effects on the nose.

HAZARDOUS REACTIONS
Aluminum: violent reaction when heated

When heated to decomposition, it emits toxic fumes of SO_x.

CAMPHOR

CAS #76-22-2 mf: $C_{10}H_{16}O$
DOT #2717 mw: 152.26

SYNONYMS
2-Bornanone
2-Camphanone
Camphor, synthetic
Camphor, natural
Formosa camphor
Gum camphor
Japan camphor
2-Keto-1,7,7-trimethylnorcamphane
Laurel camphor
Matricaria camphor
2-Oxobornane
1,7,7-trimethylbicyclo(2.2.1)-2-heptanone
1,7,7-trimethylnorcamphor

PROPERTIES   White, transparent, crystalline masses; penetrating odor; pungent, aromatic taste; melting point 180°C, boiling point 204°C; density 0.992 at 25°C/4°C; flash point 65.6°C (closed cup); lower explosive limit 0.6%, upper explosive limit 3.5%; autoignition temperature 466.1°C; vapor density 5.24

OSHA PEL   TWA 2 mg/m³; IDLH 2 ppm

ACGIH TLV   TWA 2 ppm; STEL 3 ppm

DOT Classification   Flammable solid; Label: Flammable solid

TOXIC EFFECTS   A poison by ingestion and possibly other routes. An experimental poison by inhalation, and by subcutaneous and intraperitoneal routes. A local irritant. Ingestion causes nausea, vomiting, dizziness, excitation, and convulsions. Used as a topical anti-infective and anti-itching agent.

Target Organs   Central nervous system, eye, skin, respiratory system

Routes   Ingestion, inhalation, contact

Symptoms   Irritated eyes, skin, and mucous membranes; nausea, vomiting, diarrhea; headache; dizziness; irrational behavior, epileptiform convulsions

HAZARDOUS REACTIONS
Chromic anhydride: incompatible
Chromium trioxide: explosive reaction with camphor vapor
Oxidizing materials: incompatible

FIRST AID   In case of contact with the eyes, immediately wash the eyes with large amounts of water, occasionally lifting the lower and upper lids. Get medical attention immediately. Contact lenses should not be worn when working with this chemical.

In case of contact with the skin, immediately wash the contaminated skin with soap and water. If this chemical penetrates through the clothing, immediately remove the clothing and wash the skin with soap and water. Get medical attention promptly.

If a person breathes in large amounts of this chemical, move the victim to fresh air at once. If breathing has stopped, perform artificial respiration. Keep the

affected person warm and at rest. Get medical attention as soon as possible.

If this chemical has been swallowed, get medical attention immediately.

FIRE HAZARD   Flammable when exposed to heat or flame. Vapor is explosive when exposed to heat or flame.

Use foam, carbon dioxide, or dry chemical extinguishers to fight a fire.

HANDLING   Wear appropriate clothing and equipment to prevent repeated or prolonged skin contact. Wash promptly when the skin becomes contaminated. Promptly remove nonimpervious clothing that becomes contaminated. Work clothing should be changed daily if it is reasonably probable that the clothing has been contaminated.

Wear eye protection to reduce probability of eye contact.

CARBON

CAS #7440-44-0   mf: C
   mw: 12.01

SYNONYMS
Black pearls
Carbon black
Charcoal black
Columbian carbon
Diamond
Purified charcoal

PROPERTIES   Black crystals, powder or diamond form; melting point 3652–3697°C (sublimes), boiling point approximately 4200°C; density (amorphous) 1.8–2.1, density (graphite) 2.25, density (diamond) 3.51; vapor pressure 1 mm Hg at 3586°C

TOXIC EFFECTS   Moderately toxic by intravenous route. The dust can cause irritation to the eyes and mucous membranes.

HAZARDOUS REACTIONS
Ammonium nitrate + heat: explosive reaction with carbon dust
Ammonium perchlorate: explosive reaction with carbon dust at 240°C
Bromates: explosive reaction with carbon dust
Calcium hypochlorite: explosive reaction with carbon dust
Chlorates: explosive reaction with carbon dust
Chlorine oxide: explosive reaction with carbon dust
Halogens: explosive reaction with carbon dust
Interhalogens: explosive reaction with carbon dust
Iodates: explosive reaction with carbon dust
Iodine pentoxide: explosive reaction with carbon dust
Lead nitrate: explosive reaction with carbon dust
Mercury nitrate: explosive reaction with carbon dust
Metals: incompatible
Nitric acid: explosive reaction with carbon dust
Oils + air: explosive reaction with carbon dust
Oxidants: incompatible
Oxides: explosive reaction with carbon dust
Oxosalts: explosive reaction with carbon dust
Oxygen: explosive reaction with carbon dust
Peroxides: explosive reaction with carbon dust
Potassium + air: explosive reaction with carbon dust
Sodium sulfide: explosive reaction with carbon dust
Unsaturated oils: incompatible
Zinc nitrate: explosive reaction with carbon dust

FIRE HAZARD   Combustible when exposed to heat. The dust is explosive when exposed to heat or flame.

CARBON DIOXIDE

CAS #124-38-9   mf: CO_2
DOT #1013; 1845; 2187   mw: 44.01

SYNONYMS
Carbonic acid gas
Carbonic anhydride

PROPERTIES   Colorless, odorless gas; solubility in water 0.14%; melting point, sublimes at −78.5°C (−56.6°C at 5.2 atm); vapor density 1.53

OSHA PEL   TWA 5000 ppm; IDLH 50,000 ppm

ACGIH TLV   TWA 5000 ppm; STEL 30,000 ppm

NIOSH REL   10,000 ppm; ceiling 30,000 ppm/10M

DOT Classification   Nonflammable gas; Label: Nonflammable gas; ORM-A

TOXIC EFFECTS   An asphyxiant. A teratogen. Solid carbon dioxide can cause skin burns.

Concentrations of 10% or more cause rapid unconsciousness, followed by death or severe brain damage from tissue anoxia. However, even small quantities up to 2% carbon dioxide can seriously enhance the toxicity of carbon monoxide by stimulating overbreathing.

Target Organs   Lungs, skin, cardiovascular system

Routes   Ingestion, contact

Symptoms   Headache, dizziness, restless, paresthesia; dyspnea; sweating; malaise; increased heart rate, elevated blood and pulse pressure; coma; asphyxia; convulsions; frostbite

HAZARDOUS REACTIONS
Acrylaldehyde: incompatible
Aluminum dust: ignites and then explodes when heated in CO_2
Aluminum + sodium peroxide: vigorous reaction
Aziridine: incompatible
Cesium oxide: vigorous reaction
Chromium dust: ignites and then explodes when heated in CO_2
KHC: vigorous reaction
Lithium: vigorous reaction
Magnesium + sodium peroxide: vigorous reaction
Magnesium dust: ignites and then explodes in CO_2 atmospheres
Magnesium-aluminum alloy dust: ignites and then explodes in CO_2 atmospheres
Manganese dust: ignites and then explodes when heated in CO_2
Metals: several bulk metals will burn in CO_2 once ignited
Metals (chemically active): incompatible
Metal acetylides: incompatible
$Mg(C_2H_5)_2$: vigorous reaction
Na_2C_2: vigorous reaction
Potassium: vigorous reaction
Sodium: vigorous reaction
Sodium peroxide: incompatible
Sodium-potassium alloys: vigorous reaction
Titanium: vigorous reaction
Titanium dust: ignites and then explodes in CO_2 atmospheres
Zirconium dust: ignites and then explodes in CO_2 atmospheres

CO_2 fire extinguishers can produce highly incendiary sparks of 5–15 mJ at 10–20 KV by electrostatic discharge.

FIRST AID   For dry ice accidents: In the case of contact with the skin or eyes, get medical attention for frostbite.

If a person breathes in large amounts of this chemical, move the victim to fresh air at once. If breathing has stopped, perform artificial respiration. Keep the affected person warm and at rest. Get medical attention as soon as possible.

If this chemical has been swallowed, get medical attention immediately.

HANDLING   For dry ice work: Wear appropriate clothing and equipment to prevent skin freezing.

CARBON DISULFIDE

CAS #75-15-0   mf: CS_2
DOT #1131   mw: 76.13

STRUCTURE   NFPA LABEL
S=C=S

SYNONYMS
Carbon bisulfide
Carbon sulfide
Dithiocarbonic anhydride
Sulphocarbonic anhydride
Weevitox

PROPERTIES   Colorless to faintly yellow liquid; nearly odorless when pure but may have a disagreeable or sweetish odor when impure; soluble at 0.22 g/100 mL in water at 22°C; miscible with alcohol, ether, and benzene; melting point −110.8°C, boiling point 46.5°C; density 1.261 at 20°C/20°C; flash point −30°C (closed cup); lower explosive limit 1.3%, upper explosive limit 50%; autoignition temperature 125°C; vapor pressure 400 mm Hg at 28°C; vapor density 2.64. Conversions: 1 ppm = 3.11 mg/m³

OSHA PEL   TWA 20 ppm; ceiling 30 ppm; peak 100 ppm/30 min/8 h

ACGIH TLV   TWA 10 ppm (skin)

NIOSH REL   TWA 1 ppm; ceiling 10 ppm/15 min

DOT Classification   Label: Flammable liquid

IMO Classification   Label: Flammable liquid, poison

TOXIC EFFECTS   A poison by ingestion and possibly other routes. Mildly toxic by inhalation. Skin absorption can significantly contribute to toxic effects. A teratogen.

Carbon disulfide is rapidly absorbed when inhaled, and inhalation can produce acute poisoning. Poisoning can also occur from ingestion; death has been known to occur after ingestion of as little as 15 mL.

Acute poisoning by ingestion or inhalation can produce narcosis, accompanied by delirium. This may progress to areflexia, paralysis, coma, and death. It is reported that exposure in excess of 500 ppm is required before acute effects will be noted.

Chronic exposure has resulted in neuropsychiatric manifestations with mental disorders, including psychoses, weakness, paralysis, parkinsonism, and blindness or in polyneuritis with pain along nerves, loss of strength, and parethesias.

In milder exposures, the reported effects have been attributed to cerebral vascular damage with symptoms related to central nervous system damage involving pyramidal, extrapyramidal, and pseudobulbar tracts. There are also reports of hypertension, renal damage, elevated cholesterol, and early arteriosclerosis. Some recovery from these effects is the rule, but such recovery is slow, occurring over months or years, and some paralysis may persist. There have also been reported effects on the reproductive system in both sexes, with women having menstrual disorders, abortions, and infertility, and males having spermatic disorders.

Target Organs   Central nervous system, peripheral nervous system, cardiovascular system, eyes, kidneys, liver, skin

Routes   Ingestion, absorption, inhalation, contact

Symptoms   Dizziness, headache, sleeplessness, fatigue, nervousness; anorexia, weight loss; psychosis; polyneuropathy; Parkinson-like symptoms; ocular changes; cardiovascular system changes; gastrointestinal changes; burns, dermatitis; spermatogenesis effects; death

HAZARDOUS REACTIONS
Air: incompatible
Aluminum: violent reaction
Aluminum powder: ignites in CS_2 vapor
Amines (organic): incompatible
Azides: violent reaction
Cesium azide: violent reaction
Chlorine (catalyzed by iron): potentially explosive reaction
Chlorine monoxide: violent reaction
Dinitrogen tetraoxide: mixtures are heat-, spark-, and shock-sensitive explosives
Ethylamine diamine: violent reaction
Ethylene diamine: violent reaction
Ethylene imine: violent reaction
Fluorine: CS_2 vapor ignites on contact
Iron: ignition and potentially explosive reaction when heated
Lead azide: violent reaction
Lithium azide: violent reaction
Metal azides: reacts to form shock- and heat-sensitive, explosive metal azidodithioformates
Metals (chemically active): incompatible
Nitric oxide: violent reaction
Nitrogen oxide: potentially explosive reaction
Nitrogen tetroxide: violent reaction
Oxidizing agents: violent reaction
Permanganic acid: explodes on contact
Phenylcopper-triphenylphosphine complexes: violent reaction
Potassium: violent reaction
Potassium azide: violent reaction
Potassium-sodium alloys: mixtures are powerful, shock-sensitive explosives
Rubidium azide: violent reaction
Rust: ignition and potentially explosive reaction when heated
Sodium: mixtures are powerful, shock-sensitive explosives
Sodium azide: violent reaction
Sulfuric acid + permanganates: violent reaction
Zinc: violent reaction

When carbon disulfide is used to desorb organic materials from activated charcoal, as in the case of air sample analysis, a significant amount of heat can be liberated.

When heated to decomposition, it emits toxic fumes of SO_x.

FIRST AID   In case of contact with the eyes, immediately wash the eyes with large amounts of water, occasionally lifting the lower and upper lids. Get medical attention immediately. Contact lenses should not be worn when working with this chemical.

In case of contact with the skin, immediately wash the contaminated skin with soap and water. If this chemical penetrates the clothing, immediately remove the clothing and wash the skin with soap and water. Get medical attention promptly.

If a person breathes in large amounts of this chemical, move the victim to fresh air at once. If breathing has stopped, perform artificial respiration. Keep the affected person warm and at rest. Get medical attention as soon as possible.

If this chemical has been swallowed, get medical attention immediately.

FIRE HAZARD   A dangerous fire hazard when exposed to heat, flame, sparks, friction, or oxidizing materials. Severe explosion hazard when exposed to heat or flame.

Use water, CO_2, dry chemical, fog, or mist extinguishers to fight a fire.

Carbon disulfide is a very flammable substance; a steam pipe or even a hot radiator can ignite its vapors.

It also has a very wide range of explosive concentrations and thus should not be exposed to heat, flame, sparks, or friction.

HANDLING Gloves and protective apparel should be used when handling liquid carbon disulfide. As much as possible, laboratory operations should be confined to a hood that has protection factors high enough to prevent significant exposure or to closed systems.

Wear appropriate clothing and equipment to prevent a reasonable probability of skin contact. Wash promptly when the skin becomes contaminated. Immediately remove any clothing that becomes wet to avoid flammability hazard.

Wear eye protection to reduce probability of eye contact.

CARBON MONOXIDE

CAS #630-08-0 mf: CO
DOT #1016; 9202 mw: 28.01

SYNONYMS
Carbonic oxide
Carbon monoxide, cryogenic liquid
Carbon oxide (CO)
Exhaust gas
Flue gas

PROPERTIES Colorless, odorless gas; melting point −207°C, boiling point −191.3°C; density (gas) 1.250 g/L at 0°C, (liquid) 0.793; lower explosive limit 12.5%, upper explosive limit 74.2%; autoignition temperature 608.9°C

OSHA PEL TWA 50 ppm; IDLH 500 ppm

ACGIH TLV TWA 50 ppm; STEL 400 ppm

NIOSH REL TWA 35 ppm; ceiling 200 ppm

DOT Classification Flammable gas; Label: Flammable gas and poison gas

TOXIC EFFECTS Mildly toxic by inhalation. A teratogen. A direct and cumulative poison. It combines with the hemoglobin of the blood to form a relatively stable compound, carboxyhemoglobin, and renders it useless as an oxygen carrier. When about one-third of the hemoglobin has entered into such combination, the victim dies. The gas is a treacherous poison because of its odorless character and insidious action. Exposure to 1500–2000 ppm of carbon monoxide in air for 1 h is dangerous, and exposure to 4000 ppm is fatal in less than 1 h. After removal from exposure, the half-life of its elimination from the blood is 1 h. Headache and dizziness are the usual symptoms of carbon monoxide poisoning, but occasionally the first evidence of poisoning is the collapse of the patient. Where the poisoning has been relatively long and severe, cerebral congestion and edema may occur, resulting in long-lasting mental or nervous damage.

PHYSICAL EFFECTS OF CO

ppm	Effect
50	Level allowable for several hours
400–500	No appreciable effect for times less than 1 h
600–700	First appreciable effect after 1-h exposure
1000–1200	Unpleasant but not dangerous symptoms after 1 h
1500–2000	Dangerous level for 1-h exposure
4000	Fatal in less than 1 h

Target Organs Cardiovascular system, lungs, blood, central nervous system

Routes Ingestion, contact (liquid)

Symptoms Headache, tachypnea, nausea, weakness, dizziness, confusion, hallucinations, cyanosis, depressant/depression, angina, syncope, frostbite

HAZARDOUS REACTIONS
Bromine pentafluoride: violent or explosive reaction on contact
Bromine trifluoride: violent or explosive reaction on contact
Cesium oxide + water: ignites on contact
Chlorine dioxide: violent or explosive reaction on contact
Chlorine trifluoride: exothermic reaction
Copper powder + copper(II) perchlorate + water: mixture forms an explosive complex
Dinitrogen oxide (liquid): mixture with liquid CO is a rocket propellant
Iodine heptafluoride: ignites on warming
Iron(III) oxide: potentially explosive reaction between 0 and 150°C
Lithium + water: exothermic reaction
Nitrogen trifluoride: exothermic reaction
Oxidizers (strong): incompatible
Oxygen (liquid): mixture with liquid CO is explosive
Oxygen difluoride: exothermic reaction
Peroxodisulfuryl difluoride: violent or explosive reaction on contact
Potassium: reacts to form explosive products sensitive to shock, heat, or contact with water
Potassium + oxygen: exothermic reaction
Silver oxide: exothermic reaction

Sodium: reacts to form explosive products sensitive to shock, heat, or contact with water

Sodium + ammonia: exothermic reaction

FIRST AID If a person breathes in large amounts of this chemical, move the victim to fresh air at once. If breathing has stopped, perform artificial respiration. Keep the affected person warm and at rest. Get medical attention as soon as possible.

FIRE HAZARD A dangerous fire hazard when exposed to flame. Severe explosion hazard when exposed to heat or flame.

To extinguish fires, stop flow of gas.

HANDLING Carbon monoxide should be used only in areas that have adequate ventilation at all times. A trap or vacuum break should always be used to prevent impurities from being sucked back into the cylinder.

The main danger lies in the difficulty of detection, as CO has no smell. Detection plaques are available, and anyone who has reason to believe there is a possible CO hazard in his or her work should obtain a supply of these plaques.

Gas mixtures containing CO and water vapor should not be stored in steel cylinders due to the possibility of stress corrosion.

To handle liquid: Wear appropriate clothing and equipment to prevent skin freezing. Immediately remove any clothing that becomes wet to avoid flammability hazard.

Wear eye protection to reduce probability of eye contact.

CARBON TETRACHLORIDE

CAS #56-23-5   mf: CCl_4
DOT #1846   mw: 153.81

NFPA LABEL

SYNONYMS

Benzinoform
Carbona
Carbon chloride
Carbon tet
Fasciolin
Flukoids
Halon 104
Methane tetrachloride
Necatornia
Perchloromethane
Tetrachlorocarbon
Tetrachloromethane
Tetrafinol
Tetraform
Tetrasol
Univerm
Vermoestricid

PROPERTIES Colorless liquid with a heavy, ethereal odor; solubility 0.080% in water at 25°C; miscible with most organic solvents; melting point −22.6°C, boiling point 76.8°C; density 1.597 at 20°C; flash point, none; vapor pressure 100 mm Hg at 23.0°C

OSHA PEL TWA 10 ppm; ceiling 25 ppm; peak 200 ppm/5M/4H

ACGIH TLV TWA 5 ppm; suspected carcinogen

NIOSH REL Ceiling 2 ppm/60M

DOT Classification Label: Poison

TOXIC EFFECTS A poison by ingestion and possibly other routes. Mildly toxic by inhalation. A carcinogen and teratogen. A narcotic. Contact dermatitis can result from skin contact. An eye and skin irritant. Damages liver, kidneys, and lungs. Individual susceptibility varies widely; some persons appear to be unaffected by exposures that seriously poison their fellow workers. Ingestion of alcohol has been implicated repeatedly as predisposing the worker to increased effects of liver damage from carbon tetrachloride exposure. Previous liver and kidney damage also seem to render the individual more susceptible.

Although inhalation of carbon tetrachloride can cause depression of the central nervous system, with dizziness, headaches, depression, mental confusion, and even unconsciousness, such effects probably are the result of exposure at concentrations of 100–500 ppm, and serious poisonings rarely occur. Ingestion of as little as 4 mL of carbon tetrachloride has been reported to be fatal. Many deaths have occurred from accidental ingestion; the early initial symptom is central nervous system depression, which usually clears the second day. Then, if the dose has been large enough, the victim may become jaundiced and, if the dose is sufficiently large, he or she may die in a few days. If the dose is smaller, the liver effects partially abate, only to have the victim go into renal failure with anuria, oliguria, uremia, proteinuria, and possibly death. Acute inhalation has produced almost the identical picture. In addition, occasionally, very brief inhalations of carbon tetrachloride have been reported to produce sudden death thought to be due to ventricular fibrillation.

Chronic inhalation of carbon tetrachloride at concentrations of 10–100 ppm has resulted in liver damage, which can be detected by abnormal liver function tests, and liver biopsies have disclosed centrilobular necrosis. Chronic exposure in animals has resulted in the appearance of cirrhosis and of hepatomas. There is little information regarding effects of chronic exposure in humans other than the laboratory abnormalities as described. A few cases of cirrhosis and hepatic cancer have been reported, but a causal relationship to carbon tetrachloride is difficult to confirm or deny. Prolonged exposure of the skin to the solvent can result in extreme dryness and fissuring, with redness and some secondary infection.

ACGIH states that skin contact may account for a substantial part of toxic response.

Target Organs   Central nervous system, eyes, lungs, liver, kidneys, skin

Routes   Ingestion, absorption, inhalation, contact

Symptoms   Pupillary constriction, central nervous system depressant/depression; tremors, somnolence, anorexia, coma; nausea, vomiting; liver, kidney damage; skin irritation; carcinogen

HAZARDOUS REACTIONS

$Al(C_2H_5)_3$: vigorous exothermic reaction
Allyl alcohol: vigorous exothermic reaction
Aluminum: mixture is an impact-sensitive explosive when ball milled or heated to 152°C in a closed container
Aluminum + methanol: exothermic reaction with 9:1 methanol and CCl_4
Aluminum trichloride: incompatible
Barium: mixture is an impact-sensitive explosive; bulk metal also reacts violently
Benzoyl peroxide + ethylene: vigorous exothermic reaction
Beryllium: mixture is an impact-sensitive explosive
Boranes: potentially explosive reaction on contact
Bromine trifluoride: vigorous exothermic reaction
Calcium disilicide: forms explosive mixtures (friction and pressure sensitive)
Calcium hypochlorite: forms explosive mixtures (heat sensitive)
Chlorine trifluoride: forms explosive mixtures
Decarborane(14): forms explosive mixtures (impact sensitive)
Dibenzoyl peroxide: incompatible
Diborane: vigorous exothermic reaction
Dimethyl formamide: potentially dangerous reaction when iron is present as a catalyst
Dimethylacetamide: potentially dangerous reaction when iron is present as a catalyst
Dinitrogen tetroxide: forms explosive mixture
Disilane: vigorous exothermic reaction
Ethylene: forms explosive mixture at 25–105°C and 30–80 bar
Fluorine: violent or explosive reaction on contact
1,2,3,4,5,6-Hexachlorocyclohexane: potentially dangerous reaction when iron is present as a catalyst
Lithium: mixture is an impact-sensitive explosive
Magnesium: exothermic reaction with 9:1 mixtures of methanol and CCl_4
Magnesium: incompatible
Metal (particulates): mixtures are impact-sensitive explosives
Metals (chemically active): incompatible
Oxygen (liquid): vigorous exothermic reaction
Plutonium: vigorous exothermic reaction
Potassium: mixture is an impact-sensitive explosive (200 times more shock-sensitive than mercury fulminate)
Potassium-sodium alloy: mixture is an impact-sensitive explosive (more sensitive than potassium)
Potassium-*tert*-butoxide: vigorous exothermic reaction
Silver perchlorate + hydrochloric acid: vigorous exothermic reaction
Sodium: mixture is an impact-sensitive explosive
Tetraethylaluminum: hazardous reaction
Tetraethylenepentamine: vigorous exothermic reaction
Tetrasilane: vigorous exothermic reaction
Triethyldialuminum trichloride: forms explosive mixtures (heat-sensitive)
Trisilane: vigorous exothermic reaction
Uranium: hazardous reaction
Wax (burning): hazardous reaction
Zinc: mixture is an impact-sensitive explosive, which burns readily
Zinc + methanol: exothermic reaction with 9:1 methanol and CCl_4
zirconium: vigorous exothermic reaction

CCl_4 has caused explosions when used as a fire extinguisher on wax fires and uranium fires.

When heated to decomposition it emits toxic fumes of Cl^- and phosgene.

FIRST AID   In case of contact with the eyes, immediately wash the eyes with large amounts of water, occasionally lifting the lower and upper lids. Get medical attention immediately. Contact lenses should not be worn when working with this chemical.

In case of contact with the skin, immediately wash the contaminated skin with soap and water. If this chemical penetrates the clothing, immediately remove the clothing and wash the skin with soap and water. Get medical attention promptly.

If a person breathes in large amounts of this chemical, move the victim to fresh air at once. If breathing has stopped, perform artificial respiration. Keep the affected person warm and at rest. Get medical attention as soon as possible.

If this chemical has been swallowed, get medical attention immediately.

FIRE HAZARD   Although carbon tetrachloride is nonflammable, on exposure to heat or flame it may decompose with the formation of phosgene.

HANDLING   Because the carcinogenic potency of carbon tetrachloride is low, Procedure B (see Chap. 1) provides adequate protection for laboratory operations in which it is used. All operations should be carried out in a hood, not only because of the carcinogenicity of the substance, but also because of its other toxic effects and its volatility. Nitrile rubber is the recommended material for gloves and other protective clothing.

Wear appropriate clothing and equipment to prevent repeated or prolonged skin contact. Personnel should wash promptly when skin becomes wet. Promptly remove nonimpervious clothing that becomes contaminated.

Wear eye protection to reduce probability of eye contact.

CATECHOL

CAS #120-809   mf: $C_6H_6O_2$
mw: 110.12

NFPA LABEL

SYNONYMS

o-Benzenediol	Fouramine PCH
1,2,-Benzenediol	Fourrine 68
Catechin	o-Hydroquinone
o-Dihydroxybenzene	o-Hydroxyphenol
1,2-Dihydroxybenzene	2-Hydroxyphenol
o-Dioxybenzene	Oxyphenic acid
o-Diphenol	Pelagol gray C
Durafur developer C	o-Phenylenediol
Pyrocatechin	Pyrocatechuic acid
Pyrocatechinic acid	

PROPERTIES   Colorless crystals; soluble in water, chloroform, and benzene; very soluble in alcohol and ether; melting point 105°C, boiling point 246°C; density 1.341 at 15°C; flash point 127.2°C (closed cup); vapor pressure 10 mm Hg at 118.3°C; vapor density 3.79

ACGIH TLV   TWA 5 ppm

EPA Extremely Hazardous Substances List

TOXIC EFFECTS   Poisonous by ingestion and many other routes. Moderately toxic by skin contact. Can cause dermatitis on skin contact. An allergen. Symptoms of exposure are similar to those of phenol.

HAZARDOUS REACTIONS
Nitric acid (conc.): hypergolic reaction
Oxidizing materials: vigorous reaction

When heated to decomposition, it emits acrid smoke and irritating fumes.

FIRE HAZARD   Combustible when exposed to heat or flame.

Use water, CO_2, or dry chemical extinguishers to fight a fire.

CHLORAL HYDRATE

CAS #302-17-0   mf: $C_2HCl_3O \cdot H_2O$
mw: 165.40

SYNONYMS
Aquachloral
Chloraldurat
Dormal
Felsules
Hydral
Kessodrate
Lorinal
Noctec
Nortec
Nycoton
Phaldrone
Rectules
SK-choral hydrate
Somni sed
Sontec
Tosyl
Trawotox

Trichloroacetaldehyde hydrate
Trichloroacetaldehyde monohydrate
2,2,2-Trichloro-1,1-ethanediol

PROPERTIES   Transparent, colorless crystals; aromatic, penetrating, slightly acrid odor and slightly bitter, caustic taste; melting point 52°C, boiling point 97.5°C; density 1.9

TOXIC EFFECTS   A poison by ingestion and some other routes. Poisonous experimentally by ingestion, intravenous, and rectal routes. Symptoms of ingestion: general anesthesia, cardiac arrythmias, blood pressure depression. Tumorigenic and mutagenic data. Used as a sedative, anesthetic, and narcotic.

HAZARDOUS REACTIONS   When heated to decomposition, it emits toxic fumes of Cl$^-$.

FIRE HAZARD   Combustible when exposed to heat or flame.

CHLORINE

CAS #7782-50-5   mf: Cl_2
DOT #1017   mw: 70.90

NFPA LABEL

SYNONYMS
Bertholite
Chlorine mol.
Molecular chlorine

PROPERTIES   Greenish-yellow gas, liquid, or rhombic crystals; penetrating and irritating odor, threshold = 0.3 ppm; solubility 1.46% in water at 25°C; melting point −101°C, boiling point −34.5°C; density (liquid) 1.47 at 0°C (3.65 atm); vapor pressure 4800 mm Hg at 20°C; vapor density 2.49

OSHA PEL   Ceiling 1 ppm

ACGIH TLV   TWA 0.5 ppm; STEL 1 ppm

NIOSH REL   Ceiling 0.5 ppm/15M

EPA Extremely Hazardous Substances List and Community Right to Know List

DOT Classification   Nonflammable gas; Label: Nonflammable gas and poison; Poison A; Label: Poison gas

TOXIC EFFECTS   Moderately toxic by inhalation.
Reacts with moisture to form oxygen and HCl which can cause inflammation of body tissues.
Humans can generally detect the odor of chlorine at about 0.3 ppm. Irritation of the throat and nose is first noticed at about 2.6 ppm and becomes quite strong at 41 ppm. An exposure of a large number of humans to 5–7 ppm for 1 h did not result in serious or long-term consequences. Exposure to about 17 ppm causes coughing, and levels as low as 10 ppm may cause lung edema. Human exposure to 14–21 ppm for 30 min to 1 h is regarded as dangerous and may, after a delay of 6 h or more, result in death from anoxia due to serious pulmonary edema. Concentrations of 50 ppm are dangerous even with short exposure; 1000 ppm may be fatal, even when exposure is brief. For rats, the LC_{50} (1 h) = 293 ppm.
The subjective response to chlorine is less pronounced with prolonged exposure.
Chronic effects on humans from long-term, low-level exposure have not been well documented. Animal exposures have indicated that prolonged exposure to approximately 1.7 ppm for 1 h per day may cause deterioration in the nutritional state, blood alteration, and decreased resistance to disease.

Target Organs   Respiratory system

Routes   Ingestion, contact

Symptoms   Burning of eyes, nose, mouth; lacrimation; rhinorrhea; cough; choking; nausea, vomiting; substernal pain; headache; dizziness; syncope; emphysema, pulmonary edema, pneumonia; hypoxemia; dermatitis

HAZARDOUS REACTIONS
N-aryl sulfinamides: violent reaction
Acetaldehyde: contact may cause fires or explosions
Acetylene + heat or UV light: explodes on contact
Acetylides (metal): ignites
Air: explosive mixture
Air + ethylene: explodes on contact
Alcohols: contact may cause fires or explosions
Alkylisothiourea salts: contact may cause fires or explosions
Alkylthiouronium salts: reacts to form explosive products
Aluminum: ignition or explosive reaction
Aluminum (molten): explodes on contact

132 CHLORINE

Amidosulfuric acid: explodes on contact
Ammonia: explodes on contact
Ammonium chloride solutions (acid): reacts to form explosive products
Antimony powder: ignition or explosive reaction
Antimony trichloride + tetramethyl silane: explodes on contact (at 100°C)
Arsenic: contact may cause fire or explosion
Arsenic disulfide: contact may cause fire or explosion
Arsine: contact may cause fire or explosion
$As_2(CH_3)_4$: contact may cause fire or explosion
Aziridine: reacts to form explosive products
Ba_3P_2: contact may cause fire or explosion
Benzene: contact may cause fire or explosion
Bismuth: contact may cause fire or explosion
Biuret: explodes on contact
Boron: contact may cause fire or explosion
Boron trisulfide: contact may cause fire or explosion
BPI_2: contact may cause fire or explosion
Brass: contact may cause fire or explosion
Bromine pentafluoride: contact may cause fire or explosion
tert-Butanol: explodes on contact
Butyl rubber + naphtha: explodes on contact
$C_2H_5PH_2$: contact may cause fire or explosion
Calcium powder: ignition or explosive reaction
Ca_3P_2: contact may cause fire or explosion
$Ca(ClO_2)_2$: contact may cause fire or explosion
Calcium carbide + potassium hydroxide: contact may cause fire or explosion
Calcium nitride: contact may cause fire or explosion
Carbides (metal): ignites
Carbon: contact may cause fire or explosion
Carbon disulfide: contact may cause fire or explosion
Cesium: contact may cause fire or explosion
Cesium acetylide: contact may cause fire or explosion
Cesium oxide: incandescent reaction when warmed above 150°C
Chlorinated pyridine + iron powder: explodes on contact
3-Chloropropyne: explodes on contact
Co_2O: contact may cause fire or explosion
Cobalt(II) chloride + methanol: explodes on contact
Combustible substances: incompatible
Copper: ignition or explosive reaction
Copper + oxygen difluoride: contact may cause fire or explosion
Copper hydride: ignites
Copper(II) phosphide: ignites
Cyanuric acid: reacts to form explosive products
Diborane: explodes on contact
Dibutyl phthalate: explodes at 118°C
Dichloro(methyl)arsine: explodes in a sealed container

Diethyl ether: explodes on contact
Diethyl zinc: ignition on contact
Dimethyl formamide: violent reaction
Dimethyl phosphoramidate: explodes on contact
bis(2,4-Dinitrophenyl)disulfide: reacts to form explosive products
Dioxygen difluoride: explodes on contact
Disilyl oxide: explodes on contact
4,4'-Dithiodimorpholine: explodes on contact
Ethane + activated carbon: explodes at 350°C
Ethylene: contact may cause fire or explosion
Ethylene imine: contact may cause fire or explosion
Fluorine: contact may cause fire or explosion
Gasoline: explodes on contact
Germanium: ignition or explosive reaction
Glycerol: explodes above 70°C in a sealed container
H_2O + Mg_2P_3: contact may cause fire or explosion
H_2SiO: contact may cause fire or explosion
Hexachlorodisilane: explodes above 300°C
Hg_3P_2: contact may cause fire or explosion
Hydrazine: contact may cause fire or explosion
Hydrides: ignites
Hydrocarbon oils or waxes: explodes on contact
Hydrocarbons + Lewis acids: potentially dangerous reaction releases HCl gas
Hydrochloric acid + dinitroanilines: violent reaction
Hydrogen: explosive mixture
Hydrogen chloride: explosive mixture
Hydroxylamine: ignites
Illuminating gas: contact may cause fire or explosion
Iron: ignition or explosive reaction
Iron(III) chloride + monomers (e.g., styrene): explodes on contact
Lead oxide + ethylene: mixtures are explosive (at 100°C)
Manganese: ignition or explosive reaction
Mercuric oxide: contact may cause fire or explosion
Mercuric sulfide: ignites
Mercuric oxide (yellow) + hydrogen: mixture explodes
Mercury oxide + ethylene: mixtures are explosive
Mercury: contact may cause fire or explosion
Metals: ignition or explosive reaction, reaction with some metals requires moist Cl_2 or heat
Methane: contact may cause fire or explosion
Methane + mercury oxide: explodes on contact
Methane + oxygen: ignites
Methanol: explodes on contact
Methanol + tetrapyridine cobalt(II) chloride: explodes on contact
Mn_3P_2: contact may cause fire or explosion
Naphtha + sodium hydroxide: explodes on contact
Niobium: contact may cause fire or explosion
Nitrogen trichloride + hydrogen: mixture explodes
Nitrogen triiodide: explodes on contact

Nonmetals: ignites
Oxygen: explosive mixture
Oxygen difluoride: explodes on contact
P_2O_3: contact may cause fire or explosion
$P(SNC)_3$: contact may cause fire or explosion
PCBs: contact may cause fire or explosion
Phenyl magnesium bromide: reacts to form explosive products
Phosphides (metal): ignites
Phosphine: ignites
Phosphines (alkyl): contact may cause fire or explosion
Phosphorus: contact may cause fire or explosion
Phosphorus (white) + liquid chlorine: explodes on contact
Phosphorus compounds: explodes on contact
Polychlorobiphenyl: violent reaction
Polydimethyl siloxane: explodes above 88°C in a sealed container
Polyisobutylene: ignition at 130°C
Polymethyl trifluoropropylsiloxane: explodes above 68°C in a sealed container
Polypropylene: contact may cause fire or explosion
Polypropylene + zinc oxide: explodes on contact
Potassium: contact may cause fire or explosion
Potassium acetylide: contact may cause fire or explosion
Potassium hydride: ignites
Propane: explodes at 300°C
Rubber: contact may cause fire or explosion
Rubidium: contact may cause fire or explosion
Rubidium acetylide: contact may cause fire or explosion
SbH_3: contact may cause fire or explosion
SiH_2: contact may cause fire or explosion
Silane: ignites
Silicon: ignites
Silicones: mixture explodes when heated in a sealed container
Silver oxide: contact may cause fire or explosion
Silver oxide + ethylene: mixtures are explosive
SnF_2: contact may cause fire or explosion
Sodium: contact may cause fire or explosion
Sodium acetylide: contact may cause fire or explosion
Sodium carbide: contact may cause fire or explosion
Sodium hydride: ignites
Sodium hydroxide: violent reaction
Sr_3P: contact may cause fire or explosion
Steel: ignition above 200°C or as low as 50°C when impurities are present
Stibine: explodes on contact
Sulfamic acid: contact may cause fire or explosion
Sulfides: ignites
Synthetic rubber (in liquid Cl_2): explodes on contact
Tellurium: contact may cause fire or explosion
Tetraselenium tetranitride: explodes on contact
Thorium: contact may cause fire or explosion
Tin: ignition or explosive reaction
Trialkyl boranes: ignites
Trimethyl thionophosphate: explodes on contact
Tungsten dioxide: incandescent reaction when warmed
Turpentine: contact may cause fire or explosion
Uranium: contact may cause fire or explosion
Uranium dicarbide: contact may cause fire or explosion
Vanadium: contact may cause fire or explosion
Vanadium powder: ignition or explosive reaction
WO_2: contact may cause fire or explosion
Zinc: contact may cause fire or explosion
$Zn(C_2H_5)_2$: contact may cause fire or explosion
Zirconium dicarbide: contact may cause fire or explosion

FIRST AID   In case of contact with the eyes, immediately wash the eyes with large amounts of water, occasionally lifting the lower and upper lids. Get medical attention immediately. Contact lenses should not be worn when working with this chemical.

In case of contact with the skin, immediately flush the contaminated skin with water. If this chemical penetrates the clothing, immediately remove the clothing and flush the skin with water. Get medical attention promptly.

If a person breathes in large amounts of this chemical, move the victim to fresh air at once. If breathing has stopped, perform artificial respiration. Keep the affected person warm and at rest. Get medical attention as soon as possible.

HANDLING   Chlorine should be kept away from easily oxidized materials. Chlorine reacts readily with many organic chemicals, sometimes with explosive violence. Because of the high toxicity of chlorine, laboratory operations using it should be carried out in a hood (see *Fundamentals of Laboratory Safety: Physical Hazards in the Academic Laboratory,* Chap. 9) and appropriate gloves (see *Fundamentals of Laboratory Safety: Physical Hazards,* Chap. 10).

Respiratory protective equipment of a type approved by MSHA and NIOSH for chlorine service should always be readily available in places where this substance is used and so located as to be easily reached in case of need. Proper protection should be afforded the eyes by use of goggles or large-lens spectacles to eliminate the possibility of liquid chlorine coming in contact with the eyes and causing injury. Chlorine should be used only by experienced or properly instructed persons.

Only auxiliary valves and gauges designed solely for chlorine use should be used. Stainless steel equipment should not be used.

Every precaution should be taken to avoid drawing liquids back into chlorine containers as their contents are being exhausted. To avoid this possibility, the valve should be closed immediately after the container has been emptied.

Chlorine leaks may be detected by passing a rag dampened with aqueous ammonia over the suspected valve or fitting. White fumes of hydrogen chloride indicate escaping chlorine gas.

Leaks from cylinders of chlorine in a confined space can easily lead to a concentration that could be fatal. A complete mask with air supply should be provided in rooms where chlorine cylinders are used.

If the valve cannot be closed, bubble the gas through sodium sulfite and excess sodium bicarbonate solution, including a trap in line to prevent suck-back.

Wear appropriate clothing and equipment to prevent any possibility of skin contact. Wash immediately when skin becomes contaminated. Immediately remove any nonimpervious clothing that becomes contaminated.

Wear eye protection to prevent any possibility of eye contact.

The following equipment should be available: an eyewash station and a quick-drench shower.

CHLORINE DIOXIDE

CAS #10049-04-4   mf: ClO_2
DOT #9191   mw: 67.45

SYNONYMS
Alcide
Anthium dioxide
Chlorine oxide
Chlorine peroxide
Chlorine(IV) oxide
Chloroperoxyl
Chloryl radical

PROPERTIES   Yellow-green to orange gas or liquid with a pungent, sharp odor; solubility in water 0.8%; melting point −58.8°C, boiling point 11.1°C; lower explosive limit 10%; vapor pressure >1 atm

OSHA PEL   TWA 0.1 ppm (0.3 mg/m^3); IDLH 10 ppm

ACGIH TLV   TWA 0.1 ppm; STEL 0.3 ppm

DOT Classification   Forbidden (not hydrated); Oxidizer; Label: Oxidizer and poison (hydrated, frozen)

TOXIC EFFECTS   Moderately toxic by inhalation. An eye irritant.

Routes   Inhalation, ingestion, contact

Symptoms   Irritated eyes, nose, and throat; cough, wheeze, bronchitis, pulmonary edema, chronic bronchitis

Target Organs   Respiratory system, eyes

HAZARDOUS REACTIONS   A powerful oxidizer. A powerful explosive, sensitive to spark, impact, strong light, or rapid heating. Concentrations of 10% or higher in air are explosive.

Butadiene: explodes on mixing
Carbon monoxide: explodes on mixing
Combustible substances: incompatible
Difluoramine: explodes on mixing
Dust: incompatible
Ethane: explodes on mixing
Ethylene: explodes on mixing
Fluoramines: explodes on mixing
Fluorine: violent reaction
Hydrocarbons: explodes on mixing
Hydrogen: mixture is an explosive sensitive to sparks
Mercury: explodes on contact
Methane: explodes on mixing
NHF_2: violent reaction
Nonmetals: ignites or explodes on contact
Organic matter: incompatible
Phosphorus: ignites or explodes on contact
Phosphorus pentachloride + chlorine: explodes on contact
Platinum + hydrogen: explodes
Potassium hydroxide: explodes on contact
Propane: explodes on mixing
Sugar: ignites or explodes on contact
Sulfur: ignites or explodes on contact
Trifluoramine: explodes on mixing
Water or steam: reacts to produce hydrogen chloride gas

FIRST AID   In case of contact with the eyes, immediately wash the eyes with large amounts of water, occasionally lifting the lower and upper lids. Get medical attention immediately. Contact lenses should not be worn when working with this chemical.

In case of contact with the skin, immediately wash the contaminated skin with soap and water. If this chemical penetrates the clothing, immediately remove the clothing and wash the skin with soap and water. Get medical attention promptly.

If a person breathes in large amounts of this chemical, move the victim to fresh air at once. If breathing has stopped, perform artificial respiration. Keep the affected person warm and at rest. Get medical attention as soon as possible.

If this chemical has been swallowed, get medical attention immediately.

HANDLING   Wear appropriate clothing and equipment to prevent any possibility of skin contact. Wash immediately when skin becomes contaminated. Immediately remove any clothing that becomes contaminated.

Wear eye protection to prevent any possibility of eye contact.

The following equipment should be available: an eyewash station and a quick-drench shower.

CHLORINE TRIFLUORIDE

CAS #7790-91-2   mf: ClF_3
DOT #1749   mw: 92.45

SYNONYMS
Chlorine fluoride
Chlorotrifluoride

PROPERTIES   Greenish-yellow, almost colorless gas or liquid; sweet, irritating odor; melting point −83°C, boiling point 11.8°C; density 1.77 at 13°C

OSHA PEL   Ceiling 0.1 ppm; IDLH 20 ppm

ACGIH TLV   Ceiling 0.1 ppm

DOT Classification   Oxidizer; Label: Oxidizer and poison; Poison A; Label: Poison gas, oxidizer, corrosive

TOXIC EFFECTS   Human poison by inhalation. An eye irritant.

The liquid causes deep penetrating burns on contact. The effect may be delayed and progressive, as in the case of burns by hydrogen fluoride. It may be more dangerous than chlorine.

Target Organs   Skin, eyes

Routes   Ingestion, inhalation, contact

Symptoms   Irritation of the eyes and respiratory tract; burns eyes and skin

HAZARDOUS REACTIONS   A powerful oxidant, which may react violently with oxidizable materials. A rocket propellant. The substance resembles elemental fluorine in many of its chemical properties, and procedures for handling it, as well as precautions against accidents in use, are similar.

Acetic acid: explosive reaction
Acids: violent reaction
Aluminum: violent reaction
Ammonia: explosive reaction
Ammonium fluoride: reaction forms explosive gaseous products
Ammonium hydrogen fluoride: reaction forms explosive gaseous products
Antimony: violent reaction
Arsenic: violent reaction
Asbestos: ignites on contact
Benzene: explosive reaction
Boron-containing materials: ignites on contact
Carbon monoxide: violent reaction
Carbon tetrachloride: forms shock-sensitive, explosive mixture
Chlorinated compounds: forms shock-sensitive, explosive mixture
Chromium trioxide: violent reaction
Coal gas: explosive reaction
Combustible substances: incompatible
Copper: violent reaction
Cotton: explosive reaction
Ether: explosive reaction
Fluorinated polymers: ignites on contact with flowing trifluoride
Fuels: incompatible
Glass: incompatible
Glass wool: ignites on contact
Graphite: violent reaction
Hexanitrodiphenyl amine: forms shock-sensitive, explosive mixture
Hexanitrodiphenyl ether: forms shock-sensitive, explosive mixture
Hexanitrodiphenyl sulfide: forms shock-sensitive, explosive mixture
HgI_2: violent reaction
Hydrogen: explosive reaction
Hydrogen sulfide: explosive reaction
Hydrogen-containing materials: explosive reaction
Iodine: ignites on contact
Iridium: violent reaction
Iron: violent reaction
K_2CO_3: violent reaction
Lead: violent reaction
Magnesium: violent reaction
Metal oxides: violent reaction
Metal salts: violent reaction
Metals: violent reaction

136 CHLORINE TRIFLUORIDE

Methane: explosive reaction
Molybdenum: violent reaction
Nitric acid: violent reaction
Nitro compounds: incompatible
Nitroaryl compounds: forms shock-sensitive, explosive mixture
Nonmetal salts: violent reaction
Nonmetals: violent reaction
Organic matter: violent reaction
Organic vapors: forms explosive mixtures
Osmium: violent reaction
Oxides: violent reaction
Oxidizable substances: vigorous reaction, frequently with immediate ignition
Paper: explosive reaction
Phosphorus: violent reaction
Polychlorotrifluoroethylene: explosive reaction
Potassium: violent reaction
Potassium iodide: violent reaction
Refractory materials (finely divided): ignites on contact
Rubber: violent reaction
Ruthenium: violent reaction
Sand: ignites on contact
Selenium: violent reaction
Selenium tetrafluoride: violent reaction above 106°C
Silicon: violent reaction
Silicon-containing compounds: incompatible
Silver: violent reaction
Silver azide: violent reaction
Silver nitrate: violent reaction
Sodium: violent reaction
Sodium hydroxide: violent reaction
Sulfur: violent reaction
Sulfuric acid: violent reaction
Tellurium: violent reaction
Tin: violent reaction
Trifluoromethanesulfenyl chloride: explosive reaction
bis(Trifluoromethyl)sulfide or -disulfide: explosive reaction
Trinitrotoluene hexanitrobiphenyl: forms shock-sensitive, explosive mixture
Tungsten: violent reaction
Tungsten carbide: ignites on contact
V_2P_5: violent reaction
Water: explosive reaction produces toxic gases
Water vapor: forms explosive mixtures
WO_3: violent reaction
Wood: explosive reaction
Zinc: violent reaction

When heated to decomposition, it emits toxic fumes of F^- and Cl^-.

FIRST AID   In case of contact with the eyes, immediately wash the eyes with large amounts of water, occasionally lifting the lower and upper lids. Get medical attention immediately. Contact lenses should not be worn when working with this chemical.

In case of contact with the skin, immediately flush the contaminated skin with water. If this chemical penetrates the clothing, immediately remove the clothing and flush the skin with water. Get medical attention promptly.

If a person breathes in large amounts of this chemical, move the victim to fresh air at once. If breathing has stopped, perform artificial respiration. Keep the affected person warm and at rest. Get medical attention as soon as possible.

If this chemical has been swallowed, get medical attention immediately.

FIRE HAZARD   Spontaneously flammable.

HANDLING   Because of its extreme activity, the area in the vicinity of a chlorine trifluoride cylinder and its associated apparatus should be well ventilated and cleared of easily combustible material.

Chlorine trifluoride should always be removed from the cylinder as a gas. Cylinders of it should never be connected directly to apparatus containing a liquid absorbing medium. A trap should be inserted in the line between the cylinder and the reaction vessel adequate to contain the entire volume of liquid that might be drawn into the chlorine trifluoride line.

Protective measures against chlorine trifluoride are not fully developed, and entry into zones contaminated with the substance should be avoided. Only air-line or oxygen-supplied respiratory protective equipment is advised for personal protection against atmospheres contaminated with chlorine trifluoride. Gas masks, rubber gloves, rubber aprons, and face shields give protection only at low concentrations of gas and, if brought into local contact with a leak, may inflame. Thorough flushing with an inert gas should precede any opening of apparatus that has contained chlorine trifluoride.

Wear appropriate clothing and equipment to prevent any possibility of skin contact. Wash immediately when skin becomes contaminated. Immediately remove any clothing that becomes contaminated.

Wear eye protection to prevent any possibility of eye contact.

The following equipment should be available: an eyewash station and a quick-drench shower.

CHLOROBENZENE

CAS #108-90-7 mf: C_6H_5Cl
DOT #1134 mw: 112.56

NFPA LABEL

SYNONYMS
Benzene chloride
Chlorbenzene
Chlorbenzol
Chlorobenzol
MCB
Monochlorbenzene
Monochlorobenzene
Phenyl chloride

PROPERTIES   Clear, colorless liquid with a mild aromatic odor; melting point −45°C, boiling point 131.7°C; density 1.113 at 15.5°C; flash point 29.4°C (closed cup), lower explosive limit 1.3%, upper explosive limit 7.1% at 150°C; autoignition temperature 637.8°C; vapor pressure 10 mm Hg at 22.2°C; vapor density 3.88

OSHA PEL   TWA 75 ppm; IDLH 2400 ppm

ACGIH TLV   TWA 75 ppm

Community Right to Know List

DOT Classification   Flammable or combustible liquid; Label: Flammable liquid

TOXIC EFFECTS   Poisonous by ingestion. A teratogen. Dichlorobenzols are strong narcotics.

Target Organs   Respiratory system, eyes, skin, central nervous system, liver

Routes   Ingestion, inhalation, contact

Symptoms   Irritation of the skin, eyes, and nose; drowsiness, incoordination; liver damage, narcosis

HAZARDOUS REACTIONS
Dimethyl sulfoxide: violent reaction
Oxidizers: vigorous reaction
Phosphorus trichloride + sodium: potentially explosive reaction
Silver perchlorate: violent reaction
Sodium powder: potentially explosive reaction

FIRST AID   In case of contact with the eyes, immediately wash the eyes with large amounts of water, occasionally lifting the lower and upper lids. Get medical attention immediately. Contact lenses should not be worn when working with this chemical.

In case of contact with the skin, immediately wash the contaminated skin with soap and water. If this chemical penetrates through the clothing, immediately remove the clothing and wash the skin with soap and water. Get medical attention promptly.

If a person breathes in large amounts of this chemical, move the victim to fresh air at once. If breathing has stopped, perform artificial respiration. Keep the affected person warm and at rest. Get medical attention as soon as possible.

If this chemical has been swallowed, get medical attention immediately.

FIRE HAZARD   Dangerous fire hazard when exposed to heat or flame. An explosion hazard when exposed to heat or flame.

Use foam, CO_2, dry chemical, or water extinguishers to blanket the fire.

HANDLING   Wear appropriate clothing and equipment to prevent repeated or prolonged skin contact. Wash promptly when skin becomes wet. Immediately remove any clothing that becomes wet to avoid flammability hazard.

Wear eye protection to reduce probability of eye contact.

CHLOROFORM

CAS #67-66-3 mf: $CHCl_3$
DOT #1888 mw: 119.37

NFPA LABEL

SYNONYMS
Formyl trichloride
Methane trichloride
Methenyl trichloride
Methyl trichloride
TCM
Trichloroform
Trichloromethane

PROPERTIES   Colorless liquid, heavy, ethereal odor; solubility in water 0.74% at 25°C, miscible with most organic solvents; melting point −63.5°C, boiling point 61.26°C; density 1.49845 at 15°C; flash point none; vapor pressure 100 mm Hg at 10.4°C; vapor density 4.12

138 CHLOROFORM

OSHA PEL   Ceiling 50 ppm

ACGIH TLV   TWA 10 ppm; suspected carcinogen

NIOSH REL   Waste anesthetic gases and vapors ceiling 2 ppm/1 h; chloroform ceiling 2 ppm/60 min

Community Right to Know List and EPA Extremely Hazardous Substances List

DOT Classification   ORM-A; Label: None

IMO Classification   Poison B; Label: Poison

TOXIC EFFECTS   A poison by ingestion, inhalation, and some other routes. A carcinogen and teratogen.

Prolonged inhalation will bring on paralysis accompanied by cardiac respiratory failure and finally death.

Chloroform is an anesthetic but is too toxic for use in humans. Inhalation of 70,000 ppm in air can cause death in a few minutes; 14,000 ppm for 30–60 min may cause death. Inhalation of concentrations greater than 1000 ppm can produce dizziness, nausea, and headache. Prolonged administration as an anesthetic may lead to such serious effects as profound toxemia and damage to the liver, heart, and kidneys. Chronic exposure to chloroform at concentrations of 100–200 ppm has been reported to produce enlarged livers.

High concentrations of the vapor can produce conjunctivitis. Liquid chloroform in the eye will produce a painful corneal injury that usually heals in several days. Continued contact with the skin can produce drying, fissuring, and inflammation.

Target Organs   Liver, kidneys, heart, eyes, skin

Routes   Ingestion, inhalation, contact

Symptoms   Nausea, vomiting, dizziness, mental dullness, loss of reflexes, headache; fatigue; anesthesia; hepatomegaly; irritation of the eyes, skin, and mucous membranes; hallucinations and distorted perceptions

HAZARDOUS REACTIONS
Acetone + alkali: violent reaction
Acetone + calcium hydroxide: violent reaction
Acetone + potassium hydroxide: violent reaction
Acetone + sodium hydroxide: violent reaction
Alkali (strong): incompatible
Aluminum: violent reaction
Chemically active metals: incompatible
Dinitrogen tetraoxide: incompatible
Disilane: violent reaction
Fluorine: incompatible
Lithium: violent reaction
Magnesium: violent reaction
Magnesium powder: incompatible
Metals: incompatible
Methanol + alkali: violent reaction
Nitrogen tetroxide: violent reaction
Perchloric acid + phosphorus pentoxide: violent reaction
Potassium: violent reaction
Potassium hydroxide + methanol: violent reaction
Potassium-*tert*-butoxide: violent reaction
Sodium: violent reaction
Sodium + methanol: explosive reaction
Sodium hydride: violent reaction
Sodium hydroxide + methanol: violent reaction
Sodium methoxide + methanol: explosive reaction
Sodium methylate: violent reaction
Sodium-potassium alloy: violent reaction
Triisopropylphosphine: incompatible

When heated to decomposition, it emits toxic fumes of Cl^-.

FIRST AID   In case of contact with the eyes, immediately wash the eyes with large amounts of water, occasionally lifting the lower and upper lids. Get medical attention immediately. Contact lenses should not be worn when working with this chemical.

In case of contact with the skin, promptly wash the contaminated skin with soap and water. If this chemical penetrates through the clothing, promptly remove the clothing and wash the skin with soap and water. Get medical attention promptly.

If a person breathes in large amounts of this chemical, move the victim to fresh air at once. If breathing has stopped, perform artificial respiration. Keep the affected person warm and at rest. Get medical attention as soon as possible.

If this chemical has been swallowed, get medical attention immediately.

FIRE HAZARD   Combustible when exposed to high heat. Exposure to heat or flame can generate phosgene gas.

HANDLING   Although chloroform has caused tumors in animals, its potency is low. Hence, Procedure B (see Chap. 1) provides adequate protection during laboratory operations with it. The high volatility of chloroform emphasizes the importance of a hood for such operations. Polyvinyl alcohol gloves provide the best hand protection.

Wear appropriate clothing and equipment to prevent a reasonable probability of skin contact. Wash promptly when skin becomes wet. Promptly remove any nonimpervious clothing that becomes contaminated.

Wear eye protection to reduce probability of eye contact.

CHLOROMETHANE

CAS #74-87-3   mf: CH_3Cl
DOT #1063   mw: 50.49

SYNONYMS
Methyl chloride
Monochloromethane

PROPERTIES   Colorless gas, with a slight, etherlike odor and a sweet taste; slightly soluble in water, miscible with chloroform, ether, glacial acetic acid, soluble in alcohol; melting point −97°C, boiling point −23.7°C; density 0.918 at 20°/4°C; flash point <0°C; lower explosive limit 8.1%, upper explosive limit 17%; autoignition temperature 632.2°C; vapor pressure 4.8 atm; vapor density 1.78

OSHA PEL   TWA 100 ppm; ceiling 200; peak 300 ppm/5 min/3 h

ACGIH TLV   TWA 50 ppm; STEL 100 ppm (skin)

DOT Classification   Flammable gas; Label: Flammable gas

IMO Classification   Poison A; Label: Poison gas and flammable gas

TOXIC EFFECTS   Mildly toxic by inhalation. A carcinogen and teratogen.

The odor is not noticeable at dangerous concentrations. With exposure to high concentrations, death may be immediate or delayed by several days. If exposure is not fatal, recovery is usually slow. Chronic exposure to low concentrations causes damage to the central nervous system and, less frequently, to the liver, kidneys, bone marrow, and cardiovascular system. Contact with the liquid can cause frostbite by evaporation from the skin.

Target Organs   Central nervous system, liver, kidneys, intestines, skin

Routes   Inhalation, ingestion, contact

Symptoms   Dizziness, drowsiness; nausea, vomiting; liver, kidney damage; abdominal pains; hiccoughs; hemorrhages in the lungs, intestinal tract, and dura; visual disturbances (diplopia and dimness of vision); lack of coordination, confusion, delirium, staggering, slurred speech; convulsions, coma; frostbite

HAZARDOUS REACTIONS
Aluminum: potentially explosive reaction at 152°C in a sealed container
Aluminum chloride: may ignite on contact
Aluminum chloride + ethylene: mixtures react exothermically and then explode above 30 bar
Aluminum powder: may ignite on contact
Bromine pentafluoride: explodes on contact
Bromine trifluoride: explodes on contact
Interhalogens: explodes on contact
Magnesium and alloys: explodes on contact
Metals (chemically active): incompatible
Potassium and alloys: explodes on contact
Sodium and alloys: explodes on contact
Zinc: explodes on contact

When heated to decomposition, it emits toxic fumes of Cl^-.

FIRST AID   In case of contact with the eyes, immediately wash the eyes with large amounts of water, occasionally lifting the lower and upper lids. Get medical attention immediately. Contact lenses should not be worn when working with this chemical.

In case of contact with the skin, immediately flush the contaminated skin with water. If this chemical penetrates the clothing, immediately remove the clothing and flush the skin with water. Get medical attention promptly.

If a person breathes in large amounts of this chemical, move the victim to fresh air at once. If breathing has stopped, perform artificial respiration. Keep the affected person warm and at rest. Get medical attention as soon as possible.

FIRE HAZARD   Very dangerous fire hazard when exposed to heat, flame, or powerful oxidizers. Moderate explosion hazard when exposed to flame and sparks.

To extinguish fires, stop flow of gas and use carbon dioxide, dry chemical, or water spray.

HANDLING   Because methyl chloride may, under certain conditions, react with aluminum or magnesium to form materials that ignite or fume spontaneously in contact with air, contact with these metals should be avoided.

Suitable protective respiratory equipment should be available and kept in proper working order. Canister-type gas masks, however, are unsafe at concentrations higher than 50 times the TLV; at these concentrations, a positive-pressure air-line respirator must be used and, at concentrations immediately dangerous to life and health (IDLH), a pressure-demand self-contained breathing apparatus or a positive-pressure air-line respirator that has escape-cylinder provisions must be used.

Wear appropriate clothing and equipment to prevent skin becoming wet or frozen. Immediately remove any clothing that becomes wet to avoid flammability hazard.

Wear eye protection to reduce probability of eye contact.

CHROMIC ACID

CAS #7738-94-5   mf: CrH_2O_4
mw: 118.02

NFPA LABEL

SYNONYM
Chromic(VI) acid

OSHA PEL   ceiling 0.1 mg(CrO_3)/m^3; IDLH 30 mg/m^3

ACGIH TLV   TWA 0.05 mg(Cr)/m^3

NIOSH REL   TWA 0.025 mg(Cr(VI))/m^3; ceiling 0.05/15M

Community Right to Know List (chromium and its compounds)

TOXIC EFFECTS   Poison by some routes.

Target Organs   Blood, respiratory system, liver, kidneys, eyes, skin

Routes   Ingestion, inhalation, contact

Symptoms   Irritation of the respiratory system and nose/nasal septum; leukocytosis, leukopenia, monocytosis, eosinophila; eye injury, conjunctivitis; skin ulcer, sensitization dermatitis

HAZARDOUS REACTIONS   A powerful oxidizer. A storage hazard; it may burst a sealed container due to carbon dioxide release.

Acetone: may ignite on contact
Alcohols: may ignite on contact
Aluminum: incompatible
Combustible materials: incompatible
Organic materials: incompatible
Oxidizable materials: potentially explosive reaction
Paper: incompatible
Plastics: incompatible
Sulfur: incompatible
Wood: incompatible

When heated to decomposition, it emits acrid smoke and irritating fumes.

FIRST AID   In case of contact with the eyes, immediately wash the eyes with large amounts of water, occasionally lifting the lower and upper lids. Get medical attention immediately. Contact lenses should not be worn when working with this chemical.

In case of contact with the skin, immediately flush the contaminated skin with soap and water. If this chemical penetrates through the clothing, immediately remove the clothing and flush the skin with water. If irritation persists after washing, get medical attention.

If a person breathes in large amounts of this chemical, move the victim to fresh air at once. If breathing has stopped, perform artificial respiration. Keep the affected person warm and at rest. Get medical attention as soon as possible.

If this chemical has been swallowed, get medical attention immediately.

HANDLING   Wear appropriate clothing and equipment to prevent any possibility of skin contact. Wash immediately if skin becomes contaminated. Work clothing should be changed daily if it is reasonably probable that the clothing has been contaminated. Immediately remove any nonimpervious clothing that becomes contaminated.

Wear eye protection to prevent any possibility of eye contact.

Provide an eyewash and a quick-drench shower.

CHROMIUM

CAS #7440-47-3   mf: Cr
mw: 52.00

SYNONYM
Chrome

OSHA PEL   TWA 1 mg/m^3; IDLH 500 mg/m^3

ACGIH TLV   TWA 0.5 mg/m^3

Community Right to Know List

TOXIC EFFECTS   Poison by ingestion, with gastrointestinal effects. The chromium(VI) ion is a carcinogen.

Target Organs Respiratory system, gastrointestinal system

Routes Ingestion

Symptoms Gastrointestinal effects, histologic fibrosis of lungs

HAZARDOUS REACTIONS

Air: powder will explode spontaneously

Ammonium nitrate: violent or explosive reaction when heated

Bromine pentafluoride: may ignite or react violently

Carbon dioxide (atmospheres): ignites and is potentially explosive

Nitrogen oxide: incandescent reaction

Oxidants: incompatible

Sulfur dioxide: incandescent reaction

FIRST AID In case of contact with the eyes, immediately wash the eyes with large amounts of water, occasionally lifting the lower and upper lids. Get medical attention immediately. Contact lenses should not be worn when working with this chemical.

In case of contact with the skin, immediately flush the contaminated skin with soap and water. If this chemical penetrates through the clothing, immediately remove the clothing and flush the skin with water. If irritation persists after washing, get medical attention.

If a person breathes in large amounts of this chemical, move the victim to fresh air at once. If breathing has stopped, perform artificial respiration. Keep the affected person warm and at rest. Get medical attention as soon as possible.

If this chemical has been swallowed, get medical attention immediately.

HANDLING Wear appropriate clothing and equipment to prevent repeated or prolonged skin contact. Wash promptly if the skin becomes contaminated. Promptly remove nonimpervious clothing that becomes contaminated.

Wear eye protection to reduce probability of eye contact.

CHROMIUM CHLORIDE

CAS #10025-73-7 mf: Cl_3Cr
  mw: 158.35

SYNONYMS
Chromic chloride
Chromium chloride, anhydrous
Chromium trichloride
Puratronic chromium chloride
Trichlorochromium

PROPERTIES Sublimes at 1300°C

OSHA PEL ceiling 0.5 mg(Cr)/m³; IDLH 250 mg/m³

ACGIH TLV TWA 0.5 mg(Cr)/m³

EPA Extremely Hazardous Substances List and Community Right to Know List

TOXIC EFFECTS Poisonous by skin contact, inhalation, and several other routes. A teratogen.

HAZARDOUS REACTIONS

Lithium under nitrogen atmosphere: violent reaction

When heated to decomposition, it emits toxic fumes of Cl^-.

FIRST AID In case of contact with the eyes, immediately wash the eyes with large amounts of water, occasionally lifting the lower and upper lids. Get medical attention immediately. Contact lenses should not be worn when working with this chemical.

In case of contact with the skin, immediately flush the contaminated skin with water. If this chemical penetrates the clothing, immediately remove the clothing and flush the skin with water. Get medical attention promptly.

If a person breathes in large amounts of this chemical, move the victim to fresh air at once. If breathing has stopped, perform artificial respiration. Keep the affected person warm and at rest. Get medical attention as soon as possible.

If this chemical has been swallowed, get medical attention immediately.

HANDLING Wear appropriate clothing and equipment to prevent repeated or prolonged skin contact. Wash promptly if the skin becomes contaminated. Promptly remove any nonimpervious clothing that becomes contaminated.

Wear eye protection to prevent any possibility of eye contact.

CHROMIUM TRIOXIDE

CAS #1333-82-0 mf: CrO_3
DOT #1463; 1755 mw: 100.00

SYNONYMS
Chromic acid
Chromic acid, solid
Chromic acid, solution
Chromic anhydride
Chromium oxide

CHROMIUM TRIOXIDE

Chromium(VI) oxide
Chromium(VI) oxide (1:3)
Chromium (6+) trioxide
Chromium trioxide, anhydrous
Monochromium oxide
Monochromium trioxide
Puratronic chromium trioxide

PROPERTIES   Red, rhombic, deliquescent crystals; solubility 61.7 g/100 cc at 0°C, 67.45 g/100 cc at 100°C; melting point 196°C, decomposes before boiling; density 2.70

OSHA PEL   ceiling 0.1 mg(CrO_3)/m^3

ACGIH TLV   TWA 0.05 mg(Cr)/m^3; human carcinogen

NIOSH REL   0.025 mg[Cr(VI)]/m^3; ceiling 0.05/15M

Community Right to Know List

DOT Classification   Oxidizer; Label: Oxidizer, solid: Corrosive material; Label: Corrosive solution: Oxidizer; Label: Oxidizer, corrosive, anhydrous

TOXIC EFFECTS   Poisonous by ingestion and some other routes. A carcinogen by inhalation which produces nasal and lung tumors. A teratogen.

Target Organs   Blood, respiratory system, liver, kidneys, eyes, skin

Routes   Ingestion, inhalation, contact

Symptoms   Corrosive; an irritant to the eyes, skin, respiratory system, mucous membranes, and nose/nasal septum; leukocytosis, leukopenia, monocytosis; eye injury, conjunctivitis; skin ulcer, sensitization dermatitis

HAZARDOUS REACTIONS   A powerful oxidizer.

Acetaldehyde: explosive reaction
Acetic acid + heat: explosive reaction
Acetic anhydride + 3-methylphenol: violent reaction above 75°C
Acetic anhydride + heat: explosive reaction
Acetic anhydride + tetrahydronaphthalene: ignites on contact
Acetone: ignites on contact
Acetylene: violent reaction
Alcohols: ignites on contact
Alkali metals: incandescent reaction
Aluminum: incompatible
Ammonia: incandescent reaction
Arsenic: incandescent reaction
Benzaldehyde: explosive reaction
Benzene: explosive reaction
Benzylthylaniline: explosive reaction
Bromine pentafluoride: violent reaction
Butanol: ignites on contact
Butyraldehyde: explosive reaction
Butyric acid: incandescent reaction above 100°C
Chlorine trifluoride: incandescent reaction
Chromium(II) sulfide: ignites on contact
Combustible materials: incompatible
Cyclohexanol: ignites on contact
Diethyl ether: explosive reaction
Dimethyl formamide: ignites on contact
1,3-Dimethylhexahydropyrimidone: explosive reaction
Ethanol: ignites on contact
Ethylacetate: explosive reaction
Ethylene glycol: ignites on contact
Glycerol: violent reaction
Hexamethylphosphoramide: violent reaction
Hydrogen sulfide + heat: incandescent reaction
Isopropylacetate: explosive reaction
Methanol: ignites on contact
Methyl dioxane: explosive reaction
N,N-dimethylformamide: incompatible
Organic materials or solvents: explosive reaction
Oxidizable materials: ignites on contact
Paper: incompatible
Pelargonic acid: explosive reaction
Pentyl acetate: explosive reaction
Peroxyformic acid: violent reaction
Phosphorus + heat: explosive reaction
Plastics: incompatible
Potassium: incandescent reaction
Potassium hexacyanoferrate: mixture is a friction- and heat-sensitive explosive
2-Propanol: ignites on contact
Propionaldehyde: explosive reaction
Pyridine: ignites on contact
Selenium: violent reaction
Sodium: incandescent reaction
Sodium amide: violent reaction
Sulfur: incandescent reaction
Wood: incompatible

FIRST AID   In case of contact with the eyes, immediately wash the eyes with large amounts of water, occasionally lifting the lower and upper lids. Get medical attention immediately. Contact lenses should not be worn when working with this chemical.

In case of contact with the skin, immediately flush the contaminated skin with soap and water. If this chemical penetrates through the clothing, immediately remove the clothing and flush the skin with water. If irritation persists after washing, get medical attention.

If a person breathes in large amounts of this chemical, move the victim to fresh air at once. If breathing has stopped, perform artificial respiration. Keep the affected person warm and at rest. Get medical attention as soon as possible.

If this chemical has been swallowed, get medical attention immediately.

HANDLING   Wear appropriate clothing and equipment to prevent any possibility of skin contact. Wash immediately if skin becomes contaminated. Immediately remove any nonimpervious clothing that becomes contaminated. Work clothing should be changed daily if there is any possibility that the clothing may be contaminated.

Wear eye protection to prevent any possibility of eye contact.

The following equipment should be available: an eyewash station and a quick-drench shower.

COBALT

CAS #7440-48-4   mf: Co
                              mw: 58.93

SYNONYMS
Aquacat
Cobalt-59
Super cobalt

PROPERTIES   Gray, hard, magnetic, ductile, somewhat malleable metal; readily soluble in dilute HNO_3, very slowly attacked by HCl or cold H_2SO_4; melting point 1493°C, boiling point about 3100°C; density 8.92; Brinell hardness 125; latent heat of fusion 62 cal/g, latent heat of vaporization 1500 cal/g; specific heat (15–100°C) 0.1056 cal/g/°C. Exists in two allotropic forms. At room temperature, the hexagonal form is more stable than the cubic form; both forms can exist at room temperature. Stable in air or toward water at ordinary temperatures. The hydrated salts of cobalt are red, and the soluble salts form red solutions which become blue upon adding concentrated HCl.

OSHA PEL   TWA 0.1 mg/m^3

ACGIH TLV   (metal, dust, and fume) TWA 0.05 mg(Co)/m^3

EPA Extremely Hazardous Substances List and Community Right to Know List

TOXIC EFFECTS   Moderately toxic by ingestion. Poisonous by some other routes.

Routes   Ingestion, inhalation, contact

Symptoms   Cough, dyspnea, decreased pulmonary function; weight loss; dermatitis; diffuse nodular fibrosis; respiratory hypersensitivity

Target Organs   Respiratory system, skin

HAZARDOUS REACTIONS
Acetylene: incandescent reaction
Air: cobalt powder ignites spontaneously
Ammonium nitrate: explosive reaction when heated
Bromine pentafluoride: ignites on contact
Hydrazinium nitrate: explosive reaction
Nitryl fluoride: incandescent reaction
Oxidizers (strong): incompatible
1,3,4,7-Tetramethylisoindole: explosive reaction at 390°C

FIRST AID   In case of contact with the eyes, immediately wash the eyes with large amounts of water, occasionally lifting the lower and upper lids. Get medical attention immediately. Contact lenses should not be worn when working with this chemical.

In case of contact with the skin, wash the contaminated skin with soap and water.

If a person breathes in large amounts of this chemical, move the victim to fresh air at once. If breathing has stopped, perform artificial respiration. Keep the affected person warm and at rest. Get medical attention as soon as possible.

If this chemical has been swallowed, get medical attention immediately.

FIRE HAZARD   Flammable when exposed to heat or flame.

HANDLING   Wear appropriate clothing and equipment to prevent repeated or prolonged skin contact. Wash promptly when the skin becomes contaminated. Promptly remove any nonimpervious clothing that becomes contaminated. Work clothing should be changed daily if it is reasonably probable that the clothing has been contaminated.

COBALT(II) CHLORIDE

CAS #7646-79-9   mf: Cl_2Co
                              mw: 129.83

SYNONYMS
Cobalt dichloride   Cobaltous chloride
Cobalt muriate   Cobaltous dichloride

PROPERTIES   Blue powder; melting point 724°C, boiling point 1049°C; density 3.348

OSHA PEL   TWA 0.1 mg(Co)/m³ (fume and dust)

ACGIH TLV   TWA 0.05 mg(Co)/m³

NIOSH REL   (To Cobalt) TWA 0.1 mg/m³

Community Right to Know List (cobalt and its compounds)

TOXIC EFFECTS   Poisonous by ingestion, skin contact, and most other routes. A carcinogen and teratogen.

Symptoms by Ingestion   Anorexia, goiter (increased thyroid size), weight loss

HAZARDOUS REACTIONS
Metals: incompatible
Potassium: incompatible
Sodium: incompatible

When heated to decomposition, it emits toxic fumes of Cl⁻.

COBALT(II) NITRATE

CAS #10141-05-6   mf: CoN_2O_6
                               mw: 182.95

NFPA LABEL

SYNONYMS
Cobalt dinitrate
Cobaltous nitrate
Nitric acid, cobalt(2+) salt

PROPERTIES   Melting point 55°C; density 1.87

OSHA PEL   TWA 0.1 mg(Co)/m³ (fume and dust)

ACGIH TLV   TWA 0.05 mg(Co)/m³

NIOSH REL   (To Cobalt) TWA 0.1 mg/m³

Community Right to Know List

TOXIC EFFECTS   Poisonous by ingestion, intramuscular, and subcutaneous routes. An experimental tumorigen. Experimental reproductive effects.

HAZARDOUS REACTIONS
Ammonium hexacyanoferrate(II): explosive reaction at 220°C

Carbon: potentially explosive reaction

When heated to decomposition, it emits toxic fumes of NO_x.

COBALT(II) SULFATE (1:1)

CAS #10124-43-3   mf: $O_4S \cdot Co$
                               mw: 154.99

SYNONYMS
Cobaltous sulfate
Cobalt sulfate
Cobalt(2+) sulfate
Sulfuric acid, cobalt(2+) salt (1:1)

PROPERTIES   Red to lavender dimorphic, orthorhombic crystals; dissolves slowly in boiling water; density 3.71; stable to 708°C

OSHA PEL   TWA 0.1 mg(Co)/m³ (fume and dust)

ACGIH TLV   TWA 0.05 mg(Co)/m³

NIOSH REL   0.1 mg/m³

Community Right to Know List

TOXIC EFFECTS   Poisonous by some routes. Moderately toxic by ingestion.

HAZARDOUS REACTIONS   When heated to decomposition, it emits toxic fumes of SO_x.

COLCHICINE

CAS #64-86-8   mf: $C_{22}H_{25}NO_6$
                               mw: 399.48

SYNONYMS
7-Acetamido-6, 7-dihydro-1,2,3,10-tetra-
    methoxybenzo(a)heptalen-9(5H)-one
N-Acetyl trimethylcolchicinic acid methylether
7-α-H-colchicine
Colchineos
Colchisol
Colcin
Colsaloid
Condylon
N-[5,6,7,9-tetrahydro-1,2,3,10-tetramethoxy-
    9-oxobenzo(α)heptalen-7-yl]-acetamide

PROPERTIES   Pale yellow scales or powder; melting point 142–150°C. Crystals from ethyl acetate, pale yellow needles, melting point 157°C. One gram dissolves in 22 mL water, 220 mL ether, 100 mL benzene;

freely soluble in alcohol or chloroform, practically insoluble in petroleum ether

EPA Extremely Hazardous Substances List

TOXIC EFFECTS   A poison by ingestion and most other routes. A teratogen. A severe eye irritant.

Symptoms by Ingestion   Dyspnea, gastrointestinal effects and body-temperature decrease

Symptoms by Other Routes   Kidney damage and hemorrhaging; somnolence, blood-pressure depression, and nausea or vomiting

Inhibits the formation of microtubules and thus impairs cell division at a very basic level.

HAZARDOUS REACTIONS   When heated to decomposition, it emits toxic fumes of NO_x.

COPPER SULFATE PENTAHYDRATE

CAS #7758-99-8   mf: $O_4S \cdot Cu \cdot 5H_2O$
mw: 249.70

SYNONYMS
Blue copperras
Blue vitriol
Bluestone
Copperfine-zinc
CSP
Cupric sulfate pentahydrate
Roman vitriol
Salzburg vitriol
Sulfuric acid, copper(2+) salt, pentahydrate
Triangle

PROPERTIES   Loses $4H_2O$ at 110°C

ACGIH TLV   TWA 1 mg(Cu)/m³

Community Right to Know List (copper and its compounds)

TOXIC EFFECTS   Poisonous by ingestion and some other routes.

Symptoms by Ingestion   jaundice, unspecified urinary system effects, and hemolysis. Mutagenic data.

HAZARDOUS REACTIONS   When heated to decomposition, it emits toxic fumes of SO_x.

COPPER(II) NITRATE

CAS #3251-23-8   mf: CuN_2O_6
DOT #1479   mw: 187.55

SYNONYMS
Copper dinitrate   Cupric dinitrate
Copper(2+) nitrate   Cupric nitrate

PROPERTIES   Large, blue-green, deliquescent, orthorhombic crystals; soluble in water, ethyl acetate, dioxane; dissolves in and reacts vigorously with ether; melting point 255–256°C, sublimes at 150–225°C; density 2.047

ACGIH TLV   TWA 1 mg(Cu)/m³

Community Right to Know List

DOT Classification   Oxidizer; Label: Oxidizer

TOXIC EFFECTS   Moderately toxic by ingestion. A severe eye and skin irritant.

HAZARDOUS REACTIONS
Acetic anhydride: violent reaction
Ammonia + potassium amide: reaction forms an explosive product
Ammonium hexacyanoferrate(II): potentially explosive reaction above 220°C
Paper: may ignite on prolonged contact
Potassium hexacyanoferrate(II): potentially explosive reaction above 220°C
Tin foil: concentrated solutions of the nitrate may ignite in contact

When heated to decomposition, it emits toxic fumes of NO_x.

COPPER(II) SULFATE

CAS #7758-98-7   mf: $O_4S \cdot Cu$
DOT #9109   mw: 159.60

SYNONYMS
BCS copper fungicide   CP basic sulfate
Blue copper   Cupric sulfate
Blue stone   Roman vitriol
Blue vitriol   Sulfuric acid,
Copper monosulfate   copper(2+) salt (1:1)
Copper(II) sulfate (1:1)   Triangle

PROPERTIES   Blue crystals or blue, crystalline granules or powder; density 2.284

OSHA PEL   TWA 1 mg(Cu)/m³

ACGIH TLV   TWA 1 mg(Cu)/m³

Community Right to Know List

DOT Classification   ORM-E; Label: None

TOXIC EFFECTS   A poison by ingestion and other routes.

Symptoms by Ingestion   Gastritis, diarrhea, nausea or vomiting, damage to kidney tubules, and hemolysis

HAZARDOUS REACTIONS
Hydroxylamine: violent reaction
Magnesium: violent reaction

When heated to decomposition, it emits toxic fumes of SO_x.

m-CRESOL

CAS #108-39-4   mf: C_7H_8O
DOT #2076   mw: 108.15

NFPA LABEL

SYNONYMS
3-Cresol   m-Kresol
m-Cresylic acid   m-Methylphenol
1-Hydroxy-3-methyl-   3-Methylphenol
  benzene   m-Oxytoluene
m-Hydroxytoluene   m-Toluol

PROPERTIES   Colorless solid or liquid (may turn yellowish), with a sweet, tarry odor; melting point 10.9°C, boiling point 202.8°C; density 1.034 at 20°C/4°C; flash point 94.4°C; lower explosive limit 1.1% at 150°C; autoignition temperature 558.9°C; vapor pressure 1 mm Hg at 52°C; vapor density 3.72

OSHA PEL   TWA 5 ppm; IDLH 250 ppm

NIOSH REL   TWA 2.3 ppm (10 mg/m³)

Community Right to Know List, EPA TSCA Inventory, and EPA Genetic Toxicology Program

TOXIC EFFECTS   Poisonous by ingestion and several other routes. Moderately toxic by skin contact. May cause severe eye and skin burns.

Routes   Ingestion, absorption, inhalation, contact

Symptoms   Central nervous system effects; confusion, depression; respiratory failure; dyspnea, irregular and/or rapid respiration, weak pulse; skin and eye burns; dermatitis

Target Organs   Central nervous system, respiratory system, liver, kidneys, skin, eyes

HAZARDOUS REACTIONS
Oxidizers (strong): incompatible

FIRST AID   In case of contact with the eyes, immediately wash the eyes with large amounts of water, occasionally lifting the lower and upper lids. Get medical attention immediately. Contact lenses should not be worn when working with this chemical.

In case of contact with the skin, immediately wash the contaminated skin with soap and water. If this chemical penetrates through the clothing, immediately remove the clothing and wash the skin with soap and water. Get medical attention promptly.

If a person breathes in large amounts of this chemical, move the victim to fresh air at once. If breathing has stopped, perform artificial respiration. Keep the affected person warm and at rest. Get medical attention as soon as possible.

If this chemical has been swallowed, get medical attention immediately.

FIRE HAZARD   Flammable when exposed to heat or flame. Moderately explosive in the form of vapor when exposed to heat or flame.

HANDLING   Wear appropriate clothing and equipment to prevent any possibility of skin contact. Wash immediately if skin becomes contaminated. Immediately remove any nonimpervious clothing that becomes contaminated. Work clothing should be changed daily if there is any possibility that the clothing may be contaminated.

Wear eye protection to prevent any possibility of eye contact.

The following equipment should be available: an eyewash station and a quick-drench shower.

CYCLOHEXANE

CAS #110-82-7   mf: C_6H_{12}
DOT #1145   mw: 84.18

NFPA LABEL

SYNONYMS
Hexahydrobenzene
Hexamethylene
Hexanaphthene

PROPERTIES Colorless, mobile liquid with a mild, sweet odor; melting point 6.5°C, boiling point 80.7°C; density 0.7791 at 20°C/4°C; flash point −17°C; lower explosive limit 1.3%, upper explosive limit 8.4%; autoignition temperature 245°C; vapor pressure 100 mm Hg at 60.8°C; vapor density 2.90

OSHA PEL TWA 300 ppm; IDLH 10,000

ACGIH TLV TWA 300 ppm

Community Right to Know List

DOT Classification Flammable liquid; Label: Flammable liquid

TOXIC EFFECTS Moderately toxic by ingestion. The liquid and vapor are irritants.

Routes Inhalation, ingestion, contact

Symptoms Irritation of the eyes and respiratory system; drowsiness; dermatitis; narcosis, coma; dizziness, nausea

Target Organs Eyes, respiratory system, skin, central nervous system

HAZARDOUS REACTIONS
Dinitrogen tetroxide (liquid) + heat: mixture can explode
Oxidizing materials: incompatible

When heated to decomposition, it emits acrid smoke and fumes.

FIRST AID In case of contact with the eyes, immediately wash the eyes with large amounts of water, occasionally lifting the lower and upper lids. Get medical attention immediately. Contact lenses should not be worn when working with this chemical.

In case of contact with the skin, immediately flush the contaminated skin with water promptly. If this chemical penetrates the clothing, immediately remove the clothing and flush the skin with water promptly. If irritation persists after washing, get medical attention.

If a person breathes in large amounts of this chemical, move the victim to fresh air at once. If breathing has stopped, perform artificial respiration. Keep the affected person warm and at rest. Get medical attention as soon as possible.

If this chemical has been swallowed, get medical attention immediately.

FIRE HAZARD Dangerous fire hazard when exposed to heat or flame. Moderate explosion hazard in the form of vapor or when exposed to flame.

Use foam, CO_2, dry chemical, spray, or fog extinguishers to fight a fire.

HANDLING Wear appropriate clothing and equipment to prevent repeated or prolonged skin contact. Wash promptly if skin is contaminated. Immediately remove any clothing that becomes wet to avoid flammability hazard.

Wear eye protection to reduce probability of eye contact.

CYCLOHEXANOL

CAS #108-94-1 mf: $C_6H_{12}O$
mw: 100.16

NFPA LABEL

SYNONYMS
Adronal	Hexalin
Anol	Hydralin
Cyclohexyl alcohol	Hydrophenol
Hexahydrophenol	Hydroxycyclohexane

PROPERTIES Colorless, viscous liquid or sticky solid with faint camphorlike odor; hygroscopic; solubility in water 3.6%; melting point 24°C, boiling point 161.5°C; flash point 67.8°C (closed cup); autoignition temperature 300°C; lower explosive limit 2.4%; density 0.9449 at 25°C/4°C; vapor pressure 1 mm Hg at 21.0°C; vapor density 3.45

OSHA PEL TWA 50 ppm; IDLH 3500 ppm

ACGIH TLV TWA 50 ppm (skin)

TOXIC EFFECTS Moderately toxic by ingestion and some other routes. Mildly toxic by skin contact. Has caused damage to the kidneys, liver, and blood vessels in experimental animals. A severe eye irritant.

Routes Ingestion, inhalation, contact

Symptoms Irritation of the eyes, skin, nose, and throat; narcosis; changes in the olfactory and respiratory systems

Target Organs Eyes, respiratory system, skin

HAZARDOUS REACTIONS

Chromium trioxide: ignites on contact
Nitric acid: violent reaction
Oxidizing materials: incompatible

When heated to decomposition, it emits acrid smoke and fumes.

FIRST AID
In case of contact with the eyes, immediately wash the eyes with large amounts of water, occasionally lifting the lower and upper lids. Get medical attention immediately. Contact lenses should not be worn when working with this chemical.

In case of contact with the skin, immediately flush the contaminated skin with water. If this chemical penetrates the clothing, immediately remove the clothing and flush the skin with water promptly. If irritation persists after washing, get medical attention.

If a person breathes in large amounts of this chemical, move the victim to fresh air at once. If breathing has stopped, perform artificial respiration. Keep the affected person warm and at rest. Get medical attention as soon as possible.

If this chemical has been swallowed, get medical attention immediately.

FIRE HAZARD
Flammable when exposed to heat or flame.

Use alcohol foam, foam, CO_2, or dry chemical extinguishers to fight a fire.

HANDLING
Wear appropriate clothing and equipment to prevent repeated or prolonged skin contact. Wash promptly if the skin becomes contaminated. Promptly remove any nonimpervious clothing that becomes contaminated.

Wear eye protection to reduce probability of eye contact.

CYCLOHEXANONE

CAS #108-94-1   mf: $C_6H_{10}O$
DOT #1915   mw: 98.16

NFPA LABEL

SYNONYMS

Hexanon   Pimelic ketone
Ketohexamethylene   Sextone
Nadone

PROPERTIES
Colorless liquid, acetonelike odor; solubility in water 15%; melting point −45.0°C, boiling point 115.6°C; density 0.9478 at 20°C/4°C; flash point 43.9°C; lower explosive limit 1.1% at 100°C; autoignition temperature 420°C; vapor pressure 10 mm Hg at 38.7°C; vapor density 3.4

OSHA PEL   TWA 50 ppm; IDLH 5000 ppm
ACGIH TLV   TWA 25 ppm (skin)
NIOSH REL   TWA 100 mg/m³

DOT Classification   Flammable or combustible liquid; Label: Flammable liquid

TOXIC EFFECTS
Moderately toxic by ingestion, inhalation, and several other routes. A skin and severe eye irritant.

Routes   Ingestion, inhalation, contact

Symptoms   Irritation of the skin, eyes, and mucous membranes; narcosis; changes in the sense of smell

Target Organs   Respiratory system, eyes, skin, central nervous system

HAZARDOUS REACTIONS

Hydrogen peroxide + nitric acid: reaction forms an explosive peroxide
Nitric acid: explosive reaction at 75°C
Oxidizing materials: vigorous reaction

When heated to decomposition, it emits acrid smoke and irritating fumes.

FIRST AID
In case of contact with the eyes, immediately wash the eyes with large amounts of water, occasionally lifting the lower and upper lids. Get medical attention immediately. Contact lenses should not be worn when working with this chemical.

In case of contact with the skin, immediately flush the contaminated skin with water. If this chemical penetrates the clothing, immediately remove the clothing and flush the skin with water. If irritation persists after washing, get medical attention.

If a person breathes in large amounts of this chemical, move the victim to fresh air at once. If breathing has stopped, perform artificial respiration. Keep the affected person warm and at rest. Get medical attention as soon as possible.

If this chemical has been swallowed, get medical attention immediately.

FIRE HAZARD
Flammable when exposed to heat or flame. Slight explosion hazard in the form of vapor when exposed to flame.

Use alcohol foam, dry chemical, or CO_2 extinguishers to fight a fire.

HANDLING   Wear appropriate clothing and equipment to prevent repeated or prolonged skin contact. Wash promptly if skin becomes wet. Immediately remove any clothing that becomes wet to avoid flammability hazard.

Wear eye protection to reduce probability of eye contact.

CYCLOHEXENE

CAS #110-83-8 mf: C_6H_{10}
DOT #2256 mw: 82.15

SYNONYMS
Benzenetetrahydride
1,2,3,4-Tetrahydrobenzene

PROPERTIES   Colorless liquid with a sweetish odor; very low solubility in water; freezing point −103.7°C, boiling point 83°C; density 0.8102 at 20°C/4°C; flash point <−6°C; lower explosive limit 1.2%, upper explosive limit (at 100°C) 5%; autoignition temperature 310°C; vapor pressure 160 mm Hg at 38°C; vapor density 2.8

OSHA PEL   TWA 300 ppm; 10,000 ppm

ACGIH TLV   TWA 300 ppm

DOT Classification   Flammable liquid; Label: Flammable liquid

TOXIC EFFECTS   Moderately toxic by inhalation and ingestion.

Routes   Ingestion, inhalation, contact

Symptoms   Irritation of the skin, eyes, and respiratory system; drowsiness

Target Organs   Skin, eyes, respiratory system

HAZARDOUS REACTIONS
Oxidizers: incompatible

FIRST AID   In case of contact with the eyes, immediately wash the eyes with large amounts of water, occasionally lifting the lower and upper lids. Get medical attention immediately. Contact lenses should not be worn when working with this chemical.

In case of contact with the skin, promptly wash the contaminated skin with soap and water. If this chemical penetrates through the clothing, promptly remove the clothing and wash the skin with soap and water. Get medical attention promptly.

If a person breathes in large amounts of this chemical, move the victim to fresh air at once. If breathing has stopped, perform artificial respiration. Keep the affected person warm and at rest. Get medical attention as soon as possible.

If this chemical has been swallowed, get medical attention immediately.

FIRE HAZARD   A very dangerous fire hazard when exposed to flame. The vapor is explosive when exposed to heat and flame. Use foam, CO_2, or dry chemical extinguishers to fight a fire.

HANDLING   Wear appropriate clothing and equipment to prevent repeated or prolonged skin contact. Wash promptly if skin becomes wet. Immediately remove any clothing that becomes contaminated to avoid flammability hazard.

Wear eye protection to reduce probability of eye contact.

CYCLOPENTANE

CAS #287-92-3 mf: C_5H_{10}
DOT #1146 mw: 70.15

SYNONYM
Pentamethylene

PROPERTIES   Colorless liquid; freezing point −93.7°C, boiling point 49.3°C; density 0.745 at 20°C/4°C; flash point −7°C; autoignition temperature 380°C; vapor pressure 400 mm Hg at 31.0°C; vapor density 2.42

ACGIH TLV   TWA 600 ppm

EPA Extremely Hazardous Substances List

DOT Classification   Flammable liquid; Label: Flammable liquid

TOXIC EFFECTS   Mildly toxic by ingestion and inhalation. A narcotic at high concentrations.

HAZARDOUS REACTIONS
Oxidizers: incompatible

When heated to decomposition, it emits acrid smoke and irritating fumes.

FIRE HAZARD   A very dangerous fire hazard when exposed to heat or flame. To fight a fire, use foam, dry chemicals, or CO_2 extinguishers.

ortho-DICHLOROBENZENE

CAS #95-50-1   mf: $C_6H_4Cl_2$
DOT #1591   mw: 147.00

NFPA LABEL

SYNONYMS

Chloroben
Chloroden
Cloroben
DCB
o-Dichlorobenzene
o-Dichlorobenzol
1,2-Dichlorobenzene
ODB
ODCB
Orthodichlorobenzene
Orthodichlorobenzol

PROPERTIES   Colorless to pale yellow liquid with a pleasant, aromatic odor; melting point −17.5°C, freezing point −22°C; boiling point 180–183°C; density 1.307 at 20°C/20°C; flash point 66.1°C; lower explosive limit 2.2%, upper explosive limit 9.2%; autoignition temperature 647.8°C; vapor pressure 1.2 mm Hg; vapor density 5.05

OSHA PEL   ceiling 50 ppm; IDLH 1700 ppm

ACGIH TLV   ceiling 50 ppm (skin)

Community Right to Know List

DOT Classification   ORM-A; Label: None

IMO Classification   Poison B; Label: St. Andrew's cross

TOXIC EFFECTS   Poisonous by ingestion and some other routes. Moderately toxic by inhalation. An eye, skin, and mucous membrane irritant. Causes liver and kidney injury. A teratogen and a suspected carcinogen.

Routes   Ingestion, absorption, inhalation, contact

Symptoms   Irritation of the nose, eyes, and mucous membranes; liver and kidney damage; skin blistering

Target Organs   Liver, kidneys, skin, eyes

HAZARDOUS REACTIONS
Aluminum: incompatible when heated
Aluminum alloys: incompatible
Aluminum containers: storage hazard
Oxidizing materials: vigorous reaction

When heated to decomposition, it emits very toxic fumes of Cl^-.

FIRST AID   In case of contact with the eyes, immediately wash the eyes with large amounts of water, occasionally lifting the lower and upper lids. Get medical attention immediately. Contact lenses should not be worn when working with this chemical.

In case of contact with the skin, promptly wash the contaminated skin with soap and water. If this chemical penetrates through the clothing, promptly remove the clothing and wash the skin with soap and water. Get medical attention promptly.

If a person breathes in large amounts of this chemical, move the victim to fresh air at once. If breathing has stopped, perform artificial respiration. Keep the affected person warm and at rest. Get medical attention as soon as possible.

If this chemical has been swallowed, get medical attention immediately.

FIRE HAZARD   Flammable when exposed to heat or flame.

To fight a fire, use water, foam, CO_2, or dry chemical extinguishers.

HANDLING   Wear appropriate clothing and equipment to prevent repeated or prolonged skin contact. Wash at the end of each work shift if there is a reasonable probability of contact.

Wear eye protection to reduce probability of eye contact.

See Chapter 2 for information on handling carcinogens.

DICHLORODIFLUOROMETHANE

CAS #75-71-8   mf: CCl_2F_2
DOT #1028   mw: 120.91

SYNONYMS

Difluorodichloromethane
F 12
Halon 122
Refrigerant 12
Freon 12
Propellant 12
R 12

PROPERTIES   Colorless gas with a characteristic etherlike odor at >20% by volume; solubility in water 0.008%; melting point −158°C, boiling point −29°C; vapor pressure 5 atm at 16.1°C

OSHA PEL   TWA 1000 ppm; IDLH 50,000 ppm

ACGIH TLV   TWA 1000 ppm

DOT Classification   Nonflammable gas; Label: Nonflammable gas

EPA TSCA Inventory

TOXIC EFFECTS

Routes   Ingestion, contact

Symptoms   Dizziness, tremors, unconsciousness; liver effects; narcosis; cardiac arrhythmias, cardiac arrest

Target Organs   Cardiovascular system, peripheral nervous system

HAZARDOUS REACTIONS

Aluminum: violent reaction
Aluminum powder: incompatible
Calcium: incompatible
Magnesium: incompatible
Metals (chemically active): incompatible
Potassium: incompatible
Sodium: incompatible
Zinc: incompatible

When heated to decomposition, it emits highly toxic fumes of phosgene, Cl^-, and F^-.

FIRST AID   In case of contact with the eyes, immediately wash the eyes with large amounts of water, occasionally lifting the lower and upper lids. Get medical attention immediately. Contact lenses should not be worn when working with this chemical.

In case of contact with the skin, immediately flush the contaminated skin with water. If this chemical penetrates the clothing, immediately remove the clothing and flush the skin with water. Get medical attention promptly.

If a person breathes in large amounts of this chemical, move the victim to fresh air at once. If breathing has stopped, perform artificial respiration. Keep the affected person warm and at rest. Get medical attention as soon as possible.

HANDLING   Wear appropriate clothing and equipment to prevent skin becoming wet or frozen. Immediately remove any clothing that becomes wet to avoid flammability hazard.

Wear eye protection to reduce probability of eye contact.

DICHLOROETHANE

CAS #1300-21-6   mf: $C_2H_4Cl_2$
mw: 98.96

SYNONYMS
Asymmetrical dichloroethane
Ethylidene chloride
1,1-Ethylidene dichloride

PROPERTIES   Colorless liquid with a chloroform-like odor; solubility in water <0.1%; melting point −96.7°C, boiling point 57.2°C; flash point −8.3°C; lower explosive limit 6%, upper explosive limit 16%; vapor pressure 182 mm Hg

OSHA PEL   TWA 100 ppm; IDLH 4000 ppm

DOT Classification   Flammable liquid; Label: Flammable liquid

TOXIC EFFECTS   Moderately toxic by ingestion, skin contact, and possibly other routes. Mildly toxic by inhalation. A teratogen.

Routes   Ingestion, inhalation, contact

Symptoms   Central nervous system depressant/depression; skin irritation; drowsiness; unconsciousness; liver and kidney damage

Target Organs   Skin, liver, kidneys

HAZARDOUS REACTIONS

Oxidizing agents: incompatible
Caustics (strong): incompatible

When heated to decomposition, it emits very toxic fumes of Cl^-.

FIRST AID   In case of contact with the eyes, immediately wash the eyes with large amounts of water, occasionally lifting the lower and upper lids. Get medical attention immediately. Contact lenses should not be worn when working with this chemical.

In case of contact with the skin, promptly flush the contaminated skin with soap and water. If this chemical penetrates through the clothing, promptly remove the clothing and flush the skin with water. If irritation persists after washing, get medical attention.

If a person breathes in large amounts of this chemical, move the victim to fresh air at once. If breathing has stopped, perform artificial respiration. Keep the affected person warm and at rest. Get medical attention as soon as possible.

If this chemical has been swallowed, get medical attention immediately.

HANDLING   Wear appropriate clothing and equipment to prevent repeated or prolonged skin contact. Wash immediately if the skin becomes wet. Immediately remove any clothing that becomes wet to avoid flammability hazard.

Wear eye protection to reduce probability of eye contact.

2,4-DICHLOROPHENOL

CAS #120-83-2   mf: $C_6H_4Cl_2O$
mw: 163.00

NFPA LABEL

SYNONYMS
DCP
2,4-DCP

PROPERTIES   Colorless crystals; melting point 45°C, boiling point 210°C; density 1.383 at 60°C/25°C; flash point 113.9°C; vapor pressure 1 mm Hg at 53.0°C; vapor density 5.62

Community Right to Know List

TOXIC EFFECTS   Moderately toxic by ingestion, skin contact, and some other routes. A carcinogen and teratogen. A priority pollutant.

HAZARDOUS REACTIONS
Oxidizing materials: vigorous reaction

When heated to decomposition or on contact with acid or acid fumes, it emits toxic fumes of Cl^-.

FIRE HAZARD   Combustible when exposed to heat or flame.

To fight a fire, use alcohol foam, foam, CO_2, or dry chemical extinguishers.

HANDLING   See Chap. 2 for information on handling carcinogens.

DIETHYLNITROSAMINE

CAS #55-18-5   mf: $C_4H_{10}N_2O$
mw: 102.16

SYNONYMS
DANA
DEN
DENA
Diethylnitrosoamine
N,N-Diethylnitrosamine
N-Ethyl-N-nitroso-ethan-amine
NDEA
N-Nitrosodiethylamine
Nitrosodiethylamine

PROPERTIES   A volatile yellow liquid; soluble in water (10%), alcohol, and ether; boiling point 177°C; density 0.9422 at 20°C/4°C. It is stable at room temperature for several days in aqueous solution at neutral and alkaline pH. It is less stable at strongly acid pH at room temperature, but it is not readily decomposed under the latter conditions. N-nitrosodiethylamine can be oxidized by strong oxidants to the corresponding nitramine and can be reduced by various reducing agents to the hydrazine or amine.

Community Right to Know List

Although the information given below applies specifically to N-nitrosodiethylamine [$(CH_3CH_2)_2N-N=O$], the general precautions and procedures are also applicable to other dialkylnitrosamines. It should be noted that the use of N-nitrosodimethylamine (dimethylnitrosamine) is regulated by OSHA when it is present at levels equal to or greater than 1% of a solution or mixture.

TOXIC EFFECTS   Poisonous by ingestion, inhalation, and skin contact. A carcinogen and teratogen. A transplacental carcinogen.

Symptoms   Severe kidney damage, death; cancer of the liver, lung, esophagus, trachea, and nasal cavity.

HAZARDOUS REACTIONS   When heated to decomposition, it emits toxic fumes of NO_x.

HANDLING   All work with N-nitrosodiethylamine should be carried out in a well-ventilated hood or in a glove box equipped with a HEPA filter. To the extent possible, all vessels that contain N-nitrosodiethylamine should be kept closed. All work should be carried out in apparatus that is contained in or mounted above unbreakable pans that will contain any spill. All containers should bear a label such as the following: CANCER-SUSPECT AGENT. All personnel who handle the material should wear plastic, latex, or neoprene gloves and a fully buttoned laboratory coat.

Storage   All bottles of N-nitrosodiethylamine should be stored and transported within an unbreakable outer container; storage should be in a ventilated storage cabinet (or in a hood).

See Chap. 2 for information on handling carcinogens.

SPILL CLEANUP   Because N-nitrosodiethylamine is chemically stable under usual conditions, disposal is best carried out by incineration. For incineration of liquid wastes, solutions should be neutralized if necessary, filtered to remove solids, and put in closed polyethylene containers for transport. All equipment should be thoroughly rinsed with solvent, which should then be added to the liquid waste for incineration. Great care should be exercised to prevent contamination of the outside of the solvent container. If possible, solid wastes should be incinerated; if this is not possible, solid wastes from reaction mixtures that may contain N-nitrosodiethylamine should be extracted and the extracts added to the liquid waste. Similarly, any rags, paper, and such that may be contaminated should be incinerated. Contaminated solid materials should be enclosed in sealed plastic bags labeled CANCER-SUSPECT AGENT and with the name and amount of the carcinogen. The bags should be stored in a well-ventilated area until they are incinerated.

Spills of N-nitrosodiethylamine can be absorbed by Celite or a commercial spill absorbant. After the absorbant containing the major share of the nitrosamine has been picked up (avoid dusts; do not sweep), the surface should be thoroughly cleaned with a strong detergent solution. If a major spill occurs outside a ventilated area, the room should be evacuated and the cleanup operation should be carried out by persons equipped with self-contained respirators. Those involved in this operation should wear rubber gloves, laboratory coats, and plastic aprons or equivalent protective apparel.

DIISOPROPYL ETHER

CAS #108-20-3   mf: $C_6H_{14}O$
DOT #1159   mw: 102.20

SYNONYMS
Diisopropyl oxide
2-Isopropoxypropane
Isopropyl ether

PROPERTIES   Colorless liquid; ethereal odor, miscible in water; melting point −60°C, boiling point 68.5°C; density 0.719 at 25°C; flash point −27.7°C (closed cup); lower explosive limit 1.4%, upper explosive limit 7.9%; autoignition temperature 443.3°C; vapor pressure 150 mm Hg at 25°C; vapor density 3.52

OSHA PEL   TWA 500 ppm

ACGIH TLV   TWA 250 ppm; STEL 310 ppm

DOT Classification   Flammable liquid; Label: Flammable liquid

TOXIC EFFECTS   Moderately toxic by several routes. Mildly toxic by ingestion, inhalation, and skin contact. A skin irritant.

HAZARDOUS REACTIONS   A shock-sensitive explosive under some conditions. Reacts rapidly with air to form very sensitive, explosive peroxides which precipitate as crystals.

Chlorosulfonic acid: violent reaction
Nitric acid: violent reaction
Oxidizing materials: vigorous reaction
Propionyl chloride: reaction can burst a sealed container

When heated to decomposition, it emits acrid smoke and fumes.

FIRE HAZARD   A very dangerous fire hazard and severe explosion hazard when exposed to heat, flame, sparks, or oxidizers.

To fight a fire, use alcohol foam, CO_2, foam, or dry chemical extinguishers.

DIISOPROPYL FLUOROPHOSPHATE

CAS #55-91-4   mf: $C_6H_{14}FO_3P$
  mw: 184.17

SYNONYMS
DFP
Diflupyl
Diisopropoxyphosphoryl fluoride
O,O-Diisopropyl fluorophosphate
Diisopropyl fluorophosphonate
Diisopropyl phosphofluoridate
Diisopropyl phosphorofluoridate
Diisopropylfluorophosphoric acid ester
Dyflos
Floropryl
Fluophosphoric acid, diisopropyl ester
Fluorodiisopropyl phosphate
Fluoropryl
Fluostigmine
Isofluorophate

Isopropyl fluophosphate
Neoglaucit
Phosphorofluoridic acid, diisopropyl ester

PROPERTIES   Oily liquid; melting point −82°C, boiling point 46° at 5 mm; density 1.07 (approximately); vapor density 5.24

EPA Extremely Hazardous Substances List

USE   Diisopropyl fluorophosphate (DFP) is a valuable reagent in biochemical laboratories because it is an effective in-vitro inhibitor of serine-type proteolytic enzymes useful in curtailing destruction of proteins.

TOXIC EFFECTS   A poison by ingestion, inhalation, skin contact, ocular, and most other routes. An FDA proprietary drug. Used as a miotic agent.

DFP, as well as the organophosphorus insecticides, reacts with acetylcholinesterase to form covalent derivatives, which leads to inactivation of the enzyme and the accumulation of acetylcholine. Fortunately, the toxic effects respond to treatment if the proper antidotes are immediately available.

It causes pinpoint pupils in those affected, as well as severe lacrymation and rhinitis. Contact with skin can lead to localized fasciculations (visible contractions) of the muscles nearby. Other symptoms of exposure include weakness, wheezing, and tachycardia (increased heart rate). Severe intoxication is evidenced by ataxia (failure of muscular coordination), confusion, convulsions, and respiratory paralysis leading to death.

Routes   Inhalation, ingestion, contact

Symptoms   Miosis (pupillary constriction), headache, damage to eyes; nausea, vomiting, diarrhea; central nervous system disturbances

HAZARDOUS REACTIONS   When heated to decomposition, it emits toxic fumes of F^- and PO_x.

FIRST AID   Treatment consists of immediate administration of atropine followed by a specific antidote, pyridine-2-aldoxime (2-PAM; pralidoxime chloride; available commercially as Protopam Chloride). Atropine is used to block certain receptors on which acetylcholine acts, while pyridine-2-aldoxime reactivates the covalently modified acetylcholinesterase. Both drugs should be on hand in any laboratory in which DFP is used, and medical assistance must be readily available.

HANDLING   The keys to safety in handling DFP are an appreciation of its toxicity, the presence of adequate ventilation, a solution of base in case of spillage and in which to discard all implements that have come into contact with the reagent, and the presence at the site of the two antidotes.

DFP should be obtained in ampules containing the minimum amount necessary for use; the reagent is not expensive, and excess material should be neutralized and disposed of in a safe manner. Ampules containing 0.1, 0.4, and 1.0 g are available. The ampules should be opened in a hood using great care to avoid splatter or in a glove box equipped with a HEPA filter. Protective apparel should include disposable gloves and a face shield.

A large container of 2 N NaOH should be available wherever DFP is being used so that the entire ampule, if necessary, can be discarded into it; decomposition of DFP is relatively rapid in basic solution. Once the ampule has been opened, the required amount of DFP should be withdrawn with a disposable syringe and used immediately. The opened ampule and gloves and other items that have come in contact with the DFP should be washed with base before disposal. Any spills that occur should be neutralized with excess base before they are cleaned up.

Solutions of DFP should be handled only under conditions that provide adequate ventilation. Most cold rooms, for example, are not adequately ventilated. All solutions of DFP should be kept in tightly stoppered containers. After such solutions have served their purpose, they should be made alkaline to ensure complete decomposition of any remaining DFP before disposal.

DIMETHYL ACETAMIDE

CAS #127-19-5   mf: C_4H_9NO
mw: 87.14

SYNONYMS
Acetdimethylamide
Acetic acid dimethylamide
Dimethylacetamide
Dimethylacetone amide
Dimethylamide acetate
DMA
DMAC

PROPERTIES   Colorless liquid with a faint, ammonialike odor; miscible with water; melting point −20°C, boiling point 165°C; density 0.9448 at 15.5°C; flash point 77.2°C (TOC); lower explosive limit 2.0%, upper explosive limit 11.5% at 740 mm and 160°C; vapor pressure 1.3 mm Hg at 25°C; vapor density 3.01

OSHA PEL TWA 10 ppm (skin); IDLH 400 ppm
ACGIH TLV TWA 10 ppm (skin)

TOXIC EFFECTS Moderately toxic by skin contact, inhalation, and other routes. Mildly toxic by ingestion. A teratogen. A skin irritant. Less toxic than dimethylformamide.

Routes Ingestion, absorption, inhalation, contact

Symptoms Jaundice, liver damage; depressant/depression; lethargy, hallucinations, delusions; irritation of the skin

Target Organs Liver, skin

HAZARDOUS REACTIONS A moderate explosion hazard.

Carbon tetrachloride: violent reaction when heated above 90°C, 71°C with iron powder catalyst
Halogenated compounds: violent reaction when heated above 90°C, 71°C with iron powder catalyst
Hexachlorocyclohexane: violent reaction when heated above 90°C, 71°C with iron powder catalyst

When heated to decomposition, it emits highly toxic fumes of NO_x.

FIRST AID In case of contact with the eyes, immediately wash the eyes with large amounts of water, occasionally lifting the lower and upper lids. Get medical attention immediately. Contact lenses should not be worn when working with this chemical.

In case of contact with the skin, immediately flush the contaminated skin with water. If this chemical penetrates the clothing, immediately remove the clothing and flush the skin with water. Get medical attention promptly.

If a person breathes in large amounts of this chemical, move the victim to fresh air at once. If breathing has stopped, perform artificial respiration. Keep the affected person warm and at rest. Get medical attention as soon as possible.

If this chemical has been swallowed, get medical attention immediately.

HANDLING Wear appropriate clothing and equipment to prevent any possibility of skin contact. Wash immediately if skin becomes contaminated. Immediately remove any nonimpervious clothing that becomes contaminated.

Wear eye protection to reduce probability of eye contact.

A quick-drench shower should be available.

DIMETHYLFORMAMIDE

CAS #68-12-2 mf: C_3H_7NO
DOT #2265 mw: 73.11

STRUCTURE

$$\overset{O}{\underset{}{\overset{\|}{HC-N(CH_3)_2}}}$$

NFPA LABEL

SYNONYMS
N,N-Dimethyl formamide
DMF
DMFA
N-Formyldimethylamine

PROPERTIES Colorless liquid with a faint ammonialike odor; miscible with water; freezing point −61°C, boiling point 152.8°C; density 0.9445 at 25°C/4°C; flash point 136°C; lower explosive limit 2.2% at 100°C, upper explosive limit 15.2% at 100°C; autoignition temperature 445°C; vapor pressure 3.7 mm Hg at 25°C; vapor density 2.51

OSHA PEL TWA 10 ppm (skin); IDLH 3500 ppm
ACGIH TLV TWA 10 ppm (skin)

DOT Classification Flammable or combustible liquid; Label: Flammable liquid

TOXIC EFFECTS Moderately toxic by ingestion and several other routes. Mildly toxic by skin contact and inhalation. A teratogen. A skin and eye irritant.

Routes Ingestion, absorption, inhalation, contact

Symptoms Nausea, vomiting, colic; liver damage, hepatomegaly; high blood pressure; face flush; dermatitis

Target Organs Liver, kidneys, cardiovascular system, skin

HAZARDOUS REACTIONS Explosion hazard when exposed to flame.

Aluminum alkyls: incompatible
Bromine: explosive reaction
C_6Cl_6: incompatible
Carbon tetrachloride: incompatible
Carbon tetrachloride + iron: violent reaction
Chlorine: violent reaction
Chromic acid: incompatible
Chromium trioxide: ignition on contact

156 DIMETHYL SULFATE

Diisocyanatomethane: violent reaction
2,5-Dimethyl pyrrole + phosphorus hypochlorite: incompatible
Halogenated compounds + iron: incompatible
Halogenated hydrocarbons: incompatible
1,2,3,4,5,6-Hexachlorocyclohexane + iron: violent reaction
Lithium azide: forms explosive mixture, shock sensitive above 200°C
Magnesium nitrate: vigorous exothermic reaction
Methylene diisocyanates: incompatible
Nitrates (inorganic and organic): incompatible
Oxidizing materials: incompatible
P_2O_3: incompatible
Phosphorus pentoxide: incompatible
Potassium permanganate: explosive reaction
Sodium + heat: vigorous exothermic reaction
Sodium hydride + heat: vigorous exothermic reaction
Sodium hydroborate + heat: violent reaction
Sulfinyl chloride + traces of iron or zinc: vigorous exothermic reaction
Tetraethylaluminum: incompatible
2,4,6-Trichloro-1,3,5-triazine (with gas evolution): vigorous exothermic reaction
Triethylaluminum + heat: explosive reaction
Uranium perchlorate: forms explosive mixture

When heated to decomposition, it emits toxic fumes of NO_x.

FIRST AID   In case of contact with the eyes, immediately wash the eyes with large amounts of water, occasionally lifting the lower and upper lids. Get medical attention immediately. Contact lenses should not be worn when working with this chemical.

In case of contact with the skin, immediately flush the contaminated skin with water. If this chemical penetrates the clothing, immediately remove the clothing and flush the skin with water. If irritation persists after washing, get medical attention.

If a person breathes in large amounts of this chemical, move the victim to fresh air at once. If breathing has stopped, perform artificial respiration. Keep the affected person warm and at rest. Get medical attention as soon as possible.

If this chemical has been swallowed, get medical attention immediately.

FIRE HAZARD   Flammable when exposed to heat or flame.
To fight a fire, use foam, CO_2, or dry chemical extinguishers.

HANDLING   Significant toxicity can result from skin contact. Therefore, gloves or other protective apparel (preferably of butyl rubber) should be worn when handling liquid DMF. It should also be noted that use of DMF as a solvent for toxic materials that are not ordinarily absorbed may increase their skin-contact hazard.

Wear appropriate clothing and equipment to prevent repeated or prolonged skin contact. Wash promptly if the skin becomes contaminated. Promptly remove any nonimpervious clothing that becomes contaminated.

Wear eye protection to reduce probability of eye contact.

DIMETHYL SULFATE

CAS #77-78-1   mf: $C_2H_6O_4S$
DOT #1595   mw: 126.13

STRUCTURE

$$CH_3O\overset{\overset{O}{\|}}{\underset{\underset{O}{\|}}{S}}-OCH_3$$

NFPA LABEL

4 / 2 / 0

SYNONYMS
DMS
DMS (methyl sulfate)
Dimethyl monosulfate
Methyl sulfate
Sulfuric acid, dimethyl ester

PROPERTIES   Colorless, oily liquid; slight onion odor; solubility in water 2.8 g/100 mL at 18°C; hydrolyzes to sulfuric acid and methanol in water; soluble in alcohol, ether, and benzene; melting point −31.8°C; boiling point 188.8°C; density 1.328 at 20°C; flash point 28.3°C (closed cup), 83.3°C (open cup); autoignition temperature 187.8°C; vapor pressure 0.54 mm Hg at 20°C; vapor density 4.35

OSHA PEL   TWA 1 ppm (skin); IDLH 10 ppm

ACGIH TLV   TWA 0.1 ppm (skin)

Community Right to Know List and EPA Extremely Hazardous Substances List

DOT Classification   Corrosive material; Label: Corrosive; Poison B; Label: Poison

TOXIC EFFECTS   A poison by inhalation, ingestion, and other routes. A carcinogen and teratogen. A corrosive irritant to the skin, eyes, and mucous membranes.

There is no odor or initial irritation to give warning of exposure.

Many cases of DMS poisoning have been reported. The common initial symptoms are headache and giddiness, with burning of the eyes. The patient's condition may worsen, with painful eyes, nose, and throat irritation, loss of voice, coughing, difficulty in breathing and swallowing, vomiting, and diarrhea possible. The onset of these symptoms may be delayed for up to 10 h.

Skin contact causes blistering and necrosis, and DMS can be absorbed through the skin in sufficient quantities to cause systemic intoxication. In the worst cases, there is severe inflammation of the mucous membranes and pulmonary injury that may be fatal; several deaths have occurred. For example, exposure to 97 ppm for 10 min was fatal.

Survivors of severe exposures may have serious injury to the liver and kidneys, with suppression of urine, jaundice, albuminuria, and hematuria appearing. Death from kidney or liver damage may be delayed for several weeks.

Routes   Ingestion, absorption, inhalation, contact

Symptoms   Irritation of the eyes, nose, throat; headache; giddiness; photophobia; periorbital edema; dysphonia, aphonia, dysphagia cough, chest pain; cyanosis; vomiting; diarrhea, dysuria, icterus, albuminuria, hematuria

Target Organs   Eyes, respiratory system, liver, kidneys, central nervous system, skin

HAZARDOUS REACTIONS
Ammonia solutions: violent reaction
Oxidizers (strong): incompatible
Sodium azide: violent reaction

When heated to decomposition, it emits toxic fumes of SO_x.

FIRST AID   In case of contact with the eyes, immediately wash the eyes with large amounts of water, occasionally lifting the lower and upper lids. Get medical attention immediately. Contact lenses should not be worn when working with this chemical.

In case of contact with the skin, immediately flush the contaminated skin with water. If this chemical penetrates the clothing, immediately remove the clothing and flush the skin with water. Get medical attention promptly.

If a person breathes in large amounts of this chemical, move the victim to fresh air at once. If breathing has stopped, perform artificial respiration. Keep the affected person warm and at rest. Get medical attention as soon as possible.

If this chemical has been swallowed, get medical attention immediately.

FIRE HAZARD   Flammable when exposed to heat, flame, or oxidizers.

To fight a fire, use water, foam, CO_2, or dry chemical extinguishers.

HANDLING   The ACGIH and OSHA levels include a warning of the potential contribution of skin absorption to the overall exposure.

Procedure A (see Chap. 2) should be used when handling more than a few grams of DMS in view of its fairly high carcinogenic potency in rats by inhalation and its ability to penetrate the skin. It is particularly important to avoid skin contact by appropriate use of rubber gloves, a rubber apron, and other protective apparel, and to avoid inhalation of even low concentrations of vapor by working in a hood.

Wear appropriate clothing and equipment to prevent any possibility of skin contact. Wash immediately if skin becomes contaminated. Immediately remove any nonimpervious clothing that becomes contaminated.

Wear eye protection to prevent any possibility of eye contact.

The following equipment should be available: an eyewash station and a quick-drench shower.

DIMETHYL SULFOXIDE

CAS #67-68-5   mf: C_2H_6OS
    mw: 78.14

NFPA LABEL

SYNONYMS

Deltan	DMS-70
Demasorb	DMS-90
Demavet	DMSO
Demeso	Dolicur
Demsodrox	Doligur
Dermasorb	Domoso
Dimethyl sulphoxide	Dromisol
Dimexide	Durasorb
Dipirartril-tropico	Gamasol 90

Hyadur
Infiltrina
Methylsulfinylmethane
Methyl sulfoxide
Somipront
Sulfinylbis(methane)
Syntexan
Topsym

PROPERTIES   Clear, water-white, hygroscopic liquid; melting point 18.5°C, boiling point 189°C; density 1.100 at 20°C; flash point 95°C (open cup); lower explosive limit 2.6%, upper explosive limit 28.5%; autoignition temperature 215°C; vapor pressure 0.37 mm Hg at 20°C

TOXIC EFFECTS   Poisonous by ingestion. Moderately toxic by some other routes. A teratogen. A skin and eye irritant. Readily passes through the skin, and is capable of carrying dissolved chemicals with it into the body.

Symptoms   Nausea, vomiting, jaundice, anaphylactic reaction, corneal opacity

HAZARDOUS REACTIONS
Acetyl chloride: violent or explosive reaction
Aluminum perchlorate: forms powerfully explosive mixture
Benzenesulfonyl chloride: violent or explosive reaction
Borane: violent or explosive reaction
Boron compounds: violent or explosive reaction
Bromobenzoyl acetanilide: violent or explosive reaction
Bromoform: violent or explosive reaction
Carbonyl diisothiocyanate: violent or explosive reaction
Copper + trichloroacetic acid: violent or explosive reaction
Cyanuric chloride: violent or explosive reaction
Dinitrogen tetroxide: violent or explosive reaction
Disulfur dichloride: violent or explosive reaction
Halides (acyl, aryl, and nonmetal): violent or explosive reaction
Hexachlorocyclotriphosphazine: violent or explosive reaction
Iodine pentafluoride: violent or explosive reaction
Iron(III) nitrate: forms powerfully explosive mixture
Magnesium perchlorate: violent or explosive reaction
Metal alkoxides: violent or explosive reaction
Metal oxosalts: incompatible
Metal salts of oxoacids: forms powerfully explosive mixture
NIO_4: violent or explosive reaction
Nonanhydrononaborate(2−) ion: violent or explosive reaction
Oxalyl chloride: violent or explosive reaction
Oxidizing materials: incompatible
P_2O_3: violent or explosive reaction
Perchloric acid: incompatible
Periodic acid: incompatible
Phosphorous trichloride: violent or explosive reaction
Phosphoryl chloride: violent or explosive reaction
Potassium *tert*-butoxide: violent or explosive reaction
Silver difluoride: violent or explosive reaction
Silver fluoride: violent or explosive reaction
Sodium hydride: violent or explosive reaction
Sodium isopropoxide: violent or explosive reaction
Sodium perchlorate: forms powerfully explosive mixture
Sulfur dichloride: violent or explosive reaction
Sulfur trioxide: incompatible
Sulfuryl chloride: violent or explosive reaction
Tetrachlorosilane: violent or explosive reaction
Thionyl chloride: violent or explosive reaction
Trifluoroacetic acid anhydride: violent or explosive reaction

When heated to decomposition, it emits highly toxic fumes of SO_x.

FIRE HAZARD   Combustible when exposed to heat or flame.
To fight a fire, use foam, alcohol foam, CO_2, or dry chemical extinguishers.

2,4-DINITROPHENOL

CAS #51-28-5
mf: $C_6H_4N_2O_5$
mw: 184.12

SYNONYMS
Aldifen
Chemox PE
α-Dinitrophenol
2,4-DNP
Fenoxyl carbon N
1-Hydroxy-2, 4-dinitrobenzene
Maroxol-50
Nitro kleenup
Solfo Black B
Tertrosulphur Black PB
Tertrosulphur PBR

PROPERTIES   Yellow crystals; melting point 112°C; density 1.683 at 24°C; vapor density 6.35

TOXIC EFFECTS   A deadly poison by ingestion. A poison by inhalation and most other routes. Moderately toxic by skin contact. A teratogen. A skin irritant. Phytotoxic.

HAZARDOUS REACTIONS   An explosive.

Alkalies: forms explosive salts
Ammonia: forms explosive salts

When heated to decomposition, it emits toxic fumes of NO_x.

DIOXANE

CAS #123-91-1 mf: $C_4H_8O_2$
DOT #1165 mw: 88.11

STRUCTURE

NFPA LABEL

SYNONYMS
Diethylene dioxide
1,4-Diethylene ether
Diethylene ether
Diokan
1,4-Dioxacyclohexane
1,4-Dioxane
p-Dioxane
Dioxyethylene ether
Tetrahydro-p-dioxin
Tetrahydro-1,4-dioxin

PROPERTIES   Colorless liquid, faintly alcoholic, pleasant odor, threshold = 5.7 ppm; miscible in all proportions with water, acetone, alcohols, and most organic solvents; melting point 12°C, boiling point 101.1°C; density 1.0353 at 20°C/4°C; flash point 11°C (closed cup); lower explosive limit 2.0%, upper explosive limit 22.2%; autoignition temperature 180°C; vapor pressure 40 mm Hg at 25.2°C; vapor density 3.03

OSHA PEL   TWA 100 ppm (skin)

ACGIH TLV   TWA 25 ppm (skin)

NIOSH REL   ceiling 1 ppm/30M

Community Right to Know List

DOT Classification   Flammable liquid, Label: Flammable liquid

TOXIC EFFECTS   Moderately toxic by ingestion and inhalation. Mildly toxic by skin contact. A weak carcinogen. A teratogen. An eye and skin irritant.

A worker who was exposed via the skin and inhalation routes to 500 ppm for 1 week died. Autopsy revealed damage to the kidney, liver, and brain. Symptoms of inhalation overexposure include irritation of the upper respiratory tract, coughing, eye irritation, vertigo, headache, and vomiting. An airborne concentration of 300 ppm of dioxane causes irritation of the eyes, nose, and throat. At lower levels, exposure sufficient to cause harm can occur before one realizes it through smell or irritation. Prolonged or repeated skin contact can produce drying and fissuring of the skin.

Routes   Ingestion, absorption, inhalation, contact

Symptoms   Drowsiness, headache; nausea, vomiting; irritation of the eyes, nose, and throat; lacrimation; liver damage; kidney failure; skin irritation; convulsions; high blood pressure; a carcinogen

Target Organs   Liver, kidneys, skin, eyes

HAZARDOUS REACTIONS   Dioxane forms explosive peroxides in contact with air, especially in the presence of moisture. A very dangerous explosion hazard when exposed to heat or flame.

Air: can form dangerous peroxides
Decaborane: forms impact-sensitive explosive mixture
Hydrogen + Raney nickel: violent reaction
Nitric acid + perchloric acid: potentially explosive reaction
Oxidizing materials: can react vigorously
Raney nickel catalyst: potentially explosive reaction above 210°C
Silver chloride: violent reaction
Sulfur trioxide: violent reaction
Triethynylaluminum: forms an explosive mixture sensitive to heating or drying

When heated to decomposition, it emits acrid smoke and irritating fumes.

FIRST AID   In case of contact with the eyes, immediately wash the eyes with large amounts of water, occasionally lifting the lower and upper lids. Get medical attention immediately. Contact lenses should not be worn when working with this chemical.

In case of contact with the skin, immediately flush the contaminated skin with water. If this chemical penetrates the clothing, immediately remove the clothing and flush the skin with water. If irritation persists after washing, get medical attention.

If a person breathes in large amounts of this chemical, move the victim to fresh air at once. If breathing has stopped, perform artificial respiration. Keep the affected person warm and at rest. Get medical attention as soon as possible.

If this chemical has been swallowed, get medical attention immediately.

FIRE HAZARD   A very dangerous fire hazard when exposed to heat or flame.

To fight a fire, use alcohol foam, CO_2, or dry chemical extinguishers.

HANDLING   Dioxane is the principal ingredient of Bray's solution (scintillation cocktail), and uninhabited solutions have been known to explode if left for a period of time.

The OSHA and ACGIH limits include a warning about the potential contribution of skin absorption to the overall exposure.

Although dioxane has caused tumors in animals, these have occurred only after prolonged exposure to very large amounts and, hence, it is considered a carcinogen of such low potency that no special precautions beyond normal good laboratory practice are needed for working with it. Nitrile rubber is the preferred material for gloves and other apparel used to protect against skin contact. The high volatility of the compound requires that all laboratory operations with it be carried out in a hood. Because it is miscible in water, prompt washing is an effective way to remove dioxane that has accidentally contacted the skin.

Wear appropriate clothing and equipment to prevent repeated or prolonged skin contact. Wash promptly if the skin becomes contaminated. Immediately remove any clothing that becomes wet to avoid flammability hazard.

Wear eye protection to reduce probability of eye contact.

See Chap. 2 for information on handling carcinogens.

ETHANOLAMINE

CAS #141-43-5
DOT #2491
mf: C_2H_7NO
mw: 61.10

STRUCTURE
$HOCH_2CH_2NH_2$

NFPA LABEL

2
2 0

SYNONYMS
2-Aminoethanol
β-Aminoethyl alcohol
β-Ethanolamine
Ethylolamine
Glycinol
2-Hydroxyethylamine
β-Hydroxyethylamine
MEA
Monoethanolamine
Olamine

PROPERTIES   Colorless liquid with a mild, ammonialike odor; miscible with water and alcohol, soluble in chloroform, slightly soluble in benzene; hygroscopic; boiling point 170.5°C, freezing point 10.5°C; density 1.0180 at 20°C/4°C; flash point 93.3°C (open cup); vapor pressure 6 mm Hg at 60°C; vapor density 2.11

OSHA PEL   TWA 3 ppm; IDLH 1000 ppm

ACGIH TLV   TWA 3 ppm; STEL 6 ppm

DOT Classification   Corrosive material; Label: Corrosive

TOXIC EFFECTS   Moderately toxic by ingestion, skin contact, and other routes. A corrosive irritant to skin, eyes, and mucous membranes.

Routes   Ingestion, inhalation, contact

Symptoms   Respiratory symptoms, irritation to the skin and eyes, lethargy

Target Organs   Skin, eyes, respiratory system

HAZARDOUS REACTIONS   A powerful base.

Acetic acid: violent reaction
Acetic anhydride: violent reaction
Acids: incompatible
Acrolein: violent reaction
Acrylic acid: violent reaction
Acrylonitrile: violent reaction
Cellulose: violent reaction
Chlorosulfonic acid: violent reaction
Epichlorohydrin: violent reaction
Hydrochloric acid: violent reaction
Hydrogen fluoride: violent reaction
Mesityl oxide: violent reaction
Nitric acid: violent reaction
Oleum: violent reaction
Oxidizers (strong): incompatible
Sulfuric acid: violent reaction
Vinyl acetate: violent reaction
β-Propiolactone: violent reaction

When heated to decomposition, it emits toxic fumes of NO_x.

FIRST AID   In case of contact with the eyes, immediately wash the eyes with large amounts of water, occasionally lifting the lower and upper lids. Get medical attention immediately. Contact lenses should not be worn when working with this chemical.

In case of contact with the skin, immediately flush the contaminated skin with water. If this chemical penetrates the clothing, immediately remove the clothing and flush the skin with water. If irritation persists after washing, get medical attention.

If a person breathes in large amounts of this chemical, move the victim to fresh air at once. If breathing has stopped, perform artificial respiration. Keep the affected person warm and at rest. Get medical attention as soon as possible.

If this chemical has been swallowed, get medical attention immediately.

FIRE HAZARD   Flammable when exposed to heat or flame.

To fight a fire, use foam, alcohol foam, or dry chemical extinguishers.

HANDLING   Wear appropriate clothing and equipment to prevent repeated or prolonged skin contact. Wash promptly if the skin becomes contaminated. Promptly remove any nonimpervious clothing that becomes contaminated. Work clothing should be changed daily if it is reasonably probable that the clothing has been contaminated.

Wear eye protection to prevent any possibility of eye contact.

The following equipment should be available: an eyewash station.

2-ETHOXYETHANOL

CAS #110-80-5   mf: $C_4H_{10}O_2$
DOT #1171   mw: 90.14

STRUCTURE
$CH_3CH_2OCH_2CH_2OH$

SYNONYMS
Cellosolve
Cellosolve solvent
Dowanol EE
Ektasolve EE
Ethyl cellosolve
Ethylene glycol ethyl ether
Ethylene glycol monoethyl ether
Glycol ether EE
Glycol ethyl ether
Glycol monoethyl ether
Hydroxy ether
Jeffersol EE
Oxitol
Poly-solv EE

PROPERTIES   Colorless liquid, practically odorless; boiling point 135.1°C; density 0.9360 at 15°C/15°C; flash point 94.4°C (closed cup); lower explosive limit 1.8%, upper explosive limit 14%; autoignition temperature 235°C; vapor pressure 3.8 mm Hg at 20°C; vapor density 3.10

OSHA PEL   TWA 200 ppm (skin)

ACGIH TLV   TWA 5 ppm (skin)

NIOSH REL   (Glycol ethers) Reduce to lowest level

Community Right To Know List (glycol ether compounds)

DOT Classification   Combustible liquid; Label: None

IMO Classification   Flammable or combustible liquid; Label: Flammable liquid

TOXIC EFFECTS   Moderately toxic by ingestion, skin contact, and other routes. Mildly toxic by inhalation. A teratogen. An eye and skin irritant.

HAZARDOUS REACTIONS   Moderate explosion hazard in the form of vapor when exposed to heat or flame.

Hydrogen peroxide + polyacrylamide gel + toluene: mixture is explosive when dry
Oxidizing materials: incompatible

FIRE HAZARD   Combustible when exposed to heat or flame.

To fight a fire, use alcohol foam or dry chemical extinguishers.

ETHYL ACETATE

CAS #141-78-6   mf: $C_4H_8O_2$
DOT #1173   mw: 88.12

STRUCTURE
$CH_3CH_2OCO\cdot CH_3$

NFPA LABEL

SYNONYMS
Acetic ether   Ethyl acetic ester
Acetidin   Ethyl ethanoate
Acetoxyethane   Vinegar naphtha

PROPERTIES   Colorless liquid with a pleasant, fruity odor; melting point −83.6°C, boiling point 77.15°C; density 0.8946 at 25°C; flash point −4.4°C; lower explosive limit 2.2%, upper explosive limit 11%; autoignition temperature 426.7°C; vapor pressure 100 mm Hg at 27°C; vapor density 3.04

162 ETHYL ALCOHOL

Underwriters Laboratory Classification 85–90

OSHA PEL   TWA 400 ppm; IDLH 10,000 ppm

ACGIH TLV   TWA 400 ppm

DOT Classification   Flammable liquid; Label: Flammable liquid

TOXIC EFFECTS   Poisonous by inhalation. Mildly toxic by ingestion. Human eye irritant. Irritating to mucous surfaces, particularly the eyes, gums, and respiratory passages, and also mildly narcotic.

Chronic exposure causes conjunctival irritation and corneal clouding as well as anemia, leucocytosis (transient increase in the white blood cell count), and cloudy swelling, and fatty degeneration of the viscera.

Routes   Ingestion, inhalation, contact

Symptoms   Irritation of the eyes, nose, and throat; dermatitis; narcosis; congestion of the liver and kidneys

Target Organs   Eyes, skin, respiratory system

HAZARDOUS REACTIONS   Moderate explosion hazard when exposed to flame.

Acids: incompatible
Alkalies: incompatible
Chlorosulfonic acid: violent reaction
LiAlH$_2$ + 2-chloromethyl furan: violent reaction
Lithium tetrahydroaluminate: potentially explosive reaction
Nitrates: incompatible
Oleum: violent reaction
Oxidizing materials: incompatible
Potassium *tert*-butoxide: ignites on contact

When heated to decomposition, it emits acrid smoke and irritating fumes.

FIRST AID   In case of contact with the eyes, immediately wash the eyes with large amounts of water, occasionally lifting the lower and upper lids. Get medical attention immediately. Contact lenses should not be worn when working with this chemical.

In case of contact with the skin, immediately flush the contaminated skin with water. If this chemical penetrates the clothing, immediately remove the clothing and flush the skin with water. If irritation persists after washing, get medical attention.

If a person breathes in large amounts of this chemical, move the victim to fresh air at once. If breathing has stopped, perform artificial respiration. Keep the affected person warm and at rest. Get medical attention as soon as possible.

If this chemical has been swallowed, get medical attention immediately.

FIRE HAZARD   A very dangerous fire hazard when exposed to heat or flame.

To fight a fire use CO_2, dry chemical, or alcohol foam extinguishers.

HANDLING   Wear appropriate clothing and equipment to reduce probability of skin contact. Wash immediately if skin becomes contaminated. Immediately remove any clothing that becomes wet and any nonimpervious clothing that becomes contaminated.

Wear eye protection to prevent any possibility of eye contact.

The following equipment should be available: an eyewash station and a quick-drench shower.

ETHYL ALCOHOL

CAS #64-17-5   mf: C_2H_6O
DOT #1170   mw: 46.08

NFPA LABEL

SYNONYMS

Absolute ethanol	Ethyl hydrate
Alcohol	Ethyl hydroxide
Alcohol, anhydrous	Fermentation alcohol
Alcohol dehydrated	Grain alcohol
Algrain	Jaysol
Anhydrol	Methylcarbinol
Cologne spirit	Molasses alcohol
Cologne spirits (alcohol)	Potato alcohol
Ethanol	Spirits of wine
Ethanol 200 proof	Spirit
Ethanol solution	Tecsol
Ethyl alcohol anhydrous	

PROPERTIES   Clear, colorless, fragrant liquid; burning taste; miscible in water, alcohol, chloroform, and ether; freezing point ≤130°C, boiling point 78.32°C; density 0.7893 at 20°C/4°C; flash point 13.1°C; lower explosive limit 3.3%, upper explosive limit 19%; autoignition temperature 422.8°C; vapor pressure 40 mm Hg at 19°C; vapor density 1.59

OSHA PEL   TWA 1000 ppm

ACGIH TLV   TWA 1000 ppm

DOT Classification   Flammable or combustible liquid; Label: Flammable liquid

TOXIC EFFECTS   Moderately toxic by ingestion. Mildly toxic by inhalation and skin contact. A teratogen. Affects female fertility. An eye and severe skin irritant. A central nervous system depressant.

Symptoms   Sleep disorders, hallucinations and distorted perceptions, convulsions, motor activity changes, ataxia, coma, antipsychotic, headache, pulmonary changes, alteration in gastric secretion, nausea or vomiting, other gastrointestinal changes, menstrual cycle changes and body-temperature decrease, glandular effects

HAZARDOUS REACTIONS
Acetic anhydride + sodium hydrogen sulfate: ignites and then explodes on contact
Acetyl bromide: violent reaction evolves hydrogen bromide
Acetyl chloride: incompatible
Ammonia + silver nitrate: reacts to form explosive silver nitride and silver fulminate
Bromine pentafluoride: incompatible
Calcium hypochlorite: incompatible
Chlorine trioxide: incompatible
Chromium hypochlorite: incompatible
Chromium trioxide: incompatible
Cyanuric acid + H_2O: incompatible
Dichloromethane + sulfuric acid + nitrate or nitrite: violent reaction
Disulfuric acid + nitric acid: ignites on contact
Disulfuryl difluoride: violent reaction
$HMnO_4$: incompatible
Hydrogen peroxide: incompatible
Hydrogen peroxide + sulfuric acid: incompatible
Iodine + methanol + mercuric oxide: incompatible
KO_2: incompatible
$KOC(CH_3)_3$: incompatible
Magnesium perchlorate: reaction forms explosive ethyl perchlorate
Manganese perchlorate + 2,2-dimethoxy propane: incompatible
Mercuric nitrate: incompatible
NaH_3N_2: incompatible
Nitric acid + silver: reaction forms explosive silver fulminate
Nitric acid: incompatible
Oxidizers: vigorous reaction
Oxidizing materials: violent reaction
Perchlorates: incompatible
Perchloric acid: incompatible
Phosphorous(III) oxide: ignites on contact
Platinum: ignites on contact
Potassium metal (oxidized coating): explosive reaction
Potassium-*tert*-butoxide + acids: ignites on contact
Silver perchlorate: incompatible
Sodium: reaction evolves explosive hydrogen gas
Sulfuric acid + permanganates: incompatible
Tetrachlorosilane + water: violent reaction
$UO_2(ClO_4)_2$: incompatible

FIRE HAZARD   Flammable when exposed to heat or flame.
To fight a fire, use alcohol foam, CO_2, or dry chemical extinguishers.

ETHYL ETHER

CAS #60-29-7   mf: $C_4H_{10}O$
DOT #1155   mw: 74.14

STRUCTURE   NFPA LABEL
$CH_3CH_2OCH_2CH_3$

SYNONYMS
Aether   Ether
Anesthetic ether   Ethoxyethane
Anesthesia ether   1,1'-Oxybisethane
Anesthetic ether   Solvent ether
Diethyl ether

PROPERTIES   Clear, colorless, volatile liquid, with a sweet, pungent odor, threshold = 0.2 ppm; solubility in water 6.05 wt% at 25°C, miscible in alcohol and ether, soluble in chloroform; melting point −116.2°C, boiling point 34.6°C; density 0.7135 at 20°C/4°C; flash point −45°C; lower explosive limit 1.85%, upper explosive limit 36%; autoignition temperature 180°C; vapor pressure 442 mm Hg at 20°C; vapor density 2.56

OSHA PEL   TWA 400 ppm; IDLH 19,000 ppm

ACGIH TLV   TWA 400 ppm; STEL 500 ppm

DOT Classification   Flammable liquid; Label: Flammable liquid

TOXIC EFFECTS   Moderately toxic by ingestion and other routes. Mildly toxic by inhalation. A moderate skin and eye irritant.
Not safe to inhale or ingest.
Repeated exposure of humans to diethyl ether via inhalation has caused central nervous system depres-

sion including loss of appetite, exhaustion, headache, intoxication, drowsiness, stupor, and unconsciousness. Death due to respiratory failure may result from severe and continued exposure. General anesthesia occurs at a concentration of 3.6–6.5% in air. Human subjects find diethyl ether irritating to the nose but not to the eyes or throat at a level of 200 ppm. Acute overexposure produces vomiting, paleness, irregular respiration, and low pulse rates and body temperatures. The human oral lethal dose for diethyl ether is about 420 mg/kg.

Its absorption through the skin is not usually great enough to cause a deleterious effect. Diethyl ether can cause eye irritation but no permanent damage.

Routes   Ingestion, inhalation, contact

Symptoms   Dizziness; drowsiness; headache, excitement, narcosis; nausea, vomiting; irritation of the eyes, upper respiratory tract, and skin; olfactory changes

Target Organs   Central nervous system, skin, respiratory system, eyes

HAZARDOUS REACTIONS   Ethyl ether is not corrosive or dangerously reactive. It is a storage hazard. It readily forms explosive polymeric 1-oxyperoxides on exposure to air, sometimes leading to explosive residues when distilled.

Acetyl peroxide: vigorous reaction
Air: vigorous reaction
Air (liquid): violent reaction or ignition on contact
Boron triazide: explosive reaction
Bromine: violent reaction or ignition on contact
Bromine pentafluoride: explosive reaction
Bromine trifluoride: explosive reaction
Bromoazide: vigorous reaction
Chlorine: violent reaction or ignition on contact
Chlorine trifluoride: vigorous reaction
Chromium oxide: vigorous reaction
Chromyl chloride: violent reaction or ignition on contact
Fluorine nitrate: violent reaction or ignition on contact
Halogens: violent reaction or ignition on contact
Hydrogen peroxide: violent reaction or ignition on contact
Interhalogens: violent reaction or ignition on contact
Iodine heptafluoride: violent reaction or ignition on contact
Iodine(VII) oxide: violent reaction or ignition on contact
$LiAlH_2$: vigorous reaction
Nitric acid: violent reaction or ignition on contact
Nitrosyl perchlorate: violent reaction or ignition on contact
Nitryl chloride: vigorous reaction
Nitryl perchlorate: violent reaction or ignition on contact
$NOClO_4$: vigorous reaction
Oxidants: violent reaction or ignition on contact
Oxygen: vigorous reaction
Ozone: violent reaction or ignition on contact
Perchloric acid: explosive reaction
Permanganic acid: violent reaction or ignition on contact
Peroxodisulfuric acid: violent reaction or ignition on contact
Potassium peroxide: vigorous reaction
Silver perchlorate: violent reaction or ignition on contact
Sodium peroxide: violent reaction or ignition on contact
Sulfanoyl chloride: violent reaction or ignition on contact
Sulfur and sulfur compounds: violent reaction or ignition on contact
Sulfur when dried with peroxidized ether: violent reaction or ignition on contact
Sulfuric acid + permanganates: vigorous reaction
Triethylaluminum + air: vigorous reaction
Trimethyl aluminum + air: vigorous reaction
Uranyl nitrate + light: explosive reaction
Wood pulp extracts + heat: explosive reaction

When heated to decomposition, it emits acrid smoke and irritating fumes.

FIRST AID   In case of contact with the eyes, immediately wash the eyes with large amounts of water, occasionally lifting the lower and upper lids. Get medical attention immediately. Contact lenses should not be worn when working with this chemical.

In case of contact with the skin, promptly wash the contaminated skin with water. If this chemical penetrates the clothing, promptly remove the clothing and wash the skin with water. If irritation persists after washing, get medical attention.

If a person breathes in large amounts of this chemical, move the victim to fresh air at once. If breathing has stopped, perform artificial respiration. Keep the affected person warm and at rest. Get medical attention as soon as possible.

If this chemical has been swallowed, get medical attention immediately.

FIRE HAZARD   A very dangerous fire and explosion hazard when exposed to heat or flame.

To fight a fire, use alcohol foam, CO_2, or dry chemical extinguishers.

HANDLING   Because of its high volatility and flammability, diethyl ether should be used in a hood that has a spark-proof mechanical system and be kept well away from flames or sparking devices. It should be stored in a cool place, preferably an explosion-proof refrigerator. Opened bottles of diethyl ether, even those containing an oxidation inhibitor such as BHT, should not be kept more than a few months to avoid the hazard of peroxide formation. Uninhibited ether, such as that prepared specifically for anesthesia use, should be handled with particular care.

Wear appropriate clothing and equipment to prevent repeated or prolonged skin contact. Immediately remove any clothing that becomes wet to avoid flammability hazard.

Wear eye protection to reduce probability of eye contact.

ETHYLENE DIBROMIDE

CAS #106-93-4   mf: $C_2H_4Br_2$
DOT #1605   mw: 187.88

STRUCTURE   NFPA LABEL
$BrCH_2CH_2Br$

SYNONYMS
Bromofume   Ethylene bromide
Celmide   Fumo-gas
DBE   Glycol bromide
Dibromoethane   Kopfume
1,2-Dibromoethane   Nephis
sym-Dibromoethane   Pestmaster
Dowfume 40   Soilbrom-40
Dowfume EDB   Soilfume
EDB   Unifume

PROPERTIES   Colorless, heavy liquid; sweet odor, threshold = 10 ppm; solubility 0.43 g/100 mL of water at 30°C, miscible with most organic solvents; melting point 9.79°C, boiling point 132°C; density 2.1792 at 20°C; flash point—nonflammable; vapor pressure 12 mm Hg at 25°C, 17.4 mm Hg at 30°C; vapor density 6.5

OSHA PEL   TWA 20 ppm; ceiling 30 ppm; peak 50 ppm/5 min/8 h

ACGIH TLV   Suspected carcinogen

NIOSH REL   0.045 ppm; ceiling 1 mg/m³/15 min

Community Right to Know List

DOT Classification   ORM-A; Label: None; Poison B; Label: Poison

USES   An insecticide.

TOXIC EFFECTS   A poison by inhalation, ingestion, skin contact, and other routes. A carcinogen and teratogen. A severe skin and eye irritant.

The approximate lethal dose of EDB for humans is 5 mL. Skin adsorption can also cause death, and inhalation of the vapor can produce pulmonary edema. Can cause severe irritation to all exposed tissues, respiratory tract, skin, and eyes.

Fifty percent of rats repeatedly exposed to 50 ppm EDB for 6 months died from pneumonia and upper respiratory tract infections. Slight changes in the liver and kidney were seen.

Ethylene dibromide has produced a high incidence of tumors (squamous-cell carcinomas of the forestomach) in mice and rats following oral administration. It is considered a human carcinogen.

Routes   Ingestion, absorption, inhalation, contact

Symptoms   Irritation of the respiratory system and eyes, dermatitis with vesiculation, central nervous system depression, kidney injury, severe liver necrosis, sterility; a carcinogen

Target Organs   Respiratory system, liver, kidneys, skin, eyes

HAZARDOUS REACTIONS
Aluminum powder: incompatible
Ammonia (liquid): incompatible
Calcium: incompatible
Magnesium: incompatible
Metals (chemically active): incompatible
Oxidizers (strong): incompatible
Potassium: incompatible
Sodium: incompatible
Zinc: incompatible

When heated to decomposition, it emits toxic fumes of Br^-.

FIRST AID   In case of contact with the eyes, immediately wash the eyes with large amounts of water, occasionally lifting the lower and upper lids. Get medical attention immediately. Contact lenses should not be worn when working with this chemical.

In case of contact with the skin, immediately wash the contaminated skin with soap and water. If this chemical penetrates through the clothing, immediately remove the clothing and wash the skin with soap and water. Get medical attention promptly.

If a person breathes in large amounts of this chemical, move the victim to fresh air at once. If breathing has stopped, perform artificial respiration. Keep the affected person warm and at rest. Get medical attention as soon as possible.

If this chemical has been swallowed, get medical attention immediately.

HANDLING   The ACGIH and OSHA limits include a warning about the potential contribution of skin absorption to the overall exposure.

ACGIH recommends that those working with human carcinogens should be properly equipped to ensure virtually no contact with the carcinogen.

Procedure A (see Chap. 1) should be followed when handling more than a few grams in the laboratory. Serious skin injury can occur from direct exposure to EDB. The substance can penetrate neoprene and several other types of plastic; therefore, gloves and other protective apparel made of these materials provide only temporary protection if EDB spills on them.

Wear appropriate clothing and equipment to prevent repeated or prolonged skin contact. Wash immediately if skin becomes contaminated. Immediately remove any nonimpervious clothing that becomes contaminated.

Wear eye protection to reduce probability of eye contact.

The following equipment should be available: a quick-drench shower.

See Chap. 2 for information on handling carcinogens.

ETHYLENE DICHLORIDE

CAS #107-06-2 mf: $C_2H_4Cl_2$
DOT #1184 mw: 98.96

NFPA LABEL

SYNONYMS
1,2-Bichloroethane
Borer sol
Brocide
1,2-DCE
Destruxol borer-sol
Dichloremulsion
α,β-Dichloroethane
sym-Dichloroethane
1,2-Dichloroethane
Dichloroethylene
Dutch liquid
Dutch oil
EDC
Ethane dichloride
Ethylene chloride
1,2-Ethylene dichloride
Glycol dichloride

PROPERTIES   Colorless liquid, pleasant odor, sweet taste; freezing point −35.7°C, boiling point 83.5°C; density 1.257 at 20°C/4°C; flash point 13.3°C; lower explosive limit 6.2%, upper explosive limit 15.9%; autoignition temperature 412.8°C; vapor pressure 100 mm Hg at 29.4°C; vapor density 3.35

OSHA PEL   TWA 50 ppm; ceiling 100 ppm; peak 200 ppm/5M/3H

ACGIH TLV   TWA 10 ppm

NIOSH REL   TWA 1 ppm; ceiling 2 ppm/15M

DOT Classification   Flammable liquid; Label: Flammable liquid

IMO Classification   Flammable liquid; Label: Flammable liquid, Poison

TOXIC EFFECTS   A poison by ingestion and some other routes. Moderately toxic by inhalation and skin contact. A carcinogen and teratogen. A strong narcotic. A skin and severe eye irritant.

Its smell and irritant effects warn of its presence at relatively safe concentrations.

Symptoms   Flaccid paralysis without anesthesia (usually neuromuscular blockade), somnolence, cough, jaundice, nausea or vomiting, hypermotility, diarrhea, ulceration or bleeding from the stomach, fatty liver degeneration, change in cardiac rate, cyanosis and coma, dermatitis, edema of the lungs, toxic effects on the kidneys, severe corneal effects

HAZARDOUS REACTIONS   Moderately explosive in the form of vapor when exposed to flame.

Aluminum: violent reaction
Ammonia: violent reaction
Dimethylaminopropylamine: violent reaction
N_2O_4: violent reaction
Oxidizing materials: vigorous reaction emits vinyl chloride and HCl gases

When heated to decomposition it emits toxic fumes of Cl^- and phosgene.

FIRE HAZARD   A dangerous fire hazard if exposed to heat, flame, or oxidizers.

To fight a fire, use water, foam, CO_2, or dry chemical extinguishers.

HANDLING   See Chap. 2 for information on handling carcinogens.

SPILL CLEANUP   A priority pollutant.

ETHYLENE GLYCOL

CAS #107-21-1

mf: $C_2H_6O_2$
mw: 62.08

NFPA LABEL

SYNONYMS

1,2-Dihydroxyethane	Lutrol-9
1,2-Ethanediol	Macrogol 400 BPC
Ethylene alcohol	M.E.G.
Ethylene dihydrate	Monoethylene glycol
Glycol	Norkool
Glycol alcohol	Tescol

PROPERTIES   Colorless, hygroscopic liquid; freezing point −13°C, boiling point 197.5°C; density 1.113 at 25°C; flash point 111.1°C (closed cup); lower explosive limit 3.2%; autoignition temperature 400°C; vapor pressure 0.05 mm Hg at 20°C; vapor density 2.14

ACGIH TLV   ceiling 50 ppm

Community Right to Know List

TOXIC EFFECTS   A poison by ingestion (the lethal dose for humans is reported to be 100 mL). Moderately toxic by other routes. Mildly toxic by skin contact. Particulates are very toxic when inhaled. May be a carcinogen. A teratogen.

Symptoms   Irritation of the skin, eyes, and mucous membranes; eye lacrimation; general anesthesia; headache; cough; respiratory stimulation; nausea, vomiting, pulmonary, kidney, and liver damage (kidney damage may be fatal)

HAZARDOUS REACTIONS   Moderate explosion hazard when exposed to flame.

Ammonium dichromate: ignites when heated to 100°C
Chlorosulfonic acid: can react violently
Chromium trioxide: ignites on contact
Oleum: can react violently
Oxidants: vigorous reaction
P_2S_5: can react violently
Perchloric acid: can react violently
Potassium permanganate: ignites on contact
Silver chlorate: ignites when heated to 100°C
Sodium chlorite: ignites when heated to 100°C
Sodium peroxide: ignites on contact
Sulfuric acid: can react violently
Uranyl nitrate: ignites when heated to 100°C

Aqueous solutions may ignite silvered copper wires which have an applied D.C. voltage.

FIRE HAZARD   Combustible when exposed to heat or flame.

To fight a fire, use alcohol foam, water, foam, CO_2, or dry chemical extinguishers.

HANDLING   See Chap. 2 for information on handling carcinogens.

ETHYLENE OXIDE

CAS #75-21-8
DOT #1040

mf: C_2H_4O
mw: 44.06

NFPA LABEL

SYNONYMS

Amprolene	Oxacyclopropane
Anproline	Oxane
Dihydrooxirene	Oxidoethane
Dimethylene oxide	α,β-Oxidoethane
Epoxyethane	Oxirane
1,2-Epoxyethane	Oxyfume
Ethene oxide	Sterilizing gas ethylene
ETO	oxide 100%
Merpol	T-gas

PROPERTIES   Colorless gas at room temperature, with an etherlike odor that is irritating at high concentrations; miscible in water and alcohol, very soluble in ether; melting point −111.3°C, boiling point 10.7°C; density 0.8711 at 20°C; flash point −20°C; lower ex-

168 ETHYLENE OXIDE

plosive limit 3.0%, upper explosive limit 100%; autoignition temperature 428.9°C; vapor pressure 1095 mm Hg at 20°C; vapor density 1.52

OSHA PEL   TWA 1 ppm

ACGIH TLV   TWA 1 ppm; suspected carcinogen

NIOSH REL   (Oxirane) TWA 0.1 ppm; ceiling 5 ppm/10M/D

EPA Extremely Hazardous Substances List and Community Right to Know List

DOT Classification   Flammable liquid; Label: Flammable liquid; flammable gas; Label: Poison gas and flammable gas

TOXIC EFFECTS   Poison by ingestion and several other routes. Moderately toxic by inhalation. A carcinogen and teratogen. A skin, eye, and mucous membrane irritant.

Routes   Ingestion, inhalation, contact

Symptoms   Irritation (and possibly burns) of the eyes, nose, and throat; peculiar taste; headache; nausea, vomiting, diarrhea; dyspnea; cyanosis; pulmonary edema; drowsiness, weakness, incoordination; electrocardiogram abnormalities, convulsions; frostbite; carcinogen

HAZARDOUS REACTIONS   Severe explosion hazard when exposed to flame. Rapid compression of the vapor with air causes explosions.

Acids: violent polymerization reaction on contact
Air: incompatible
Alcohols: incompatible
Alkali hydroxides: violent polymerization reaction on contact
Alkane thiols: incompatible
Aluminum: incompatible
Aluminum chloride: violent polymerization reaction on contact
Aluminum chlorides (anhydrous): incompatible
Aluminum oxide: violent polymerization reaction on contact
Amines: violent polymerization reaction on contact
Ammonia: violent polymerization reaction on contact
Bases: incompatible
Bromoethane: incompatible
Contaminants: incompatible
Copper: incompatible
Covalent halides: violent polymerization reaction on contact
Glycerol: explosive reaction at 200°C
Iron chlorides: incompatible
Iron oxide: violent polymerization reaction on contact
Iron(III) chloride: violent polymerization reaction on contact
m-Nitroaniline: incompatible
Magnesium perchlorate: incompatible
Mercaptans: incompatible
Oxidizers: incompatible
Potassium (metallic): violent polymerization reaction on contact
Rust: violent polymerization reaction on contact
Tin chlorides: incompatible
Tin(IV) chloride: violent polymerization reaction on contact
Trimethyl amine: incompatible

When heated to decomposition, it emits acrid smoke and irritating fumes.

FIRST AID   In case of contact with the eyes, immediately wash the eyes with large amounts of water, occasionally lifting the lower and upper lids. Get medical attention immediately. Contact lenses should not be worn when working with this chemical.

In case of contact with the skin, immediately flush the contaminated skin with water. If this chemical penetrates the clothing, immediately remove the clothing and flush the skin with water. Get medical attention promptly.

If a person breathes in large amounts of this chemical, move the victim to fresh air at once. If breathing has stopped, perform artificial respiration. Keep the affected person warm and at rest. Get medical attention as soon as possible.

If this chemical has been swallowed, get medical attention immediately.

FIRE HAZARD   A very dangerous fire hazard when exposed to heat or flame.

To fight a fire, use alcohol foam, CO_2, or dry chemical extinguishers.

HANDLING   Wear appropriate clothing and equipment to prevent any possibility of skin contact. Wash immediately if skin becomes contaminated. Immediately remove any clothing that becomes wet to avoid flammability hazard.

Wear eye protection to reduce probability of eye contact.

The following equipment should be available: an eyewash station.

See Chap. 2 for information on handling carcinogens.

FERRIC CHLORIDE

CAS #7705-08-0
DOT #1773; 2582
mf: Cl_3Fe
mw: 162.20

NFPA LABEL

SYNONYMS
Ferric chloride, solid, anhydrous
Flores martis
Iron chloride
Iron sesquichloride, solid
Iron trichloride

PROPERTIES   Black-brown solid; melting point 292°C, boiling point 319°C, density 2.90 at 25°C; vapor pressure 1 mm Hg at 194°C

ACGIH TLV   TWA 1 mg(Fe)/m³

DOT Classification   ORM-B; Label: None, anhydrous; Corrosive material; Label: Corrosive

TOXIC EFFECTS   Moderately toxic by ingestion. Corrosive to tissue.

HAZARDOUS REACTIONS
Allyl chloride: violent reaction
Chlorine + monomers (e.g., styrene): potentially explosive polymerization reaction
Ethylene oxide: potentially explosive polymerization reaction
Metals (chemically active): mixtures are shock-sensitive explosives
Potassium: mixtures are shock-sensitive explosives
Sodium: mixtures are shock-sensitive explosives
Water: reaction produces toxic and corrosive fumes

When heated to decomposition, it emits toxic fumes of HCl.

FERROUS SULFATE

CAS #7720-78-7
DOT #9125
mf: $O_4S \cdot Fe$
mw: 151.91

SYNONYMS
Copperas
Duretter
Duroferon
Exsiccated ferrous sulfate
Feosol
Feospan
Fer-in-sol
Fero-gradumet
Ferralyn
Ferrosulfate
Ferro-theron
Fersolate
Green vitriol
Iron monosulfate
Iron protosulfate
Iron vitriol
Irospan
Irosul
Slow-Fe
Sulferrous
Sulfuric acid, iron(2+) salt (1:1)

ACGIH TLV   TWA 1 mg(Fe)/m³

DOT Classification   ORM-E; Label: None

TOXIC EFFECTS   A poison by ingestion and some other routes. A systemic toxin.

Symptoms by Ingestion   Aggression, somnolence, brain-recording changes, diarrhea, nausea, or vomiting, bleeding from the stomach, coma

HAZARDOUS REACTIONS
Arsenic trioxide + sodium nitrate: may ignite on contact
Methyl isocyanoacetate: potentially explosive reaction at 25°C

When heated to decomposition, it emits toxic fumes of SO_x.

FLUORIDES

OSHA PEL   TWA 2.5 mg(F)/m³; IDLH 500 mg(F)/m³

ACGIH TLV   TWA 2.5 mg(F)/m³

NIOSH REL   TWA 2.5 mg(F)/m³

TOXIC EFFECTS   Inorganic fluorides are generally highly irritating and toxic. The estimated lethal dose for humans is 2.5 to 5.9 g of F. Large doses can cause very severe nausea, vomiting, diarrhea, abdominal burning, and cramplike pains. Chronic exposure can cause loss of weight, anorexia, anemia, and dental defects.

Organic fluorides are generally less toxic than other halogenated hydrocarbons.

Routes   Ingestion, inhalation, contact

Symptoms   Irritation of the eyes, respiratory system; nausea, abdominal pain, diarrhea; excess salivation, thirst, sweating; stiff spine; dermatitis; calcification of ligaments of ribs, pelvis; skin rashes, pulmonary fibrosis

Target Organs   Eyes, respiratory system, central nervous system, skeleton, kidneys, skin

HAZARDOUS REACTIONS   When heated to decomposition, they emit toxic fumes of F⁻.

FIRST AID   In case of contact with the eyes, immediately wash the eyes with large amounts of water, occasionally lifting the lower and upper lids. Get medical attention immediately. Contact lenses should not be worn when working with this chemical.

In case of contact with the skin, promptly wash the contaminated skin with soap and water. If this chemical penetrates through the clothing, promptly remove the clothing and wash the skin with soap and water. Get medical attention promptly.

If a person breathes in large amounts of this chemical, move the victim to fresh air at once. Other measures are usually unnecessary; however, artificial respiration may be required.

If this chemical has been swallowed, get medical attention immediately.

HANDLING   Wear appropriate clothing and equipment to prevent repeated or prolonged skin contact. Wash promptly if the skin becomes contaminated. Promptly remove any nonimpervious clothing that becomes contaminated. Work clothing should be changed daily if there is any possibility that the clothing has been contaminated.

Wear eye protection to prevent any possibility of eye contact.

FLUORINE

CAS #7782-41-4   mf: F_2
DOT #1045   mw: 38.00

NFPA LABEL

SYNONYM
Fluorine, compressed

PROPERTIES   Pale yellow gas with a pungent, irritating odor; melting point −218°C, boiling point −187°C; density 1.14 at −200°C; vapor density 1.695

OSHA PEL   TWA 0.1 ppm

ACGIH TLV   TWA 1 ppm; STEL 2 ppm

DOT Classification   Nonflammable gas; Label: Poison and oxidizer; Poison A; Label: Poison gas, Oxidizer

TOXIC EFFECTS   A poisonous gas. Moderately toxic by inhalation. A skin, eye, and mucous membrane irritant. A powerful, caustic irritant to tissue.

Fluorine causes deep, penetrating burns on contact with the body, an effect that may be delayed and progressive, as in the case of burns by hydrogen fluoride. The hazard of exposure to fluorine is at least as great as that of chlorine.

Routes   Ingestion, contact

Symptoms   Irritation of the eyes, nose, respiratory tract; laryngeal spasm, bronchitis spasm, pulmonary edema; burns of the eyes and skin; liver and kidney damage in animals

Target Organs   Respiratory system, eyes, skin; liver and kidneys in animals

HAZARDOUS REACTIONS   A powerful oxidizer. Reacts violently with many materials.

Acetaldehyde: ignition or violent reaction on contact
Acetone: ignition or violent reaction on contact
Acetonitrile: forms explosive mixtures
Acetylene: ignition or violent reaction on contact
Air (liquid): violent reaction
Alkali metal oxides: incandescent reaction
Alkali metals: incompatible
Alkaline earth oxides: incandescent reaction
Alkaline earths: incompatible
Alkanes + oxygen: reaction forms explosive products (peroxides)
Alkenes: violent reaction
Alkyl benzenes: violent reaction
Aluminum: violent reaction
Ammonia: explosive or potentially explosive reaction
Aniline: incandescent reaction
Anthracene: explosive or potentially explosive reaction
Antimony: violent reaction
Antimony trisulfide: ignition or violent reaction on contact
Arsenic: violent reaction
Arsenic trioxide: ignition or violent reaction on contact
Asbestos: supports continued combustion
Barium sulfide: ignition or violent reaction on contact
Benzene: ignition or violent reaction on contact
Benzoic acid: ignition or violent reaction on contact
Boron: ignition or violent reaction on contact
Boron nitride: incandescent reaction
Boron trichloride: ignition or violent reaction on contact
Bromine: ignition or violent reaction on contact
Calcium disilicide: ignition or violent reaction on contact

Calcium iodide: ignition or violent reaction on contact
Carbon disulfide: violent reaction
Carbon disulfide vapor: ignition or violent reaction on contact
Carbon monoxide: explosive or potentially explosive reaction
Carbon tetrachloride: explosive or potentially explosive reaction
Ceramic materials: ignition or violent reaction on contact
Cesium acetylide: ignition or violent reaction on contact
Cesium fluoride + fluorocarboxylic acids: explosive or potentially explosive reaction
Cesium heptafluoropropoxide: explosive or potentially explosive reaction
Charcoal: ignition or violent reaction on contact
Chlorine: ignition or violent reaction when exposed to sparks or heating to 100°C
Chlorine dioxide: violent reaction
Chlorine fluoride: forms explosive mixtures
Chloroform: explosive or potentially explosive reaction
Chromium hypochlorite: violent reaction
Chromium(II) sulfide: ignition or violent reaction on contact
Chromyl chloride: ignition or violent reaction on contact
Coke: violent reaction
Copper hydride: ignition or violent reaction on contact
Cyanamide: violent reaction
Cyanides: violent reaction
Cyanoguanidine: reaction forms explosive products
1,2-Dichlorotetrafluoroethane: explosive or potentially explosive reaction
Dicyanogen: ignition or violent reaction on contact
Dimethylamine: incandescent reaction
Dinitrogen tetroxide: ignition or violent reaction on contact
Ditungsten carbide: ignition or violent reaction on contact
Ethanol: ignition or violent reaction on contact
Ethyl acetate: ignition or violent reaction on contact
1- or 2-Fluoriminoperfluoropropane: explosive or potentially explosive reaction
Gallic acid: incandescent reaction
Glass: supports continued combustion
Graphite: explosive or potentially explosive reaction
Halides (covalent): ignition or violent reaction on contact
Halocarbons: explosive or potentially explosive reaction
Halogen acids: violent reaction
Halogens: ignition or violent reaction on contact
Hexalithium disilicide + heat: incandescent reaction
Hydrazine: violent reaction
Hydrocarbons (gaseous): ignition or violent reaction on contact
Hydrocarbons (liquid): explosive or potentially explosive reaction
Hydrogen: explosive or potentially explosive reaction
Hydrogen + oxygen: explosive or potentially explosive reaction
Hydrogen bromide: ignition or violent reaction on contact
Hydrogen chloride: ignition or violent reaction on contact
Hydrogen-containing molecules: incompatible
Hydrogen fluoride: ignition or violent reaction on contact
Hydrogen fluoride + seleninyl fluoride + heat: explosive or potentially explosive reaction
Hydrogen halide gases or concentrated solutions: ignition or violent reaction on contact
Hydrogen iodide: ignition or violent reaction on contact
Hydrogen sulfide: ignition or violent reaction on contact
Ice: forms explosive mixtures
Iodine: ignition or violent reaction on contact
Iodoform: explosive or potentially explosive reaction
Lactic acid: ignition or violent reaction on contact
Lead hexacyanoferrate(III): ignition or violent reaction on contact
Lead iodide: ignition or violent reaction on contact
Lead monoxide + glycerol: violent reaction
Lithium acetylide: ignition or violent reaction on contact
Lithium hexasilicide: ignition or violent reaction on contact
Mercury iodide: ignition or violent reaction on contact
Metal acetylides and carbides: ignition or violent reaction on contact
Metal borides: incandescent reaction
Metal cyano complexes: ignition or violent reaction on contact
Metal hydrides: ignition or violent reaction on contact
Metal iodides: ignition or violent reaction on contact
Metal oxides: incandescent reaction
Metal salts: ignition or violent reaction on contact
Metal silicides: ignition or violent reaction on contact
Metals: ignition or violent reaction on contact
Methane: ignition or violent reaction on contact
Methanol: ignition or violent reaction on contact
Methyl borate: ignition or violent reaction on contact
3-Methyl butanol: ignition or violent reaction on contact
Molybdenum sulfide: ignition or violent reaction on contact

Monocesium acetylide: ignition or violent reaction on contact
Natural gas: violent reaction
Neoprene: ignition or violent reaction on contact
Nickel(II) oxide: incandescent reaction
Nickel(III) oxide: ignition or violent reaction on contact
Nickel(IV) oxide: ignition or violent reaction on contact
Nitric acid: explosive or potentially explosive reaction
Nitrogen oxide: ignition or violent reaction on contact
Nitrogenous bases: incandescent reaction
Nonmetal oxides: ignition or violent reaction on contact
Nonmetals: ignition or violent reaction on contact
Nylons: ignition or violent reaction on contact
Organic vapors: forms explosive mixtures
Oxides of sulfur, nitrogen, phosphorus: incompatible
Oxidizable substances: vigorous reaction, frequently with immediate ignition
Oxygen: incompatible
Oxygen + polymers (e.g., phenol-formaldehyde resins, Bakelite): ignition or violent reaction on contact
Oxygenated organic compounds: ignition or violent reaction on contact
Perchloric acid: reaction forms explosive products (fluorine perchlorates)
Perfluorocyclobutane: explosive or potentially explosive reaction
Perfluoropropionyl fluoride: violent reaction
Phosphorus (yellow or red): ignition or violent reaction on contact
Phosphorus pentachloride: ignition or violent reaction on contact
Phosphorus trichloride: ignition or violent reaction on contact
Phosphorus trifluoride: ignition or violent reaction on contact
Polyacrylonitrile-butadiene: ignition or violent reaction on contact
Polyamides (nylons): ignition or violent reaction on contact
Polychloropene (neoprene): ignition or violent reaction on contact
Polyethylene: ignition or violent reaction on contact
Polymethyl methacrylate (Perspex): ignition or violent reaction on contact
Polytetrafluoroethylene (Teflon): ignition or violent reaction on contact
Polytrifluoropropylmethylsiloxane: ignition or violent reaction on contact
Polyurethane foam: ignition or violent reaction on contact
Polyvinyl chloride acetate: violent reaction
Polyvinylchloride-vinyl acetate (Tygon): ignition or violent reaction on contact
Polyvinylidene fluoride-hexafluoropropylene (Viton): ignition or violent reaction on contact
Potassium chlorate: reaction forms explosive fluorine perchlorate gas
Potassium hexacyanoferrate(II): ignition or violent reaction on contact
Potassium hexacyanoferrate(III): ignition or violent reaction on contact
Potassium hydride: ignition or violent reaction on contact
Potassium hydroxide: reaction forms explosive potassium trioxide
Potassium iodide: ignition or violent reaction on contact
Potassium nitrate: violent reaction
Potassium sulfide: ignition or violent reaction on contact
Pyridine: incandescent reaction
Rubidium acetylide: ignition or violent reaction on contact
Salicylic acid: ignition or violent reaction on contact
Selenium: ignition or violent reaction on contact
Seleninyl fluoride: incompatible
Silicates: violent reaction
Silicides: violent reaction
Silicon carbon: ignition or violent reaction on contact
Silicon tetrachloride: ignition or violent reaction on contact
Silicon-containing compounds: vigorous reaction
Silver cyanide: explosive or potentially explosive reaction
Silver difluoride: ignition or violent reaction on contact
Silver(I) oxide: ignition or violent reaction on contact
Sodium acetate: explosive or potentially explosive reaction
Sodium bromate: explosive or potentially explosive reaction
Sodium dicyanamides: incompatible
Sodium hydride: ignition or violent reaction on contact
Stainless steel: explosive or potentially explosive reaction
Sulfides: ignition or violent reaction on contact
Sulfur: ignition or violent reaction on contact
Sulfur dioxide: explosive or potentially explosive reaction
Teflon: ignition or violent reaction on contact
Tellurium: ignition or violent reaction on contact
Thallium: violent reaction
Tin: violent reaction
Town gas: ignition or violent reaction on contact

Trichloroacetaldehyde: ignition or violent reaction on contact

Trinitromethane: violent reaction

Tungsten carbide: ignition or violent reaction on contact

Turpentine: explosive or potentially explosive reaction

Uranium dicarbide: ignition or violent reaction on contact

Water: explosive or potentially explosive reaction

Water or steam: reaction produces heat and toxic and corrosive fumes

Water vapor: forms explosive mixtures

Xenon + catalysts (e.g., nickel fluoride): ignition or violent reaction on contact

Zinc sulfide: ignition or violent reaction on contact

Zirconium dicarbide: ignition or violent reaction on contact

When heated to decomposition, it emits highly toxic fumes of F^-.

FIRST AID   In case of contact with the eyes, immediately wash the eyes with large amounts of water, occasionally lifting the lower and upper lids. Get medical attention immediately. Contact lenses should not be worn when working with this chemical.

In case of contact with the skin, immediately flush the contaminated skin with water. If this chemical penetrates the clothing, immediately remove the clothing and flush the skin with water. Get medical attention promptly.

If a person breathes in large amounts of this chemical, move the victim to fresh air at once. If breathing has stopped, perform artificial respiration. Keep the affected person warm and at rest. Get medical attention as soon as possible.

FIRE HAZARD   A very dangerous fire hazard.

HANDLING   Because of the high activity of fluorine, the area in the vicinity of a fluorine-containing cylinder and its associated apparatus should be well ventilated and cleared of easily combustible material.

When a cylinder of fluorine is to be opened, the user should be protected by a suitable shield, and the valve should be opened by remote control. Any apparatus that is to contain fluorine under pressure should be surrounded by a protective barrier. Fluorine cylinder valves are not adapted to fine adjustment, and the flow of fluorine from a cylinder should therefore be controlled by a needle valve located close to the cylinder and operated by remote control.

All equipment that may be in contact with fluorine should be completely dry.

Protective measures against fluorine are not fully developed, and entry into zones contaminated with this substance should be avoided. Only positive-pressure, atmosphere-supplying respiratory protective equipment is advised and, if IDLH concentrations are reached, a pressure-demand, self-contained breathing apparatus or a positive-pressure, airline respirator that has escape-cylinder provisions must be used. Gauntlet-type rubber gloves, rubber aprons, and face shields give only temporary protection against fluorine and, if brought into local contact with a fluorine leak, may inflame. A thorough flushing of fluorine lines with an inert gas should precede any opening of the lines for any reason.

The reaction of fluorine with some metals is slow and results in the formation of a protective metallic fluoride film. Brass, iron, aluminum, and copper, as well as certain of their alloys, react in this way at standard temperatures and atmospheric pressure. Thus, these metals can be passivated by passing fluorine gas highly diluted with argon, neon, or nitrogen through tubing or over the surface with proper precautions and gradually increasing the concentration of fluorine. As long as the protective coating is not cracked or dislodged to create a "hot spot," passivated apparatus is safe to use.

Once a fire that involves fluorine as an oxidizer has started, there is no effective way stopping it other than shutting off the source of fluorine. The area should be cleared and the fire allowed to burn itself out. No attempts should be made to extinguish the fire by using water or chemicals, as these act as additional fuel. Anyone in the vicinity of such a fire should wear impervious clothing and supplied-air or self-contained breathing apparatus.

Wear appropriate clothing and equipment to prevent any possibility of skin contact. Wash immediately if skin becomes contaminated. Immediately remove any nonimpervious clothing that becomes contaminated.

Wear eye protection to prevent any possibility of eye contact.

The following equipment should be available: an eyewash station and a quick-drench shower.

FLUOROTRICHLOROMETHANE

CAS #75-69-4 mf: CCl_3F
DOT #1198 mw: 136.36

SYNONYMS
Freon 11
Monofluorotrichloromethane
Refrigerent 11

Trichlorofluoromethane
Trichloromonofluoromethane

PROPERTIES   Colorless liquid or gas with a chlorinated solvent odor which is detectable >20% by volume; solubility in water 0.11%; melting point −111.1°C, boiling point 23.9°C; density 1.484 at 17.2°C; vapor pressure 690 mm Hg

OSHA PEL   TWA 1000 ppm; IDLH 10,000 ppm

ACGIH TLV   ceiling 1000 ppm

TOXIC EFFECTS   Poisonous by inhalation. High concentrations cause narcosis and anesthesia.

Routes   Ingestion, inhalation, contact

Symptoms   Incoordination, tremors, dermatitis, frostbite, cardiac arrhythmias, cardiac arrest, conjunctiva irritation, fibrosing alveolitis, and liver changes

Target Organs   Cardiovascular system, skin, lungs

HAZARDOUS REACTIONS
Aluminum: violent reaction
Barium: violent reaction
Calcium: incompatible
Lithium: violent reaction
Magnesium: incompatible
Metals (chemically active): incompatible
Potassium: incompatible
Powdered aluminum: incompatible
Sodium: incompatible
Zinc: incompatible

When heated to decomposition, it emits highly toxic fumes of F$^-$ and Cl$^-$.

FIRST AID   In case of contact with the eyes, immediately wash the eyes with large amounts of water, occasionally lifting the lower and upper lids. Get medical attention immediately. Contact lenses should not be worn when working with this chemical.

In case of contact with the skin, immediately flush the contaminated skin with water. If this chemical penetrates the clothing, immediately remove the clothing and flush the skin with water. Get medical attention promptly.

If a person breathes in large amounts of this chemical, move the victim to fresh air at once. If breathing has stopped, perform artificial respiration. Keep the affected person warm and at rest. Get medical attention as soon as possible.

If this chemical has been swallowed, get medical attention immediately.

HANDLING   Wear appropriate clothing and equipment to prevent repeated or prolonged skin contact. Promptly remove any nonimpervious clothing that becomes wet.

Wear eye protection to prevent any possibility of eye contact.

The following equipment should be available: an eyewash station and a quick-drench shower.

FORMALDEHYDE

CAS #50-00-0   mf: CH$_2$O
DOT #1198; 2209   mw: 30.03

STRUCTURE   NFPA LABEL

O
‖
HC—H

SYNONYMS
BFV   Methanal
Fannoform   Methyl aldehyde
Formaldehyde, solution Methylene glycol
Formalin   Methylene oxide
Formalith   Morbocid
Formic aldehyde   Oxomethane
Formol   Oxymethylene
Fyde   Paraform
Hoch   Polyoxymethylene
Ivalon   glycol
Karsan   Superlysoform
Lysoform

PROPERTIES   Clear, water-white, very slightly acid gas or liquid (the solid polymer is called paraformaldehyde), pungent and irritating odor, threshold = 1 ppm; very soluble in water; soluble in ether, alcohol, and most organic solvents; melting point −92°C, boiling point −19°C; density (liquid) 0.815 at 20°C; flash points: 6% methanol 72.2°C (closed cup), 10% methanol 63.8°C, 15% methanol 50°C; lower explosive limit 7.0%, upper explosive limit 73.0%; autoignition temperature 430°C; vapor pressure 10 mm Hg at −88°C; vapor density 1.075. Pure formaldehyde is not available commercially because of its tendency to polymerize. It is sold as aqueous solutions containing from 37% to 50% formaldehyde by weight and varying amounts of methanol. Some alcoholic solutions are used industrially, and the physical properties and hazards may be greatly influenced by the solvent.

OSHA PEL   TWA 1 ppm

ACGIH TLV   TWA 1 ppm; suspected carcinogen

NIOSH REL   TWA 0.016 ppm

DOT Classification   Flammable or Combustible liquid; Label: Flammable liquid

TOXIC EFFECTS   A poison by ingestion, skin contact, inhalation, and some other routes. A carcinogen and teratogen. A severe eye and skin irritant. Eye contact with the liquid causes delayed effects that are not appreciably eased by eye washing.

If swallowed, it causes violent vomiting and diarrhea, which can lead to collapse. Ingestion of as little as 30 mL of formalin has been fatal. Frequent or prolonged exposure can cause hypersensitivity leading to contact dermatitis, possibly of an eczematoid nature.

Inhalation of vapors may result in severe irritation and edema of the upper respiratory tract, burning and stinging of the eyes, and headache, and has been known to cause death. Workers exposed to 2–10 ppm have experienced headaches, nausea, dizziness, and vomiting; lacrimation occurs at 4–5 ppm. For several minutes of exposure, 10 ppm or more is intolerable. Central nervous system effects were seen among rats exposed to 0.8 ppm for 3 months.

Routes   Ingestion, inhalation, contact

Symptoms   Irritation of the eyes, nose, and throat; lacrimation; burns nose; cough, bronchitis spasm, pulmonary irritation; dermatitis; nausea, vomiting; loss of consciousness; carcinogen

Target Organs   Respiratory system, eyes, skin

HAZARDOUS REACTIONS
Acids: incompatible
Alkalies: incompatible
Hydrogen peroxide: violent reaction
Magnesium carbonate: violent reaction
Nitromethane: violent reaction
Oxides of nitrogen: the reaction becomes explosive at about 180°C
Oxidizers: vigorous reaction
Perchloric acid + aniline: violent reaction
Performic acid: violent reaction
Phenols: incompatible
Urea: incompatible

When heated to decomposition, it emits acrid smoke and irritating fumes.

FIRST AID   In case of contact with the eyes, immediately wash the eyes with large amounts of water, occasionally lifting the lower and upper lids. Get medical attention immediately. Contact lenses should not be worn when working with this chemical.

In case of contact with the skin, immediately flush the contaminated skin with water. If this chemical penetrates the clothing, immediately remove the clothing and flush the skin with water. Get medical attention promptly.

If a person breathes in large amounts of this chemical, move the victim to fresh air at once. If breathing has stopped, perform artificial respiration. Keep the affected person warm and at rest. Get medical attention as soon as possible.

If this chemical has been swallowed, get medical attention immediately.

FIRE HAZARD   Flammable when exposed to heat or flame. The gas is a more dangerous fire hazard than the vapor. When aqueous formaldehyde solutions are heated above their flash points, a potential for explosion hazard exists. High formaldehyde concentration or methanol content lowers the flash point. A moderate explosion hazard when exposed to heat or flame. To extinguish fires, stop flow of gas for pure form; use alcohol foam for the 37% methanol form.

HANDLING   Laboratory operations with formalin in open vessels should be carried out in a hood or other local-exhaust device (formaldehyde has such an objectionable odor that there may be little need for this admonition). Because repeated exposure to formaldehyde can lead to a formaldehyde allergy, it is well to avoid skin contact with aqueous solutions by appropriate use of neoprene, butyl rubber, or polyvinyl chloride gloves and other protective apparel. Splash-proof goggles should be worn if there is any possibility of splashing formaldehyde in the eyes.

On the basis of carcinogenicity data, Procedure A (see Chap. 1) should be followed when handling more than a few grams in the laboratory.

Wear appropriate clothing and equipment to reduce probability of skin contact. Wash immediately if skin becomes contaminated. Immediately remove any nonimpervious clothing that becomes contaminated.

Wear eye protection to prevent any possibility of eye contact.

The following equipment should be available: an eyewash station and a quick-drench shower.

See Chap. 2 for information on handling carcinogens.

FORMIC ACID

CAS #64-18-6   mf: CH_2O_2
DOT #1779   mw: 46.03

NFPA LABEL

SYNONYMS
Aminic acid
Formic acid, solution
Formylic acid
Hydrogen carboxylic acid
Methanoic acid

PROPERTIES   Colorless, fuming liquid with a pungent, penetrating odor; miscible in water and alcohol; freezing point 8.2°C, boiling point 100.8°C; density 1.2267 at 15°C/4°C, 1.220 at 20°C/4°C; flash point 68.9°C (open cup); lower explosive limit (90% solution) 18%, upper explosive limit (90% solution) 57%; autoignition temperature (90% solution) 433.9°C; vapor pressure 40 mm Hg at 24°C; vapor density 1.59

OSHA PEL   TWA 5 ppm; IDLH 100 ppm

ACGIH TLV   TWA 5 ppm

DOT Classification   Corrosive material; Label: Corrosive; Corrosive material; Label: Corrosive, solution

TOXIC EFFECTS   Moderately toxic by ingestion. Mildly toxic by inhalation. A suspected carcinogen. Corrosive to the skin and eyes.

Routes   Ingestion, inhalation, contact

Symptoms   Irritation of the eyes, throat, and skin; nasal discharge; cough, dyspnea; nausea; skin burns, dermatitis

Target Organs   Respiratory system, skin, kidneys, liver, eyes

HAZARDOUS REACTIONS
Caustics (strong): incompatible
Furfuryl alcohol: explosive reaction
Hydrogen peroxide: explosive reaction
Nitromethane: explosive reaction
Oxidizing materials: vigorous reaction
P_2O_5: explosive reaction
Sulfuric acid (concentrated): incompatible
Thallium nitrate·$3H_2O$: explosive reaction

When heated to decomposition, it emits acrid smoke and irritating fumes.

FIRST AID   In case of contact with the eyes, immediately wash the eyes with large amounts of water, occasionally lifting the lower and upper lids. Get medical attention immediately. Contact lenses should not be worn when working with this chemical.

In case of contact with the skin, immediately flush the contaminated skin with water. If this chemical penetrates the clothing, immediately remove the clothing and flush the skin with water. Get medical attention promptly.

If a person breathes in large amounts of this chemical, move the victim to fresh air at once. If breathing has stopped, perform artificial respiration. Keep the affected person warm and at rest. Get medical attention as soon as possible.

If this chemical has been swallowed, get medical attention immediately.

FIRE HAZARD   Flammable when exposed to heat or flame.

To fight a fire, use CO_2, dry chemical, or alcohol foam extinguishers.

HANDLING   Wear appropriate clothing and equipment to prevent any possibility of skin contact. Wash immediately if skin becomes contaminated. Immediately remove any nonimpervious clothing that becomes contaminated.

Wear eye protection to prevent any possibility of eye contact.

The following equipment should be available: an eyewash station and a quick-drench shower.

See Chap. 2 for information on handling carcinogens.

FUMARIC ACID

CAS #110-17-8   mf: $C_4H_4O_4$
DOT #9126   mw: 116.08

SYNONYMS
Allomaleic acid
Boletic acid
trans-Butenedioic acid
(E)-Butenedioic acid
trans-1,2-Ethylenedicarboxylic acid
(E)1,2-Ethylenedicarboxylic acid
Lichenic acid

PROPERTIES   Colorless, odorless crystals; melting point 287°C, density 1.635 at 20°C/4°C

DOT Classification   ORM-E; Label: None

TOXIC EFFECTS   Mildly toxic by ingestion and skin contact. A skin and eye irritant.

HAZARDOUS REACTIONS   Oxidizing materials: vigorous reaction

When heated to decomposition, it emits acrid smoke and irritating fumes.

FIRE HAZARD   Combustible when exposed to heat or flame.

GALLIC ACID

CAS #149-91-7   mf: $C_7H_6O_5$
mw: 170.13

SYNONYM
3,4,5-Trihydroxybenzoic acid

PROPERTIES   White or off-white, odorless crystals; slightly water-soluble; decomposes at 225–250°C, loses H_2O at 100–120°C; density 1.694

EPA TSCA Inventory and EPA Genetic Toxicology Program

TOXIC EFFECTS   Poisonous by intravenous route. Mildly toxic by ingestion. Experimental reproductive effects. Mutagenic data.

HAZARDOUS REACTIONS   When heated to decomposition, it emits acrid smoke and irritating fumes.

GIBBERELLIC ACID

CAS #77-06-5   mf: $C_{19}H_{22}O_6$
mw: 346.41

SYNONYMS
Berelex
Brellin
Cekugib
Floraltone
GA
Gibberellin
Gibbrel
Gib-sol
Gib-tabs
Grocel
Pro-gibb
2,4a,7-Trihydroxy-1-methyl-8-methylenegibb-3-ene-1,10-carboxylic acid 1-4-lactone

PROPERTIES   A crystalline, plant-growth-promoting hormone; slightly soluble in water and ether; soluble in methanol, ethanol, acetone, and aqueous solutions of sodium bicarbonate and sodium acetate; moderately soluble in ethyl acetate; melting point 233–235°C

EPA Genetic Toxicology Program and EPA TSCA Inventory

TOXIC EFFECTS   Mildly toxic by ingestion. An experimental tumorigen. Mutagenic data.

HAZARDOUS REACTIONS   When heated to decomposition, it emits acrid smoke and irritating fumes.

d-GLUCOSE

CAS #50-99-7   mf: $C_6H_{12}O_6$
mw: 180.18

SYNONYMS
Anhydrous dextrose
Cartose
Cerelose
Corn sugar
Dextropur
Dextrose
Dextrose, anhydrous
Dextrosol
Glucolin
Glucose
d-Glucose, anhydrous
Glucose liquid
Grape sugar
Sirup

PROPERTIES   Colorless crystals or white crystalline or granular powder, odorless; soluble in water, slightly soluble in alcohol; melting point 146°C; density 1.544. α-Form monohydrate, crystals from water, melting point 83°C. α-Form anhydrous, crystals from hot ethanol or water, melting point 146°C. Very sparingly soluble in absolute alcohol, ether, acetone; soluble in hot glacial acetic acid, pyridine, aniline. β-Form crystals from hot H_2O + ethanol, from dilute acetic acid, or from pyridine; melting point 148–155°C.

EPA TSCA Inventory and EPA Genetic Toxicology Program

TOXIC EFFECTS   Mildly toxic by ingestion. Experimental reproductive effects. Mutagenic data.

GLYCEROL

HAZARDOUS REACTIONS
Alkali: mixtures release carbon monoxide when heated
Potassium nitrate + sodium peroxide: potentially explosive reaction when heated in a sealed container

When heated to decomposition, it emits acrid smoke and irritating fumes.

GLYCEROL

CAS #56-81-5   mf: $C_3H_8O_3$
    mw: 92.09

STRUCTURE
$HOCH_2CHOHCH_2OH$

TOXIC EFFECTS
Poisonous by subcutaneous route. Mildly toxic by ingestion. Experimental reproductive effects. Human mutagenic data. A skin and eye irritant.

Symptoms   Headache and nausea or vomiting when ingested

HAZARDOUS REACTIONS
Acetic anhydride: violent reaction
Aniline + nitrobenzene: violent reaction
Calcium hypochlorite: ignites on contact
Chlorine: confined mixture explodes if heated to 70–80°C
Chromium trioxide: violent reaction
Ethylene oxide + heat: violent reaction
Fluorine + lead monoxide: violent reaction
Hydrogen peroxide: mixtures are highly explosive
Nitric acid + hydrofluoric acid: storage hazard due to gas evolution
Nitric acid + sulfuric acid: mixture forms explosive glyceryl nitrate
Perchloric acid + lead oxide: mixtures form explosive perchlorate esters
Phosphorus triiodide: violent reaction
Potassium permanganate: ignites on contact
Potassium peroxide: violent reaction
Silver perchlorate: violent reaction
Sodium hydride: violent reaction
Sodium peroxide: violent reaction

When heated to decomposition, it emits acrid smoke and fumes.

FIRE HAZARD
Combustible when exposed to heat, flame, or powerful oxidizers. To fight a fire, use alcohol foam, CO_2, or dry chemical extinguishers.

GLYME

CAS #110-71-4   mf: $C_4H_{10}O_2$
    mw: 90.14

STRUCTURE
$CH_3OC_2H_4OCH_3$

SYNONYMS
Dimethoxyethane
α,β-Dimethoxyethane
1,2-Dimethoxyethane
Dimethylcellosolve
2,5-Dioxahexane
EDGME
Ethylene dimethyl ether
Ethylene glycol dimethyl ether
Glycol dimethyl ether
Monoethylene glycol dimethyl ether
Monoglyme

PROPERTIES
Liquid; sharp, ethereal odor; miscible with water, alcohol; soluble in hydrocarbon solvents; melting point −58°C, boiling point 82–83°C; density 0.86877; flash point 4.5°C

Community Right to Know List

DOT Classification   Flammable liquid; Label: Flammable liquid

TOXIC EFFECTS
Experimental reproductive effects.

HAZARDOUS REACTIONS
Readily forms an explosive peroxide.

Lithium tetrahydroaluminate: mixture may ignite or explode if heated.

When heated to decomposition, it emits acrid smoke and fumes.

FIRE HAZARD
A very dangerous fire hazard when exposed to heat, flame, or oxidizers.

HAFNIUM AND COMPOUNDS

CAS #7440-58-6   mf: Hf
DOT #1326   mw: 178.49

PROPERTIES
Silvery metal; melting point 2227°C, boiling point 4602°C; density 13.31 at 20°C

OSHA PEL   TWA 0.5 mg/m³; IDLH 250 mg/m³

ACGIH TLV   TWA

DOT Classification   Flammable solid; Label: Flammable solid, dry and wet

IMO Classification   Flammable solid; Label: Spontaneously combustible, dry

TOXIC EFFECTS   A poison. Low solubility in water prevents its efficient absorption by ingestion. Many of its compounds are poisons.

Routes   Ingestion, inhalation, contact

Symptoms   Irritation of the eyes, skin, and mucous membranes

Target Organs   Eyes, skin, mucous membranes

HAZARDOUS REACTIONS
Chlorine: incompatible
Halogens: mixture with hafnium powder may explode when heated
Nitric acid: may explode on contact
Nitrogen: mixture with hafnium powder may explode when heated
Nonmetals: mixture with hafnium powder may explode when heated
Oxidants: may explode on contact
Oxygen: mixture with hafnium powder may explode when heated
Phosphorus: mixture with hafnium powder may explode when heated
Sulfur: mixture with hafnium powder may explode when heated

FIRST AID   In case of contact with the eyes, immediately wash the eyes with large amounts of water, occasionally lifting the lower and upper lids. Get medical attention immediately. Contact lenses should not be worn when working with this chemical.

In case of contact with the skin, promptly wash the contaminated skin with soap and water. If this chemical penetrates through the clothing, promptly remove the clothing and wash the skin with soap and water. Get medical attention promptly.

If a person breathes in large amounts of this chemical, move the victim to fresh air at once. If breathing has stopped, perform artificial respiration. Keep the affected person warm and at rest. Get medical attention as soon as possible.

If this chemical has been swallowed, get medical attention immediately.

FIRE HAZARD   A dangerous fire hazard. The powder ignites with friction, heat, sparks, or exposure to air and may explode. The damp powder burns explosively.

HANDLING   Wear appropriate clothing and equipment to prevent any possibility of skin contact. Wash promptly if skin becomes contaminated and at the end of each work shift. Work clothing should be changed daily if there is any possibility that the clothing has been contaminated. Promptly remove any nonimpervious clothing that becomes contaminated.

Wear eye protection to prevent any possibility of eye contact.

The following equipment should be available: an eyewash station and a quick-drench shower.

HEPTANE

CAS #142-82-5   mf: C_7H_{16}
DOT #1206   mw: 100.23

NFPA LABEL

SYNONYMS
Dipropyl methane   n-Heptane
Gettysolve-C   Heptyl hydride

PROPERTIES   Colorless liquid with a mild, gasolinelike odor; slightly soluble in alcohol, miscible in ether and chloroform; insoluble in H_2O; freezing point −90.5°C, boiling point 98.52°C; density 0.684 at 20°C/4°C; flash point 3.9°C (closed cup); lower explosive limit 1.05%, upper explosive limit 6.7%; autoignition temperature 223°C; vapor pressure 40 mm Hg at 22.3°C; vapor density 3.45

OSHA PEL   TWA 500 ppm; IDLH 5000 ppm

ACGIH TLV   TWA 400 ppm; STEL 500 ppm

NIOSH REL   TWA 85 ppm; ceiling 440 ppm/15 min

DOT Classification   Flammable liquid; Label: Flammable liquid

TOXIC EFFECTS   Poisonous. Narcotic in high concentrations.

Routes   Ingestion, inhalation, contact

Symptoms   Hallucinations, lightheadedness, giddiness, stupor, no appetite, nausea; dermatitis; chemical pneumonia; unconsciousness

Target Organs   Skin, respiratory system, peripheral nervous system

HAZARDOUS REACTIONS

Oxidizing materials: vigorous reaction
Phosphorus + chlorine: violent reaction

When heated to decomposition, it emits acrid smoke and irritating fumes.

FIRST AID
In case of contact with the eyes, immediately wash the eyes with large amounts of water, occasionally lifting the lower and upper lids. Get medical attention immediately. Contact lenses should not be worn when working with this chemical.

In case of contact with the skin, promptly wash the contaminated skin with soap and water. If this chemical penetrates through the clothing, promptly remove the clothing and wash the skin with soap and water. Get medical attention promptly.

If a person breathes in large amounts of this chemical, move the victim to fresh air at once. If breathing has stopped, perform artificial respiration. Keep the affected person warm and at rest. Get medical attention as soon as possible.

If this chemical has been swallowed, get medical attention immediately.

FIRE HAZARD
A volatile, flammable liquid when exposed to heat or flame. Moderately explosive when exposed to heat or flame.

To fight a fire, use foam, CO_2, or dry chemical extinguishers.

HANDLING
Wear appropriate clothing and equipment to prevent repeated or prolonged skin contact. Wash promptly if skin becomes wet. Immediately remove any clothing that becomes wet to avoid flammability hazard.

Wear eye protection to reduce probability of eye contact.

HEXANE

CAS #110-54-3   mf: C_6H_{14}
DOT #1208   mw: 86.20

NFPA LABEL

SYNONYMS

Gettysolve-B   Hexyl hydride
n-Hexane   Normal hexane

PROPERTIES
Colorless liquid with a mild, gasolinelike odor; insoluble in water; miscible in chloroform, ether, alcohol; very volatile liquid; freezing point −95.6°C, boiling point 69°C; density 0.6603 at 20°C/4°C; flash point −23°C; lower explosive limit 1.2%, upper explosive limit 7.5%; autoignition temperature 225°C; vapor pressure 100 mm Hg at 15.8°C; vapor density 2.97

OSHA PEL   TWA 500 ppm; IDLH 5000 ppm

ACGIH TLV   TWA 50 ppm

NIOSH REL   TWA 100 ppm; ceiling 510 ppm/15 min

DOT Classification   Flammable liquid; Label: Flammable liquid

TOXIC EFFECTS
Slightly toxic by ingestion and inhalation. A teratogen. An eye irritant. High concentrations may be narcotic and irritate the respiratory tract. Can cause a motor neuropathy in exposed workers. Inhalation of 5000 ppm for 10 min causes disorientation; 1000–2500 ppm for 12 h produces drowsiness, fatigue and loss of appetite, paresthesia of the hands and feet; 500–2500 ppm produces muscle weakness, blurred vision, headache, and anorexia; 2000 ppm for 10 minutes produces no symptoms. Chronic exposure to 500–1000 ppm for 3–6 months produces fatigue, loss of appetite, and distal paresthesia.

Routes   Ingestion, inhalation, contact

Symptoms   Lightheadedness; nausea; headache; numbness, muscle weakness; irritation of the eyes and nose; dermatitis; chemical pneumonia; giddiness; hallucinations

Target Organs   Skin, eyes, respiratory system, lungs

HAZARDOUS REACTIONS

Dinitrogen tetroxide: mixtures may explode at 28°C
Oxidizing materials: vigorous reaction

When heated to decomposition, it emits acrid smoke and fumes.

FIRST AID
In case of contact with the eyes, immediately wash the eyes with large amounts of water, occasionally lifting the lower and upper lids. Get medical attention immediately. Contact lenses should not be worn when working with this chemical.

In case of contact with the skin, immediately wash the contaminated skin with soap and water. If this chemical penetrates through the clothing, immediately remove the clothing and wash the skin with soap and water. Get medical attention promptly.

If a person breathes in large amounts of this chemical, move the victim to fresh air at once. If breathing has stopped, perform artificial respiration. Keep the affected person warm and at rest. Get medical attention as soon as possible.

If this chemical has been swallowed, get medical attention immediately.

FIRE HAZARD   A dangerous fire hazard and a very dangerous explosion hazard when exposed to heat or flame.

To fight a fire, use CO_2 or dry chemical extinguishers.

HANDLING   Wear appropriate clothing and equipment to prevent repeated or prolonged skin contact. Wash promptly if there has been any probability of contact. Immediately remove any clothing that becomes wet to avoid flammability hazard.

Wear eye protection to reduce probability of eye contact.

2-HEXANONE

CAS #591-78-6   mf: $C_6H_{12}O$
    mw: 100.18

STRUCTURE
$CH_3CO(CH_2)_3CH_3$

SYNONYMS
Butyl methyl ketone   Methyl butyl ketone
Hexanone-2   Methyl *n*-butyl ketone
MBK   MNBK

PROPERTIES   Colorless liquid with a characteristic odor; slightly soluble in water; soluble in alcohol and ether; melting point −56.9°C, boiling point 127.2°C; lower explosive limit 1.22%, upper explosive limit 8%; flash point 35°C; autoignition temperature 532.8°C; density 0.830 at 0°C/4°C; vapor pressure 10 mm Hg at 38.8°C; vapor density 3.45

OSHA PEL   TWA 100 ppm; IDLH 5000 ppm

ACGIH TLV   TWA 5 ppm

NIOSH REL   1 ppm

TOXIC EFFECTS   Moderately toxic by ingestion. Mildly toxic by inhalation and skin contact. A teratogen. An eye irritant.

Routes   Ingestion, absorption, inhalation, contact

Symptoms   Irritation of the eyes and nose; peripheral neuropathy; weakness, paresthesia; dermatitis; headache; drowsiness; nausea or vomiting

Target Organs   Central nervous system, skin, respiratory system

HAZARDOUS REACTIONS
Oxidizers (strong): incompatible

FIRST AID   In case of contact with the eyes, immediately wash the eyes with large amounts of water, occasionally lifting the lower and upper lids. Get medical attention immediately. Contact lenses should not be worn when working with this chemical.

In case of contact with the skin, immediately wash the contaminated skin with soap and water. If this chemical penetrates through the clothing, immediately remove the clothing and wash the skin with soap and water. Get medical attention promptly.

If a person breathes in large amounts of this chemical, move the victim to fresh air at once. If breathing has stopped, perform artificial respiration. Keep the affected person warm and at rest. Get medical attention as soon as possible.

If this chemical has been swallowed, get medical attention immediately.

FIRE HAZARD   Dangerous fire and explosion hazard when exposed to heat or flame.

To fight a fire, use alcohol foam, CO_2, or dry chemical extinguishers.

HANDLING   Wear appropriate clothing and equipment to reduce probability of skin contact. Wash promptly if the skin becomes contaminated. Immediately remove any clothing that becomes wet to avoid flammability hazard.

Wear eye protection reduce probability of eye contact.

HEXONE

CAS #108-10-1   mf: $C_6H_{12}O$
DOT #1245   mw: 100.18

SYNONYMS
Isobutyl methyl ketone   4-Methyl-2-pentanone
Isopropylacetone   MIBK
Methyl isobutyl ketone   MIK
2-Methyl-4-pentanone   Shell MIBK

PROPERTIES   Clear liquid with a pleasant odor; freezing point −80.2°C, boiling point 118°C; density

0.803; flash point 17°C; lower explosive limit 1.4%, upper explosive limit 7.5%; autoignition temperature 458.9°C; vapor pressure 16 mm Hg at 20°C; density 3.45

OSHA PEL   TWA 100 ppm; IDLH 3000 ppm

ACGIH TLV   TWA 50 ppm; STEL 75 ppm

NIOSH REL   (Ketones) TWA 200 mg/m^3

Community Right to Know List

TOXIC EFFECTS   Moderately toxic by ingestion. Mildly toxic by inhalation. Very irritating to the skin, eyes, and mucous membranes. Narcotic in high concentration.

Routes   Ingestion, inhalation, contact

Symptoms   Irritation of the eyes and mucous membranes; headache; narcosis, coma; dermatitis

Target Organs   Respiratory system, eyes, skin, central nervous system

HAZARDOUS REACTIONS   May form explosive peroxides upon exposure to air.

Air: incompatible
Oxidizers (strong): incompatible
Potassium-*tert*-butoxide: ignites on contact
Reducing materials: vigorous reaction

FIRST AID   In case of contact with the eyes, immediately wash the eyes with large amounts of water, occasionally lifting the lower and upper lids. Get medical attention immediately. Contact lenses should not be worn when working with this chemical.

In case of contact with the skin, immediately flush the contaminated skin with water. If this chemical penetrates the clothing, immediately remove the clothing and flush the skin with water. If irritation persists after washing, get medical attention.

If a person breathes in large amounts of this chemical, move the victim to fresh air at once. If breathing has stopped, perform artificial respiration. Keep the affected person warm and at rest. Get medical attention as soon as possible.

If this chemical has been swallowed, get medical attention immediately.

FIRE HAZARD   Dangerous fire hazard when exposed to heat, flame, or oxidizers. The vapor is explosive when exposed to heat or flame.

To fight a fire, use alcohol foam, CO_2, or dry chemical extinguishers.

HANDLING   Wear appropriate clothing and equipment to prevent repeated or prolonged skin contact. Wash promptly if skin becomes wet. Immediately remove any clothing that becomes wet to avoid flammability hazard.

Wear eye protection to reduce probability of eye contact.

HYDRAZINE AND ITS SALTS

CAS #302-01-2　　　　　　　　　　mf: H_4N_2
DOT #2029; 2030　　　　　　　　　mw: 32.06

NFPA LABEL

SYNONYMS

Anhydrous hydrazine
Diamide
Diamine
Hydrazine, aqueous
　solution
Hydrazine base

The salts include:
Hydrazine
　hydrochloride
Hydrazine sulfate
Hydrazine hydrate

PROPERTIES   Colorless, oily fuming liquid or white crystals; ammonialike, fishy odor, threshold 3–4 ppm; miscible with water and ethanol, insoluble in hydrocarbons; melting point 1.4°C, boiling point 113.5°C; flash point 52°C (open cup), 37.8°C (closed cup); lower explosive limit 4.7%, upper explosive limit 100%; density 1.1011 at 15° (liquid); autoignition temperature can vary from 23.3°C in contact with iron rust to 132.2°C in contact with black iron to 156.1°C in contact with stainless steel to 270°C in contact with glass; vapor pressure 10.4 mm Hg at 20°C; vapor density 1.1

IARC Experimental Carcinogen, EPA Extremely Hazardous Substances List, Community Right to Know List, and EPA TSCA Inventory

OSHA PEL   TWA 1 ppm

ACGIH TLV   TWA 0.1 ppm; suspected carcinogen

NIOSH REL   ceiling 0.04 mg/m^3/120 min

DOT Classification   Label: Flammable liquid and poison; Corrosive material; Label: Corrosive, aqueous solution

HYDRAZINE AND ITS SALTS

TOXIC EFFECTS   A poison by ingestion, skin contact, and several other routes. Moderately toxic by inhalation. A carcinogen which produces tumors of the lung, nervous system, liver, kidney, hematopoietic organs, breast, and subcutaneous tissue. A teratogen. Corrosive to the eyes, skin, and mucous membranes.

Acute exposure to hydrazine vapors can cause respiratory-tract irritation, excitement, convulsions, cyanosis, and a decrease in blood pressure. The liquid can severely burn the eyes and skin. Hydrazine can cause fatty degeneration of the liver, lung damage, nephritis (kidney damage), and hemolysis (destruction of red blood cells).

A strong eye, skin, and mucous membrane irritant, and a strong skin sensitizer. Hydrazine hydrate produces severe irritation when applied to rabbit eyes.

After repeated oral, skin, or injection exposure, the effects noted include the following: Among guinea pigs and dogs exposed to hydrazine in the air 5–47 times, the dogs showed liver damage, with lesser damage to the kidneys and lungs, while the guinea pigs had pneumonitis and partial lung collapse.

Hydrazine and hydrazine salts are carcinogenic.

Routes   Ingestion, absorption, inhalation, contact

Symptoms   Irritation of the eyes, nose, and throat; burning of the skin and eyes; dizziness; nausea; dermatitis; weight loss; weakness; vomiting; convulsions; a carcinogen

Target Organs   Central nervous system, respiratory system, skin, eyes, kidneys, liver

HAZARDOUS REACTIONS   A sensitive and powerful explosive. A powerful reducing agent.

Acids (strong): incompatible
Air: forms sensitive, explosive mixture
Alkali metals: potentially explosive reaction
Ammonia: potentially explosive reaction
Asbestos: mixture ignites spontaneously in air
Barium oxide: explodes on contact
Benzene-seleninic acid: vigorous reaction
Benzene-seleninic anhydride: vigorous reaction
Cadmium perchlorate: forms sensitive, explosive mixture
Calcium oxide: explodes on contact
Carbon dioxide + stainless steel: vigorous reaction
Catalysts: ignites on contact
Chlorates: violent reaction
Chlorine: potentially explosive reaction
1-Chloro-2,4-dinitrobenzene: violent reaction
2-Chloro-5-methylnitrobenzene: forms sensitive, explosive mixture
Chromate salts: explodes on contact
Chromates: potentially explosive reaction
Chromium dioxide: explodes on contact
Cloth: mixture ignites spontaneously in air
Copper chlorate: forms sensitive, explosive mixture (heat-sensitive)
Copper oxide: vigorous reaction
Copper oxide (black): potentially explosive reaction
Copper(II) salts: potentially explosive reaction
Copper-iron oxide: decomposes on contact to ammonia, hydrogen, and nitrogen gases, which may ignite or explode
Cotton waste + heavy metals: ignites on contact
Dicyanofurazin: explodes on contact
Diethyl zinc: potentially explosive reaction
Dinitrogen oxide: ignites on contact
Dinitrogen tetroxide: hypergolic reaction
Earth: mixture ignites spontaneously in air
Fluorine: potentially explosive reaction
N-haloimides: explodes on contact
Hydrogen peroxide: ignites on contact
Iridium: decomposes on contact to ammonia, hydrogen, and nitrogen gases, which may ignite or explode
Iron oxide: violent reaction
Iron rust: potentially explosive reaction
Lead oxide: vigorous reaction
Lithium perchlorate: forms sensitive, explosive mixture
Manganese nitrate: forms sensitive, explosive mixture (heat-sensitive)
Mercury oxide: explodes on contact
Mercury(I) chloride: forms sensitive, explosive mixture
Mercury(I) nitrate: forms sensitive, explosive mixture
Mercury(II) chloride: forms sensitive, explosive mixture
Mercury(II) nitrate: forms sensitive, explosive mixture
Metal catalysts: decomposes on contact to ammonia, hydrogen, and nitrogen gases, which may ignite or explode
Metal salts: forms sensitive, explosive mixture
Metallic oxides: potentially explosive reaction
Methanol + nitromethane: forms sensitive, explosive mixture
Molybdenum: decomposes on contact to ammonia, hydrogen, and nitrogen gases, which may ignite or explode
Molybdenum oxides: decomposes on contact to ammonia, hydrogen, and nitrogen gases, which may ignite or explode
Nitrous oxide: potentially explosive reaction
Nickel: potentially explosive reaction

Nickel perchlorate: potentially explosive reaction
Nitric acid: ignites on contact
Nitric oxide: potentially explosive reaction
Oxidants: violent reaction
Oxygen: potentially explosive reaction
Oxygen (liquid): potentially explosive reaction
Peroxides: violent reaction
Platinum black: decomposes on contact to ammonia, hydrogen, and nitrogen gases, which may ignite or explode
Porous materials: incompatible
Potassium: explodes on contact
Potassium dichromate: potentially explosive reaction
Potassium peroxodisulfate: vigorous reaction
Raney nickel: decomposes on contact to ammonia, hydrogen, and nitrogen gases, which may ignite or explode
Rhenium + alumina: ignites on contact
Rust + heat: ignites on contact
Ruthenium(III) chloride: vigorous reaction
Silver compounds: explodes on contact
Sodium: forms sensitive, explosive mixture
Sodium dichromate: potentially explosive reaction
Sodium hydroxide: explodes on contact
Sodium perchlorate: forms sensitive, explosive mixture
N,2,4,6-Tetranitroaniline: ignites on contact
Tetryl: potentially explosive reaction
Thiocarbonyl azide thiocyanate: violent reaction
Tin(II) chloride: forms sensitive, explosive mixture
Titanium compounds: explodes on contact at 130°C
Trioxygen difluoride: explodes on contact
Wood: mixture ignites spontaneously in air
Zinc diamide: potentially explosive reaction

When heated to decomposition, it emits highly toxic nitrogen compounds.

FIRST AID   In case of contact with the eyes, immediately wash the eyes with large amounts of water, occasionally lifting the lower and upper lids. Get medical attention immediately. Contact lenses should not be worn when working with this chemical.

In case of contact with the skin, immediately flush the contaminated skin with water. If this chemical penetrates the clothing, immediately remove the clothing and flush the skin with water. Get medical attention promptly.

If a person breathes in large amounts of this chemical, move the victim to fresh air at once. If breathing has stopped, perform artificial respiration. Keep the affected person warm and at rest. Get medical attention as soon as possible.

If this chemical has been swallowed, get medical attention immediately.

FIRE HAZARD   Hydrazine poses a dangerous fire and explosion risk and can explode during distillation if traces of air are present. Severe explosion hazard when exposed to heat or flame. The vapor will burn without air.

HANDLING   Do not use without reading instructions from the manufacturer for handling, storage, and disposal.

The OSHA limits include a warning about the potential contribution of skin absorption to the overall exposure.

When more than a few grams of hydrazine are to be used in the laboratory, Procedure A (see Chap. 1) should be used because hydrazine is carcinogenic in animal tests, quite volatile, and readily absorbed through the skin. Moreover, it is a serious risk as an acute poison and a skin and eye irritant. Nitrile rubber is recommended for gloves and other protective apparel. Prompt washing with water effectively removes hydrazine from skin that it has splashed on.

Wear appropriate clothing and equipment to prevent any possibility of skin contact. Wash immediately if skin becomes contaminated. Immediately remove any clothing that becomes wet to avoid flammability hazard.

Wear eye protection to prevent any possibility of eye contact.

The following equipment should be available: an eyewash station and a quick-drench shower.

See Chap. 2 for information on handling carcinogens.

HYDROGEN

CAS #1333-74-0   mf: H_2
DOT #1049; 1966   mw: 2.02

SYNONYMS
Hydrogen, compressed
Hydrogen, refrigerated liquid

PROPERTIES   Colorless, odorless, tasteless gas; melting point −259.18°C, boiling point −252.8°C; density 0.0899 g/L; lower explosive limit 4.1%, upper explosive limit 74.2%; autoignition temperature 400°C; vapor density 0.069

DOT Classification   Flammable gas; Label: Flammable gas

TOXIC EFFECTS   A simple asphyxiant in high concentration. Otherwise practically no toxicity.

HAZARDOUS REACTIONS
Acetylene + ethylene: explodes on contact
Air: mixture is explosive
Barium: vigorous exothermic reaction
Benzene + Raney nickel catalyst: vigorous exothermic reaction
Bromine: violent reaction or ignition
Bromine fluoride: ignites on contact
Bromine trifluoride: explodes on contact
Calcium: vigorous exothermic reaction
Calcium carbonate + magnesium: mixture explodes when heated
Catalysts + air: violent reaction or ignition
Chlorine: forms sensitive explosive mixtures
Chlorine dioxide: forms sensitive explosive mixtures
Chlorine trifluoride: explodes on contact
Copper(II) oxide: mixture explodes when heated
Dichlorine oxide: forms sensitive explosive mixtures
3,4-Dichloronitrobenzene + catalysts: mixture explodes when heated
Difluorodiazene: mixture explodes when heated above 90°C
Dinitrogen oxide: forms sensitive explosive mixtures
Dinitrogen tetraoxide: forms sensitive explosive mixtures
Dioxane + nickel: violent reaction or ignition
Ethylene + nickel catalysts: mixture explodes when heated
Fluorine: explodes on contact
Fluorine perchlorate: ignition on contact
Hydrogen peroxide + catalysts: explodes on contact
Iodine: violent reaction or ignition
Iodine heptafluoride: forms heat- or spark-sensitive explosive mixtures
Lead trifluoride: violent reaction or ignition
Lithium: violent reaction or ignition
Metals: vigorous exothermic reaction
3-Methyl-2-penten-4-yn-1-ol: violent reaction or ignition
Nickel + oxygen: violent reaction or ignition
2-Nitroanisole: mixture explodes when heated above 250°C/34 bar + 12% catalyst
Nitrogen (liquid): mixtures react with heat or form an explosive product
Nitrogen oxide + oxygen: ignition above 360°C
Nitrogen trifluoride: violent reaction or ignition
Nitryl fluoride: mixture explodes when heated above 200°C
Oxygen (gas): forms sensitive explosive mixtures
Oxygen difluoride: violent reaction or ignition
Palladium + isopropyl alcohol: violent reaction or ignition
Palladium powder + 2-propanol + air: spontaneous ignition
Palladium trifluoride: vigorous exothermic reaction
Palladium(II) oxide: vigorous exothermic reaction
Platinum + air: violent reaction or ignition
Platinum catalyst: violent reaction or ignition
Polycarbon monofluoride: ignition or explosion above 400°C
Potassium: vigorous exothermic reaction above 300°C
Sodium: vigorous exothermic reaction
Strontium: vigorous exothermic reaction
1,1,1-Trisazidomethylethane + palladium catalyst: forms sensitive explosive mixtures
1,1,1-tris(Hydroxymethyl)-nitromethane + nickel catalyst: vigorous exothermic reaction
Vegetable oils + catalysts: mixture explodes when heated
Xenon hexafluoride: violent reaction

FIRE HAZARD   Highly dangerous fire hazard and severe explosion hazard when exposed to heat, flame, or oxidizers.

To extinguish fires, stop flow of gas.

HANDLING   The past history of explosions and accidents involving hydrogen emphasizes that it is unpredictable in its behavior and should be treated with a great deal of respect.

When constructing and operating apparatus for hydrogen, the following facts must always be borne in mind:

1. Mixtures of hydrogen and air containing from 4% to 75% hydrogen by volume are explosive. When working with hydrogen, the most elementary precaution is to avoid the formation of explosive mixtures.

2. The ignition temperature of hydrogen-air mixtures is 573°C.

3. Hydrogen is 15 times lighter than air.

Hydrogen should not be allowed to pass into the air inside buildings. When hydrogen has to be exhausted from apparatus, this should be done via a vent tube which will allow the hydrogen to pass safely to the outside atmosphere.

It is recommended that any part of an apparatus which might develop an excessive pressure be connected to a vent tube via a pressure-relief valve, bursting disk, or other safety device as appropriate to each individual case. The operation of these devices should not in itself contribute to any undue rise in pressure.

If an apparatus which has contained hydrogen is due for alteration or repair, it must be purged with an inert gas.

Mixtures of hydrogen and air must not be pumped. Vacuum pumps which pass hydrogen should have their exhausts connected to a vent tube.

Catalytic oxygen removal units must be used intelligently. They are perfectly safe for removing traces of oxygen from hydrogen, or vice versa, but if a hydrogen/oxygen mixture in the explosive range enters the unit, it will explode.

HYDROGEN BROMIDE

CAS #10035-10-6 mf: BrH
DOT #1048; 1788 mw: 80.92

SYNONYMS
Anhydrous hydrobromic acid
Hydrobromic acid
Hydrobromic acid, anhydrous

PROPERTIES   Colorless gas or pale yellow liquid (under pressure) with an irritating, sharp odor; miscible with water, alcohol; keep protected from light; melting point −87°C, boiling point −66.5°C; density 3.50 g/L at 0°C

OSHA PEL   TWA 3 ppm; IDLH 50 ppm

ACGIH TLV   Ceiling 3 ppm

DOT Classification   Corrosive material; Label: Corrosive; Nonflammable gas; Label: Nonflammable gas, anhydrous

IMO Classification   Poison A; Label: Poison gas, corrosive

TOXIC EFFECTS   Mildly toxic by inhalation. A corrosive irritant to the eyes, skin, and mucous membranes.

Routes   Inhalation, contact

Symptoms   Irritation of the eyes, nose, and throat; burns the eyes and skin

Target Organs   Respiratory system, eyes, skin

HAZARDOUS REACTIONS
Ammonia: violent reaction
Caustics (strong): incompatible
Fe_2O_3: violent reaction
Fluorine: violent reaction
Metals: incompatible
Moisture: incompatible
Oxidizers (strong): incompatible
Ozone: violent reaction

When heated to decomposition or in reaction with water or steam, it emits toxic and corrosive fumes of Br^- and HBr.

FIRST AID   In case of contact with the eyes, immediately wash the eyes with large amounts of water, occasionally lifting the lower and upper lids. Get medical attention immediately. Contact lenses should not be worn when working with this chemical.

In case of contact with the skin, immediately flush the contaminated skin with water. If this chemical penetrates the clothing, immediately remove the clothing and flush the skin with water. Get medical attention promptly.

If a person breathes in large amounts of this chemical, move the victim to fresh air at once. If breathing has stopped, perform artificial respiration. Keep the affected person warm and at rest. Get medical attention as soon as possible.

If this chemical has been swallowed, get medical attention immediately.

HANDLING   Wear appropriate clothing and equipment to prevent any possibility of skin contact. Wash immediately if skin becomes contaminated. Immediately remove any nonimpervious clothing that becomes contaminated.

Wear eye protection to prevent any possibility of eye contact.

The following equipment should be available: an eyewash station and a quick-drench shower.

HYDROGEN CHLORIDE

CAS #7647-01-0 mf: ClH
DOT #1050 mw: 36.46

NFPA LABEL

SYNONYMS
Anhydrous hydrogen chloride
Hydrochloric acid

PROPERTIES   Colorless, corrosive, nonflammable gas; pungent odor; fumes in air; boiling point −154.37°C at 1.0 mm; density 1.639 at −137.77°C

OSHA PEL   Ceiling 5 ppm

ACGIH TLV   TWA

EPA Extremely Hazardous Substances List

TOXIC EFFECTS   Mildly toxic by inhalation. A highly corrosive irritant to the eyes, skin, and mucous membranes.

Routes   Ingestion, inhalation, contact

Symptoms   Inflammation of the nose, throat, and larynx; cough; burns the throat, eyes, and skin; choking; dermatitis

Target Organs   Respiratory system, skin, eyes

HAZARDOUS REACTIONS
Alcohols + hydrogen cyanide: explosive reaction
Alkali or active metals: incompatible
Aluminum: vigorous reaction
Aluminum-titanium alloys: ignites on contact with HCl vapor
Cesium acetylide: ignites on contact
Chlorine + dinitroanilines: vigorous reaction evolves gas
1,1-Difluoroethylene: violent reaction
Fluorine: ignites on contact
Hexalithium disilicide: ignites on contact
Metal acetylides: ignites on contact
Metal carbides: ignites on contact
Metals: incompatible
Potassium permanganate: explosive reaction
Rubidium acetylide: ignites on contact
Silicon dioxide: adsorption of the acid is exothermic
Sodium: mixture with aqueous HCl explodes
Sulfuric acid: potentially dangerous reaction releases HCl gas
Tetraselenium tetranitride: explosive reaction

FIRST AID   In case of contact with the eyes, immediately wash the eyes with large amounts of water, occasionally lifting the lower and upper lids. Get medical attention immediately. Contact lenses should not be worn when working with this chemical.

In case of contact with the skin, immediately flush the contaminated skin with water. If this chemical penetrates the clothing, immediately remove the clothing and flush the skin with water. Get medical attention promptly.

If a person breathes in large amounts of this chemical, move the victim to fresh air at once. If breathing has stopped, perform artificial respiration. Keep the affected person warm and at rest. Get medical attention as soon as possible.

If this chemical has been swallowed, get medical attention immediately.

HANDLING   Wear appropriate equipment to prevent any possibility of skin contact with HCl solutions with a pH<3, and repeated or prolonged skin contact with liquids of pH>3. Wash immediately if skin is contaminated by liquids of pH<3, or after repeated or prolonged contact with liquids of pH>3. Immediately remove nonimpervious clothing that becomes contaminated with liquids of pH<3, and promptly remove clothing contaminated with liquids of pH >3.

Wear eye protection to prevent any possibility of eye contact.

Provide an eyewash station and a quick-drench shower if hydrochloric acid solutions of pH<3 are present.

HYDROGEN CYANIDE

CAS #74-90-8   mf: CHN
DOT #1614; 1051   mw: 27.03

NFPA LABEL

SYNONYMS
Aero liquid HCN
Cyclon
Cyclone B
Fluohydric acid gas
Formonitrile
HCN
Hydrocyanic acid, liquefied
Hydrocyanic acid (prussic), unstabilized
Hydrogen cyanide, anhydrous, stabilized
Prussic acid
Prussic acid, unstabilized
Zaclon discoids

PROPERTIES   Colorless or pale blue liquid or gas with a bitter almond odor, threshold 2–5 ppm; miscible in water, alcohol, and ether; melting point −13.2°C, boiling point 25.7°C; density (liquid) 0.6876 at 20°C/4°C; flash point −17.8°C (closed cup); lower explosive limit 5.6%, upper explosive limit 40%; autoignition temperature 537.8°C; vapor pressure 400 mm Hg at 9.8°C, 807 mm Hg at 27.2°C; vapor density 0.932

OSHA PEL   TWA 10 ppm (skin); IDLH 50 ppm

ACGIH TLV   Ceiling 10 ppm (skin)

NIOSH REL   (Cyanide) ceiling 5 mg(CN)/m^3/10 min

EPA Extremely Hazardous Substances List and Community Right to Know List

DOT Classification   Poison A; Label: Poison gas and flammable gas

IMO Classification   Poison B; Label: Poison (UN 1614); Flammable liquid and poison; Forbidden, unstabilized

TOXIC EFFECTS   A deadly human and experimental poison by all routes. Hydrocyanic acid and its salts inactivate the enzymes associated with cellular oxidation. This reaction makes oxygen unavailable to body tissues and thus causes death through asphyxia. If the cyanide is removed before death occurs, normal function is restored.

Hydrogen cyanide is among the most toxic and rapidly acting of all known substances. Exposure to high doses may be followed by almost instantaneous collapse, cessation of respiration, and death. At lower dosages, the early symptoms include weakness, headache, confusion, nausea, and vomiting. In humans, the approximate fatal dose is 40 mg via the oral route. Exposure to 3000 ppm HCN is immediately fatal, while 200–480 ppm can be fatal after 30 min. Exposure to 18–36 ppm HCN causes slight symptoms after several hours. The liquid is rapidly absorbed through the skin or the eyes.

The presence of cherry-red venous blood in cases of cyanide poisoning is due to the inability of the tissues to remove the oxygen from the blood. If the patient recovers, there is rarely any disability.

Dogs subjected repeatedly to 30-min exposures of 45 ppm HCN exhibited cumulative effects, particularly central nervous system lesions, hemorrhages, and vasodilation.

Routes   Ingestion, absorption, inhalation, contact

Symptoms   Asphyxia and death at high levels; weakness; headache; confusion; nausea, vomiting; increase rated and depth of respiration or respiratory slowing and gasping, a feeling of suffocation, cyanosis; unsteadiness of gait

Target Organs   Central nervous system, cardiovascular system, liver, kidneys

HAZARDOUS REACTIONS   Can polymerize explosively at 50–60°C or in the presence of traces of alkali. The anhydrous liquid is stabilized at or below room temperature by the addition of acid.

Because of its low flash point and wide range of explosive mixtures, HCN presents a serious fire and explosion hazard. Another often-overlooked hazard is spontaneous polymerization, which results from the attack of cyanide ion on HCN. The reaction is unpredictable, but may occur in pure unstabilized HCN if a base is present. Hence, amines, hydroxides, and cyanide salts that are capable of producing the cyanide ion should not be added to liquid HCN without suitable precautions. Once started, the polymerization is likely to become violent and be accompanied by sharp increases in temperature and pressure. If liquid HCN is heated above 115°C in a sealed vessel, a violent exothermic reaction generally occurs.

Commercial HCN is normally stabilized by the addition of a little phosphoric acid, although sulfuric acid or sulfur dioxide may also be used. Either acid may be removed readily by distillation, but sulfur dioxide is extremely difficult to remove. Distilled HCN constitutes a greater explosion hazard than stabilized materials, so use of the distilled materials should be avoided.

Acetaldehyde: violent reaction
Acid or acid fumes: reacts to form toxic cyanide gas
Air: mixture is explosive
Amines: incompatible
Bases: incompatible
Caustics: incompatible
Water or steam: reacts to form toxic cyanide gas

When heated to decomposition, it forms highly toxic CN^- gas.

FIRST AID   The following equipment and procedures are recommended:

Equipment—An HCN first-aid kit and an oxygen cylinder equipped with pressure gauge and needle valve should be available on any floor of a building on which work with cyanides is in progress. In special cases, first-aid kits and oxygen cylinders may be located near the work area, but they should not be in the same room. Except when the cylinder is being used or checked, the main cylinder valve should be kept closed and the pressure kept off the gauge. A tag should be attached to the oxygen cylinder indicating that it is reserved for emergency HCN first aid.

The HCN first-aid kit should contain a box of amyl nitrite pearls, a facepiece and length of rubber tubing for administering oxygen, and a bottle of 1% sodium thiosulfate solution.

Procedures—Anyone who has been exposed to HCN should be removed from the contaminated atmosphere immediately. Any contaminated clothing should be removed and the affected area deluged with water. The person should be kept warm. An amyl nitrite pearl should be held under the person's nose for not more than 15 s out of each minute (excess nitrite will reduce the blood pressure), and oxygen should be administered in the intervals. If the person is not breathing, artificial resuscitation should be started; when breathing starts, amyl nitrite and oxygen should be administered immediately. If HCN has been ingested, the person should be given 1 pint of 1% sodium thiosulfate and then soapy water or mustard water to induce vomiting. Such cases must be transported to definitive medical care immediately.

In case of contact with the eyes, immediately wash the eyes with large amounts of water, occasionally lifting the lower and upper lids. Get medical attention immediately. Contact lenses should not be worn when working with this chemical.

In case of contact with the skin, immediately flush the contaminated skin with water. If this chemical penetrates the clothing, immediately remove the clothing and flush the skin with water. Get medical attention promptly.

FIRE HAZARD   Very dangerous fire hazard when exposed to heat, flame, or oxidizers. Severe explosion hazard when exposed to heat or flame or by chemical reaction with oxidizers.

To fight a fire, use CO_2, nonalkaline dry chemical, or foam extinguishers.

HANDLING   The procedures given below are appropriate for the safe use of hydrogen cyanide (HCN) and related compounds (e.g., cyanogen and cyanogen halides) that may release cyanide ion or cyanogen or generate HCN when acidified (e.g., sodium or potassium cyanide). The use of the term HCN in these procedures is intended to be specific for hydrogen cyanide and to indicate general applicability to related compounds. All users of HCN should study or be instructed in HCN procedures before starting to work with this material.

All work with HCN must be confined to hoods, which should have a minimum face velocity of 60 lfm. Care must be exercised to prevent the contact of either liquid HCN or its vapors with the skin. Neoprene or rubber gloves should be worn at all times when working with HCN. Whenever work with HCN or related compounds is being carried out in a laboratory, there should be at least two people present in the area at all times.

Signs warning that HCN is in use should be posted at each entrance to the laboratory area whenever work is being done with HCN. WARNING or NO ADMITTANCE signs should be posted on the doors to fan lofts and roofs whenever cyanides are being used or stored in hoods.

All reaction equipment in which cyanides are used or produced should be placed in or over shallow pans so that spills or leaks will be contained. In the event of spills of HCN or cyanide solutions, the contaminated area should be evacuated promptly, and it should be determined immediately if anyone has been exposed to cyanide vapors or liquid splash. Consideration should be give to the need for evacuating other parts of the building or notifying other occupants that the spill has occurred. In general, it is usually best not to attempt to dilute or absorb such spills if they occur in well-ventilated areas.

Detection—Hydrogen cyanide has a characteristic odor that resembles that of bitter almonds; however, many people cannot smell it in low concentrations, and this method of detection should not be relied on. Vapor-detector tubes sensitive to 1 ppm of HCN are available commercially. The presence of free cyanide ion in aqueous solution may be detected by treating an aliquot of the sample with ferrous sulfate and an excess of sulfuric acid. A precipitate of Prussian blue indicates that free cyanide ion is present.

Sodium cyanide and acids should not be stored or transported together. An open bottle of NaCN can generate HCN in humid air, and HCN may be liberated from spills of sodium cyanide solutions.

Storage and dispensing—Storage of liquid HCN (except in commercial cylinders) in laboratory areas should be prohibited without special permission. If such storage is necessary, it must be in an exhaust hood or a barricaded area that has independent ventilation facilities. Liquid HCN is dispensed from cylinders. Only trained personnel should be permitted to operate the dispensing equipment.

Waste disposal—Hydrogen cyanide and waste solutions containing cyanides must not be emptied into the sewer or left to evaporate in an exhaust hood. The most effective way to dispose of such material is to dilute it with an equal amount of alcohol and burn it in a solvent incinerator.

OSHA and ACGIH limits include a warning against the potential contribution of skin absorption to the overall exposure.

In addition to its high toxicity, HCN has a low flash point and forms an explosive mixture with air over a wide range of concentrations. Moreover, traces of base can cause rapid spontaneous polymerization, sometimes resulting in detonation. Hence HCN is very dangerous; anyone working with it should wear goggles and impervious gloves, and no one should work alone with it. In cases of overexposure to HCN, quick action is called for in removing the victim from the contaminated area, using amyl nitrate ampules to restore consciousness, or, if breathing has stopped, beginning artificial respiration. Medical assistance should be summoned as soon as possible, but the victim should not be left unattended. Speed in providing treatment is of the utmost importance.

Wear appropriate clothing and equipment to prevent any possibility of skin contact. Wash promptly if the skin becomes contaminated. Immediately remove any nonimpervious clothing that becomes contaminated.

Wear eye protection to prevent any possibility of eye contact.

Provide an eyewash and quick-drench shower.

HYDROGEN FLUORIDE

CAS #7664-39-3 mf: HF
DOT #1790; 1052 mw: 20.01

NFPA LABEL

SYNONYMS
Anhydrous hydrofluoric acid
Hydrofluoric acid
Hydrofluoride

PROPERTIES   Clear, colorless, fuming, corrosive liquid or gas.

Anhydrous HF is a clear, colorless liquid that is miscible with water in all proportions and forms an azeotrope (38.3% HF) that boils at 112°C; melting point −83.1°C, boiling point 19.54°C; density 0.901 g/L (gas), 0.699 at 22° (liquid); vapor pressure 40 mm Hg at 2.5°C. Because of its low boiling point and high vapor pressure, anhydrous HF must be stored in pressure containers. A 70% aqueous solution is a common form of HF.

OSHA PEL   TWA 3 ppm

ACGIH TLV   Ceiling 3 ppm (F)

NIOSH REL   (HF) TWA 2.5 mg(F)/m^3; ceiling 5.0 mg(F)/m^3/15M

EPA Extremely Hazardous Substances List and Community Right to Know List

DOT Classification   Corrosive material; Label: Corrosive

IMO Classification   Poison A; Label: Poison gas, corrosive; Corrosive material; Label: Corrosive, poison

TOXIC EFFECTS   A poison by inhalation and other routes. A corrosive irritant to skin, eyes (at 0.05 mg/L), and mucous membranes. A teratogen.

Burns caused by solutions of up to 20% concentration are not dangerous and occur only after a few hours exposure. Burns due to solutions of 20–50% concentration occur after a few minutes and are more serious. Burns due to solutions of 50–100% concentration are immediate and dangerous. They are slow to heal, and the subcutaneous tissues may be affected. Gangrene of the affected areas may follow.

The effects of the fumes and of the acid itself on the eyes are very serious. Acid splashed into the eyes may cause blindness or at least acute irritation.

Inhalation of anhydrous HF or HF mists or vapors can cause severe respiratory-tract irritation, which may be fatal. Concentrations of 50–250 ppm are dangerous, even in brief exposures. The distinctive smell of this acid facilitates the quick detection of dangerous concentrations and prevents the inhaling of large amounts.

When ingested, this acid causes acute irritation of the digestive tract and of the esophagus, although it is generally considered not to be a source of permanent poisoning.

Wearing clothing (including leather shoes and gloves) that has absorbed small amounts of HF can result in serious delayed effects such as painful, slow-healing skin ulcers.

Routes   Ingestion, absorption, inhalation, contact

Symptoms   Irritation of the eyes, nose, and throat; pulmonary edema; burns of the skin and eyes; nasal congestion; bronchitis

Target Organs   Eyes, respiratory system, skin

HAZARDOUS REACTIONS
Acetic anhydride: violent reaction
2-Amino ethanol: violent reaction
Ammonium hydroxide: violent reaction
Arsenic trioxide: violent reaction
Bismuthic acid: violent reaction evolves oxygen
Calcium oxide: violent reaction

Ceramics: incompatible
Chlorosulfonic acid: violent reaction
Concrete: incompatible
Cyanogen fluoride: explosive reaction
Ethylene diamine: violent reaction
Ethylene imine: violent reaction
Fluorine: violent reaction
Glass: incompatible
Glycerol + nitric acid: explosive reaction
$HBiO_3$: violent reaction
Mercuric oxide: violent reaction
Mercury(II) oxide + organic materials: violent reaction above 0°C
Metals: incompatible
Methanesulfonic acid: reaction evolves oxygen difluoride, which explodes
Nitric acid + lactic acid: violent reaction
Nitric acid + propylene glycol: dangerous storage hazard—mixtures evolve gas which may burst a sealed container
Oleum: violent reaction
Oxides: incandescent reaction
n-Phenylazopiperidine: violent reaction
P_2O_5: violent reaction
Potassium permanganate: violent reaction
Potassium tetrafluorosilicate (2−): violent reaction evolves silicon tetrafluoride gas
β-Propiolactone: violent reaction
Propylene oxide: violent reaction
Sodium: explosive reaction with aqueous hydrofluoric acid
Sodium hydroxide: violent reaction
Sodium tetrafluorosilicate: violent reaction
Sulfuric acid: violent reaction
Vinyl acetate: violent reaction
Water or steam: reaction produces toxic and corrosive fumes

When heated to decomposition, it emits highly corrosive fumes of F^-.

FIRST AID   Anyone who knows or even suspects that he or she has come into direct contact with HF should immediately flush the exposed area with large quantities of cool water. Exposed clothing should be removed as quickly as possible while flushing. Medical attention should be obtained promptly, even if the injury appears slight. On the way to the physician, the burned area should be immersed in a mixture of ice and water or, if readily available, an iced solution of benzalkonium chloride. (It may be necessary to remove the area being soaked from the solution periodically to relieve discomfort caused by the cold.) If immersion is impractical, a compress made by inserting ice cubes between layers of gauze should be used.

In case of contact with the eyes, immediately wash the eyes with large amounts of water, occasionally lifting the lower and upper lids. Continue for at least 15 min. Get medical attention immediately. Contact lenses should not be worn when working with this chemical.

Anyone who has inhaled HF vapor should be removed *immediately* to fresh air and kept warm. If a person breathes in large amounts of this chemical, move the victim to fresh air at once. If breathing has stopped, perform artificial respiration. Keep the affected person warm and at rest. Get medical attention as soon as possible.

Anyone who has ingested HF should drink a large quantity of water as quickly as possible. Do not induce vomiting. Again, medical help should be obtained promptly. After the acid has been thoroughly diluted with water, if medical attention is delayed, the person should be given milk or 2 fluid ounces of milk of magnesia to drink to soothe the burning effect.

FIRE HAZARD   Hydrofluoric acid is nonflammable. Because aqueous HF can cause formation of hydrogen in metallic containers and piping, which presents a fire and explosion hazard, potential sources of ignition (sparks and flames) should be excluded from areas having equipment containing HF.

HANDLING   It is crucial to ensure adequate ventilation by working only in a hood so that safe levels (3 ppm) are not exceeded. All contact of the vapor or the liquid with eyes, skin, respiratory system, or digestive system must be avoided by using protective equipment such as face shields and neoprene or polyvinyl chloride gloves. The protective equipment should be washed after each use to remove any HF on it. Safety showers and eyewash fountains should be nearby. Anyone working with HF should have received prior instruction about its hazards and proper protective measures and should know the recommended procedure for treatment in the event of exposure.

To avoid burns and poisoning, wear anti-acid clothing, neoprene gloves, goggles or a plastic mask, and rubber shoes. If acid fumes are suspected, the mask should be equipped with a fresh-air supply.

HF is difficult to contain because it attacks glass, concrete, and some metals—especially cast iron and alloys that contain silica. It also attacks such organic materials as leather, natural rubber, and wood.

Wear appropriate clothing and equipment to prevent any possibility of skin contact. Wash immediately if skin becomes contaminated. Immediately remove any nonimpervious clothing that becomes contaminated.

Wear eye protection to prevent any possibility of eye contact.

SPILL CLEANUP   The vapors of both anhydrous HF and aqueous 70% HF produce visible fumes if they contact moist air. This characteristic can be useful in detecting leaks but cannot be relied on because of atmospheric variations. Spills of HF must be treated immediately to minimize the dangers of vapor inhalation, body contact, corrosion of equipment, and possible generation of hazardous gases. Spills should first be contained. Next, cover the contaminated surface with lime or sodium bicarbonate. Mix and add water to form a slurry, which should be washed down the drain with excess water.

Dispose of HF solutions by adding slowly with stirring to a large volume of soda ash and slaked lime and flush with excess water.

WASTE DISPOSAL   Waste HF may be added slowly to a larger volume of an agitated solution of slaked lime. This neutralized solution is then added to excess running water before final disposal. Because sodium fluoride is highly soluble and toxic to warm-blooded animals, lime is the preferred neutralizing agent.

HYDROGEN PEROXIDE

CAS #7722-84-1   mf: H_2O_2
DOT #2015   mw: 34.02

NFPA LABEL

SYNONYMS

Albone
Dihydrogen dioxide
Hioxyl
Hydrogen dioxide
Hydrogen peroxide solution (over 52% peroxide)
Hydrogen peroxide, stabilized (over 60% peroxide)
Hydroperoxide
Inhibine
Oxydol
Perhydrol
Perone
PEROXAN
Peroxide
Superoxol
T-stuff

PROPERTIES   Colorless, heavy liquid with a slightly sharp odor; at low temperatures it forms a crystalline solid; miscible with water; soluble in ether, insoluble in petroleum ether; bitter taste; melting point −0.43°C, boiling point 152°C; density 1.71 at −20°C, 1.46 at 0°C; vapor pressure 1 mm Hg at 15.3°C. Unstable, can be decomposed by many organic solvents.

OSHA PEL   TWA 1 ppm; IDLH 75 ppm

ACGIH TLV   TWA 1 ppm

DOT Classification   Oxidizer; Label: Oxidizer and corrosive

IARC Cancer Review: Animal Limited Evidence, EPA TSCA Inventory, and EPA Extremely Hazardous Substances List

TOXIC EFFECTS   Moderately toxic by inhalation, ingestion, and skin contact. A corrosive irritant to skin, eyes, and mucous membranes. A suspected carcinogen. A very powerful oxidizer.

Solutions of H_2O_2 of 35 wt% and over can easily cause blistering of the skin. Irritation caused by H_2O_2 which does not subside upon flushing the affected part with water should be treated by a physician. The eyes are particularly sensitive to this material.

Routes   Ingestion, inhalation, contact

Symptoms   Irritation of the eyes, nose, and throat; corneal ulcer; erythema, vesicles on skin; bleaches hair

Target Organs   Eyes, skin, respiratory system

HAZARDOUS REACTIONS H_2O_2 is a powerful oxidizer, particularly in the concentrated state. It is important to keep containers of this material covered, because uncovered containers are much more prone to react with flammable vapors, gases, etc.; and, if uncovered, the water from an H_2O_2 solution can evaporate, concentrating the material and thus increasing the fire hazard of the remainder. For instance, solutions of H_2O_2 of concentration in excess of 65 wt% heat up spontaneously when decomposed to $H_2O + 1/2 O_2$. Thus 90 wt% solutions, when caused to decompose rapidly due to the introduction of a catalytic decomposition agent, can get quite hot and perhaps start fires.

Aqueous solutions of 96 wt% and 100 wt% can be powerful explosives. When stored in sealed containers, gradual decomposition of H_2O_2 to $H_2O + 1/2 O_2$ can cause large pressures to build up.

There is a severe explosion hazard when highly concentrated or when pure H_2O_2 is exposed to heat, mechanical impact, detonation of a blasting cap, or is caused to decompose catalytically by catalytic metals (in order of decreasing effectiveness: osmium, palladium, platinum, iridium, gold, silver, manganese, cobalt, copper, lead). Many mixtures of H_2O_2 and organic materials can be detonated by flame or impact.

Acetal + acetic acid: mixture explodes when heated
Acetaldehyde + desiccants: reacts to form explosive polyethylidine peroxide
Acetaldehyde + water: explodes on contact
Acetic acid: reacts to form explosive peracetic acid
Acetic acid + n-heterocycles: explodes above 50°C
Acetic acid + water: explodes on contact
Acetic anhydride: reacts to form unstable explosive products
Acetone: reaction forms explosive peroxides
Acetone + water: explodes on contact
Alcohols: reacts to form shock- and heat-sensitive explosive products
Alcohols + sulfuric acid: explodes on contact
Alcohols + tin chloride: explodes on contact
Aluminum isopropoxide + heavy metal salts: violent reaction
2-Amino-4-methyloxazole + iron(II) catalyst: explodes on contact
Ammonia: explodes on contact
Aniline: explodes on contact
Antimony trisulfide: explodes on contact
Aromatic hydrocarbons + trifluoroacetic acid: explodes on contact
Arsenic trisulfide: explodes on contact
Azeliac acid + sulfuric acid: explodes above 45°C
Benzenesulfonic anhydride: explodes on contact
Brass: incompatible
Bronze: incompatible
tert-Butanol + sulfuric acid: explodes on contact
2-Butanone: explodes on contact
Calcium permanganate: violent reaction
Carboxylic acids: reacts to form unstable explosive products
Cellulose: explodes on contact
Charcoal: explodes on contact
Chlorine + potassium hydroxide: explodes on contact
Chromium: incompatible
Coal: violent reaction
Cobalt oxide: violent reaction
Copper: incompatible
Copper sulfide: explodes on contact
Cyclohexanone: explodes on contact
Cyclopentanone: explodes on contact
Diethyl ether: reacts to form unstable explosive products
3,5-Dimethyl-3-hexanol + sulfuric acid: explodes on contact

uns-Dimethylhydrazine: explodes on contact
1,1-Dimethylhydrazine: explodes on contact
Dimethylphenylphosphine: violent reaction
Diphenyl diselenide: explodes above 53°C
Ethanol: explodes on contact
Ethanol + water: explodes on contact
2-Ethoxyethanol + polyacrylamide gel + toluene: explodes when heated
Ethyl acetate: reacts to form unstable explosive products
Formaldehyde + water: explodes on contact
Formic acid: reacts to form unstable explosive products
Formic acid + metaboric acid: reacts to form unstable explosive products
Formic acid + organic matter: explodes on contact
Formic acid + water: explodes on contact
Furfuryl alcohol: ignites on contact
Gadolinium hydroxide: explodes above 80°C
Gallium + hydrochloric acid: explodes on contact
Glycerol: explodes on contact
Hydrazine: explodes on contact
Hydrazine hydrate: explodes on contact
Hydrogen + palladium catalysts: explodes on contact, reaction has caused major industrial explosions
Hydrogen selenide: violent reaction
Iron oxide: violent reaction
Iron powder: ignites on contact
Iron(II) sulfate + 2-methylpyridine + sulfuric acid: explodes on contact
Iron(II) sulfate + nitric acid + sodium carboxymethylcellulose: explodes when evaporated
Iron sulfide: explodes on contact
Ketene: reaction forms explosive diacetyl peroxide
Lead: incompatible
Lead dioxide: explodes on contact
Lead hydroxide: violent reaction
Lead monoxide: explodes on contact
Lead oxide: violent reaction
Lead sulfide: explodes on contact
Lithium: violent reaction
Lithium tetrahydroaluminate: violent reaction
Magnesium powder: ignites on contact
Manganese: incompatible
Manganese dioxide: explodes on contact
Manganese oxide: violent reaction
Mercuric oxide: explodes on contact
Mercuric oxide + nitric acid: explodes on contact
Mercurous oxide: explodes on contact
Mercury(II) oxide + nitric acid: reaction forms explosive mercury(II) peroxide
Metals: violent reaction
Metal oxides: violent reaction
Metal powders: ignites on contact

Metal salts: violent reaction
Methanol + *tert*-amines + platinum catalysts: explodes on contact
Methanol + phosphoric acid: violent reaction
Methanol + water: explodes on contact
3-Methylcyclohexanone: explodes on contact
2-Methyl-1-phenyl-2-propanol + sulfuric acid: explodes on contact
4-Methyl-2,4,6-triazatricyclo[5.2.2.0^{2,6}] undeca-8-ene-3,5-dione + potassium hydroxide: violent reaction
Molybdenum disulfide: explodes on contact
NaIO$_3$: explodes on contact
Nickel oxide: violent reaction
Nitric acid: explodes on contact
Nitric acid + ketones: explodes on contact
Nitric acid + soils: explodes on contact
Nitrogenous bases: explodes on contact
Organic compounds: explodes on contact
Organic compounds (unsaturated): violent reaction
Organic materials + sulfuric acid: explodes, especially if confined
Organic matter: explodes on contact
Oxidizable materials: incompatible
P$_2$O$_5$: explodes on contact
3-Pentanone: explodes on contact
2-Phenyl-1,1-dimethylethanol: explodes on contact
α-Phenylselenoketones: violent reaction
Phosphorus: explodes on contact
Phosphorus(V) oxide: violent reaction
Polyacetoxyacrylic acid lactone + poly(2-hydroxyacrylic acid) + sodium hydroxide: reacts to form unstable explosive products
Potassium: violent reaction
Potassium permanganate: explodes on contact
2-Propanol + water: explodes on contact
Propionaldehyde + water: explodes on contact
Quinoline: explodes on contact
Selenium hydride: explodes on contact
Silver: incompatible
Sodium: violent reaction
Sulfuric acid: explodes during evaporation
Tartaric acid: reacts to form unstable explosive products
Tetrahydrothiophene: explodes on contact
Thiodiglycol: explodes on contact
Thiourea + nitric acid: reacts to form unstable explosive products
Tin(II) chloride: violent reaction
Trioxane: mixture is an explosive sensitive to heat, shock, or contact with lead
Vinyl acetate: explodes on contact
Water: explodes on contact
Water + oxygenated compounds: explodes on contact
Wood: ignites on contact
Zinc: incompatible

FIRE HAZARD   A dangerous fire hazard by chemical reaction with flammable materials.

FIRST AID   In case of contact with the eyes, immediately wash the eyes with large amounts of water, occasionally lifting the lower and upper lids. Get medical attention immediately. Contact lenses should not be worn when working with this chemical.

In case of contact with the skin, immediately flush the contaminated skin with water. If this chemical penetrates the clothing, immediately remove the clothing and flush the skin with water. Get medical attention promptly.

If a person breathes in large amounts of this chemical, move the victim to fresh air at once. If breathing has stopped, perform artificial respiration. Keep the affected person warm and at rest. Get medical attention as soon as possible.

If this chemical has been swallowed, get medical attention immediately.

HANDLING   Wear appropriate clothing and equipment to prevent any possibility of skin contact. Wash promptly if the skin becomes contaminated. Immediately remove any nonimpervious clothing that becomes contaminated.

Wear eye protection to prevent any possibility of eye contact.

The following equipment should be available: an eyewash station and a quick-drench shower.

See Chap. 2 for information on handling carcinogens.

HYDROGEN SULFIDE

CAS #7783-06-4   mf: H$_2$S
DOT #1053   mw: 34.08

NFPA LABEL

SYNONYMS
Hepatic gas
Hydrogen sulfuric acid
Hydrosulfuric acid
Stink damp
Sulfurated hydrogen
Sulfur hydride

PROPERTIES   Colorless, flammable gas with a strong odor of rotten eggs, threshold 0.2–0.003 ppm; odor seems sweet at 30 ppm and above, high concentrations deaden the sense of smell; solubility 437 mL/100 mL in water at 0°C and 186 mL/100 mL at 40°C; also soluble in alcohol and petroleum solvents; melting point −82.9°C, liquid at high pressure, boiling point −61.8°C; density 1.539 g/L at 0°C; lower explosive limit 4%, upper explosive limit 46%; autoignition temperature 260°C; vapor pressure 8.77 atm at 20°C, 20 atm at 25.5°C; vapor density 1.189 at 15°C

OSHA PEL   Ceiling 20 ppm; peak 50/10 min; IDLH 300 ppm

ACGIH TLV   TWA 10 ppm; STEL 15 ppm

NIOSH REL   Ceiling 15 mg/m^3/10M

DOT Classification   Flammable gas; Label: Poison gas and Flammable gas

EPA Extremely Hazardous Substances List

TOXIC EFFECTS   A poison by inhalation. A severe irritant to eyes and mucous membranes. An asphyxiant.

Extremely dangerous. Human exposure to relatively low concentrations has caused corneal damage, headache, sleepiness disturbances, nausea, weight loss, and other symptoms suggestive of brain damage. Higher concentrations can cause irritation of the lungs and respiratory passages and even pulmonary edema. Exposure to 210 ppm for 20 min has caused unconsciousness, arm cramps, and low blood pressure. Coma may occur within seconds after one or two breaths at high concentrations and be followed rapidly by death. For example, workers exposed to 930 ppm for less than 1 min died.

The irritant action has been explained on the basis that H_2S combines with the alkali present in moist surface tissues to form sodium sulfide, a caustic. With higher concentration, the action of the gas on the nervous system, including paralysis of the respiratory center, becomes more prominent. Fatal hydrogen sulfide poisoning can be faster than that due to a similar concentration of HCN.

It is an insidious poison since sense of smell may be fatigued. The odor and irritating effects do not offer a dependable warning to workers who may be exposed to gradually increasing amounts and therefore become used to it.

Skin absorption is slight and not considered significant. However, prolonged or repeated skin contact might cause mild irritation.

Routes   Ingestion, inhalation, contact

Symptoms   Irritation of the eyes, conjunctivitis keratitis; excitement; headache, staggering gait, diarrhea, colored urine, dysuria, bronchitis, nausea; dizziness; suffocation, rapid breathing; chronic pulmonary edema, muscle twitch, delirium; collapse; coma

Target Organs   Respiratory system, eyes

HAZARDOUS REACTIONS
Acetaldehyde: violent reaction
Barium oxide + mercurous oxide + air: violent reaction
Barium oxide + nickel oxide + air: violent reaction
Barium peroxide: ignites on contact
Bromine pentafluoride: explodes on contact
4-Bromobenzenediazonium chloride: reaction forms an explosive product
Calcium oxide: ignites on contact
Chlorine oxide: violent reaction
Chlorine trifluoride: explodes on contact
Chromium trioxide: ignites on contact
Copper: violent reaction
Copper + oxygen: potentially explosive reaction
Copper chromate: ignites on contact
Copper oxide: ignites on contact
Dibismuth dichromium nonaoxide: ignites on contact
Dichlorine oxide: explodes on contact
Fluorine: ignites on contact
Heptasilver nitrate octaoxide: ignites on contact
Iron oxide (hydrated): violent reaction
Lead dioxide: ignites on contact
Lead(II) hypochlorite: ignites on contact
Lead(IV) oxide: ignites on contact
Manganese dioxide: ignites on contact
Mercury oxide: ignites on contact
Mercury(I) bromate: ignites on contact
Metal powders: vigorous reaction
Metal oxides: ignites on contact
Metals: incompatible
Nickel oxide: ignites on contact
Nitric acid: ignites on contact
Nitrogen trichloride: explodes on contact
Nitrogen trifluoride: violent reaction
Nitrogen triiodide: violent reaction
Oxidants: ignites on contact
Oxygen: explosive reaction above 280°C
Oxygen difluoride: explodes on contact
Perchloryl fluoride: explosive reaction above 100°C
Phenyl diazonium chloride: violent reaction
Rust: ignites on contact
Silver bromate: ignites on contact
Silver fulminate: explodes on contact
Silver(I) oxide: ignites on contact

Silver(II) oxide: ignites on contact
Soda-lime + air: ignites on contact
Sodium hydroxide + calcium oxide + air: violent reaction
Sodium peroxide: ignites on contact
Thallium(III) oxide: ignites on contact
Tungsten: vigorous reaction

When heated to decomposition, it emits highly toxic fumes of SO_x.

FIRST AID   In case of contact with the eyes, immediately wash the eyes with large amounts of water, occasionally lifting the lower and upper lids. Get medical attention immediately. Contact lenses should not be worn when working with this chemical.

In case of contact with the skin, immediately flush the contaminated skin with water. If this chemical penetrates the clothing, immediately remove the clothing and flush the skin with water. Get medical attention promptly.

If a person breathes in large amounts of this chemical, move the victim to fresh air at once. If breathing has stopped, perform artificial respiration. Keep the affected person warm and at rest. Get medical attention as soon as possible.

FIRE HAZARD   Very dangerous fire hazard when exposed to heat, flame, or oxidizers. Moderately explosive when exposed to heat or flame.

To extinguish fires, stop flow of gas.

HANDLING   Partly because of its disagreeable odor, but also because of its toxicity, laboratory operations with H_2S should be carried out in a hood. Cylinders of it should not be stored in small, unventilated rooms, as deaths have resulted from people entering such a room containing a leaking cylinder.

Wear appropriate clothing and equipment to prevent skin freezing. Immediately remove any clothing that becomes wet to avoid flammability hazard.

Wear eye protection to reduce probability of eye contact.

SPILL CLEANUP   If the valve on a hydrogen sulfide cylinder is leaking, the gas can be bubbled through $FeCl_3$ solution. Include a trap in the line to prevent suck-back.

HYDROQUINONE

CAS #123-31-9   mf: $C_6H_6O_2$
DOT #2662   mw: 110.12

SYNONYMS
Arctuvin
p-Benzenediol
1,4-Benzenediol
Benzohydroquinone
Benzoquinol
Black-and-white bleaching cream
Dihydroxybenzene
p-Dihydroxybenzene
1,4-Dihydroxybenzene
p-Dioxobenzene
Eldopaque
Eldoquin
Hydroquinol
α-Hydroquinone
p-Hydroquinone
p-Hydroxyphenol
β-Quinol

PROPERTIES   Light tan, light gray, or colorless hexagonal crystals; solubility in water 2.9%; very soluble in alcohol and ether, slightly soluble in benzene; melting point 170.5°C, boiling point 286.2°C; density 1.358 at 20°C/4°C; flash point 165°C (closed cup); autoignition temperature 515.6°C (closed cup); vapor pressure 1 mm Hg at 132.4°C; vapor density 3.81. Keep well closed and protected from light.

IARC Cancer Review: Animal Inadequate Evidence

OSHA PEL   TWA 2 mg/m³; IDLH 200 mg/m³

ACGIH TLV   TWA 2 mg/m³

NIOSH REL   ceiling 2.0 mg/m³/15M

IMO Classification   Poison B; Label: St. Andrew's cross

EPA Extremely Hazardous Substances List and Community Right to Know List

TOXIC EFFECTS   A poison by ingestion and several other routes. Symptoms by ingestion: pulse rate increase without fall in blood pressure; cyanosis; coma. A suspected carcinogen. An active allergen and a severe irritant. A severe human skin irritant.

Fatalities have been caused by the ingestion of 5–12 g. More toxic than phenol. The inhalation of vapors of this material, particularly when liberated at high temperatures, must be avoided. Exposure to concentrations ranging from 10 to 30 mg of the vapor or dust per cubic meter of air causes keratitis and discoloration of the conjunctiva.

Routes   Ingestion, inhalation, contact

Symptoms   Irritation of the eyes, conjunctivitis keratitis; excitement; colored urine; nausea; dizziness;

suffocation, rapid breathing; muscle twitch, delirium; collapse, dermatitis, tinnitus, vomiting, pallor, headache, dyspnea, cyanosis

Target Organs   Eyes, respiratory system, skin, central nervous system

HAZARDOUS REACTIONS
Slight explosion hazard when exposed to heat.

Oxidizing materials: incompatible
Oxygen: potentially explosive reaction at 90°C/100 bar
Sodium hydroxide: violent reaction

FIRST AID   When personnel working with this material exhibit some of the symptoms listed above, they should immediately be removed to fresh air. If the symptoms do not subside quickly, consult a physician. In cases of dermatitis due to this material, removal from exposure will quickly clear up the symptoms. If this material accidentally comes into contact with the skin, it should be removed at once and the affected area washed with plenty of soap and water.

In case of contact with the eyes, immediately wash the eyes with large amounts of water, occasionally lifting the lower and upper lids. Get medical attention immediately. Contact lenses should not be worn when working with this chemical.

In case of contact with the skin, immediately flush the contaminated skin with water. If this chemical penetrates the clothing, immediately remove the clothing and flush the skin with water. Get medical attention promptly.

If a person breathes in large amounts of this chemical, move the exposed person to fresh air at once. Other measures are usually unnecessary.

If this chemical has been swallowed, get medical attention immediately.

FIRE HAZARD   Combustible when exposed to heat or flame.

To fight a fire, use water, CO_2, or dry chemical extinguishers.

HANDLING   Wear appropriate clothing and equipment to prevent repeated or prolonged skin contact. Wash promptly if the skin becomes contaminated. Promptly remove any nonimpervious clothing that becomes contaminated. Work clothing should be changed daily if it is possible that the clothing has been contaminated.

Wear eye protection to prevent any possibility of eye contact with materials containing >7% hydroquinone, or reasonable probability of contact with <7%.

An eyewash station should be available if materials containing >7% of hydroquinone are present.

See Chap. 2 for information on handling carcinogens.

HYPOCHLORITES

PROPERTIES   Salts of hypochlorous acid.

TOXIC EFFECTS   Toxic by ingestion and inhalation. Powerful irritants to the skin, eyes, and mucous membranes.

HAZARDOUS REACTIONS   Flammable by chemical reaction with reducing agents. Powerful oxidizers, particularly at higher temperatures, when chlorine and then oxygen are evolved, or in the presence of moisture or carbon dioxide.

Urea: reactions form highly explosive nitrogen trichloride
Acid or acid fumes: reaction forms toxic chlorine gas
Water or steam: reaction forms hydrogen chloride and chlorine gases

When heated to decomposition, they emit highly toxic fumes of Cl^-.

IODINE

CAS #7553-56-2   mf: I_2
mw: 253.80

SYNONYMS
Iodine crystals
Iodine sublimed

PROPERTIES   Violet-black solid (rhombic crystals) with a metallic luster and a sharp, characteristic odor; sharp, acrid taste; solubility in water 0.03%; soluble in aqueous solutions of HI and iodides; melting point 113.5°C, boiling point 185.24°C; density 4.93 (solid at 25°C); vapor pressure 1 mm Hg at 38.7°C (solid), 0.030 mm Hg at 0°C.

198 IODINE

OSHA PEL   ceiling 0.1 ppm; IDLH 10 ppm

ACGIH TLV   ceiling 0.1 ppm

EPA TSCA Inventory

TOXIC EFFECTS   A poison by ingestion and some other routes. Moderately toxic by inhalation.

Doses of 2–3 g have been fatal. Chronic ingestion of large amounts (200 mg/day) results in thyroid disease.

The vapor has an effect similar to that of chlorine and bromine, but with greater lung irritation. Generally the vapor is not a problem, because of the low volatility of the solid at ordinary room temperatures.

Routes   Ingestion, inhalation, contact

Symptoms   Irritation of the eyes and nose; burns of the skin and mouth; lacrimation; headache; tight chest; rash; cutaneous hypersensitivity; vomiting; abdominal pain; diarrhea

Target Organs   Respiratory system, eyes, skin, central nervous system, cardiovascular system

HAZARDOUS REACTIONS

Acetaldehyde: violent reaction
Acetylene: explosive reaction
Aluminum: incompatible
Aluminum + diethyl ether: violent reaction
Aluminum-titanium alloys: ignites when heated
Ammonia: reacts to form explosive products
Ammonia + lithium 1-heptynide: reacts to form explosive products
Ammonia + potassium: reacts to form explosive products
Antimony powder: explosive reaction
Barium acetylide: incandescent reaction above 122°C
Boron: mixture ignites at 700°C
Bromine pentafluoride: ignites (or violent reaction) on contact
Bromine trifluoride: incandescent reaction
Butadiene + ethanol + mercuric oxide: reacts to form explosive products
Calcium acetylide: incandescent reaction above 305°C
Cesium acetylide: ignites on contact
Cesium oxide: incandescent reaction above 150°C
Chlorine: incompatible
Chlorine trifluoride: ignites on contact
Copper(I) acetylide: ignites on contact
Dipropylmercury: violent reaction
Ethanol: incompatible
Ethanol + butadiene: incompatible
Ethanol + methanol + mercuric oxide: incompatible
Ethanol + phosphorus: incompatible
Fluorine: ignites on contact
Formamide: incompatible
Formamide + pyridine + sulfur trioxide: incompatible
Hafnium powder: explosive reaction when heated
Halogens: incompatible
Hydrogen: incompatible
Interhalogens: incompatible
Lithium: incompatible
Lithium acetylide: ignites on contact
Lithium carbide: incompatible
Magnesium: incompatible
Mercuric oxide: incompatible
Metal acetylides: ignites on contact
Metal carbides: incandescent reaction
Metals (active): incompatible
Metals (powdered) + water: ignites on contact
Nonmetals: ignites on contact
Oxygen: incompatible
Oxygen difluoride: mixture is a heat-sensitive explosive
Phosphorus: ignites on contact
Polyacetylene: explosive reaction at 113°C
Potassium: mixture is an impact- and heat-sensitive explosive
Pyridine: incompatible
Reducing materials: vigorous reaction
Rubidium acetylide: ignites on contact
Silver azide: reacts to form explosive products
Sodium: mixture is a shock-sensitive explosive
Sodium hydride: incompatible
Sodium phosphinate: ignites on contact
Strontium acetylide: incandescent reaction above 182°C
Sulfides: incompatible
Tetraamine copper(II) sulfate + ethanol: explosive reaction
Titanium: violent reaction above 113°C
Trioxygen difluoride: explosive reaction, possibly ignition
Zirconium acetylide: incandescent reaction above 400°C
Zirconium carbide: incompatible

When heated to decomposition, it emits toxic fumes of I^- and various iodine compounds.

FIRST AID   In case of contact with the eyes, immediately wash the eyes with large amounts of water, occasionally lifting the lower and upper lids. Get medical attention immediately. Contact lenses should not be worn when working with this chemical.

In case of contact with the skin, immediately wash the contaminated skin with soap and water. If this chemical penetrates through the clothing, immediately remove the clothing and wash the skin with soap and water. Get medical attention promptly.

If a person breathes in large amounts of this chemical, move the victim to fresh air at once. If breathing has stopped, perform artificial respiration. Keep the affected person warm and at rest. Get medical attention as soon as possible.

If this chemical has been swallowed, get medical attention immediately.

HANDLING   Wear appropriate equipment to prevent any possibility of skin contact with liquids containing >7% iodine, and repeated or prolonged skin contact with liquids containing <7%. Wash immediately if skin is contaminated with liquids containing >7% iodine, or after prolonged contact with liquids containing <7%. Remove clothing that has been contaminated with liquids containing >7% iodine immediately, and promptly remove clothing contaminated with liquids containing <7%.

Wear eye protection to prevent any possibility of contact with materials containing >7% iodine, or if there is a reasonable possibility of contact with materials containing <7%.

Provide an eyewash station and a quick-drench shower if liquids containing >7% iodine are present.

ISOBUTYL ALCOHOL

CAS #78-83-1   mf: $C_4H_{10}O$
DOT #1212   mw: 74.14

STRUCTURE

$OHCH_2CH_2CH_2CH$

SYNONYMS

2-Methyl propanol	Isobutanol
Alcohol isobutylique (French)	Isobutylalkohol (Czech)
	Isopropylcarbinol
Fermentation butyl alcohol	2-methyl-1-propanol
	2-methylpropan-1-ol
1-hydroxymethylpropane	2-methylpropyl alcohol

PROPERTIES   Colorless liquid with a mild, sweet odor; solubility in water 8.7%; miscible with alcohol and ether; boiling point 107.9°C; flash point 27.8°C; Underwriters Laboratory code: 40–45; lower explosive limit 1.2%, upper explosive limit 10.9% at 100°C; freezing point −108°C; density 0.805 at 20°/4°C; autoignition temperature 426.7°C; vapor pressure 10 mm Hg at 21.7°C; vapor density 2.55

OSHA PEL   TWA 100 ppm; IDLH 8000 ppm

ACGIH TLV   TWA 50 ppm

DOT Classification   Flammable or combustible liquid; Label: Flammable liquid

TOXIC EFFECTS   Moderately toxic by ingestion and skin contact. Mildly toxic by inhalation. A carcinogen. A severe skin and eye irritant.

Routes   Ingestion, inhalation, contact

Symptoms   Irritation of the eyes, skin, and throat, headache, drowsiness, skin cracking

Target Organs   Eyes, skin, respiratory system

HAZARDOUS REACTIONS

Aluminum: reaction at 100°C forms explosive hydrogen gas
Chromium trioxide: ignites on contact
Oxidizers (strong): incompatible

FIRST AID   In case of contact with the eyes, immediately wash the eyes with large amounts of water, occasionally lifting the lower and upper lids. Get medical attention immediately. Contact lenses should not be worn when working with this chemical.

In case of contact with the skin, immediately flush the contaminated skin with water. If this chemical penetrates the clothing, immediately remove the clothing and flush the skin with water. If irritation persists after washing, get medical attention.

If a person breathes in large amounts of this chemical, move the victim to fresh air at once. If breathing has stopped, perform artificial respiration. Keep the affected person warm and at rest. Get medical attention as soon as possible.

If this chemical has been swallowed, get medical attention immediately.

FIRE HAZARD   A dangerous fire hazard when exposed to heat or flame. Moderately explosive in the form of vapor when exposed to heat, flame, or oxidizers. Keep away from open flame.

To fight a fire, use alcohol foam, CO_2, or dry chemical extinguishers.

HANDLING   Wear appropriate clothing and equipment to prevent repeated or prolonged skin contact. Wash promptly if skin becomes wet. Immediately remove any clothing that becomes wet to avoid flammability hazard.

Wear eye protection to reduce probability of eye contact.

See Chap. 2 for information on handling carcinogens.

ISOPROPYL ALCOHOL

CAS #67-63-0   mf: C_3H_8O
DOT #1219   mw: 60.11

STRUCTURE   NFPA LABEL
$(CH_3)_2CHOH$

SYNONYMS

Dimethylcarbinol	Petrohol
IPA	Propan-2-ol
Isohol	2-Propanol
Isopropanol	*sec*-Propyl alcohol

PROPERTIES   Colorless liquid with an odor of rubbing alcohol; slightly bitter taste; miscible with water, alcohol, ether, chloroform; insoluble in salt solutions; melting point −89°C; boiling point 82.5°C; density 0.7854 at 20°C/4°C; flash point 11.7°C (closed cup); lower explosive limit 2%, upper explosive limit 12%; autoignition temperature 455.6°C; vapor density 2.07; Underwriters Laboratory code 70

OSHA PEL   TWA 400 ppm; IDLH 12,000 ppm

ACGIH TLV   TWA 400 ppm

NIOSH REL   TWA 400 ppm; ceiling 800 ppm/15 min

TOXIC EFFECTS   Poisonous by ingestion and some other routes. Mildly toxic by skin contact. A teratogen. An eye and skin irritant.

Ingestion of 100 mL can be fatal. Ingestion of as little as 10 mL may cause serious illness. It can cause corneal burns and eye damage. Acts as a local respiratory irritant and in high concentration as a narcotic. It gives good warning because it causes a mild irritation of the eyes, nose, and throat at a concentration level of 400 ppm. Less toxic than the normal isomer, but twice as volatile. It acts very much like ethanol with regard to absorption, metabolism, and elimination but with a stronger narcotic action.

Routes   Ingestion, inhalation, contact

Symptoms   Mild irritation of the eyes, nose, throat; drowsiness; dizziness; headache; dry, cracking skin; gastrointestinal cramps; nausea, diarrhea, flushing, pulse rate decrease, blood-pressure lowering, anesthesia, narcosis, mental depression, hallucinations, distorted perceptions, dyspnea, respiratory depression, vomiting, coma

Target Organs   Eyes, skin, respiratory system

HAZARDOUS REACTIONS   Reacts with air to form dangerous peroxides. The presence of 2-butanone increases the reaction rate for peroxide formation. Hydrogen peroxide sharply reduces the autoignition temperature.

Air: reaction forms dangerous peroxides
Aluminum: vigorous reaction after a delay period
Aluminum isopropoxide + crotonaldehyde: violent explosive reaction when heated
Aluminum triisopropoxide: violent reaction
Barium perchlorate: reacts to form the highly explosive propyl perchlorate
2-Butanone + air: rapidly forms dangerous peroxides
Chromium trioxide: ignites on contact
$COCl_2$: violent reaction
Dioxgenyl tetrafluoroborate: ignites on contact
Hydrogen + palladium: violent reaction
Hydrogen peroxide: forms explosive mixture similar in power and sensitivity to glyceryl nitrate
Nitroform: violent reaction
Oleum: violent reaction
Oxidants: violent reaction
Oxygen: reaction forms dangerously unstable peroxides
Potassium *tert*-butoxide: ignites after a delay
Sodium dichromate + sulfuric acid: vigorous reaction
Trinitromethane: forms explosive mixture

When heated to decomposition, it emits acrid smoke and fumes.

FIRST AID   In case of contact with the eyes, immediately wash the eyes with large amounts of water, occasionally lifting the lower and upper lids. Get medical attention immediately. Contact lenses should not be worn when working with this chemical.

In case of contact with the skin, flush with water.

If a person breathes in large amounts of this chemical, move the victim to fresh air at once. If breathing has stopped, perform artificial respiration. Keep the affected person warm and at rest. Get medical attention as soon as possible.

If this chemical has been swallowed, get medical attention immediately.

FIRE HAZARD   A very dangerous fire hazard when exposed to heat, flame, or oxidizers. Moderately explosive when exposed to heat or flame.

To fight a fire, use CO_2, dry chemical, or alcohol foam extinguishers.

HANDLING   Wear appropriate clothing and equipment to prevent repeated or prolonged skin contact. Wash promptly if skin becomes wet. Immediately remove any clothing that becomes wet to avoid flammability hazard.

Wear eye protection to reduce probability of eye contact.

LEAD

CAS #7439-92-1   mf: Pb
                       mw: 207.19

SYNONYMS
Glover   Omaha
KS-4   Omaha & Grant
Lead flake   SI
Lead S2   SO

PROPERTIES   Bluish-gray, soft metal; melting point 327.43°C, boiling point 1740°C; density 11.34 at 20°C/4°C; vapor pressure 1 mm Hg at 973°C

OSHA PEL   TWA 0.05 mg(Pb)/m^3

ACGIH TLV   TWA 0.15 mg(Pb)/m^3

NIOSH REL   TWA (Inorganic lead) 0.10 mg(Pb)/m^3

IARC Cancer Review   Animal Inadequate Evidence, Community Right to Know List, EPA TSCA Inventory, and EPA Genetic Toxicology Program

TOXIC EFFECTS   Poison by ingestion. Moderately toxic by some other routes. A suspected carcinogen of the lungs and kidneys. A teratogen.

For the general population, 50% of the exposure to lead occurs through inhalation of dusts of various types and 50% through ingestion of food and water. Lead is found in water in either dissolved or particulate form. At low pH, lead is more easily dissolved. Chemical softening of water increases the solubility of lead. Adults absorb about 5–15% of ingested lead and retain less than 5%. Children absorb about 50% and retain about 30%. High levels of lead have been found in water that has been sitting in pipes that have lead solder. Some water coolers have this kind of solder.

Severe lead poisoning will generally show over 0.07 mg/lead per 100 mL of whole blood, and more than 0.1 mg/lead per liter of urine.

Experimental evidence now suggests that blood levels of lead below 10 µg/dL can have the effect of diminishing the IQ scores of children. Low levels of lead impair neurotransmission and immune system function.

The toxicity of various lead compounds is a function of the solubility of the compound in water, the particle size of the material, and environmental conditions in the area. Lead carbonate, lead monoxide, and lead sulfate are more toxic than the metal. Many organic lead compounds such as tetraethyl lead can be absorbed through the skin, lungs, and gastrointestinal system. Tetraethyl lead is metabolized to triethyl lead, which is a severe neurotoxin. Some lead compounds are carcinogens of the lungs and kidneys.

Routes   Inhalation, ingestion, contact

Symptoms   Lassitude, insomnia, pallor, anorexia, weight loss, loss of appetite, malnutrition, constipation, abdominal pain, colic, hypotension, anemia, gingival lead line, tremors, paralysis of the wrist, malaise, headache, irritability, muscle and joint pains, tremors, hallucinations and distorted perceptions, muscle weakness, gastritis

Target Organs   Gastrointestinal tract, central nervous system, kidneys, blood, gingival tissue

HAZARDOUS REACTIONS
Ammonium nitrate: violent reaction or ignition below 200°C with lead powder
Chlorine trifluoride: violent reaction or ignition
Disodium acetylide: incompatible
Hydrogen peroxide (concentrated): violent reaction or ignition
Hydrogen peroxide + trioxane: explodes on contact
Metals (active): incompatible
Oxidants: incompatible
Oxidizing materials: vigorous reaction
Potassium: incompatible
Rubber gloves containing lead + nitric acid: ignition
Sodium acetylide: violent reaction or ignition with powdered lead
Sodium: incompatible
Sodium azide: incompatible
Zirconium: incompatible

When heated to decomposition, it emits highly toxic lead fumes.

FIRST AID   In case of contact with the eyes, immediately wash the eyes with large amounts of water, occasionally lifting the lower and upper lids. Get medical attention immediately. Contact lenses should not be worn when working with this chemical.

In case of contact with the skin, promptly flush the contaminated skin with soap and water. If this chemical penetrates through the clothing, promptly remove the clothing and flush the skin with water. If irritation persists after washing, get medical attention.

If a person breathes in large amounts of this chemical, move the victim to fresh air at once. If breathing has stopped, perform artificial respiration. Keep the affected person warm and at rest. Get medical attention as soon as possible.

If this chemical has been swallowed, get medical attention immediately.

FIRE HAZARD   Flammable and moderately explosive in the form of dust when exposed to heat or flame.

HANDLING   Poisoning by lead depends essentially on the way it is worked. There is more risk from the inhalation of lead dust or fumes than in handling solid lead. Lead poisoning can be avoided if suitable precautions are taken:

1. Use adequate ventilation
2. Prohibit eating, drinking, and smoking in lead-containing environments.

Molten lead should be handled in a separate shop especially equipped for this work. The risk is minimized by reducing the temperature of the molten lead to the lowest possible level.

After handling lead or lead compounds, it is important to wash thoroughly.

Wear appropriate clothing and equipment to prevent repeated or prolonged skin contact. Wash at the end of each work shift. Promptly remove any nonimpervious clothing that becomes contaminated.

Wear eye protection to reduce probability of eye contact.

See Chap. 2 for information on handling carcinogens.

LITHIUM AND ITS COMPOUNDS

CAS #7439-93-2 mf: Li
DOT #1415 mw: 6.94

PROPERTIES   A very light, silvery metal; soluble in liquid ammonia; melting point 179°C, boiling point 1317°C; density 0.534 at 25°C; vapor pressure 1 mm at 723°C

TOXIC EFFECTS   The lithium ion has a toxic effect on the central nervous system. Aqueous solutions of lithium oxide, lithium hydroxide, and lithium carbonate are very caustic.

The toxicity of lithium compounds is a function of their solubility in water (most soluble—halide salts, carbonate, phosphate, oxalate, and fluoride—least soluble).

HAZARDOUS REACTIONS
Acetonitrile: violent reaction
Arsenic: violent reaction
Atmospheric gases: incompatible
Beryllium: violent reaction
Bromine: mixture is a friction- and impact-sensitive explosive
Bromine pentafluoride: violent reaction (may ignite with lithium powder)
Bromobenzene: explosive reaction
Bromoform: mixture is a friction- and impact-sensitive explosive
Carbides: violent reaction
Carbon + lithium tetrachloroaluminate + sulfinyl chloride: explosive reaction
Carbon + sulfinyl chloride: ignites when ground
Carbon dioxide: violent reaction
Carbon monoxide + water: violent reaction
Carbon tetrabromide: mixture is a friction- and impact-sensitive explosive
Carbon tetrachloride: mixture is a friction- and impact-sensitive explosive
Carbon tetraiodide: mixture is a friction- and impact-sensitive explosive
CHI_3: violent reaction
Chlorine: violent reaction
Chlorine tri- and pentafluorides: hypergolic reaction
Chloroform: mixture is a friction- and impact-sensitive explosive
Chromium: violent reaction
Chromium trichloride: violent reaction
Chromium(III) oxide: violent reaction at 185°C
Cobalt alloys: violent reaction
Diazomethane: explosive reaction
Diborane: violent reaction
Dichlorormethane: mixture is a friction- and impact-sensitive explosive
Diiodomethane: mixture is a friction- and impact-sensitive explosive
Ethylene + heat: incandescent reaction
Fluorotrichloromethane: mixture is a friction- and impact-sensitive explosive
Halocarbons: mixture is a friction- and impact-sensitive explosive
Halogens: mixture is a friction- and impact-sensitive explosive
Hydrogen: ignition above 300°C
Iodine: mixture is a friction- and impact-sensitive explosive above 200°C
Iron alloys: violent reaction
Iron sulfide: violent reaction

Iron(II) sulfide: violent reaction at 260°C
Maleic anhydride: violent reaction
Manganese alloys: violent reaction
Manganese telluride: violent reaction at 230°C
Mercury: violent, potentially explosive reaction
Metal chlorides: incompatible
Metal oxides: violent reaction
Molybdenum trioxide: violent reaction at 180°C
Nickel alloys: violent reaction
Niobium pentoxide: violent reaction at 320°C
Nitric acid: reaction becomes violent
Nitrogen: violent reaction
Nitrogen + metal chlorides: incandescent reaction
Nitryl fluoride: incandescent reaction at 200°C
Nonmetal oxides: incompatible
Organic matter: violent reaction
Oxygen: violent reaction
Phosphorus: violent reaction
Platinum: violent reaction at about 540°C
Rubber: violent reaction
Silicates: violent reaction
Sodium nitrite: violent reaction
Sulfur: violent reaction
Ta_2O_5: violent reaction
Tetrachloroethylene: mixture is a friction- and impact-sensitive explosive
Titanium dioxide: violent reaction at 200–400°C
Trichloroethylene: mixture is a friction- and impact-sensitive explosive
1,1,2-Trichloro-trifluoroethane; mixture is a friction- and impact-sensitive explosive
Trifluoromethylhypofluorite: violent reaction at about 170°C
Tungsten trioxide: violent reaction at 200°C
Vanadium: violent reaction
Vanadium pentoxide: violent reaction at 394°C
Viton [poly(1,1-difluoroethylene-hexafluoro-propylene]: ignites on contact
Water: vigorous reaction produces heat and hydrogen gas
Zirconium tetrachloride: violent reaction

FIRE HAZARD A very dangerous fire hazard when exposed to heat or flame. The powder may ignite spontaneously in air. The solid metal ignites in air at 180°C.

Use fire extinguishers designed specifically for metal fires. Most types of fire extinguishers (water, foam, carbon dioxide, halocarbons, sodium carbonate, sodium chloride, or other dry powders) will only cause the burning metal to explode. Fires cannot be extinguished by sand.

HANDLING Keep under mineral oil or other liquid free of oxygen and water.

MAGNESIUM

CAS #7439-95-4 mf: Mg
DOT #1418/1869/2950 mw: 24.31

SYNONYMS
Magnesio (Italian)
Magnesium metal (DOT)
Magnesium pellets
Magnesium powdered
Magnesium ribbons
Magnesium turnings

PROPERTIES Silvery metal; melting point 651°C, boiling point 1100°C; density 1.74 at 5°C, 1.738 at 20°C; vapor pressure 1 mm Hg at 621°C

EPA TSCA Inventory

DOT Classification Label: Flammable solid and dangerous when wet

TOXIC EFFECTS Poisonous by ingestion. Particles embedded in the skin can produce slow-healing wounds. Inhalation of dust and irritating fumes can cause metal-fume fever.

HAZARDOUS REACTIONS
Acetylenic compounds: reacts to form explosive magnesium acetylide
Aluminum + potassium chlorate: incompatible
Ammonium nitrate: explosive reaction or ignition
Ammonium salts: violent reaction
Barium carbonate (molten) + water: mixture ignites
Barium nitrate + barium oxide + zinc: incompatible
Barium peroxide: violent reaction
Beryllium fluoride: violent reaction
Beryllium oxide: violent reaction
Boron diiodophosphide: violent reaction
Bromine: violent reaction
Bromobenzyl trifluoride: incompatible
Cadmium cyanide: incompatible
Cadmium oxide: violent reaction
Calcium carbide: incompatible
Calcium carbonate + hydrogen: explosive reaction or ignition when heated
Carbon dioxide: mixture ignites at 780°C
Carbon tetrachloride: mixture ignites, powder may explode on contact
Carbon tetrachloride + methanol: violent reaction
Carbonates: incompatible
Chlorate salts: violent reaction
Chlorine: violent reaction
Chlorine trifluoride: violent reaction
Chloroform: magnesium powder may explode on contact

Chloroformamidinium nitrate + water: explosive reaction or ignition when ignited with magnesium powder
Chloromethane: explodes on contact
Cobalt cyanide: incompatible
Copper cyanide: incompatible
Copper oxide: violent reaction
Copper sulfate + ammonium nitrate + potassium chlorite + water: incompatible
1,2-Dibromoethane: violent reaction
Dichlorodifluoromethane: mixture is a spark-sensitive explosive
Dinitrogen tetraoxide: violent reaction when ignited
Ethylene gas: reacts with traces of acetylene to form explosive magnesium acetylide
Ethylene oxide: incompatible
Fluorine: violent reaction
Fluorocarbon polymers: ignites when heated
Gold cyanide: explosive reaction or ignition
Halocarbons: magnesium powder may explode on contact
Halogens: violent reaction
Hydrogen iodide: violent reaction
Hydrogen peroxide: violent reaction
Interhalogens: violent reaction
Iodine heptafluoride: violent reaction
Iodine vapor: violent reaction
Lead cyanide: incompatible
Lead dioxide: violent reaction
Mercury cyanide: explosive reaction when heated
Mercury oxide: violent reaction
Metal cyanides: incompatible
Metal nitrates: explosive reaction or ignition
Metal oxides + heat: violent reaction
Metal oxosalts: incompatible
Methanol: mixtures with magnesium powder are powerful explosives
Molybdenum oxide: violent reaction
Nickel cyanide: incompatible
Nitrates (fused): explosive reaction or ignition
Nitric acid + 2-nitroaniline: hypergolic reaction
Nitric acid vapor: violent reaction
Nitrogen: violent reaction when ignited
Nitrogen dioxide: incompatible
Oxidants: incompatible
Oxygen (liquid): incompatible
Performic acid: incompatible
Phosphates: explosive reaction or ignition
Polytetrafluoroethylene powder: violent reaction when heated
Potassium carbonate: incompatible
Potassium chlorate: incompatible
Potassium chlorite: incompatible
Silicon dioxide powder: violent reaction when heated
Silver nitrate: incompatible
Silver oxide: explosive reaction or ignition when heated
Sodium chlorate: incompatible
Sodium iodate: violent reaction when heated
Sodium nitrate: violent reaction when heated
Sodium peroxide: incompatible
Sodium peroxide + carbon dioxide: incompatible
Sulfates: explosive reaction or ignition
Sulfur: violent reaction when heated
Tellurium: violent reaction when heated
Tin oxide: violent reaction
1,1,1-Trichloroethane: violent reaction
Trichloroethylene: mixture ignites on impact
Water: mixtures with magnesium powder can be detonated
Zinc cyanide: incompatible
Zinc oxide: violent reaction

FIRE HAZARD A dangerous fire hazard in the form of dust or flakes when exposed to flame or oxidizing agents. Solid magnesium is difficult to ignite. It must be heated above its melting point before it will burn. Finely divided magnesium will ignite by a spark or the flame of a match. Water will cause magnesium fires to flare up violently. Therefore, it must be kept away from water, moisture, etc. It may be ignited by a spark, match flame, spontaneously when the material is finely divided and damp, particularly with water-oil emulsion. Reacts with moisture, acids, etc., to evolve hydrogen gas. The dust is moderately explosive when exposed to flame.

To extinguish fires, operators and fire fighters can approach a magnesium fire to within a few feet if no moisture is present. Use fire extinguishers designed specifically for metal fires. Water and ordinary fire extinguishers (e.g., CO_2, powder) should not be used.

MERCURY

CAS #7439-97-6 mf: Hg
DOT #2809 mw: 200.59

SYNONYMS
Colloidal mercury
Kwik (Dutch)
Mercure (French)
Mercurio (Italian)
Mercury, metallic (DOT)
NCI-C60399
Quecksilber (German)
Quick silver
Rtec (Polish)

PROPERTIES Heavy, silvery liquid metal; solubility in water 0.002%; melting point −38.89°C; boiling point 356.9°C; specific gravity 13.595 at 4°C; vapor pressure 1 mm Hg at 126.2°C, 0.002 mm Hg at 25°C. Mercury vapor is colorless, odorless, and tasteless.

OSHA PEL   Ceiling 0.1 mg/m³; IDLH 28 mg/m³

ACGIH TLV   TWA 0.05 mg(Hg)/m³

NIOSH REL   TWA 0.05 mg(Hg)/m³

TOXIC EFFECTS   Poisonous by inhalation. Mercury compounds can be as dangerous as mercury itself.

Mercury and its compounds can be easily absorbed into the body by inhalation, ingestion, or skin contact.

Mercury is a subtle poison, the effects of which are cumulative and not readily reversible. After exposure by chronic inhalation, the symptoms of poisoning gradually disappear when the source of exposure is removed. However, improvement may be slow, and complete recovery may take years. Skin contact with mercury compounds produces irritation and various degrees of corrosion. Soluble mercury salts can be absorbed through the intact skin and produce poisoning. Kidney damage may result from poisoning by mercurial salts.

When there is equilibrium between the liquid mercury and the mercury vapor in the air, the concentration may reach 200 times the allowance figure at room temperature.

There is a risk that droplets of the metal may get lodged in cracks in the flooring or be hidden in corners or beneath the skirting of a room where it is habitually used. If this is allowed to happen, it is very easy for the air concentration of mercury vapor to reach or exceed the permitted maximum, and to remain so for prolonged periods. Similarly, care must be taken to see that any clothing or footwear which has been in contact with mercury is thoroughly cleaned.

Routes   Inhalation, absorption, contact

Symptoms   Cough, dyspnea, bronchial pneumonia, tremor, insomnia, irritability, indecision, headache, fatigue, memory loss, weakness, stomatitis, salivation, anorexia, weight loss, proteinuria, irritated eyes and skin, wakefulness, muscle weakness, anorexia, tinnitus, hypermotility, diarrhea, liver changes, dermatitis, fever

Target Organs   Skin, respiratory system, central nervous system, kidneys, eyes

HAZARDOUS REACTIONS
Acetylene: violent reaction
Acetylenic compounds: violent reaction
Alkynes + silver perchlorate: explodes on contact
Aluminum: violent reaction, exothermic formation of amalgams
Ammonia: violent reaction
Ammonia gases: incompatible
Boron diiodophosphide: mercury vapor ignites on contact
BPI_2: violent reaction
3-Bromopropyne: explodes on contact
2-Butyne-1,4-diol + acid: violent reaction
Calcium: violent reaction, exothermic formation of amalgams
CH_3N_3: violent reaction
Chlorine: violent reaction
Chlorine dioxide: explodes on contact
Ethylene oxide: explodes on contact
Lithium: explodes on contact
Metals: violent reaction, exothermic formation of amalgams
Methyl azide: mixtures are shock- and spark-sensitive explosives
Methylsilane: incompatible
Methylsilane + oxygen: mixture explodes when shaken
Na_2C_2: violent reaction
Nitromethane: violent reaction
Oxidants: incompatible
Oxygen: incompatible
Peroxyformic acid: explodes on contact
Potassium: violent reaction, exothermic formation of amalgams
Rubidium: violent reaction, exothermic formation of amalgams
Sodium: violent reaction, exothermic formation of amalgams
Sodium acetylide: violent reaction
Tetracarbonylnickel + oxygen: explodes on contact

FIRST AID   In case of contact with the eyes, immediately wash the eyes with large amounts of water, occasionally lifting the lower and upper lids. Get medical attention immediately. Contact lenses should not be worn when working with this chemical.

In case of contact with the skin, promptly wash the contaminated skin with soap and water. If mercury penetrates through the clothing, promptly remove the clothing and wash the skin with soap and water. Get medical attention promptly.

If a person breathes in large amounts of this chemical, move the victim to fresh air at once. If breathing has stopped, perform artificial respiration. Keep the affected person warm and at rest. Get medical attention as soon as possible.

If this chemical has been swallowed, get medical attention immediately.

MERCURY

USES   Metallic mercury (Hg) is widely used in laboratory instruments, and mercury compounds are used in many laboratory experiments.

HANDLING   The danger to health that can exist if proper precautions are not taken when handling this substance cannot be overemphasized. Every effort should be made to prevent spills of metallic Hg, because the substance is extremely difficult and time-consuming to pick up. Droplets get into cracks and crevices, under table legs, and under and into equipment. If spills are frequent and Hg is added to the general air level, the combined concentration may exceed the allowable limits.

Storage—Containers of Hg should be kept closed and stored in secondary containers in a well-ventilated area. When breakage of instruments or apparatus containing Hg is a possibility, the equipment should be placed in an enameled or plastic tray or pan that can be cleaned easily and is large enough to contain the Hg. Transfers of Hg from one container to another should be carried out in a hood, over a tray or pan to confine any spills.

Mercury should be stored in a well-ventilated and cool place, sheltered as far as possible from all fire hazards. The containers should be kept closed and clearly marked so as to be easily recognizable.

1. The rooms where mercury is used should be well ventilated.
2. When handling of mercury or its compounds, the hands should be protected by gloves or by a barrier cream. After working with mercury, it is essential that the hands be thoroughly cleaned, so as to remove all traces of the liquid that may be on the skin and under the nails. Smoking while using mercury should be avoided. If the work is carried on frequently, or for prolonged periods, the people handling mercury should have a set of overalls to be worn only during this work; these should be washed frequently, as should the floors and working surfaces of the rooms.
3. Mercury under water emits no vapor. This property should be remembered when decontaminating a laboratory or preventing the evaporation of mercury. Fragile instruments containing mercury should be placed above containers full of water, and large rigs should be built over water-tight trays.
4. Mercury salts should always be placed in a ventilation hood when heated.
5. Since mercury has no odor, it is difficult to detect. The best detection method is air analysis.
6. If because of an accident the concentration of mercury in the air exceeds the maximum figure allowed, staff must wear breathing apparatus, and the room must be thoroughly ventilated.
7. The exhaust of rotary vacuum pumps backing mercury diffusion pumps should be connected to the extract ventilation system.

The toxicity of mercury is such that it must not be disposed of via the drains in any form. It must be recovered. Collect all droplets and pools at once with a capillary tube attached to a suction pump and respirator bottle, or a commercial mercury-spill cleanup package. Water-soluble mercury compounds should be converted to insoluble mercuric sulfide by treatment with hydrogen sulfide.

Wear appropriate clothing and equipment to prevent repeated or prolonged skin contact. Wash promptly if the skin becomes contaminated. Promptly remove any nonimpervious clothing that becomes contaminated. Work clothing should be changed daily if it is at all probable that the clothing has been contaminated.

SPILL CLEANUP   Pools and droplets of metallic Hg can be pushed together and then collected by suction by using an aspirator bulb or a vacuum device made from a filtering flask, a rubber stopper, and several pieces of flexible and glass tubing. Alternatively, mercury-spill cleanup kits are available commercially, or small drops can be picked up on cellophane tape. A mercury-vapor analyzer should be available for determining the effectiveness of the cleanup operation.

If Hg has been spilled on the floor, the workers involved in cleanup and decontamination activities should wear plastic shoe covers. When the cleanup is complete, the shoe covers should be disposed of and the workers should thoroughly wash their hands, arms, and faces several times.

Spilled mercury compounds or solutions can be cleaned up by any method that does not cause excessive airborne contamination or skin contact.

Waste disposal—Significant quantities of metallic Hg from spills or broken thermometers or other equipment and contaminated Hg from laboratory activities should be collected in thick-walled, high-density-polyethylene bottles for reclamation.

Rags, sponges, shoe covers, and such used in cleanup activities, and broken thermometers containing small amounts of residual mercury, should be placed in a sealed plastic bag, labeled, and disposed of in a safe manner.

METHANOL

CAS #67-56-1　　　　　　　　　　mf: CH_4O
DOT #1230　　　　　　　　　　　mw: 32.05

NFPA LABEL

SYNONYMS
Carbinol
Colonial spirit
Columbian spirits
Methyl alcohol
Methyl hydroxide
Monohydroxymethane
Pyroxylic spirit
Wood alcohol
Wood naphtha
Wood spirit

PROPERTIES　　Clear liquid; mild alcoholic odor, threshold 3–8 ppm; completely miscible with water, ether, and most organic solvents; freezing point −97.8°C, boiling point 65°C; flash point 12°C; lower explosive limit 6.0%, upper explosive limit 36.5%; Underwriters Laboratory classification 70%; autoignition temperature 470°C; density 0.7915 at 20°C/4°C; vapor pressure 100 mm Hg at 21.2°C; vapor density 1.11

OSHA PEL　　TWA 200 ppm; IDLH 10,000 ppm

ACGIH TLV　　TWA 200 ppm; STEL 250 ppm (skin)

NIOSH REL　　TWA 200 ppm; ceiling 800 ppm/15 min

DOT Classification　　Flammable liquid; Label: Flammable liquid, poison

EPA TSCA Inventory and Community Right to Know List

TOXIC EFFECTS　　A poison by ingestion and skin contact. Mildly toxic by inhalation. A teratogen. An eye and skin irritant. A narcotic.

The main toxic effect is on the nervous system, particularly the optic nerves and possibly the retinas. Permanent blindness can result. Once absorbed, it is eliminated slowly. Coma lasting from 2 to 4 days can result from massive exposures. Death from ingestion of less than 30 mL has been reported.

Methanol is metabolized to formaldehyde and formic acid, both of which are toxic. Because of the slow elimination, methanol should be regarded as a cumulative poison. Chronic inhalation to the fumes may accumulate sufficient methanol to cause illness.

Routes　　Inhalation, ingestion, contact

Symptoms　　Eye irritation, headache, drowsiness, light-headedness, nausea, vomiting, eye burns, digestive disturbances, failure of vision, optic nerve neuropathy, visual-field changes, lacrimation, cough, dyspnea, other respiratory effects

Target Organs　　Eyes, skin, central nervous system, gastrointestinal tract

HAZARDOUS REACTIONS
Acetylene bromide: violent reaction
Alkyl aluminum salts: violent reaction
Aluminum: incompatible
Barium perchlorate: incompatible
Beryllium dihydride: incompatible
Bromine: incompatible
Carbon tetrachloride + metals: incompatible
Chlorine: incompatible
Chloroform: explosive reaction when heated
Chloroform + sodium hydroxide: violent reaction
Chromium trioxide: violent reaction
Cyanuric chloride: violent reaction
Dichloromethane: incompatible
Diethyl zinc: explosive reaction
Hydrogen peroxide: incompatible
Iodine + ethanol + mercuric oxide: violent reaction
Lead perchlorate: violent reaction
Magnesium: incompatible
Metals: incompatible
Nitric acid: violent reaction
Oxidants: incompatible
P_2O_3: violent reaction
Perchloric acid: violent reaction
Potassium: incompatible
Potassium hydroxide + chloroform: violent reaction
Potassium *tert*-butoxide: incompatible
Sodium hypochlorite: incompatible
Zinc: incompatible

When heated to decomposition, it emits acrid smoke and irritating fumes.

FIRST AID　　In case of contact with the eyes, immediately wash the eyes with large amounts of water, occasionally lifting the lower and upper lids. Get medical attention immediately. Contact lenses should not be worn when working with this chemical.

If this chemical comes in contact with the skin, promptly wash the contaminated skin with water. If it penetrates the clothing, promptly remove the clothing and wash the skin with water. If irritation persists after washing, get medical attention.

If a person breathes in large amounts of this chemical, move the victim to fresh air at once. If breathing

has stopped, perform artificial respiration. Keep the affected person warm and at rest. Get medical attention as soon as possible.

If this chemical has been swallowed, get medical attention immediately.

FIRE HAZARD Dangerous fire hazard when exposed to heat, flame, or oxidizers. Explosive in the form of vapor when exposed to heat or flame.

To fight a fire, use alcohol foam extinguishers.

HANDLING OSHA limits include a caution against skin contact.

Although methanol is one of the safest solvents, it is best to carry out operations in a hood if significant amounts could escape into the laboratory atmosphere—for example, during recrystallization from boiling methanol in an open flask. If there are opportunities for significant hand contact, neoprene gloves should be worn. If a still safer solvent of similar properties seems called for, ethanol is often a good choice.

Wear appropriate clothing and equipment to prevent repeated or prolonged skin contact. Wash promptly if skin becomes wet. Immediately remove any clothing that becomes wet to avoid flammability hazard.

Wear eye protection to reduce probability of eye contact.

METHYL CELLOSOLVE

CAS #109-86-4 mf: $C_3H_8O_2$
DOT #1188 mw: 76.11

STRUCTURE
$CH_3OCH_2CH_2OH$

SYNONYMS
EGM   Jeffersol EM
EGME   MECS
Ektasolve   2-Methoxyethanol
Ethylene glycol   Methyl ethoxol
  monomethyl ether   Methyl glycol
Glycol ether EM   Methyl oxitol
Glycol methyl ether   Monomethyl ether of
Glycol monomethyl   ethylene glycol
  ether   Poly-solv EM

PROPERTIES Colorless liquid with a mild, nonresidual odor; miscible in water, alcohol, ether, benzene; boiling point 124.5°C, freezing point −86.5°C; flash point 46.1°C (open cup); lower explosive point 2.5%, upper explosive point 14%; density 0.9660 at 20°C/4°C; autoignition temperature 285°C; vapor pressure 6.2 mm Hg at 20°C; vapor density 2.62

OSHA PEL   TWA 25 ppm
ACGIH TLV   TWA 5 ppm
NIOSH REL   (Glycol ethers): Reduce to lowest level

DOT Classification   Combustible liquid; Label: Flammable or combustible liquid

TOXIC EFFECTS Moderately toxic by ingestion, inhalation, skin contact, and other routes. A suspected carcinogen. A teratogen. A skin and eye irritant.

When used under conditions which do not require the application of heat, this material probably presents little hazard to health.

Some of the glycol ethers are potentially dangerous teratogens.

Routes Inhalation, absorption, ingestion, skin contact

Symptoms Headache, drowsiness, weakness, ataxia, tremors, somnolence, anemic pallor, irritated eyes, change in motor activity, tremors, convulsions

Target Organs Central nervous system, blood, skin, eyes, kidneys

HAZARDOUS REACTIONS Oxidizing materials: reaction can form explosive peroxides

FIRST AID In case of contact with the eyes, immediately wash the eyes with large amounts of water, occasionally lifting the lower and upper lids. Get medical attention immediately. Contact lenses should not be worn when working with this chemical.

In case of contact with the skin, flush promptly with water.

If a person breathes in large amounts of this chemical, move the victim to fresh air at once. If breathing has stopped, perform artificial respiration. Keep the affected person warm and at rest. Get medical attention as soon as possible.

If this chemical has been swallowed, get medical attention immediately.

FIRE HAZARD Flammable when exposed to heat or flame. A moderate explosion hazard in the form of vapor.

To fight a fire, use alcohol foam, CO_2, or dry chemical extinguishers.

HANDLING Wear appropriate clothing and equipment to prevent repeated or prolonged skin contact. Wash immediately if skin becomes contaminated. Immediately remove any nonimpervious clothing that becomes contaminated.

Wear eye protection to reduce probability of eye contact.

The following equipment should be available: a quick-drench shower.

See Chap. 2 for information on handling carcinogens.

METHYL ETHYL KETONE

CAS #78-93-3   mf: C_4H_8O
DOT #1193/1232   mw: 72.12

STRUCTURE   NFPA LABEL
$CH_3COCH_2CH_3$

SYNONYMS
2-Butanone   MEK
Ethyl methyl ketone   Methyl acetone

PROPERTIES   Clear, colorless liquid with a fragrant, mintlike, moderately sharp odor; solubility in water 27%; boiling point 79.57°C, freezing point −85.9°C; lower explosive limit 1.8%, upper explosive limit 11.5%; flash point −5.6°C (TOC); density 0.80615 at 20°C/4°C; vapor pressure 71.2 mm Hg at 20°C; autoignition temperature 515.6°C; vapor density 2.42; Underwriters Laboratories code 85–90

OSHA PEL   TWA 200 ppm (590 mg/m³); IDLH 3000 ppm

ACGIH TLV   TWA 200 ppm; short-term exposure limit 300 ppm

NIOSH REL   200 ppm (590 mg/m³)

DOT Classification   Flammable liquid; Label: Flammable liquid

TOXIC EFFECTS   Moderately toxic by ingestion, skin contact, and intraperitoneal routes. A teratogen. A strong irritant. An eye irritant at 350 ppm.

Routes   Inhalation, ingestion, contact

Symptoms   Irritated eyes and nose, headache, dizziness, vomiting

Target Organs   Central and peripheral nervous systems, lungs

HAZARDOUS REACTIONS
Chloroform + alkali: vigorous reaction
Chlorosulfonic acid: incompatible
Hydrogen peroxide + nitric acid: reaction forms a heat- and shock-sensitive explosive product
Oleum: incompatible
Oxidizers (very strong): incompatible
Potassium *tert*-butoxide: ignites on contact
2-Propanol: mixtures form explosive peroxides during storage

FIRST AID   In case of contact with the eyes, immediately wash the eyes with large amounts of water, occasionally lifting the lower and upper lids. Get medical attention immediately. Contact lenses should not be worn when working with this chemical.

In case of contact with the skin, immediately flush the contaminated skin with water. If this chemical penetrates the clothing, immediately remove the clothing and flush the skin with water. Get medical attention promptly.

If a person breathes in large amounts of this chemical, move the victim to fresh air at once. If breathing has stopped, perform artificial respiration. Keep the affected person warm and at rest. Get medical attention as soon as possible.

If this chemical has been swallowed, get medical attention immediately.

FIRE HAZARD   Dangerous fire hazard when exposed to heat or flame. Moderately explosive when exposed to flame.

To fight a fire, use alcohol foam, CO_2, or dry chemical extinguishers.

HANDLING   Wear appropriate clothing and equipment to prevent repeated or prolonged skin contact. Promptly remove any nonimpervious clothing that becomes contaminated.

Wear eye protection to reduce probability of eye contact.

The following equipment should be available: an eyewash station.

METHYL ETHYL KETONE PEROXIDE

CAS #1338-23-4   mf: $C_8H_{16}O_4$
    mw: 176.24

SYNONYMS
Methylethylketonhydroperoxide
NCI-C55447

ACGIH TLV   Ceiling 0.2 ppm

EPA TSCA Inventory

USES   Used as a catalyst for hardening polyester resins, especially in fiberglass manufacture.

TOXIC EFFECTS   Poisonous by inhalation, ingestion, and intraperitoneal routes. A skin and eye irritant. Gastrointestinal tract effects by ingestion. When heated to decomposition, it emits acrid smoke and fumes.

METHYLENE CHLORIDE

CAS #75-09-2   mf: CH_2Cl_2
DOT #1593   mw: 84.93

SYNONYMS
Aerothene MM
DCM
Dichloromethane
Freon 30
Methane dichloride
Methylene bichloride
Methylene dichloride
Solmethine

PROPERTIES   Colorless, volatile liquid with a chloroformlike odor; solubility in water 1.3%; boiling point 39.8°C, freezing point −96.7°C; density 1.326 at 24°C/4°C; lower explosive limits 12% in O_2, upper explosive limit 66.4% in O_2; autoignition temperature 615°C; vapor pressure 380 mm Hg at 22°C; vapor density 2.93

OSHA PEL   TWA 500 ppm; ceiling 1000 ppm; peak 2000/5 min/2 h

ACGIH TLV   TWA 50 ppm; suspected carcinogen

NIOSH REL   TWA 75 ppm; peak 500 ppm/15 min; reduce to lowest feasible level

DOT Classification   Poison B; Label: St. Andrew's cross

TOXIC EFFECTS   Moderately toxic by ingestion and by subcutaneous and intraperitoneal routes. Mildly toxic by inhalation. A carcinogen and teratogen. An eye and severe skin irritant. Human mutagenic data.

Routes   Ingestion, contact

Symptoms   Fatigue, weakness, sleepiness, lightheadedness; limb numbness, tingling; nausea; irritation of the eyes and skin, vertigo, worsened angina, paresthesia, somnolence, altered sleep time, convulsions, euphoria, and change in cardiac rate

Target Organs   Skin, cardiovascular system, eyes, central nervous system

HAZARDOUS REACTIONS
Will not form explosive mixtures with air at ordinary temperatures.
Air + methanol vapor: mixtures are flammable
Aluminum: incompatible
Caustics: incompatible
Dinitrogen tetraoxide: mixture is explosive
Lithium: violent reaction
Magnesium powder: incompatible
Metals (chemically active): incompatible
Oxidizers (strong): incompatible
Oxygen (high-content atmospheres): mixture is explosive
Oxygen (liquid): mixture is explosive
Potassium: mixture is explosive
Potassium hydroxide + *n*-methyl-*n*-nitrosourea; violent reaction
Potassium *tert*-butoxide: violent reaction
Sodium: mixture is explosive
Sodium-potassium alloy: mixture is explosive

Mild heating causes decomposition to toxic fumes of Cl^- and phosgene.

FIRST AID   In case of contact with the eyes, immediately wash the eyes with large amounts of water, occasionally lifting the lower and upper lids. Get medical attention immediately. Contact lenses should not be worn when working with this chemical.

In case of contact with the skin, promptly wash the contaminated skin with soap and water. If this chemical penetrates through the clothing, promptly remove the clothing and wash the skin with soap and water. Get medical attention promptly.

If a person breathes in large amounts of this chemical, move the victim to fresh air at once. If breathing has stopped, perform artificial respiration. Keep the affected person warm and at rest. Get medical attention as soon as possible.

If this chemical has been swallowed, get medical attention immediately.

FIRE HAZARD   Flammable in the range 12–19% in air, but ignition is difficult. Explosive when vapor is exposed to heat or flame.

HANDLING   Wear appropriate clothing and equipment to prevent repeated or prolonged skin contact. Wash promptly if skin becomes wet. Promptly remove any nonimpervious clothing that becomes wet.

Wear eye protection to reduce probability of eye contact.

See Chap. 2 for information on handling carcinogens.

5-METHYL-2-HEXANONE

CAS #110-12-3   mf: $C_7H_{14}O$
mw: 114.21

SYNONYMS
Isoamyl methyl ketone   Methyl isoamyl ketone
Isopentyl methyl ketone   MIAK
2-Methyl-5-hexanone

PROPERTIES   Colorless, stable liquid with a pleasant odor; slightly soluble in water, miscible with most organic solvents; boiling point 144°C, freezing point −73.9°C; flash point 43.3°C (open cup); density 0.8132 at 20°C/20°C

EPA TSCA Inventory

ACGIH TLV   TWA 50 ppm

NIOSH REL   TWA 230 mg/m³

TOXIC EFFECTS   Moderately toxic by ingestion and inhalation. Mildly toxic by skin contact.

FIRE HAZARD   Flammable when exposed to heat, flame, or oxidizers.
To fight a fire, use dry chemical, CO_2, foam, or fog extinguishers.

MORPHOLINE

CAS #110-91-8   mf: C_4H_9NO
DOT #2054; 1760   mw: 87.12

STRUCTURE

SYNONYMS
Diethyleneimide oxide
Diethylene imidoxide
Morpholine, aqueous mixture
1-oxo-4-Azacyclohexane
Tetrahydro-*p*-isoxazine
Tetrahydro-1,4-oxazine
Tetrahydro-1,4-isoxazine
Tetrahydro-2H-1,4-oxazine

PROPERTIES   Colorless, mobile, hygroscopic liquid; amine odor; miscible with water and most organic solvents (e.g., acetone, benzene, ether, castor oil, turpentine, pine oil); immiscible with concentrated NaOH solutions; volatile with steam; melting point −4.9°C, boiling point 128.9°C; density 1.002 at 20°C; flash point 37.7°C (open cup); lower explosive limit 1.8%, upper explosive limit 11%; vapor pressure 6.6 mm Hg at 20°C; vapor density 3.0

OSHA PEL   TWA 20 ppm (skin); IDLH 8000 ppm

ACGIH TLV   TWA 20 ppm; STEL 30 ppm (skin)

DOT Classification   Flammable liquid; Label: Flammable liquid; Corrosive material; Label: Corrosive, aqueous solution; Flammable or combustible liquid; Label: Flammable liquid

TOXIC EFFECTS   Moderately toxic by ingestion, inhalation, skin contact, and some other routes. A corrosive irritant to skin, eyes, and mucous membranes.
In laboratory experiments, repeated exposure of rats at high concentrations (18,000 ppm) produced death, with damage to the lung, liver, and kidneys.

Routes   Inhalation, absorption, ingestion, contact

Symptoms   Corneal edema, visual disturbances; irritation of the eye, skin, nose, and respiratory system; cough; kidney damage

Target Organs   Respiratory system, eyes, skin

HAZARDOUS REACTIONS
Acids (strong): incompatible
Cellulose nitrate (with high surface area): may ignite spontaneously
Nitromethane: mixtures are explosive
Oxidizing materials: incompatible

When heated to decomposition, it emits toxic fumes of NO_x.

FIRST AID   In case of contact with the eyes, immediately wash the eyes with large amounts of water, occasionally lifting the lower and upper lids. Get medical attention immediately. Contact lenses should not be worn when working with this chemical.
In case of contact with the skin, immediately flush the contaminated skin with water. If this chemical penetrates the clothing, immediately remove the clothing and flush the skin with water. Get medical attention promptly.

If a person breathes in large amounts of this chemical, move the victim to fresh air at once. If breathing has stopped, perform artificial respiration. Keep the affected person warm and at rest. Get medical attention as soon as possible.

If this chemical has been swallowed, get medical attention immediately.

FIRE HAZARD   A very dangerous fire hazard when exposed to heat, flames, or oxidizing agents.

To fight a fire, use alcohol foam, CO_2, or dry chemical extinguishers.

HANDLING   Skin contact can be a significant contributor to toxic effects.

Wear appropriate equipment to prevent any possibility of skin contact with liquids containing >25% morpholine, and repeated or prolonged skin contact with liquids containing <25%. Wash promptly if the skin becomes contaminated. Immediately remove any clothing that becomes wet to avoid flammability hazard.

Wear eye protection to prevent any possibility of eye contact.

Provide an eyewash station if liquids containing >15% morpholine are present, and a quick-drench shower for >25%.

NAPHTHALENE

CAS #91-20-3   mf: $C_{10}H_8$
DOT #1334/2304   mw: 128.18

NFPA LABEL

SYNONYMS
Camphor tar
Mighty 150
Moth balls
Moth flakes
Naftalen (Polish)
Naphthalene, crude or refined (DOT)
Naphthalin (DOT)
Naphthaline
Naphthene
Napthalene, molten (DOT)
NCI-C52904
RCRA waste number U165
Tar camphor
White tar

PROPERTIES   Colorless to brown solid with an odor of mothballs; soluble in alcohol, benzene; solubility in water 0.003%; very soluble in ether, carbon tetrachloride, carbon disulfide, hydronaphthalenes; melting point 80.1°C, boiling point 217.9°C; density 1.162; flash point 78.9°C (open cup); lower explosive limit 0.9%, upper explosive limit 5.9%; autoignition temperature 567°C; vapor pressure 1 mm Hg at 52.6°C; vapor density 4.42

EPA TSCA Inventory, EPA Genetic Toxicology Program, and Community Right to Know List

OSHA PEL   TWA 10 ppm; IDLH 500 ppm

ACGIH TLV   TWA 10 ppm; STEL 15 ppm

DOT Classification   ORM-A; Label: None; Flammable solid; Label: Flammable solid

TOXIC EFFECTS   A poison by ingestion and several other routes. An eye and skin irritant. Poisoning may occur by ingestion of large doses, inhalation, or skin absorption.

Routes   Inhalation, ingestion, contact

Symptoms   Eye irritation; headache; confusion, excitement; malaise, nausea, vomiting, abdominal pain; irritated bladder; profuse sweating; jaundice; hematuria; hemoglobinuria; renal shutdown; dermatitis; headache; diaphoresis; fever; anemia; liver damage; convulsions; coma

Target Organs   Eyes, blood, liver, kidneys, skin, red blood cells, central nervous system

HAZARDOUS REACTIONS
Chromium trioxide: violent reaction
Dinitrogen pentoxide: incompatible
Oxidizers (strong): incompatible

When heated to decomposition, it emits acrid smoke and irritating fumes.

FIRST AID   In case of contact with the eyes, immediately wash the eyes with large amounts of water, occasionally lifting the lower and upper lids. Get medical attention immediately. Contact lenses should not be worn when working with this chemical.

In case of contact by molten naphthalene with the skin, flush immediately with water; in case of contact with solid/liquid, wash promptly with soap.

If a person breathes in large amounts of this chemical, move the victim to fresh air at once. If breathing has stopped, perform artificial respiration. Keep the affected person warm and at rest. Get medical attention as soon as possible.

If this chemical has been swallowed, get medical attention immediately.

FIRE HAZARD   Flammable when exposed to heat or flame; reacts with oxidizing materials. Moderate explosion hazard when dust is exposed to heat or flame.

To fight a fire, use water, CO_2, or dry chemical extinguishers.

HANDLING   Wear appropriate clothing and equipment to prevent repeated or prolonged skin contact. Wash promptly if the skin becomes contaminated. Promptly remove any nonimpervious clothing that becomes contaminated. Work clothing should be changed daily if there is reasonable probability that the clothing has been contaminated.

Wear eye protection to reduce probability of eye contact.

NICKEL CARBONYL

CAS #13463-39-3   mf: C_4NiO_4
DOT #1259   mw: 170.75

STRUCTURE   NFPA LABEL
$Ni(CO)_4$

SYNONYMS
Nickel tetracarbonyl

PROPERTIES   Colorless, volatile liquid or needles; musty odor; solubility in water 0.018%; soluble in alcohol, benzene, chloroform, acetone, carbon tetrachloride; melting point $-19.3°C$, boiling point $43°C$; density 1.3185 at $17°C$; flash point $\leq -4°C$, lower explosive limit 2% at $20°C$; vapor pressure 400 mm Hg at $25.8°C$. Oxidizes in air, explodes at about $60°C$.

OSHA PEL   TWA 0.007 mg/m^3

ACGIH TLV   TWA 0.05 ppm (Ni)

NIOSH REL   TWA (nickel carbonyl) 0.001 ppm

IARC Cancer Review: Animal Limited Evidence, EPA TSCA Inventory, and, EPA Extremely Hazardous Substances List

DOT Classification   Poison B; Label: Poison, flammable liquid; Flammable liquid; Label: Flammable liquid and poison

TOXIC EFFECTS   Poisonous by inhalation and some other routes. A carcinogen and teratogen. Vapors can cause irritation, congestion, and edema of lungs. Chronic exposure may cause cancer of lungs and nasal sinuses. Sensitization dermatitis is common.

Toxic effects are probably caused by nickel and carbon monoxide liberated in the lungs and then spread throughout the system. The nickel can be gradually eliminated by normal body processes.

Routes   Inhalation, ingestion, contact

Symptoms   Giddiness, headache, vertigo, nausea, epigastric pain, substernal pain, cough, hyperpnea, cyanosis, leukicytosis, pneumonitis, paranasal sinus, delirium, convulsions, somnolence, body-temperature increase; a carcinogen

Target Organs   Lungs, paranasal sinus, central nervous system

HAZARDOUS REACTIONS
Air: violent reaction
Bromine: violent reaction
n-Butane + oxygen: violent reaction
Chlorine: incompatible
Combustible vapors: incompatible
Nitric acid: incompatible
Oxidizers: incompatible
Oxygen: violent reaction

When heated to decomposition or on contact with acid or acid fumes, it emits highly toxic fumes of carbon monoxide.

FIRST AID   In case of contact with the eyes, immediately wash the eyes with large amounts of water, occasionally lifting the lower and upper lids. Get medical attention immediately. Contact lenses should not be worn when working with this chemical.

In case of contact with the skin, immediately wash the contaminated skin with soap and water. If this chemical penetrates through the clothing, immediately remove the clothing and wash the skin with soap and water. Get medical attention promptly.

If a person breathes in large amounts of this chemical, move the victim to fresh air at once. If breathing has stopped, perform artificial respiration. Keep the affected person warm and at rest. Get medical attention as soon as possible.

If this chemical has been swallowed, get medical attention immediately.

FIRE HAZARD   A very dangerous fire hazard when exposed to heat, flame, or oxidizers. Moderately explosive when exposed to heat or flame.

To fight a fire, use water, foam, CO_2, or dry chemical extinguishers.

HANDLING   Nickel carbonyl is both toxic and flammable. No one should be permitted to work with the material unless fully familiar with it. In direct sunlight, both the liquid and the gas will flash.

Although nickel carbonyl has an odor sometimes described as being like "brick dust," little confidence should be placed in detection of the odor, because some people are unable to recognize it.

Excellent ventilation, through either hoods, fans, or strong drafts, is of utmost importance when nickel carbonyl is used.

Wear appropriate clothing and equipment to prevent any possibility of skin contact. Wash immediately if the skin becomes wet. Immediately remove any clothing that becomes wet.

Wear eye protection to prevent any possibility of eye contact.

See Chap. 2 for information on handling carcinogens.

NITRATES

NFPA LABEL

PROPERTIES   Inorganic nitrates consist of the mono-valent —NO_3 radical combined with a metal. Organic nitrates (nitro compounds) consist of the nitro (—NO_2) group with an organic radical. Nitrate often denotes a nitric acid ester of an organic material.

TOXIC EFFECTS   Ingestion of large amounts can cause dizziness, abdominal cramps, vomiting, bloody diarrhea, weakness, convulsions, collapse, and possibly death. Chronic ingestion may lead to weakness, general depression, headache, and mental impairment.

HAZARDOUS REACTIONS   Inorganic nitrates can give up oxygen to other materials, which may explode. Ammonium nitrate can self-detonate under certain conditions and thus is a high explosive, although it is very insensitive to impact and difficult to detonate. Nitrates should be considered storage hazards.

Aluminum: violent reaction
Boron phosphide: violent reaction
Cyanide: violent reaction
Esters: violent reaction
Phosphorus: violent reaction
PN_2H: violent reaction
Reducing materials: incompatible
Sodium cyanide: violent reaction
Sodium hypophosphite: violent reaction
Strontium chloride: violent reaction
Thiocyanates: violent reaction

When heated to decomposition, they emit acrid smoke and irritating fumes.

FIRE HAZARD   Flammable by spontaneous chemical reaction. Nitrates may explode when shocked, exposed to heat or flame, or by spontaneous chemical reaction.

NITRIC ACID

CAS #7697-37-2   mf: HNO_3
DOT #2031   mw: 63.02

NFPA LABEL

SYNONYMS

Acide nitrique (French)	Hydrogen nitrate
Acido nitrico (Italian)	Nitric acid, over 40% (DOT)
Aqua fortis	
Azotic acid	Salpetersaure (German)
Azotowy kwas (Polish)	Salpeterzuuroplossingen (Dutch)

PROPERTIES   Colorless, yellow, or red fuming liquid with an acrid, suffocating odor; caustic and corrosive; miscible with water; melting point −42°C, boiling point 86°C; density 1.50269 at 25°C/4°C

EPA TSCA Inventory, EPA Genetic Toxicology Program, and Community Right to Know List

OSHA PEL   TWA 2 ppm; IDLH 100 ppm

ACGIH TLV   TWA 2 ppm; STEL 4 ppm

NITRIC ACID 215

NIOSH REL   TWA (nitric acid) 2 ppm

DOT Classification   Corrosive material; Label: Corrosive; Oxidizer; Label: Oxidizer and corrosive

TOXIC EFFECTS   A poison. A teratogen. Corrosive to eyes, skin, mucous membranes, and teeth. Upper respiratory irritation may seem to clear up and then become more severe in a few hours.

The vapor consists of a mixture of the various oxides of nitrogen and nitric acid.

Routes   Inhalation, ingestion, contact

Symptoms   Irritation of the eyes, mucous membrane, skin; delayed pulmonary edema; pneumonitis; bronchitis; dental erosion

Target Organs   Eyes, respiratory system, skin, teeth

HAZARDOUS REACTIONS
Acetic acid: forms explosive mixtures
Acetic acid + sodium hexahydroxyplatinate(IV): forms explosive mixtures
Acetic anhydride: explosive reaction
Acetic anhydride + hexamethylenetetramine acetate: forms explosive mixtures
Acetone: ignites on contact
Acetone + acetic acid: explosive reaction in storage
Acetone + hydrogen peroxide: explosive reaction
Acetone + sulfuric acid: explosive reaction if confined
Acetoxyethylene glycol: forms explosive mixtures
4-Acetoxy-3-methoxybenzaldehyde: incompatible
5-Acetylamine-3-bromobenzo(b)thiophene: explosive reaction
Acetylene: incompatible
Acetylene derivatives: explosive or hypergolic reaction
Acrolein: incompatible
Acrylonitrile: incompatible
Acrylonitrile + methacrylate copolymer: incompatible
Alcohols: explosive reaction
Alcohols + disulfuric acid: ignites on contact
Alcohols + potassium permanganate: ignites on contact
Aliphatic amines: ignites on contact
Alkane thiols: explosive reaction
o-Alkyl ethylene dithiophosphate: ignites on contact
Allyl alcohol: incompatible
Allyl chloride: incompatible
2-Aminoethanol: incompatible
2-Aminothiazole: explosive reaction
2-Aminothiazole + sulfuric acid: explosive reaction
Ammonia: ignites on contact
Ammonium dichromate + organic fuels: ignites on contact
Ammonium hydroxide: incompatible
Ammonium metavanadate + anilinium nitrate: ignites on contact
Ammonium metavanadate + aromatic amines: ignites on contact
Ammonium nitrate: forms explosive mixtures
Aniline: incompatible
Aniline + dinitrogen tetraoxide or sulfuric acid: explosive reaction
Anilinium nitrate: forms explosive mixtures
Anilinium nitrate + inorganic materials: ignites on contact
Anion-exchange resins: incompatible
Antimony: incompatible
Antimony hydride: incompatible
Aromatic amines + metal compounds: ignites on contact
Arsenic (powder): incompatible
Arsenic hydride: incompatible
Arsine: explosive reaction
Arsine + boron tribromide: incompatible
B_4H_{10}: incompatible
Bases (strong): incompatible
Benzene: explosive or hypergolic reaction
Benzidine: hypergolic reaction
Benzo(b)thiophene derivatives: incompatible
Benzonitrile + sulfuric acid: explosive reaction
N-Benzyl-N-ethylaniline + sulfuric acid: incompatible
1,4-Bis(methoxy-methyl)-2,3,5,5-tetramethyl-benzene: explosive reaction at 80°C
Bismuth: incompatible
1,3-Bis(trifluoromethyl)benzene + sulfuric acid: vapors are initiated by spark
Boron: incompatible
Boron decahydride: incompatible
Boron phosphide: incompatible
Boron powder: incompatible
Bromine pentafluoride: incompatible
Butanethiol: incompatible
tert-Butyl-m-xylene: explosive reaction
n-Butyraldehyde: incompatible
Cadmium phosphide: explosive reaction
Calcium hypophosphite: incompatible
Carbon: incompatible
3-Carene: explosive or hypergolic reaction
Cashew nut shell oil: explosive or hypergolic reaction
Cellulose: incompatible
Cesium acetylide: explosive reaction
Cesium carbide: incompatible
Chlorine trifluoride: incompatible

216 NITRIC ACID

Chlorobenzene: explosive reaction
4-Chloro-2-nitroaniline: incompatible
Chlorosulfonic acid: incompatible
Coal: incompatible
Combustible organics: incompatible
Copper azide: incompatible
Copper(I) chloride + anilinium nitrate: ignites on contact
Copper(I) nitride: incompatible
Copper(I) oxide + aromatic amines: ignites on contact
Copper(II) oxide + aromatic amines: ignites on contact
Cu_3N_2: incompatible
Cotton + rubber + sulfuric acid + water: explosive reaction has caused industrial accidents
Cresol: incompatible
2-Cresol: ignites on contact
3-Cresol: ignites on contact
Crotonaldehyde: hypergolic reaction
Cumene: incompatible
Cyanides: incompatible
Cyclohexanol: ignites on contact
Cyclohexanol + cyclohexanone: incompatible
Cyclohexylamine: explosive reaction
Cyclopentadienes: explosive or hypergolic reaction
$C_2H_5PH_2$: incompatible
1,3-Diaminoethanebis(trimethylgold): explosive reaction
Diborane: incompatible
Di-2-6-butoxyethylether (butex): incompatible
2,6-Di-*tert*-butyl phenol: incompatible
1,2-Dichloroethane: forms explosive mixtures
Dichloroethylene: forms explosive mixtures
Dichloromethane: forms explosive mixtures
Dichromate + anion-exchange resins: incompatible
Dichromates + organic fuels: ignites on contact
Dicyclopentadiene: explosive or hypergolic reaction
Dienes: explosive or hypergolic reaction
Diethylaminoethanol: forms explosive mixtures
Diethyl ether: explosive reaction
Diethyl ether + sulfuric acid: explosive reaction
3,6-Dihydro-1,2,2H-oxazine: forms explosive mixtures
Diisopropylether: incompatible
Dimethyl ether: forms explosive mixtures
Dimethyl hydrazine: explosive reaction
Dimethyl hydrazine (uns-): incompatible
1,1-Dimethyl hydrazine: reaction hypergolic
Dimethyl sulfide: explosive reaction
Dimethyl sulfide + 1,4-dioxane: explosive reaction
Dimethyl sulfoxide + water: explosive reaction
Dimethylaminomethylferrocene + water: incompatible
Dinitrobenzenes: forms explosive mixtures
1,3-Dinitrobenzene: forms explosive mixtures
Dinitrogen tetraoxide + aromatic amines: explosive reaction
Dinitrogen tetraoxide + nitrogenous fuels: explosive reaction
Dinitrogen tetroxide + triethylamine: hypergolic reaction
Dinitrotoluene: forms explosive mixtures
Dioxane + perchloric acid: explosive reaction
Diphenyldistibene: explosive reaction
Diphenylmercury: incompatible
Diphenyl tin: ignites on contact
Disodium phenyl orthophosphate: forms explosive mixtures
Divinyl ether: hypergolic reaction
Epichlorohydrin: incompatible
Ethane sulfonamide: explosive reaction
Ethanol: incompatible
n-Ethylamine + dinitrogen tetraoxide or sulfuric acid: explosive reaction
m-Ethylaniline: incompatible
Ethylene diamine: incompatible
Ethylene imine: incompatible
5-Ethyl-2-methylpyridine: explosive reaction
Ethyl phosphine: explosive reaction
5-Ethyl-2-picoline: incompatible
Fat + sulfuric acid: explosive reaction when confined
Fluorine: explosive reaction
Formaldehyde + impurities: incompatible
Formic acid + heat: incompatible
Formic acid + urea: incompatible
2-Formylamino-1-phenyl-1,3-propanediol: explosive reaction
Furfural: ignites on contact
Furfuryl alcohol: incompatible
Furfurylidene ketones: hypergolic reaction
Germanium: incompatible
Glycerol + hydrofluoric acid: incompatible
Glycerol + sulfuric acid: explosive reaction
Glyoxal: incompatible
Hexalithium disilicide: explosive reaction
Hexamethylbenzene: explosive or hypergolic reaction
2,2,4,4,6,6-Hexamethyl-trithiane: explosive reaction
2-Hexenal: forms heat-sensitive explosive mixtures
Hydrazine: hypergolic reaction
Hydrocarbons: explosive or hypergolic reaction
Hydrocarbons + 1,1-dimethylhydrazine: explosive reaction
Hydrofluoric acid + lactic acid: forms explosive mixtures
Hydrofluoric acid + propylene glycol + silver nitrate: forms explosive mixtures
Hydrogen iodide: ignites on contact
Hydrogen peroxide: incompatible

Hydrogen peroxide + 2-butanone: forms explosive mixtures
Hydrogen peroxide + cyclohexanone: forms explosive mixtures
Hydrogen peroxide + cyclopentanone: forms explosive mixtures
Hydrogen peroxide + ketones: forms explosive mixtures
Hydrogen peroxide + mercury(II) oxide: forms explosive mixtures
Hydrogen peroxide + 3-methylcyclohexanone: forms explosive mixtures
Hydrogen peroxide + 3-pentanone: forms explosive mixtures
Hydrogen peroxide + soils: explosive reaction
Hydrogen selenide: ignites on contact
Hydrogen sulfide: ignites on contact
Hydrogen telluride: ignites on contact
Hydrozoic acid: incompatible
Indane + sulfuric acid: incompatible
Ion-exchange resins: explosive reaction
Iron(III) chloride + aromatic amines: ignites on contact
Iron(III) oxide + aromatic amines: ignites on contact
Iron(II) oxide powder: incompatible
Iron monoxide: incompatible
Isoprene: incompatible
Ketones (cyclic): incompatible
Ketones + hydrogen epoxide: incompatible
Lactic acid + hydrogen fluoride: incompatible
Lead-containing rubber: ignites on contact
Li_6Si_2: incompatible
Lithium: ignites on contact
Magnesium: ignites on contact
Magnesium + 2-nitroaniline: hypergolic reaction
Magnesium phosphide: incompatible
Magnesium silicide: incompatible
Magnesium-titanium alloy: incompatible
Manganese: incompatible
Mesityl oxide: incompatible
Mesitylene: explosive or hypergolic reaction
Metal acetylides: explosive reaction
Metal cyanides: explosive reaction
Metal hexacyanoferrates (3^-) or (4^-): explosive reaction
Metal powders: incompatible
Metal salicylates: forms explosive mixtures
Metal thiocyanates: explosive reaction
Metals: forms explosive mixtures
4-Methylcyclohexanone: explosive reaction above 75°C
2-Methyl-5-ethyl pyridine: incompatible
Methylthiophene: explosive reaction
NdP: incompatible
Nickel tetraphosphide: ignites on contact
Nitroaromatics: forms explosive mixtures
Nitrobenzene (mono-): forms explosive mixtures
Nitrobenzene + sulfuric acid: explosive reaction
Nitrobenzene + water: forms explosive mixtures
Nitromethane: forms explosive mixtures
1-Nitronaphthalene + sulfuric acid: explosive reaction
Nonmetal hydrides: explosive reaction
Nonmetal powders: incompatible
Oleum: incompatible
Organic materials + oxidizers: explosive reaction
Oxidizable matter: incompatible
Perchloric acid + organic materials: explosive reaction
Petroleum products (burning): explosive or hypergolic reaction
Phenylacetylene + 1,1-dimethylhydrazine: explosive reaction
P-Phenylenediamine + dinitrogen tetraoxide or sulfuric acid: explosive reaction
Phosphine: explosive reaction
Phosphine derivatives: explosive reaction
Phosphonium iodide: explosive reaction
Phosphorus: explosive reaction
Phosphorus halides: incompatible
Phosphorus trichloride: explosive reaction
Phosphorus vapor: ignites on contact
Phthalic acid: incompatible
Phthalic anhydride: incompatible
Polyalkenes: incompatible
Polydibromosilane: incompatible
Polyethylene: incompatible
Polyethylene oxide derivatives: incompatible
Polypropylene: incompatible
Polysilylene: ignites on contact
Polyurethane foam: explosive reaction
Potassium chlorate + organic materials: explosive reaction
Potassium chromate + aromatic amines: ignites on contact
Potassium chromate + organic fuels: ignites on contact
Potassium dichromate + aromatic amines: ignites on contact
Potassium dichromate + organic fuels: ignites on contact
Potassium hexacyanoferrate(I) + aromatic amines: ignites on contact
Potassium hexacyanoferrate(III) + aromatic amines: ignites on contact
Potassium permanganate + anilinium nitrate: ignites on contact
Potassium phosphate, monobasic: incompatible
Potassium phosphinate + heat: explosive reaction
β-Propiolactone: incompatible

Propiophenone + sulfuric acid: explosive reaction
Propylene oxide: incompatible
Pyridine: incompatible
Pyrocatechol: hypergolic reaction
Reductants: incompatible
Resorcinol: explosive reaction
Rubber: explosive reaction
Rubidium acetylide: explosive reaction
Rubidium carbide: incompatible
Salicylic acid: forms explosive mixtures
Selenium: incompatible
Selenium iodophosphide: incompatible
Silicon (powder): incompatible
Silicone oil: explosive reaction
Silver buten-3-ynide: explosive reaction
Silver + ethanol: incompatible
Sodium: ignites on contact
Sodium azide: incompatible
Sodium hydroxide: incompatible
Sodium metavanadate + anilinium nitrate: ignites on contact
Sodium metavanadate + aromatic amines: ignites on contact
Sodium pentacyanonitrosylferrate + anilinium nitrate: ignites on contact
Sodium pentacyanonitrosylferrate(II) + aromatic amines: ignites on contact
Stibene: explosive reaction
Sucrose: incompatible
Sulfamic acid: incompatible
Sulfur dioxide: explosive reaction
Sulfuric acid: incompatible
Sulfuric acid + aromatic amines: explosive reaction
Sulfuric acid + organic materials: explosive reaction
Sulfuric acid + terephthalic acid: incompatible
Terpenes: incompatible
Tetraborane(10): explosive reaction
Tetraphosphorus diiodo triselenide: explosive reaction
Tetraphosphorus iodide: ignites on contact
Thioaldehydes: incompatible
Thiocyanates: incompatible
Thioketones: incompatible
Thiophene: incompatible
Tin: forms explosive mixtures
Titanium: forms explosive mixtures
Titanium alloy: incompatible
Titanium-magnesium alloy: incompatible
Toluene: explosive or hypergolic reaction
Toluidine: incompatible
o-Toluidine + dinitrogen tetraoxide or sulfuric acid: explosive reaction
1,3,5-Triacetylhexahydro-1,3,5-triazine + trifluoroacetic anhydride: explosive reaction
Triazine: incompatible
Triazine + trifluoroacetic anhydride: explosive reaction
Tricadmium diphosphide: incompatible
Triethylgallium monoetherate: incompatible
Trimagnesium diphosphide: incompatible
2,4,6-Trimethyltrioxane: incompatible
Trinitrotoluenes: forms explosive mixtures
Tris(iodomercury) phosphine: explosive reaction
Turpentine: explosive or hypergolic reaction
Turpentine + ammonium metavanadate: ignites on contact
Turpentine + catalysts: ignites on contact
Turpentine + copper(II) chloride: ignites on contact
Turpentine + iron(III) chloride: ignites on contact
Turpentine + sulfuric acid (concentrated): ignites on contact
Uranium: forms explosive mixtures
Uranium disulfide: incompatible
Uranium-neodymium alloy: incompatible
Uranium-neodymium-zirconium alloy: incompatible
Vanadium(V) oxide + anilinium nitrate: ignites on contact
Vanadium(V) oxide + aromatic amines: ignites on contact
Vinylacetate: incompatible
Vinylidene chloride: incompatible
Wood: ignites on contact
p-Xylene: explosive or hypergolic reaction
mixo-Xylidine: explosive reaction
Xylidine + dinitrogen tetraoxide or sulfuric acid: explosive reaction
Zinc: incompatible
Zinc ethoxide: explosive reaction
Zirconium-uranium alloys: incompatible

When heated to decomposition, it emits highly toxic fumes of NO_x and hydrogen nitrate.

FIRST AID   In case of contact with the eyes, immediately wash the eyes with large amounts of water, occasionally lifting the lower and upper lids. Get medical attention immediately. Contact lenses should not be worn when working with this chemical.

In case of contact with the skin, immediately flush the contaminated skin with water. If this chemical penetrates the clothing, immediately remove the clothing and flush the skin with water. Get medical attention promptly.

If a person breathes in large amounts of this chemical, move the victim to fresh air at once. If breathing has stopped, perform artificial respiration. Keep the affected person warm and at rest. Get medical attention as soon as possible.

If this chemical has been swallowed, get medical attention immediately.

FIRE HAZARD   Flammable by chemical reaction with reducing agents.

To extinguish fires, use water.

HANDLING   Wear appropriate equipment to prevent any possibility of skin contact with acid of pH <2.5, and repeated or prolonged skin contact with acid of pH >2.5. Wash immediately if skin becomes contaminated. Immediately remove any nonimpervious clothing that becomes contaminated.

Wear eye protection to prevent any possibility of eye contact.

Provide an eyewash station and a quick-drench shower if acid with a pH <2.5 is present.

NITRIC OXIDE

CAS #10102-43-9   mf: NO
DOT #1660   mw: 30.01

NFPA LABEL

SYNONYM
Nitrogen monoxide

PROPERTIES   Colorless gas, blue liquid and solid; melting point −163.6°C, boiling point −151.7°C, density (liquid) 1.269 at −150°C; density (gas) 1.04; vapor pressure 26,000 mm Hg

OSHA PEL   TWA 25 ppm; IDLH 100 ppm

ACGIH TLV   TWA

NIOSH REL   TWA 25 ppm

TOXIC EFFECTS   A poison gas. A severe eye, skin, and mucous membrane irritant. A systemic irritant by inhalation.

In the laboratory and workshop, nitric oxide is generated by the reaction of nitric acid with organic materials (e.g., wood, sawdust, and refuse); heating of nitric acid; burning of organic nitro compounds [e.g., celluloid, cellulose nitrate (guncotton), and dynamite]; reaction of nitric acid with metals, as in metal etching and pickling; and in high-temperature welding, as with oxyacetylene or electric torches, where the nitrogen and oxygen in the air unite to form oxides of nitrogen. Automobile exhaust and power plant emissions are the most prolific sources of NO_x.

Nitric oxide and other oxides of nitrogen react with water in the presence of oxygen to form nitric and nitrous acids. Upon inhalation of the gas, these acids are formed in the respiratory system, where they can cause congestion in the throat and bronchi and edema of the lungs. The initial response to inhalation is a slight irritation of the mucous membranes of the upper respiratory tract. Because of this, dangerous concentrations of the fumes may build up without notice.

Exposure to 60–150 ppm causes immediate irritation of the nose and throat, with coughing and burning in the throat and chest. Upon removal to fresh air, the symptoms will often clear for a period of several hours. But 6–24 h after exposure, more serious symptoms of respiratory system damage may return. Exposure to 100–150 ppm for 30–60 min is dangerous; 200–700 ppm may be fatal after even very short exposures.

Chronic exposure to low concentrations can cause irritation of the respiratory tract.

Exposure to NO_x is always potentially serious, and persons so exposed should be kept under close observation for at least 48 h.

Routes   Ingestion, contact, inhalation

Symptoms   Irritation of the eyes, nose, throat, and respiratory system, a sensation of tightness and burning in the chest, shortness of breath, sleeplessness and restlessness, dyspnea and air hunger, cyanosis, drowsiness, unconsciousness, cough, headache, loss of appetite, dyspepsia, corrosion of the teeth, gradual loss of strength

Target Organs   Respiratory system

HAZARDOUS REACTIONS   An oxidizer. The liquid is a sensitive explosive.

Acetic anhydride: violent reaction
Aluminum: violent reaction
Ammonia: violent reaction
Barium oxide: violent reaction
Boron: incompatible
Boron (amorphous): violent reaction
Boron trichloride: violent reaction
1,3-Butadiene: reaction forms explosive products
Calcium: violent reaction
Carbon disulfide: explosive reaction when ignited
Carbon + potassium hydrogen tartrate: violent reaction
Charcoal: violent reaction
Chlorinated hydrocarbons: incompatible

Chlorine monoxide: violent reaction
Chloroform: violent reaction
Chromium: incompatible
Combustible matter: incompatible
$CsHC_2$: violent reaction
Cyclopentadiene: reaction forms explosive products
Dichlorine oxide: explosive reaction
1,2-Dichloroethane: violent reaction
Dichloroethylene: violent reaction
Dienes: reaction forms explosive products
Dimethyl hydrazine (uns-): violent reaction
Ethylene: violent reaction
Fluorine: explosive reaction
Fuels: violent reaction
Hydrocarbons: violent reaction
Hydrogen + oxygen: violent reaction
Iron: violent reaction
Magnesium: violent reaction
Manganese: violent reaction
Metals: incompatible
Methanol: explosive reaction when ignited
Methylene chloride: violent reaction
Na_2O: violent reaction
Nitrogen trichloride: explosive reaction
Olefins: violent reaction
Oxygen: reacts to form toxic nitrogen dioxide
Ozone: explosive reaction
Pentacarbonyl iron: explosive reaction at 50°C
Perchloryl fluoride: explosive reaction at 100–300°C
Phosphine: violent reaction
Phosphine + oxygen: explosive reaction
Phosphorus: violent reaction
PNH_2: violent reaction
Potassium: violent reaction
Potassium sulfide: violent reaction
Propadiene: reaction forms explosive products
Propylene: violent reaction
Pyrophoric chromium: violent reaction
Reducing materials: vigorous reaction
Rubidium acetylide: violent reaction
Rubidium carbide: incompatible
Sodium: violent reaction
Sodium diphenylketyl: explosive reaction
Sulfur: violent reaction
Tetrachloroethane (uns-): violent reaction
1,1,1-Trichloroethane: violent reaction
Trichloroethylene: violent reaction
Tungsten carbide: incompatible
Tungsten hydride: violent reaction
Uranium: violent reaction
Uranium dicarbide: violent reaction
Vinyl chloride: explosive reaction
Water or steam: reaction produces heat and corrosive fumes

When heated to decomposition, it emits highly toxic fumes of NO_x.

FIRST AID If a person breathes in large amounts of this chemical, move the victim to fresh air at once. If breathing has stopped, perform artificial respiration. Keep the affected person warm and at rest. Get medical attention as soon as possible.

NITRITES

PROPERTIES Salts of nitrous acid.

TOXIC EFFECTS Poisonous. Ingestion of large amounts may produce nausea, vomiting, cyanosis (due to methemoglobin formation), collapse, and coma. Chronic ingestion of small doses cause a fall in blood pressure, rapid pulse, headache, and visual disturbances.

At least some nitrites are carcinogens.

HAZARDOUS REACTIONS Generally powerful oxidizers. Some may be explosives.

Oxidizable materials: may ignite or explode on contact
  Organic nitrite reactions:
    NH_4 may react violently
    Salts may react violently
    Cyanide may react violently
    Potassium cyanide may react violently

When heated to decomposition, they emit highly toxic fumes of NO_x.

FIRE HAZARD Some are fire hazards.

NITROBENZENE

CAS #98-95-3   mf: $C_6H_5NO_2$
DOT #1662   mw: 123.11

STRUCTURE   NFPA LABEL

SYNONYMS
Essence of mirbane   Mirbane oil
Essence of myrbane   Nitrobenzene, liquid

Nitrobenzol
Nitrobenzol, liquid
Oil of mirbane
Oil of myrbane

PROPERTIES   Bright yellow crystals or yellow, oily liquid; odor of almond oil, threshold 0.5 ppm; solubility 0.2 g/100 mL water at 20°C; miscible with most organic solvents; volatile with steam; melting point 5.7°C, boiling point 211°C; density 1.19867 at 24°C/4°C; flash point 87.8°C (closed cup); lower explosive limit 1.8%; autoignition temperature 482.2°C; vapor pressure 10 mm Hg at 85.4°C, 100 mm Hg at 139.9°C; vapor density 4.24

OSHA PEL   TWA 1 ppm (skin); IDLH 200 ppm

ACGIH TLV   TWA 1 ppm (skin)

Community Right to Know List and EPA Extremely Hazardous Substances List

DOT Classification   Poison B; Label: Poison

TOXIC EFFECTS   A poison by ingestion. Moderately toxic by skin contact. Absorbed rapidly through the skin. An eye and skin irritant. The vapors are hazardous. May cause cyanosis due to formation of methemoglobin.

Skin absorption is the greatest hazard of nitrobenzene. However, the oral lethal dose of nitrobenzene for humans is about 5 mg/kg. No effects have been seen in humans after 30–60 min of exposure to 200–300 ppm of nitrobenzene. Exposure to 40–80 ppm for several hours does cause slight symptoms. Onset of symptoms may be delayed for up to 4 h.

Rats exposed to about 0.01–0.02 ppm for 70–82 days experienced adverse central nervous system effects and inflammation of internal organs.

Routes   Inhalation, absorption, ingestion, contact

Symptoms   Anorexia, irritation of the eyes, dermatitis, anemia, dizziness, nausea, vomiting, dyspnea, a general anesthetic, respiratory stimulation, vascular changes, cyanosis due to methemoglobin formation, anoxia, weakness, shock

Target Organs   Blood, liver, kidneys, cardiovascular system, skin

HAZARDOUS REACTIONS   An oxidant.

Alkali (solid or concentrated): explosive reaction when heated
Aluminum chloride: forms explosive mixtures
Aluminum chloride + phenol: explosive reaction at 120°C
Aniline + glycerin: violent reaction
Aniline + glycerol + sulfuric acid: explosive reaction
Caustics: incompatible
Dinitrogen tetraoxide: forms explosive mixtures
Fluorodinitromethane: forms explosive mixtures
Metals (chemically active): incompatible
Nitric acid: forms explosive mixtures
Nitric acid + water: forms explosive mixtures
Nitric acid + sulfuric acid: explosive reaction when heated
Nitrogen tetroxide: incompatible
Nitrous oxide: violent reaction
Oxidants: forms explosive mixtures
Peroxodisulfuric acid: forms explosive mixtures
Phosphorous pentachloride: forms explosive mixtures
Potassium: forms explosive mixtures
Potassium hydroxide: explosive reaction when heated
Silver perchlorate: violent reaction
Sodium chlorate: forms explosive mixtures
Sodium hydroxide: explosive reaction when heated
Sulfuric acid: forms explosive mixtures
Tetranitromethane: forms explosive mixtures
Tin: incompatible
Uranium perchlorate: forms explosive mixtures
Zinc: incompatible

When heated to decomposition, it emits toxic fumes of NO_x.

FIRST AID   In case of contact with the eyes, immediately wash the eyes with large amounts of water, occasionally lifting the lower and upper lids. Get medical attention immediately. Contact lenses should not be worn when working with this chemical.

In case of contact with the skin, immediately wash the contaminated skin with soap and water. If this chemical penetrates through the clothing, immediately remove the clothing and wash the skin with soap and water. Get medical attention promptly.

If a person breathes in large amounts of this chemical, move the victim to fresh air at once. If breathing has stopped, perform artificial respiration. Keep the affected person warm and at rest. Get medical attention as soon as possible.

If this chemical has been swallowed, get medical attention immediately.

FIRE HAZARD   Flammable when exposed to heat and flame. Moderate explosion hazard when exposed to heat or flame.

To fight a fire, use water, foam, CO_2, or dry chemical extinguishers.

HANDLING   OSHA limits include a warning about the potential contribution of skin absorption to the overall exposure.

Nitrobenzene, like most aromatic nitro compounds, readily penetrates the skin to cause the serious toxic effects described above. Hence, anyone using it in the laboratory should take care to avoid skin contact. If more than a few grams are being used, rubber gloves and other protective apparel may be needed. Although nitrobenzene is only moderately volatile, it is advisable to handle it in a hood.

Wear appropriate clothing and equipment to prevent any possibility of skin contact. Immediately remove any nonimpervious clothing that becomes contaminated. Work clothing should be changed daily if there is any possibility that the clothing has been contaminated.

Wear eye protection to reduce probability of eye contact.

The following equipment should be available: a quick-drench shower.

NITROGEN

CAS #7727-37-9   mf: N_2
DOT #1066; 1977   mw: 28.02

PROPERTIES   Colorless gas, colorless liquid or cubic crystals at low temperature; slightly soluble in water; soluble in liquid ammonia, alcohol; melting point −210.0°C; density 1.2506 g/L at 0°C, density (liquid) 0.808 at −195.8°C; condenses to a liquid

SYNONYMS
Nitrogen, compressed (DOT)
Nitrogen gas
Nitrogen, refrigerated liquid (DOT)

EPA TSCA Inventory

DOT Classification   Nonflammable gas; Label: Nonflammable gas

TOXIC EFFECTS   Low toxicity except at elevated pressures. A simple asphyxiant at high concentrations by the exclusion of oxygen.

The most common illnesses resulting from exposure to the gas are nitrogen narcosis and the bends (caisson disease). They generally result from SCUBA-diving accidents. The bends is caused by the release of nitrogen from solution in the blood, with formation of small bubbles. Nitrogen narcosis is a result of nitrogen dissolved in the cell membrane of neurons at high pressure.

HAZARDOUS REACTIONS   Nitrogen (liquid) can explode during use.

Lithium: potentially violent reaction under certain conditions
Neodymium: potentially violent reaction under certain conditions
Titanium: potentially violent reaction under certain conditions

HANDLING   Liquid nitrogen is continually evaporating, and a vent must always be provided in any vessel in which it is stored or used. If the vent is closed, bursting of the vessel will result. The use of any type of plug, e.g., cotton or glass wool, is dangerous because it may become coated with ice owing to freezing of moisture from the atmosphere. For the same reason, it is dangerous to leave a glass or metal funnel in the neck of the vessel.

Freezing of the exhaust may be nearly completely prevented by placing a short piece of tube, about 8 in. long, of pure rubber without cotton lining, over the end of the container exhaust tube. This allows a temperature gradient to build up along the length of the rubber tube, the free end being at room temperature. The rubber tube should also be split lengthwise for about 4 in. near the middle; if the end of the rubber tube and the outer surfaces freeze up, the mounting pressure will simply burst the tube.

If the outlet of a liquid gas container becomes obstructed by ice, the following procedure should be used:

1. Be on the alert for the possibility of a burst. The face and hands should be protected by a mask and gloves, and other persons cleared from the vicinity.
2. Puncture the ice with a clean piece of stiff wire. Do not attempt to use heat unless from an infrared source.

See *Fundamentals of Laboratory Safety: Physical Hazards in the Academic Laboratory*, Chap. 8, for a description of the handling of cryogenic liquids.

NITROGEN DIOXIDE

CAS #10102-44-0   mf: NO_2
DOT #1067   mw: 46.01

NFPA LABEL

SYNONYMS
Dinitrogen tetroxide   Nitrito
Nitrogen peroxide   NTO
Nitrogen tetroxide

PROPERTIES   Dark brown to yellow, fuming liquid or gas with a pungent, acrid odor; insoluble in water; soluble in concentrated sulfuric acid, nitric acid, corrosive to steel when wet; melting point −9.3°C (yellow liquid), boiling point 21°C (red-brown gas with decomposition); density 1.491 at 0°C; vapor pressure 400 mm Hg at 80°C

OSHA PEL   Ceiling 5 ppm

ACGIH TLV   TWA 3 ppm; STEL 5 ppm

NIOSH REL   Ceiling (oxides of nitrogen) 1 ppm/15 min

EPA TSCA Inventory and EPA Genetic Toxicology Program

DOT Classification   Poison A; Label: Poison gas and oxidizer

TOXIC EFFECTS   A poison by inhalation. A primary irritant to the lungs and the upper respiratory tract. A teratogen.

A very insidious gas. Lung inflammation from inhalation is only slightly painful, but the edema that results may easily cause death. Even short exposures to 100 ppm in air is dangerous; 200 ppm may be fatal in a short time.

Routes   Inhalation, contact

Symptoms   Cough, mucoid frothy sputum, dyspnea, chest pain, pulmonary edema, cyanosis, tachypnea, tachycardia, eye irritation, pulmonary vascular resistance changes

Target Organs   Respiratory system, cardiovascular system

HAZARDOUS REACTIONS
Alcohols: violent reaction
Ammonia: incompatible
Carbon disulfide: incompatible
Chlorinated hydrocarbons: incompatible
Combustible matter: incompatible
Cyclohexane: violent reaction
Fluorine: violent reaction
Formaldehyde: violent reaction
Nitrobenzene: violent reaction
Petroleum: violent reaction
Toluene: violent reaction

When heated to decomposition, it emits toxic fumes of NO_x.

FIRST AID   In case of contact with the eyes, immediately wash the eyes with large amounts of water, occasionally lifting the lower and upper lids. Get medical attention immediately. Contact lenses should not be worn when working with this chemical.

In case of contact with the skin, immediately flush the contaminated skin with water. If this chemical penetrates the clothing, immediately remove the clothing and flush the skin with water. Get medical attention promptly.

If a person breathes in large amounts of this chemical, move the victim to fresh air at once. If breathing has stopped, perform artificial respiration. Keep the affected person warm and at rest. Get medical attention as soon as possible.

If this chemical has been swallowed, get medical attention immediately.

HANDLING   Nitrogen dioxide is a deadly poison, and no one should work with a cylinder of this substance who is not fully familiar with its handling and its toxic effect. Ventilation is extremely important, and respiratory protective equipment should always be available.

Exposure must be avoided by the use of an air-purifying respirator equipped with an acid-gas cartridge or canister. At concentrations greater than 50 times the TLV, a positive-pressure, atmosphere-supplying respirator must be used. In IDLH atmospheres, a pressure-demand, self-contained breathing apparatus or a positive-pressure airline respirator that has escape-cylinder provisions is required.

Only stainless steel fittings should be used.

Wear appropriate clothing and equipment to prevent any possibility of skin contact. Wash immediately if skin becomes contaminated. Immediately remove any nonimpervious clothing that becomes contaminated.

Wear eye protection to prevent any possibility of eye contact.

The following equipment should be available: an eyewash station and a quick-drench shower.

OCTANE

CAS #111-65-9   mf: C_8H_{18}
DOT #1262   mw: 114

SYNONYM
Normal octane

224 OXALIC ACID

PROPERTIES   Colorless liquid with a gasolinelike odor; solubility in water 0.04%; slightly soluble in alcohol, ether; miscible in benzene; boiling point 125.8°C, freezing point −56.5°C; density 0.7036 at 20°C/4°C; flash point 13.3°C; lower explosive limit 1.0%, upper explosive limit 6.5%; autoignition temperature 220°C; vapor pressure 10 mm at 19.2°C; vapor density 3.86

OSHA PEL   TWA 500 ppm; IDLH 5000 ppm

ACGIH TLV   TWA 300 ppm

NIOSH REL   TWA 75 ppm; ceiling 385 ppm/15 min

DOT Classification   Flammable liquid; Label: Flammable liquid

TOXIC EFFECTS   A simple asphyxiant. A narcotic at high concentrations. Skin contact with undiluted octane for 5 h causes blister formation.

Routes   Inhalation, contact, ingestion

Symptoms   Irritation of the eyes, nose; drowsiness; dermatitis; chemical pneumonia

Target Organs   Skin, eyes, respiratory system

HAZARDOUS REACTIONS   Oxidizers (strong): incompatible

FIRST AID   In case of contact with the eyes, immediately wash the eyes with large amounts of water, occasionally lifting the lower and upper lids. Get medical attention immediately. Contact lenses should not be worn when working with this chemical.

In case of contact with the skin, promptly wash the contaminated skin with soap and water. If this chemical penetrates through the clothing, promptly remove the clothing and wash the skin with soap and water. Get medical attention promptly.

If a person breathes in large amounts of this chemical, move the victim to fresh air at once. If breathing has stopped, perform artificial respiration. Keep the affected person warm and at rest. Get medical attention as soon as possible.

If this chemical has been swallowed, get medical attention immediately.

FIRE HAZARD   A very dangerous fire and explosion hazard when exposed to heat, flame, or oxidizers.

HANDLING   Wear appropriate clothing and equipment to prevent repeated or prolonged skin contact. Wash promptly if skin becomes wet. Immediately remove any clothing that becomes wet.

Wear eye protection to reduce probability of eye contact.

OXALIC ACID

CAS #144-62-7   mf: $C_2H_2O_4$
DOT #2449   mw: 90.04

NFPA LABEL

SYNONYMS

Acide oxalique (French)
Acido ossalico (Italian)
Ethanedioic acid
Kyselina stavelova (Czech)
NCI-C55209
Oxaalzuur (Dutch)
Oxalic acid dihydrate
Oxalsaeure (German)

PROPERTIES   Colorless, odorless solid (rhombic crystals); solubility in water 10%; soluble in alcohol and ether; melting point 101°C, (189°C anhydrous); sublimes 148.9–160°C; density 1.653

OSHA PEL   TWA 1 mg/m³; IDLH 500 mg/m³

ACGIH TLV   TWA 1 mg/m³; STEL 2 mg/m³

EPA TSCA Inventory

TOXIC EFFECTS   Poisonous by ingestion, skin contact, and some other routes. A skin and severe eye irritant.

Ingestion causes corrosion to the mouth, esophagus, and stomach, with symptoms of vomiting, burning and abdominal pain, collapse and sometimes convulsions. Death may follow quickly. Oxalic acid in the blood reacts with circulating calcium to form insoluble calcium oxalate, which blocks the renal tubules and results in kidney damage.

Inhalation of the dust or vapor can cause chronic irritation of the upper respiratory tract and eyes, and gastrointestinal disturbances. More severe cases may show albuminuria, chronic cough, vomiting, pain in the back, and gradual emaciation and weakness.

Skin contact can cause dermatitis or, in more severe cases, cracking and slow-healing ulcers.

Oxalic acid is a component of the saps of many poisonous plants.

Routes   Inhalation, absorption, ingestion, contact

Symptoms   Irritation of the eyes, upper respiratory tract, and skin; burning of the eyes; local pain; cyanosis; albuminuria; gradual loss of weight; weakness; ulceration of the mucous membranes of the nose and throat; epistaxis; headache; irritation and nervousness; shock; collapse; convulsions

Target Organs   Respiratory system, skin, kidneys, eyes

HAZARDOUS REACTIONS
Furfuryl alcohol: violent reaction
Mercury: incompatible
Oxidizers (strong): incompatible
Silver: violent reaction
Sodium chlorite: incompatible
Sodium hypochlorite: violent reaction

When heated to decomposition, it emits acrid smoke and irritating fumes.

FIRST AID   In case of contact with the eyes, immediately wash the eyes with large amounts of water, occasionally lifting the lower and upper lids. Get medical attention immediately. Contact lenses should not be worn when working with this chemical.

In case of contact with the skin, immediately flush the contaminated skin with water. If this chemical penetrates the clothing, immediately remove the clothing and flush the skin with water. If irritation persists after washing, get medical attention.

If a person breathes in large amounts of this chemical, move the victim to fresh air at once. If breathing has stopped, perform artificial respiration. Keep the affected person warm and at rest. Get medical attention as soon as possible.

If this chemical has been swallowed, get medical attention immediately.

HANDLING   Wear appropriate clothing and equipment to prevent repeated or prolonged skin contact. Wash promptly if skin becomes contaminated. Promptly remove any nonimpervious clothing that becomes contaminated. Work clothing should be changed daily if it is possible that the clothing has been contaminated.

Wear eye protection to prevent any possibility of eye contact.

OXYGEN

CAS #7782-44-7 mf: O_2
DOT #1072/1073 mw: 32.00

NFPA LABEL

SYNONYM
LOX

PROPERTIES   Colorless, odorless, tasteless gas or liquid or hexagonal crystals. Supports combustion; one volume of the gas dissolves in 32 volumes of water at 20°C; dissolves in 7 volumes of alcohol at 20°C; soluble in other organic liquids to a greater extent than water; melting point −218.4°C, boiling point −182.96°C; density (liquid) 1.14 at −183.0°C, density (solid) 1.426 at −252.5°C; density (gas) 1.429 g/L at 0°C

EPA TSCA Inventory and EPA Genetic Toxicology Program

DOT Classification   Nonflammable gas; Label: Nonflammable gas, oxidizer

TOXIC EFFECTS   The liquid can cause severe burns and tissue damage by skin contact due to extreme cold. A teratogen by inhalation; causes developmental abnormalities of the cardiovascular system. The gas is not toxic.

Symptoms   Cough and other pulmonary changes

HAZARDOUS REACTIONS   Very reactive. An oxidant. A slight increase in the oxygen content of the air above the normal 21% greatly increases the oxidation or burning rate (and the hazard) of many materials.

Acetaldehyde: potentially violent reaction under certain conditions
Acetone: potentially violent reaction under certain conditions
Acetylene: potentially violent reaction under certain conditions
Alcohols (secondary): potentially violent reaction under certain conditions
Alkali metals: potentially violent reaction under certain conditions
Aluminum: potentially violent reaction under certain conditions
Aluminum borohydride: potentially violent reaction under certain conditions
Aluminum hydride: potentially violent reaction under certain conditions
Aluminum tetrahydroborate: potentially violent reaction under certain conditions
Aluminum-titanium alloys: potentially violent reaction under certain conditions
Ammonia: potentially violent reaction under certain conditions
Ammonia + platinum: potentially violent reaction under certain conditions

226 OXYGEN

Asphalt: potentially violent reaction under certain conditions

Barium: potentially violent reaction under certain conditions

BaS_2Br_3: potentially violent reaction under certain conditions

Benzene: potentially violent reaction under certain conditions

1,4-Benzenediol + 1-propanol: potentially violent reaction under certain conditions

Benzoic acid: potentially violent reaction under certain conditions

Beryllium borohydride: potentially violent reaction under certain conditions

Biological materials + ether: potentially violent reaction under certain conditions

Bis(phenylhydrazone): potentially violent reaction under certain conditions

B_2H_6: potentially violent reaction under certain conditions

B_2H_{10}: potentially violent reaction under certain conditions

Boron tribromide: potentially violent reaction under certain conditions

Boron trichloride: potentially violent reaction under certain conditions

Bromine + chlorotrifluoroethylene: potentially violent reaction under certain conditions

Bromotrifluoroethylene: potentially violent reaction under certain conditions

2-Butanol: potentially violent reaction under certain conditions

Buten-3-yne: potentially violent reaction under certain conditions

Calcium: potentially violent reaction under certain conditions

Calcium phosphide: potentially violent reaction under certain conditions

Carbon disulfide: potentially violent reaction under certain conditions

Carbon disulfide + mercury + anthracene: potentially violent reaction under certain conditions

Carbon monoxide: potentially violent reaction under certain conditions

Carbon tetrachloride: potentially violent reaction under certain conditions

Cesium: potentially violent reaction under certain conditions

Cesium hydride: potentially violent reaction under certain conditions

Charcoal: potentially violent reaction under certain conditions

Chlorinated hydrocarbons: potentially violent reaction under certain conditions

Chlorotrifluoroethylene: potentially violent reaction under certain conditions

Chlorotrifluoroethylene + bromine: potentially violent reaction under certain conditions

Copper + hydrogen sulfide: potentially violent reaction under certain conditions

Cumene: potentially violent reaction under certain conditions

Cyanogen: potentially violent reaction under certain conditions

Cyclohexane: potentially violent reaction under certain conditions

Cyclohexane-1,3-dione: potentially violent reaction under certain conditions

Cyclooctatetraene: potentially violent reaction under certain conditions

$C_{10}H_{14}$: potentially violent reaction under certain conditions

Decaborane(14): potentially violent reaction under certain conditions

Diborane: potentially violent reaction under certain conditions

Diboron tetrafluoride: potentially violent reaction under certain conditions

Diethyl ether: potentially violent reaction under certain conditions

Diisopropyl ether: potentially violent reaction under certain conditions

Dimethoxymethane: potentially violent reaction under certain conditions

Dimethyl sulfide: potentially violent reaction under certain conditions

Dimethylketene: potentially violent reaction under certain conditions

Dioxane: potentially violent reaction under certain conditions

1,1-Diphenylethylene: potentially violent reaction under certain conditions

Diphenylethylene: potentially violent reaction under certain conditions

Disilane: potentially violent reaction under certain conditions

Ethers: potentially violent reaction under certain conditions

Ethyl ether: potentially violent reaction under certain conditions

Ethyl nitrate + hydrocarbons: potentially violent reaction under certain conditions

Ethylene: potentially violent reaction under certain conditions

Fibrous fabrics: potentially violent reaction under certain conditions

Fluorine + hydrogen: potentially violent reaction under certain conditions

Foam rubber: potentially violent reaction under certain conditions
Fuels: potentially violent reaction under certain conditions
Gasoline: potentially violent reaction under certain conditions
Germanium: potentially violent reaction under certain conditions
Glycerol: potentially violent reaction under certain conditions
Halocarbons: potentially violent reaction under certain conditions
Hydrazine: potentially violent reaction under certain conditions
Hydrocarbons: potentially violent reaction under certain conditions
Hydrocarbons + promoters: potentially violent reaction under certain conditions
Hydrogen: potentially violent reaction under certain conditions
Hydrogen sulfide: potentially violent reaction under certain conditions
Lithiated dialkylnitrosoamines: potentially violent reaction under certain conditions
Lithium: potentially violent reaction under certain conditions
Lithium hydride: potentially violent reaction under certain conditions
Magnesium: potentially violent reaction under certain conditions
Magnesium hydride: potentially violent reaction under certain conditions
Metal hydrides: potentially violent reaction under certain conditions
Metals: potentially violent reaction under certain conditions
Methane: potentially violent reaction under certain conditions
Methoxycyclooctatetraene: potentially violent reaction under certain conditions
4-Methoxytoluene: potentially violent reaction under certain conditions
Methyl nitrate + hydrocarbons: potentially violent reaction under certain conditions
Methylene chloride: potentially violent reaction under certain conditions
Neoprene: potentially violent reaction under certain conditions
Nickel carbonyl + butane: potentially violent reaction under certain conditions
Nitromethane + hydrocarbons: potentially violent reaction under certain conditions
Nonmetal hydrides: potentially violent reaction under certain conditions
Oil: potentially violent reaction under certain conditions
Organic matter: potentially violent reaction under certain conditions
Oxygen difluoride + water: potentially violent reaction under certain conditions
P_2O_3: potentially violent reaction under certain conditions
Paraformaldehyde: potentially violent reaction under certain conditions
Pentaborane(9): potentially violent reaction under certain conditions
Pentaborane(11): potentially violent reaction under certain conditions
Phosphine: potentially violent reaction under certain conditions
Phosphorus: potentially violent reaction under certain conditions
Phosphorus(III) oxide: potentially violent reaction under certain conditions
Phosphorus tribromide: potentially violent reaction under certain conditions
Phosphorus trifluoride: potentially violent reaction under certain conditions
Polymers: potentially violent reaction under certain conditions
Polytetrafluoroethylene (Teflon): potentially violent reaction under certain conditions
Polytetrafluoroethylene + stainless steel: potentially violent reaction under certain conditions
Polyurethane: potentially violent reaction under certain conditions
Polyvinyl chloride: potentially violent reaction under certain conditions
Potassium: potentially violent reaction under certain conditions
Potassium hydride: potentially violent reaction under certain conditions
Potassium peroxide: potentially violent reaction under certain conditions
2-Propanol: potentially violent reaction under certain conditions
Propylene oxide: potentially violent reaction under certain conditions
Rhenium: potentially violent reaction under certain conditions
Rubber + ozone: potentially violent reaction under certain conditions
Rubberized fabric: potentially violent reaction under certain conditions
Rubidium: potentially violent reaction under certain conditions
Rubidium hydride: potentially violent reaction under certain conditions

Selenium: potentially violent reaction under certain conditions

Sodium hydride: potentially violent reaction under certain conditions

Sodium hydroxide + tetramethyldisiloxane: potentially violent reaction under certain conditions

Strontium: potentially violent reaction under certain conditions

Teflon: potentially violent reaction under certain conditions

Tetraborane(10): potentially violent reaction under certain conditions

Tetracarbonylnickel: potentially violent reaction under certain conditions

Tetracarbonylnickel + mercury: potentially violent reaction under certain conditions

Tetrafluoroethylene: potentially violent reaction under certain conditions

Tetrafluorohydrazine: potentially violent reaction under certain conditions

Tetrafluorohydrazine + hydrocarbons: potentially violent reaction under certain conditions

Tetrahydrofuran: potentially violent reaction under certain conditions

Tetrasilane: potentially violent reaction under certain conditions

Titanium and alloys: potentially violent reaction under certain conditions

1,1,1-trichloroethane: potentially violent reaction under certain conditions

Trichloroethylene: potentially violent reaction under certain conditions

Trirhenium nonachloride: potentially violent reaction under certain conditions

Trisilane: potentially violent reaction under certain conditions

Uranium hydride: potentially violent reaction under certain conditions

Wood: potentially violent reaction under certain conditions

p-Xylene: potentially violent reaction under certain conditions

FIRE HAZARD   Though itself nonflammable, it is essential to combustion. Exclusion of oxygen from the neighborhood of a fire is one of the principal methods of extinguishment. Avoid smoking, flames, or electric sparks. Liquid O_2 can explode on contact with a readily oxidizable material, especially at high temperature.

HANDLING   Oils, greases, and other readily combustible substances should never come in contact with oxygen cylinders, valves, regulators, gauges, and fittings. Oil and oxygen may combine with explosive violence. Therefore, valves, regulators, gauges, and fittings used in oxygen service must not be lubricated with oil or any other combustible substance. Oxygen cylinders or apparatus should not be handled with oily hands or gloves.

Oxygen regulators, hoses, and other appliances should not be interchanged with similar equipment intended for use with other gases. Cylinders of oxygen should not be stored near flammable materials, especially oils, greases, or any substance likely to cause or accelerate fire. Oxygen is not flammable, but it supports combustion. Once a pure-oxygen fire begins, almost anything, including metal, will burn.

Compressed O_2 is shipped in steel cylinders under high pressure. If these containers are broken as a result of shock or exposure to high temperature, an explosion and fire may result.

Liquid oxygen should only be used or stored under conditions which ensure adequate dilution of the escaping gas; otherwise the increased oxygen concentration near the apparatus or container causes an increased fire and explosion risk, particularly since liquid oxygen can itself cause spontaneous ignition if it comes into contact with oil, grease, or other combustible material. Liquid oxygen must never be used in place of liquid nitrogen to cool apparatus, metals, or any combustible matter.

PENTANE

CAS #109-66-0   mf: C_5H_{12}
DOT #1265   mw: 72.17

NFPA LABEL

SYNONYMS

Pentan (Polish)   Pentani (Italian)
Pentanen (Dutch)   Normal pentane

PROPERTIES   Colorless liquid with a gasolinelike odor; solubility in water 0.002%; miscible in alcohol, ether, and other organic solvents; boiling point 36.1°C, freezing point −129.8°C; density 0.626 at 20°C/4°C; flash point <−40°C; lower explosive limit 1.5%, upper explosive limit 7.8%; autoignition temperature 153.8°C; vapor pressure 400 mm Hg at 18.5°C; vapor density 2.48

OSHA PEL   TWA 1000 ppm; IDLH 15,000 ppm

ACGIH TLV   TWA 600 ppm; STEL 750 ppm

NIOSH REL   TWA 350 mg/m³; ceiling 1800 mg/m³/15 min

Dot Classification   Label: Flammable liquid

EPA TSCA Inventory

TOXIC EFFECTS   Narcotic in high concentration. The liquid can cause blisters on contact.

Routes   Inhalation, ingestion, contact

Symptoms   Drowsiness; irritation of the eyes, skin, and nose; dermatitis; chemical pneumonia

Target Organs   Skin, eyes, respiratory system

HAZARDOUS REACTIONS
Oxidizers (strong): incompatible

FIRST AID   In case of contact with the eyes, immediately wash the eyes with large amounts of water, occasionally lifting the lower and upper lids. Get medical attention immediately. Contact lenses should not be worn when working with this chemical.

In case of contact with the skin, immediately flush the contaminated skin with water. If this chemical penetrates the clothing, immediately remove the clothing and flush the skin with water. If irritation persists after washing, get medical attention.

If a person breathes in large amounts of this chemical, move the victim to fresh air at once. If breathing has stopped, perform artificial respiration. Keep the affected person warm and at rest. Get medical attention as soon as possible.

If this chemical has been swallowed, get medical attention immediately.

FIRE HAZARD   Highly dangerous fire hazard when exposed to heat, flame, or oxidizers. Severe explosion hazard when exposed to heat or flame.

To fight a fire, use foam, CO_2, or dry chemical extinguishers.

HANDLING   Wear appropriate clothing and equipment to prevent repeated or prolonged skin contact. Wash promptly if skin becomes wet. Immediately remove any clothing that becomes wet.

Wear eye protection to reduce probability of eye contact.

2-PENTANONE

CAS #107-87-9   mf: $C_5H_{10}O$
DOT #1249   mw: 86.15

SYNONYMS
Ethyl acetone
Methyl-propyl-cetone (French)
Methylopropyloketon (Polish)
Methyl-*n*-propyl ketone
MPK
Methyl propyl ketone (DOT)

PROPERTIES   Water-white liquid with a characteristic ketone odor; solubility in water 4.3%; density 0.8; boiling point 102.2°C; flash point 7.2°C; lower explosive limit 1.5%, upper explosive limit 8.2%; autoignition temperature 505°C; vapor density 3.0

OSHA PEL   TWA 200 ppm: IDLH 5000 ppm

ACGIH TLV   TWA 200 ppm; STEL 250 ppm

NIOSH REL   TWA 530 mg/m³ (150 ppm)

DOT Classification   Label: Flammable liquid

TOXIC EFFECTS   Moderately toxic by ingestion. Mildly toxic by skin contact and inhalation. A skin and eye irritant.

Routes   Inhalation, ingestion, contact

Symptoms   Irritation of the eyes, skin, and mucous membrane; headache; dermatitis; nausea; narcosis, coma

Target Organs   Respiratory system, eyes, skin, central nervous system

HAZARDOUS REACTIONS
Bromine trifluoride: mixtures may explode during evaporation
Oxidizing agents: incompatible

FIRST AID   In case of contact with the eyes, immediately wash the eyes with large amounts of water, occasionally lifting the lower and upper lids. Get medical attention immediately. Contact lenses should not be worn when working with this chemical.

In case of contact with the skin, flush with water promptly.

If a person breathes in large amounts of this chemical, move the victim to fresh air at once. If breathing

has stopped, perform artificial respiration. Keep the affected person warm and at rest. Get medical attention as soon as possible.

If this chemical has been swallowed, get medical attention immediately.

FIRE HAZARD   Dangerous fire hazard when exposed to heat or flame.
To extinguish fires, use alcohol foam.

HANDLING   Wear appropriate clothing and equipment to prevent repeated or prolonged skin contact. Wash promptly if skin becomes wet. Immediately remove any clothing that becomes wet.

Wear eye protection to reduce probability of eye contact.

PERACETIC ACID

CAS #79-21-0   mf: $C_2H_4O_3$
DOT #2131   mw: 76.05

STRUCTURE

$$\begin{array}{c} O \\ \| \\ CH_3C-O-OH \end{array}$$

NFPA LABEL

SYNONYMS
Acetyl hydroperoxide
Ethaneperoxoic acid
Hydroperoxide, acetyl
Peroxyacetic acid
Peroxyacetic acid solution

PROPERTIES   Colorless liquid with an acrid odor; most commonly available as 40% solution in acetic acid; water-soluble; powerful oxidizer; melting point 0.1°C, freezing point approximately −30°C, boiling point 105°C; explodes at 110°C; density 1.15 at 20°C; flash point 40.6°C (open cup)

EPA Extremely Hazardous Substances List and Community Right to Know List

DOT Classification   Organic peroxide; Label: Organic peroxide, corrosive

TOXIC EFFECTS   Poisonous by ingestion. Moderately toxic by inhalation and skin contact. A corrosive irritant to the eyes, skin, and upper respiratory tract. A potent tumor promoter, and a weak carcinogen.

Symptoms by Ingestion   Severe irritation of the stomach and intestinal linings

HAZARDOUS REACTIONS   Unstable. Severe explosion hazard when exposed to heat or by spontaneous chemical reaction. Peracetic acid explodes at 100°C and decomposes at lower temperatures with the generation of oxygen. A powerful oxidizing agent.

Acetic anhydride: explosive reaction
Calcium chloride: violent reaction
5-p-Chlorophenyl-2,2-dimethyl-3-hexanone: explosive reaction
Combustible materials: incompatible
Diethyl ether: violent reaction
Ether solvents: violent reaction
Metal chloride solutions: violent reaction
Olefins: violent reaction
Organic materials: vigorous reaction
Potassium chloride: violent reaction
Sodium chloride: violent reaction
Tetrahydrofuran: violent reaction

When heated to decomposition, it emits acrid smoke and irritating fumes.

FIRE HAZARD   Flammable when exposed to heat or flames. Severe explosion hazard when exposed to heat or by spontaneous chemical reaction.

To fight a fire, use water, foam, or CO_2 extinguishers.

HANDLING   Because explosion is the greatest hazard of peracetic acid, the substance should be protected from sparks or physical shock, and laboratory operations with it should be carried out behind a shield and in a hood that has explosion-proof equipment.

Because of its high irritancy, care should be taken to prevent contact of peracetic acid with the skin, eyes, or upper respiratory tract. This is another reason why work with it should be carried out in a hood, and it also means that rubber gloves and a rubber apron should be worn as appropriate.

See Chap. 3 for information on handling carcinogens.

PERCHLORATES

mf: $-ClO_4$

TOXIC EFFECTS   Irritants due to their chemical reactivity.

HAZARDOUS REACTIONS   Unstable. Powerful oxidizers. Moderate explosion hazard when shocked, exposed to heat, or by chemical reaction. Many perchlorates of nitrogenous bases (e.g., hydroxylamine, urea, methylamine, ethylamine, isopropylamine, 4-ethylpyridine, diaminoethane) and organic perchlorates are explosives. Diazonium perchlorates are very dangerous.

Aluminum: mixture is a fire and explosion hazard
Benzene: violent reaction
Calcium hydride: violent reaction
Carbon-containing compounds: mixture is explosive
Charcoal: violent reaction
Ethanol: violent reaction
Magnesium powder: mixture is a fire and explosion hazard
Olefins: violent reaction
Reducing materials: violent reaction
Strontium hydride: violent reaction
Sulfur: violent reaction
Sulfuric acid: violent reaction
Zinc: mixture is a fire and explosion hazard

FIRE HAZARD   Flammable by chemical reaction with carbon-containing and other materials.
To extinguish fires, use water or foam.

PERCHLORIC ACID

CAS #7601-90-3   mf: ClHO$_4$
                                        mw: 100.46

NFPA LABEL

PROPERTIES   Colorless, fuming, unstable liquid; melting point −112°C, boiling point 19°C at 11 mm Hg; density 1.768 at 22°C

EPA TSCA Inventory

DOT Classification   Label: Oxidizer

TOXIC EFFECTS   Poisonous by ingestion. A severe irritant to the eyes, skin, and mucous membranes.

HAZARDOUS REACTIONS   A powerful oxidizer. A severe explosion hazard. The anhydrous form can explode spontaneously.

Acetic acid: violent reaction or ignition
Acetic acid + acetic anhydride: violent reaction or ignition
Acetic anhydride: violent reaction or ignition
Acetic anhydride + acetic acid + organic materials: potentially explosive reaction
Acetic anhydride + carbon tetrachloride + 2-methyl cyclohexanone: violent reaction or ignition
Acetic anhydride + organic materials + transition metals: potentially explosive reaction
Acetonitrile: potentially explosive reaction
Alcohols: potentially explosive reaction
Aniline + formaldehyde: reacts to form explosive products
Antimony: potentially explosive reaction above 110°C
Antimony compounds: violent reaction or ignition
Azo pigments: violent reaction or ignition
Azo dyes + orthoperiodic acid: potentially explosive reaction
Bis-1,2-diaminopropane-cis-dichlorochromium(III) perchlorate: violent reaction or ignition
1,3-Bis(di-n-cyclopentadienyl iron)-2-propen-1-one: violent reaction or ignition
Bis(2-hydroxyethyl) terephthalate + ethanol + ethylene glycol: potentially explosive reaction
Bismuth: potentially explosive reaction above 110°C
Carbon: violent reaction or ignition
Carbon tetrachloride: violent reaction or ignition
Cellulose and derivatives + heat: potentially explosive reaction
Charcoal + chromium trioxide + heat: potentially explosive reaction
Chromium + acetic anhydride + organic materials: potentially explosive reaction
Combustible materials: potentially explosive reaction
Copper dichromium tetraoxide: violent reaction or ignition at 120°C
Dehydrating agents: potentially explosive reaction
Dibutyl sulfoxide: violent reaction or ignition
Dichloromethane + dimethylsulfoxide: potentially explosive reaction
Diethyl ether: potentially explosive reaction
Dimethyl ether: potentially explosive reaction
Dimethyl sulfoxide: violent reaction or ignition
Dioxane + nitric acid + heat: potentially explosive reaction
DNA: violent reaction or ignition
Ethylbenzene: violent reaction or ignition
Ethylbenzene + thallium triacetate: reacts to form explosive products at 65°C

Fecal material + nitric acid: potentially explosive reaction
Fluorine: reacts to form explosive fluorine perchlorate
Glycerol + lead oxide: reacts to form explosive products
Glycol ethers: violent reaction or ignition
Glycols: violent reaction or ignition
Graphitic carbon + nitric acid: potentially explosive reaction
Hydrochloric acid: violent reaction or ignition
Hydrofluoric acid + structural materials: potentially explosive reaction
Hydrogen + heat: reacts to form explosive products
Hydrogen halides: reacts to form explosive products
Hypophosphites: violent reaction or ignition
Iodides: violent reaction or ignition
Iron + acetic anhydride + organic materials: potentially explosive reaction
Iron sulfate: violent reaction or ignition
Iron(II) sulfate: potentially explosive reaction
Ketones: violent reaction or ignition
Lead oxide + glycerin: violent reaction or ignition
Methanol: violent reaction or ignition
Methanol + triglycerides: violent reaction or ignition
2-Methylcyclohexanone: violent reaction or ignition
2-Methylpropene + metal oxides: violent reaction or ignition
Nickel + acetic anhydride + organic materials: potentially explosive reaction
Nitric acid: violent reaction or ignition
Nitric acid + organic matter + heat: potentially explosive reaction
Nitric acid + pyridine + sulfuric acid: potentially explosive reaction
Nitrogen triiodide: violent reaction or ignition
Nitrogenous epoxides: violent reaction or ignition
Nitrosophenol: violent reaction or ignition
Oleic acid: violent reaction or ignition
Organic materials + sodium hydrogen carbonate: potentially explosive reaction above 200°C
Organophosphorus compounds: violent reaction or ignition
Paper: violent reaction or ignition
o-Periodic acid: violent reaction or ignition
Phenyl acetylene: potentially explosive reaction at −78°C
Phosphine: reacts to form explosive products
P_2O_5: violent reaction or ignition
P_2O_5 + $CHCl_3$: violent reaction or ignition
Pyridine: reacts to form explosive products
Sodium iodide + hydroiodic acid: violent reaction or ignition
Sodium phosphinate: violent reaction or ignition
Steel: violent reaction or ignition
Sulfinyl chloride: violent reaction or ignition
Sulfoxides: reacts to form explosive products
Sulfur trioxide: violent reaction or ignition
Sulfuric acid: violent reaction or ignition
Sulfuric acid + organic materials: potentially explosive reaction
Trichloroethylene: violent reaction or ignition
Vegetable matter: violent reaction or ignition
Wood: violent reaction or ignition
Zinc phosphide: violent reaction or ignition

FIRE HAZARD   A severe explosion hazard; the anhydrous form can explode spontaneously.

PHENOL

CAS #108-95-2 mf: C_6H_6
DOT #1671; 2312; 2821 mw: 94.12

STRUCTURE

NFPA LABEL

SYNONYMS

Baker's P and S liquid and ointment
Benzenol
Carbolic acid
Hydroxybenzene
Monohydroxybenzene
Monophenol
Oxybenzene
Phenic acid
Phenyl hydrate
Phenyl hydroxide
Phenylic acid
Phenylic alcohol

PROPERTIES   Colorless to pink solid or thick liquid with a characteristic sweet, tarry odor, threshold 0.3 ppm; burning taste; solubility 6.7 g/100 mL in water at 16°C; miscible at 66°C, easily soluble in alcohol and other organic solvents; melting point 40.6°C, boiling point 181.9°C; density (liquid) 1.049 at 50°C/4°C; flash point 85°C (closed cup); lower explosive limit 1.7%, upper explosive limit 8.6%; autoignition temperature 715.5°C; vapor pressure 0.3513 mm Hg at 25°C, 1 mm Hg at 40.1°C; vapor density 3.24

OSHA PEL   TWA 5 ppm (skin); IDLH 250 ppm

ACGIH TLV   TWA 5 ppm (skin)

NIOSH REL   TWA 20 mg/m³; ceiling 60 mg/m³/15 min

DOT Classification   Poison B; Label: Poison

EPA Extremely Hazardous Substances List and Community Right to Know List

TOXIC EFFECTS   A poison by ingestion and several other routes. Moderately toxic by skin contact. A severe eye and skin irritant. A carcinogen.

In humans, lethal oral doses of phenol have ranged from 1 to 10 g. Severe phenol poisoning by ingestion is characterized by burns of the mouth and throat and rapid development of digestive disturbances, headache, fainting, vertigo, mental disturbances, collapse, and coma. Exposure to the vapor can produce marked irritation of the eyes, nose, and throat.

Phenol rapidly penetrates the skin, and can cause death within 30 min to several hours by exposure of as little as 64 in.^{2} of skin. Lesser exposures can cause damage to the kidneys, liver, pancreas, and spleen, and edema of the lungs. Ingestion can cause corrosion of the lips, mouth, throat, esophagus, and stomach, and gangrene. Chronic exposures can cause death from liver and kidney damage.

Phenol is corrosive to the eyes and skin and may cause irritant dermatitis.

Guinea pigs were severely injured by inhalation of 25–50 ppm phenol for 20 days. Damage was seen in the lungs, liver, kidneys, and heart.

Routes   Inhalation, absorption, ingestion, contact

Symptoms   Irritation of the eyes, nose, and throat; skin burns; dermatitis; anorexia, weight loss, weakness, muscle ache, pain; dark urine; cyanosis; liver, kidney damage; discoloration of body tissues; tremor, convulsions; twitching

Target Organs   Liver, kidneys, skin

HAZARDOUS REACTIONS
Aluminum chloride + nitrobenzene: violent reaction at 120°C
Aluminum chloride + nitromethane: potentially explosive reaction at 110°C/100 bar
Butadiene: violent reaction
Calcium hypochlorite: incompatible
Formaldehyde: potentially explosive reaction
Oxidizing materials: incompatible
Peroxydisulfuric acid: potentially explosive reaction
Peroxymonosulfuric acid: potentially explosive reaction
Sodium nitrite + heat: potentially explosive reaction
Sodium nitrate + trifluoroacetic acid: violent reaction

When heated to decomposition, it emits toxic fumes.

FIRST AID   In case of contact with the eyes, immediately wash the eyes with large amounts of water, occasionally lifting the lower and upper lids. Get medical attention immediately. Contact lenses should not be worn when working with this chemical.

In case of contact with the skin, immediately wash the contaminated skin with soap and water. If this chemical penetrates through the clothing, immediately remove the clothing and wash the skin with soap and water. Get medical attention promptly.

If a person breathes in large amounts of this chemical, move the victim to fresh air at once. If breathing has stopped, perform artificial respiration. Keep the affected person warm and at rest. Get medical attention as soon as possible.

If this chemical has been swallowed, get medical attention immediately.

FIRE HAZARD   Combustible when exposed to heat, flame, or oxidizers.

To fight a fire, use alcohol foam, CO_2, or dry chemical extinguishers.

HANDLING   The OSHA limit includes a warning about the potential contribution of skin absorption to the overall exposure.

Because phenol is a potent skin irritant, rubber gloves should be worn when there is opportunity for significant skin contact.

Wear appropriate clothing and equipment to prevent any possibility of skin contact. Wash promptly if the skin becomes contaminated. Promptly remove any nonimpervious clothing that becomes contaminated. Work clothing should be changed daily if it is at all probable that the clothing has been contaminated.

Wear eye protection to prevent any possibility of eye contact.

See Chap. 3 for information on handling carcinogens.

PHENOLPHTHALEIN

CAS #77-09-8   mf: $C_{20}H_{14}O_4$
mw: 318.34

SYNONYM
3,3-Bis (*p*-hydroxyphenol)-phthalide

PROPERTIES   Small crystals; insoluble in water, very soluble in chloroform; melting point 258–262°C; density 1.299

EPA TSCA Inventory

TOXIC EFFECTS   Moderately toxic by the intraperitoneal route.

USES   Used in medicine as a laxative; in chemistry as an indicator.

HAZARDOUS REACTIONS   When heated to decomposition, it emits acrid smoke and fumes.

PHOSGENE

CAS #75-44-5   mf: CCl_2O
DOT #1076   mw: 98.91

STRUCTURE

$$\underset{ClCCl}{\overset{O}{\|}}$$

NFPA LABEL

0 / 4 / 4

SYNONYMS
Carbon oxychloride
Carbonyl chloride
Chloroformyl chloride
Diphosgene

PROPERTIES   Colorless, poisonous gas or volatile liquid, sweet (geraniumlike) odor at low levels, pungent, irritating at higher levels (threshold about 0.5–1.0 ppm); very slightly soluble in water, very soluble in most organic solvents and oils; decomposes slightly in water with formation of CO_2 and HCl; melting point −118°C, boiling point 8.3°C; density 1.37 at 20°C; vapor pressure 1180 mm Hg at 20°C; vapor density 3.4

OSHA PEL   TWA 0.1 ppm; IDLH 2 ppm

ACGIH TLV   TWA 0.1 ppm

NIOSH REL   TWA 0.1 ppm; ceiling 0.2 ppm/15 min

DOT Classification   Poison A; Label: Poison gas

EPA Extremely Hazardous Substances List, Community Right to Know List, and EPA TSCA Inventory

TOXIC EFFECTS   A poison by inhalation. A severe eye, skin, and mucous membrane irritant.

Reacts with water in the lungs to form hydrochloric acid and carbon monoxide, which can cause pulmonary edema, bronchopneumonia, and sometimes lung abscess. There is little warning of exposure because of the lack of irritating effect, and dangerous concentrations can be inhaled without notice.

Symptoms of lung damage can appear after 2 to 24 h. The patient complains of burning in the throat and chest, shortness of breathing, and increasing dyspnea. In severe cases, death can occur from pulmonary edema within 36 h. In cases where the exposure has been less, pneumonia may develop several days after the occurrence of the accident.

An airborne concentration of 5 ppm may cause eye irritation and coughing in a few minutes. The substance can cause severe lung injury in 1–2 min at a level of 20 ppm. Exposure to concentrations above 50 ppm is likely to be fatal.

Pulmonary edema, bronchopneumonia, and emphysema were found in cats and guinea pigs exposed to 2.5–6.25 ppm of phosgene/day for 2–41 days. A variety of animals exposed to 0.2 or 1.1 ppm for 5 h/day for 5 days also had pulmonary edema.

Liquid phosgene is likely to cause severe skin burns and eye irritation.

Routes   Inhalation, ingestion, contact

Symptoms   Irritation of the eyes; dry, burning throat; vomiting; cough, foamy sputum, dyspnea, chest pain, bronchitis; cyanosis; skin burns; numbness

Target Organs   Respiratory system, skin, eyes

HAZARDOUS REACTIONS
Aluminum: incompatible under some conditions
tert-Butyl azido formate: incompatible under some conditions
Hexafluoroisopropylideneamino lithium: incompatible under some conditions
2,4-Hexadiyne-1,6-diol: violent reaction
Isopropyl alcohol: incompatible under some conditions
Lithium: incompatible under some conditions
Potassium: incompatible under some conditions
Sodium: incompatible under some conditions
Sodium azide: incompatible under some conditions
Water: reaction forms carbon dioxide and hydrogen chloride

When heated to decomposition, it produces toxic and corrosive fumes of Cl^-.

FIRST AID   In case of contact with the eyes, immediately wash the eyes with large amounts of water, occasionally lifting the lower and upper lids. Get medical attention immediately. Contact lenses should not be worn when working with this chemical.

In case of contact with the skin, immediately flush the contaminated skin with water. If this chemical penetrates the clothing, immediately remove the clothing and flush the skin with water. Get medical attention promptly.

If a person breathes in large amounts of this chemical, move the victim to fresh air at once. If breathing has stopped, perform artificial respiration. Keep the

affected person warm and at rest. Get medical attention as soon as possible.

If this chemical has been swallowed, get medical attention immediately.

HANDLING   No one should work with this substance who is not fully familiar with proper handling procedures and its toxic effect. Ventilation is extremely important, and respiratory protective equipment should be available. Corrosion problems are not serious, and brass fittings may be used.

In case of a leak in a phosgene cylinder, the brass cylinder cap should be affixed as tightly as possible and the cylinder placed in the coolest spot that is available. The manufacturer should be notified at once.

Work with phosgene should always be carried out within a hood. Unused quantities of phosgene greater than 1 g should be destroyed by reaction with water or dilute alkali.

Wear appropriate clothing and equipment to avoid contact with the skin. Wash immediately if skin becomes contaminated. Immediately remove any nonimpervious clothing that becomes contaminated.

Wear eye protection to prevent any possibility of eye contact.

The following equipment should be available: a quick-drench shower.

PHOSPHORIC ACID

CAS #7664-38-2   mf: H_3O_4P
DOT #1805   mw: 98.00

NFPA LABEL

SYNONYMS
Orthophosphoric acid
White phosphoric acid
Metaphosphoric acid

PROPERTIES   Viscous, colorless, odorless liquid; miscible with water; melting point 42.35°C, freezing point 42.4°C; loses 1/2 H_2O at 213°C; density 1.864 at 25°C; vapor pressure 0.0285 mm Hg at 20°C

OSHA PEL   TWA 1 mg/m^3; IDLH 1 mg/m^3

ACGIH TLV   TWA 1 mg/m^3; STEL 3 mg/m^3

DOT Classification   Label: Corrosive

EPA TSCA Inventory

TOXIC EFFECTS   Moderately toxic by ingestion and skin contact. A poison by some other routes. A corrosive irritant to eyes, skin, and mucous membranes. A systemic irritant by inhalation.

Routes   Inhalation, ingestion, contact

Symptoms   Irritation of the upper respiratory tract, eyes, and skin; burning of the skin and eyes; dermatitis

Target Organs   Respiratory system, eyes, skin

HAZARDOUS REACTIONS   A strong acid.

Caustics (strong): incompatible
Chlorides + stainless steel: reaction forms explosive hydrogen gas
Metals: incompatible
Nitromethane: forms explosive mixtures
Sodium tetrahydroborate: potentially violent reaction

When heated to decomposition, it emits toxic fumes of PO_x.

FIRST AID   In case of contact with the eyes, immediately wash the eyes with large amounts of water, occasionally lifting the lower and upper lids. Get medical attention immediately. Contact lenses should not be worn when working with this chemical.

In case of contact with the skin, immediately flush the contaminated skin with water. If this chemical penetrates the clothing, immediately remove the clothing and flush the skin with water. Get medical attention promptly.

If a person breathes in large amounts of this chemical, move the victim to fresh air at once. If breathing has stopped, perform artificial respiration. Keep the affected person warm and at rest. Get medical attention as soon as possible.

If this chemical has been swallowed, get medical attention immediately.

HANDLING   Wear appropriate equipment to prevent any possibility of skin contact with liquids containing >1.6% of the acid, and repeated or prolonged skin contact with liquids containing <1.6%. Wash immediately if skin becomes contaminated. Immediately remove any nonimpervious clothing that becomes contaminated. Work clothing should be changed daily if it is at all probable that it has been contaminated.

Wear eye protection to prevent any possibility of eye contact.

Provide an eyewash station and a quick-drench shower if liquids containing >1.6% of the acid are present.

PHOSPHORUS (RED)

CAS #7723-14-0 mf: P
DOT #1381 mw: 30.97

NFPA LABEL

SYNONYM
Phosphorus, amorphous, red (DOT)

PROPERTIES Reddish-brown powder. Boiling point 280°C (with ignition), melting point 590°C at 43 atm; density 2.34; autoignition temperature 260°C in air; vapor density 4.77

ACGIH TLV TWA 0.1 mg/m^3

DOT Classification Label: Flammable solid and poison

EPA Extremely Hazardous Substances List

TOXIC EFFECTS A poison. May contain white phosphorus as an impurity.

HAZARDOUS REACTIONS May explode on impact. Generally less reactive than white phosphorus. Moderate explosion hazard by chemical reaction.

Alkalies + heat: violent reaction or ignition
Ammonium nitrate: mixtures are sensitive explosives
Antimony pentachloride: violent reaction or ignition
Barium iodate: mixtures are sensitive explosives
Barium sulfate: incompatible
Beryllium: incompatible
Boron triiodide: incompatible
Bromates: mixtures are sensitive explosives
Bromine pentafluoride: incompatible
Bromine trifluoride: incompatible
Calcium iodate: mixtures are sensitive explosives
Calcium sulfate: incompatible
Cerium: incompatible
Cesium acetylide: incompatible
Chlorates: mixtures are sensitive explosives
Chlorine: violent reaction or ignition
Chlorine dioxide: incompatible
Chlorine trifluoride: incompatible
Chlorosulfuric acid: explosive reaction
Chromium trioxide: incompatible
Chromyl chloride: explosive reaction
Copper: incompatible
Copper oxide: incompatible
Cyanogen iodide: incompatible
Dichlorine oxide: incompatible
Dinitrogen pentoxide: incompatible
Dinitrogen tetraoxide: incompatible
Disulfur dibromide: incompatible
Disulfuryl chloride: incompatible
Fluorine: violent reaction or ignition
Halogen azides: incompatible
Halogen oxides: incompatible
Hexalithium disilicide: incompatible
Hydrogen peroxide: incompatible
Hydroiodic acid: explosive reaction
Interhalogens: incompatible
Iodine pentafluoride: incompatible
Iodine trichloride: incompatible
Lanthanum: incompatible
Lead oxide: incompatible
Lead peroxide: incompatible
Liquid bromine: violent reaction or ignition
Lithium acetylide: incompatible
Magnesium iodate: mixtures are sensitive explosives
Magnesium perchlorate: explosive reaction
Manganese: incompatible
Manganese dioxide: incompatible
Mercury oxide: incompatible
Mercury(I) nitrate: mixtures are sensitive explosives
Metal acetylides: incompatible
Metal halogenates: mixtures are sensitive explosives
Metal oxides: incompatible
Metal peroxides: incompatible
Metal sulfates: incompatible
Metals: incompatible
Neodymium: incompatible
Nitric acid: incompatible
Nitrogen halides: incompatible
Nitrogen oxide: incompatible
Nitrosyl fluoride: incompatible
Nitryl fluoride: incompatible
Nonmetal halides: incompatible
Nonmetal oxides: incompatible
Organic materials: incompatible
Osmium: incompatible
Oxidizing materials: incompatible
Oxygen: incompatible
Oxygen difluoride: incompatible
Peroxides: incompatible
Peroxyformic acid: incompatible
Platinum: incompatible
Potassium acetylide: incompatible

Potassium chlorite: incompatible
Potassium iodate: mixtures are sensitive explosives
Potassium nitride: incompatible
Potassium permanganate: mixtures are sensitive explosives
Potassium peroxide: incompatible
Praseodymium: incompatible
Reducing materials: incompatible
Rubidium acetylide: incompatible
Selenium: incompatible
Seleninyl chloride: incompatible
Silver nitrate: mixtures are sensitive explosives
Silver oxide: incompatible
Sodium acetylide: incompatible
Sodium chlorite: incompatible
Sodium iodate: mixtures are sensitive explosives
Sodium nitrate: mixtures are sensitive explosives
Sodium peroxide: incompatible
Sulfur: incompatible
Sulfur trioxide: incompatible
Sulfuric acid: incompatible
Sulfuryl chloride: incompatible
Thorium: incompatible
Trioxygen difluoride: incompatible
Zinc iodate: mixtures are sensitive explosives
Zirconium: incompatible

When heated to decomposition, it emits highly toxic fumes of PO_x.

FIRE HAZARD   Dangerous fire hazard when exposed to heat or by chemical reaction with oxidizers.
To extinguish fires, use water.

PHOSPHORUS (WHITE)

CAS #7723-14-0   mf: P_4
DOT #2447   mw: 123.88

NFPA LABEL

SYNONYMS
Common-sense cockroach and rat preparations
Fosforo bianco (Italian)
Gelber phosphor (German)
Phosphorus (yellow)
Tetrafosfor (Dutch)
Tetraphosphor (German)
Weiss phosphor (German)
White phosphorus
Yellow phosphorus

PROPERTIES   White to yellow, soft, waxy solid with acrid fumes in air; solubility in water 0.0003%; melting point 44.1°C, boiling point 280°C; spontaneously flammable in air; density 1.82; autoignition temperature 29.4°C; vapor pressure 1 mm Hg at 76.6°C; vapor density 4.42

OSHA PEL   TWA 100 µg/m³

ACGIH TLV   TWA 0.1 mg/m³

DOT Classification   Label: Flammable solid and poison

EPA TSCA Inventory

TOXIC EFFECTS   Poisonous by inhalation, ingestion, skin contact, and subcutaneous routes. Can cause severe eye damage by inhalation or ingestion.

Routes   Inhalation, ingestion, contact

Symptoms   Irritation of the eyes and respiratory tract; abdominal pain, nausea; jaundice, liver damage; severe generalized weakness, emaciation; dental pain, excess salivation, jaw pain; swelling; burns of the skin and eyes; sweating, nausea, diarrhea; cyanosis; cardiomyopathy; anemia; brittleness of the long bones; photophobia with myosis, dilation of the pupils, retinal hemorrhage, congestion of the blood vessels

Target Organs   Respiratory system, liver, kidneys, jaw, teeth, blood, eyes, skin

HAZARDOUS REACTIONS   More reactive than red phosphorus.

Air: incompatible
Alkaline hydroxides: mixture is a dangerous explosion hazard
Ammonium nitrite: mixture is a dangerous explosion hazard
Antimony pentafluoride: mixture is a dangerous explosion hazard
Barium bromate: mixture is a dangerous explosion hazard
Barium chlorate: mixture is a dangerous explosion hazard
Barium iodate: mixture is a dangerous explosion hazard
Beryllium: mixture is a dangerous explosion hazard
Bromine: mixture is a dangerous explosion hazard
Bromine azide: mixture is a dangerous explosion hazard
Bromine trifluoride: mixture is a dangerous explosion hazard
Calcium bromate: mixture is a dangerous explosion hazard

Calcium chlorate: mixture is a dangerous explosion hazard
Calcium iodate: mixture is a dangerous explosion hazard
Caustics (strong): incompatible
Cerium: mixture is a dangerous explosion hazard
Cesium: mixture is a dangerous explosion hazard
$CsHC_2$: mixture is a dangerous explosion hazard
Cesium nitride: mixture is a dangerous explosion hazard
Charcoal + air: mixture is a dangerous explosion hazard
Chlorine + heptane: mixture is a dangerous explosion hazard
Chlorine dioxide: mixture is a dangerous explosion hazard
Chlorine monoxide: mixture is a dangerous explosion hazard
Chlorine trifluoride: mixture is a dangerous explosion hazard
Chlorine trioxide: mixture is a dangerous explosion hazard
Chlorosulfonic acid: mixture is a dangerous explosion hazard
Chlorosulfuric acid: mixture is a dangerous explosion hazard
Chromium trioxide: mixture is a dangerous explosion hazard
Chromyl chloride: mixture is a dangerous explosion hazard
Copper: mixture is a dangerous explosion hazard
FNO_2: mixture is a dangerous explosion hazard
Halogen azides: mixture is a dangerous explosion hazard
Halogens: mixture is a dangerous explosion hazard
Hexalithium disilicide: mixture is a dangerous explosion hazard
Iodine monobromide: mixture is a dangerous explosion hazard
Iodine monochloride: mixture is a dangerous explosion hazard
Iodine pentafluoride: mixture is a dangerous explosion hazard
Iron: mixture is a dangerous explosion hazard
K_3N: mixture is a dangerous explosion hazard
Lanthanum: mixture is a dangerous explosion hazard
Lead dioxide: mixture is a dangerous explosion hazard
Lithium: mixture is a dangerous explosion hazard
Lithium carbide: mixture is a dangerous explosion hazard
Li_6CS: mixture is a dangerous explosion hazard
Magnesium bromate: mixture is a dangerous explosion hazard
Magnesium chlorate: mixture is a dangerous explosion hazard
Magnesium iodate: mixture is a dangerous explosion hazard
Manganese: mixture is a dangerous explosion hazard
Manganese perchlorate: mixture is a dangerous explosion hazard
Mercuric oxide: mixture is a dangerous explosion hazard
Mercury nitrate: mixture is a dangerous explosion hazard
Neodymium: mixture is a dangerous explosion hazard
Nickel: mixture is a dangerous explosion hazard
Nitrates: mixture is a dangerous explosion hazard
Nitrogen bromide: mixture is a dangerous explosion hazard
Nitrogen chloride: mixture is a dangerous explosion hazard
Nitrogen dioxide: mixture is a dangerous explosion hazard
Nitrogen tribromide: mixture is a dangerous explosion hazard
Nitrogen trichloride: mixture is a dangerous explosion hazard
NOF: mixture is a dangerous explosion hazard
Oxidizing agents: incompatible
Oxygen: mixture is a dangerous explosion hazard
Performic acid: mixture is a dangerous explosion hazard
Peroxyformic acid: mixture is a dangerous explosion hazard
Platinum: mixture is a dangerous explosion hazard
Potassium: mixture is a dangerous explosion hazard
Potassium bromate: mixture is a dangerous explosion hazard
Potassium chlorate: mixture is a dangerous explosion hazard
Potassium hydroxide: mixture is a dangerous explosion hazard
Potassium iodate: mixture is a dangerous explosion hazard
Potassium permanganate: mixture is a dangerous explosion hazard
Rubidium: mixture is a dangerous explosion hazard
$RbHC_2$: mixture is a dangerous explosion hazard
Selenium chloride: mixture is a dangerous explosion hazard
Selenium tetrafluoride: mixture is a dangerous explosion hazard
Selenium hypochlorite: mixture is a dangerous explosion hazard
$SeOF_2$: mixture is a dangerous explosion hazard
Silver nitrate: mixture is a dangerous explosion hazard

Silver oxide: mixture is a dangerous explosion hazard
Sodium: mixture is a dangerous explosion hazard
Sodium bromate: mixture is a dangerous explosion hazard
Na_2C_2: mixture is a dangerous explosion hazard
Sodium chlorate: mixture is a dangerous explosion hazard
Sodium hydroxide: mixture is a dangerous explosion hazard
Sodium hypochlorite: mixture is a dangerous explosion hazard
Sodium iodate: mixture is a dangerous explosion hazard
Sodium peroxide: mixture is a dangerous explosion hazard
Sulfur: mixture is a dangerous explosion hazard
Sulfur trioxide: mixture is a dangerous explosion hazard
Sulfuric acid: mixture is a dangerous explosion hazard
Thorium: mixture is a dangerous explosion hazard
Vanadium oxychloride: mixture is a dangerous explosion hazard
Zinc bromate: mixture is a dangerous explosion hazard
Zinc chlorate: mixture is a dangerous explosion hazard
Zinc iodate: mixture is a dangerous explosion hazard
Zirconium: mixture is a dangerous explosion hazard

When heated to decomposition, it emits highly toxic fumes of PO_x.

FIRST AID   In case of contact with the eyes, immediately wash the eyes with large amounts of water, occasionally lifting the lower and upper lids. Get medical attention immediately. Contact lenses should not be worn when working with this chemical.

In case of contact with the skin, immediately flush the contaminated skin with water. If this chemical penetrates the clothing, immediately remove the clothing and flush the skin with water. Get medical attention promptly.

If a person breathes in large amounts of this chemical, move the victim to fresh air at once. If breathing has stopped, perform artificial respiration. Keep the affected person warm and at rest. Get medical attention as soon as possible.

If this chemical has been swallowed, get medical attention immediately.

FIRE HAZARD   Ignites spontaneously in air. Dangerous fire hazard when exposed to heat or by chemical reaction with oxidizers. If combustion occurs in a confined space, it will remove the oxygen and cause asphyxiation.

To extinguish fires, use water.

HANDLING   This very flammable solid must be kept under water.

Wear appropriate clothing and equipment to prevent any possibility of skin contact. Wash immediately if skin becomes contaminated. Immediately remove any clothing that becomes contaminated.

Wear eye protection to prevent any possibility of eye contact.

The following equipment should be available: an eyewash station and a quick-drench shower.

PHOSPHORUS PENTOXIDE

CAS #1314-56-3   mf: O_5P_2
DOT #1807   mw: 141.94

SYNONYMS
Diphosphorus pentoxide
Phosphoric anhydride
Phosphorus(V) oxide
POX

PROPERTIES   Deliquescent crystals; density 2.30; melting point 340°C; sublimes at 360°C

DOT Classification   Corrosive material; Label: Corrosive

EPA Extremely Hazardous Substances List and EPA TSCA Inventory

TOXIC EFFECTS   Poisonous by inhalation. A corrosive irritant to the eyes, skin, and mucous membranes.

HAZARDOUS REACTIONS
Bases (inorganic): incompatible
Bromine pentafluoride: incompatible
Chlorine trifluoride: incompatible
Formic acid: incompatible
Hydrogen fluoride: incompatible
Hydrogen peroxide: incompatible
Iodides: incompatible
Metals: incompatible
Methyl hydroperoxide: incompatible
Oxidants: incompatible
Oxygen difluoride: incompatible
Perchloric acid: incompatible
3-Propynol: incompatible
Water: incompatible

PICRIC ACID

CAS #88-89-1
DOT #0154
mf: $C_6H_3N_3O_7$
mw: 229.12

STRUCTURE
$HOC_6H_2(NO_2)_3$

NFPA LABEL

4 / 2 / 4

SYNONYMS
Carbazoic acid
2-Hydroxy-1,3,5-trinitrobenzene
Melinite
Nitroxanthic acid
Phenol trinitrate
Picronitric acid
1,3,5-Trinitrophenol
2,4,6-Trinitrophenol

PROPERTIES   Colorless to yellow solid or liquid; very bitter taste; solubility in water 1.4%; melting point 121.8°C; explodes >300°C; flash point 150°C; autoignition temperature 300°C; density 1.763; vapor density 7.90

OSHA PEL   TWA 100 µg/m³ (skin); IDLH 100 mg/m³

ACGIH TLV   TWA 0.1 mg/m³

DOT Classification   Class A Explosive

Community Right to Know List and EPA TSCA Inventory

TOXIC EFFECTS   Poisonous by ingestion. An irritant and an allergen. Skin contact can cause local and systemic allergic reactions.

Routes   Inhalation, contact, ingestion

Symptoms   Irritated eyes; sensitization dermatitis; yellow-stained hair and teeth; weakness, muscle pain; low urine output, high urine output, blood in the urine, albumin in the urine, nephritis; bitter taste, gastrointestinal effects; hepatitis

Target Organs   Kidneys, liver, blood, skin, eyes

HAZARDOUS REACTIONS   Very unstable. A severe explosion hazard when shocked or exposed to heat. It easily forms picrate salts, which are more sensitive explosives than picric acid.

Aluminum + water: mixtures ignite after a delay period
Ammonia: forms unstable, possibly explosive salts
Bases: forms unstable, possibly explosive salts
Concrete: forms unstable, possibly explosive salts
Copper: forms unstable, possibly explosive salts
Lead: forms unstable, possibly explosive salts
Mercury: forms unstable, possibly explosive salts
Metals: forms unstable, possibly explosive salts
Plaster: incompatible
Reducing materials: vigorous reaction
Salts: incompatible
Uranium perchlorate: mixtures are extremely powerful explosives
Zinc: forms unstable, possibly explosive salts

Upon decomposition, it emits highly toxic fumes and explodes.

FIRST AID   In case of contact with the eyes, immediately wash the eyes with large amounts of water, occasionally lifting the lower and upper lids. Get medical attention immediately. Contact lenses should not be worn when working with this chemical.

In case of contact with the skin, promptly wash the contaminated skin with soap and water. If this chemical penetrates through the clothing, promptly remove the clothing and wash the skin with soap and water. Get medical attention promptly.

If a person breathes in large amounts of this chemical, move the victim to fresh air at once. If breathing has stopped, perform artificial respiration. Keep the affected person warm and at rest. Get medical attention as soon as possible.

If this chemical has been swallowed, get medical attention immediately.

FIRE HAZARD   Combustible when exposed to heat or flame.

HANDLING   Wear appropriate clothing and equipment to prevent skin contact. Wash promptly if skin becomes contaminated and at the end of each work shift. Promptly remove any nonimpervious clothing that becomes contaminated. Work clothing should be changed daily if it is at all probable that the clothing has been contaminated.

Wear eye protection to reduce probability of eye contact.

POTASSIUM

CAS #7440-09-7 mf: K
DOT #2257 mw: 39.10

NFPA LABEL

PROPERTIES   Soft, ductile, silvery-white, very reactive metal; melting point 63.65°C, boiling point 774°C; density 0.862 at 20°C

DOT Classification   Label: Flammable solid and dangerous when wet

EPA TSCA Inventory

HAZARDOUS REACTIONS   A storage hazard. Reaction with moisture leads to ignition and can result in an explosion. Forms an unstable, explosive peroxide and superoxide coating when stored under mineral oil. Handling the stored metal can lead to explosions.

Acid fumes: violent reaction
Air: potential explosive reaction
Alcohols: potential explosive reaction
Aluminum bromides: potential explosive reaction
Aluminum chlorides: potential explosive reaction
Aluminum fluorides: potential explosive reaction
Aluminum tribromide: potential explosive reaction
Ammonia + sodium nitrite: potential explosive reaction
Ammonium bromide: potential explosive reaction
Ammonium chlorocuprate: potential explosive reaction
Ammonium iodide: potential explosive reaction
Ammonium nitrate: potential explosive reaction
Ammonium nitrate + ammonium sulfate: potential explosive reaction
Ammonium tetrachlorocuprate: potential explosive reaction
Antimony halides: potential explosive reaction
Antimony oxide: potential explosive reaction
Antimony tribromides: potential explosive reaction
Antimony trichlorides: potential explosive reaction
Antimony triiodides: potential explosive reaction
Arsenic halides: potential explosive reaction
Arsenic trichloride: potential explosive reaction
Arsenic triiodide: potential explosive reaction
Arsine + ammonia: potential explosive reaction
Benzyl alcohol: potential explosive reaction
Bismuth tribromides: potential explosive reaction
Bismuth trichlorides: potential explosive reaction
Bismuth triiodides: potential explosive reaction
Bismuth trioxide: potential explosive reaction
Boric acid: potential explosive reaction
Boron tribromide: potential explosive reaction
Bromine vapor: incompatible
Bromoform: potential explosive reaction
Cadmium bromides: potential explosive reaction
Cadmium chlorides: potential explosive reaction
Cadmium iodides: potential explosive reaction
Calcium bromide: potential explosive reaction
Carbon: incompatible
Carbon dioxide (solid): potential explosive reaction
Carbon disulfide: mixture is an impact-sensitive explosive
Carbon monoxide: potential explosive reaction
Carbon tetrachloride: potential explosive reaction
Charcoal (activated): incompatible
Chlorinated hydrocarbons: potential explosive reaction
Chlorine: incompatible
Chlorine oxide: potential explosive reaction
Chlorine trifluoride: potential explosive reaction
Chloroethane: potential explosive reaction
Chloroform: potential explosive reaction
Chromium tetrachloride: potential explosive reaction
Chromium trioxide: potential explosive reaction
Cobalt(II) chloride: potential explosive reaction
$COCl_2$: potential explosive reaction
Copper hypochlorite: potential explosive reaction
Copper oxide: potential explosive reaction
Copper(I) chloride: potential explosive reaction
Copper(I) iodide: potential explosive reaction
Copper(II) bromide: potential explosive reaction
Copper(II) chloride: potential explosive reaction
Copper(II) oxide: incompatible
Cyclohexanol: potential explosive reaction
Dibromomethane: potential explosive reaction
Dichlorine oxide: potential explosive reaction
Dichloroethane: potential explosive reaction
Dichloromethane: potential explosive reaction
Diiodomethane: potential explosive reaction
Dimethyl sulfoxide: incompatible
Dinitrogen pentaoxide: potential explosive reaction
Dinitrogen tetraoxide: potential explosive reaction
Diselenium dichloride: potential explosive reaction
Disulfur dichloride: potential explosive reaction
Ethylene oxide: potential explosive reaction
Fluorine: potential explosive reaction
Graphite: potential explosive reaction

Graphite + air: potential explosive reaction
Graphite + potassium peroxide: potential explosive reaction
Halocarbons: potential explosive reaction
Hydrazine: potential explosive reaction
Hydrogen bromide: incompatible
Hydrogen chloride: potential explosive reaction
Hydrogen iodide: potential explosive reaction
Hydrogen peroxide: potential explosive reaction
Interhalogens: potential explosive reaction
Iodine: mixture is an impact-sensitive explosive
Iodine bromide: potential explosive reaction
Iodine chloride: potential explosive reaction
Iodine pentafluoride: potential explosive reaction
Iodine trichloride: potential explosive reaction
Iron(II) bromide: potential explosive reaction
Iron(II) chloride: potential explosive reaction
Iron(II) iodide: potential explosive reaction
Iron(III) bromide: potential explosive reaction
Iron(III) chloride: potential explosive reaction
Lead dioxide: potential explosive reaction
Lead hypochlorite: potential explosive reaction
Lead peroxide: potential explosive reaction
Lead sulfate: potential explosive reaction
Maleic anhydride: potential explosive reaction
Manganese(II) chloride: potential explosive reaction
Mercury: incompatible
Mercury(I) oxide: potential explosive reaction
Mercury(II) bromide: potential explosive reaction
Mercury(II) chloride: potential explosive reaction
Mercury(II) fluoride: potential explosive reaction
Mercury(II) iodide: potential explosive reaction
Mercury(II) oxide: incompatible
Metal halides: potential explosive reaction
Metal oxides: potential explosive reaction
Moisture: violent reaction
Molybdenum trioxide: potential explosive reaction
Molybdenum(III) oxide: incompatible
Nickel bromide: potential explosive reaction
Nickel chloride: potential explosive reaction
Nickel iodide: potential explosive reaction
Nitric acid: potential explosive reaction
Nitrobenzene: potential explosive reaction
Nitrogen dioxide: potential explosive reaction
Nitrogen-containing explosives: potential explosive reaction
Nonmetal halides: potential explosive reaction
Nonmetal oxides: potential explosive reaction
n-Octanol: potential explosive reaction
Oxalyl dibromide: potential explosive reaction
Oxalyl dichloride: potential explosive reaction
Oxidizers: violent reaction
Pentachloroethane: potential explosive reaction
Peroxides: potential explosive reaction
Phosgene: potential explosive reaction
Phosphine + ammonia: potential explosive reaction
Phosphorus: potential explosive reaction
Phosphorus pentachloride: potential explosive reaction
Phosphorus tribromide: potential explosive reaction
Phosphorus trichloride: potential explosive reaction
P_2NF: potential explosive reaction
P_2O_5: potential explosive reaction
Picric acid: potential explosive reaction
Potassium chlorocuprate: potential explosive reaction
Potassium iodide + magnesium bromide, chloride, or iodide: incompatible
Potassium oxides: potential explosive reaction
n-Propanol: potential explosive reaction
Seleninyl bromide: potential explosive reaction
Seleninyl chloride: potential explosive reaction
Selenium: potential explosive reaction
Selenium hypochlorite: potential explosive reaction
Silicon tetrachloride: potential explosive reaction
Silver fluoride: potential explosive reaction
Silver iodate: potential explosive reaction
Sodium iodate: potential explosive reaction
Sodium peroxide: potential explosive reaction
Soot: incompatible
Sulfur: potential explosive reaction
Sulfur dibromide: potential explosive reaction
Sulfur dichloride: potential explosive reaction
Sulfur dioxide: incompatible
Sulfuric acid: potential explosive reaction
Tellurium: potential explosive reaction
Tetrachloroethane: potential explosive reaction
Thallium(I) bromide: potential explosive reaction
Thiophosphoryl fluoride: potential explosive reaction
Tin chlorides: potential explosive reaction
Tin iodide: potential explosive reaction
Tin tetraiodide + sulfur: potential explosive reaction
Tin(IV) oxide: incompatible
Trichloroethane: potential explosive reaction
Vanadium hypochlorite: potential explosive reaction
Vanadium(V) chloride: potential explosive reaction
Water: potentially explosive reaction forms potassium hydroxide and hydrogen
Zinc bromide: potential explosive reaction
Zinc chloride: potential explosive reaction
Zinc iodide: potential explosive reaction

FIRE HAZARD   In the presence of moist air, potassium can spontaneously catch fire and burn intensely or even explode.

Burning potassium is difficult to extinguish; dry powdered soda ash or graphite or special mixtures of dry chemicals are recommended.

POTASSIUM BICHROMATE

CAS #7778-50-9   mf: $Cr_2K_2O_7$
DOT #1479   mw: 294.20

NFPA LABEL

SYNONYMS
Bichromate of potash
Dipotassium dichromate
Iopezite
Kaliumdichromat (German)
Potassium dichromate (VI)

PROPERTIES   Bright, yellowish-red, transparent crystals, bitter, metallic taste; melting point 398°C; decomposes at 500°C; density 2.69

NIOSH REL   TWA 25 µg [Cr(VI)]/m^3; ceiling 50 µg/m^3/15 min

DOT Classification   ORM-A; Label: None

EPA TSCA Inventory

TOXIC EFFECTS   Poisonous by ingestion and some other routes. A human carcinogen.

HAZARDOUS REACTIONS   A powerful oxidizer.

Sulfuric acid + acetone: violent reaction
Hydrazine: violent reaction
Hydroxylamine: violent reaction

FIRE HAZARD   Flammable by chemical reaction.

HANDLING   See Chap. 3 for information on handling carcinogens.

POTASSIUM CHLORATE

CAS #3811-04-9   mf: $ClO_3 \cdot K$
DOT #2427/1485   mw: 122.55

NFPA LABEL

SYNONYMS
Chlorate de potassium (French)
Chlorate of potash
Fekabit
Kaliumchloraat (Dutch)
Kaliumchlorat (German)
Oxymuriate of potash
Potash chlorate (DOT)
Potassio (chlorato di) (Italian)
Potassium chlorate (DOT)
Potassium (chlorate de) (French)
Potassium oxymuriate

PROPERTIES   Transparent, colorless crystals or white powder with a cooling, saline taste; melting point 368.4°C; decomposes at 400°C; density 2.32

DOT Classification   Label: Oxidizer

EPA TSCA Inventory

TOXIC EFFECTS   Moderately toxic by ingestion. A gastrointestinal tract and kidney irritant. Can cause hemolysis of red blood cells and methemoglobinemia.
The toxic dose by ingestion is about 5 g.

HAZARDOUS REACTIONS   May explode on heating. A powerful oxidizer. Very reactive material. It has caused many industrial explosions.

Agricultural materials: forms sensitive explosive mixtures
Aluminum dust: forms sensitive explosive mixtures
Aluminum + antimony trisulfide powders: forms sensitive explosive mixtures
Ammonia: violent or explosive reaction
Ammonia + salts: violent or explosive reaction
Ammonium chloride: violent or explosive reaction
Ammonium sulfate: violent or explosive reaction
Ammonium thiocyanate: forms sensitive explosive mixtures
Antimony trisulfide: violent or explosive reaction
Aqua regia + ruthenium: explosive reaction
Arsenic: violent or explosive reaction
Arsenic trisulfide: forms sensitive explosive mixtures
Barium hypophosphite: violent or explosive reaction
Barium phosphinate: forms sensitive explosive mixtures
Barium sulfide: violent or explosive reaction
Barium thiocyanate: forms sensitive explosive mixtures
Boron powder: forms sensitive explosive mixtures
Calcium hydride: forms sensitive explosive mixtures
Calcium hypophosphite: violent or explosive reaction

Calcium phosphinate: forms sensitive explosive mixtures
Calcium sulfide: violent or explosive reaction
Carbon: forms sensitive explosive mixtures
Charcoal: violent or explosive reaction
Charcoal + potassium nitrate + sulfur: forms sensitive explosive mixtures
Charcoal + sulfur: forms sensitive explosive mixtures
Chromium dust: forms sensitive explosive mixtures
Cobalt + sulfur: forms sensitive explosive mixtures
Cobalt oxide + sulfur: forms sensitive explosive mixtures
Copper dust: forms sensitive explosive mixtures
Copper chlorate + sulfur: forms sensitive explosive mixtures
Copper nitride + sulfur: forms sensitive explosive mixtures
Copper sulfate + sulfur: forms sensitive explosive mixtures
Cu_3P_2: violent or explosive reaction
Cyanides: forms sensitive explosive mixtures
Cyanoguanidine: forms sensitive explosive mixtures
Dinickel trioxide: violent or explosive reaction
Fabrics: violent or explosive reaction
Fluorine: reaction forms explosive fluorine perchlorate gas
Gallic acid: violent or explosive reaction
Germanium dust: forms sensitive explosive mixtures
Glucose: forms sensitive explosive mixtures
Hydrocarbons: forms sensitive explosive mixtures
Hydrogen iodide: violent or explosive reaction
Lactose: violent or explosive reaction
Magnesium dust: forms sensitive explosive mixtures
Magnesium + anhydrous copper sulfate + ammonium nitrate + water: violent or explosive reaction
Manganese dioxide: violent or explosive reaction
Manganese dioxide + potassium hydroxide: forms sensitive explosive mixtures
Manganese dioxide + traces of organic matter: forms sensitive explosive mixtures
Metal (finely divided): forms sensitive explosive mixtures
Metal + wood: forms sensitive explosive mixtures
Metal hypophosphites: violent or explosive reaction
Metal phosphides: forms sensitive explosive mixtures
Metal phosphinates: forms sensitive explosive mixtures
Metal sulfides: forms sensitive explosive mixtures
Metal thiocyanates: forms sensitive explosive mixtures
Nitric acid + organic materials: forms sensitive explosive mixtures
Nonmetals (powdered): forms sensitive explosive mixtures
Organic acids (dibasic): violent or explosive reaction
Organic matter: violent or explosive reaction
Peat: forms sensitive explosive mixtures
Phosphorus powder: forms sensitive explosive mixtures
Potassium hydroxide: violent or explosive reaction
Reducing agents: forms sensitive explosive mixtures
Sawdust: forms sensitive explosive mixtures
Silver sulfide: forms sensitive explosive mixtures
Sodium amide: violent or explosive reaction
Sodium phosphinate: forms sensitive explosive mixtures
Steel dust: forms sensitive explosive mixtures
Strontium hydride: forms sensitive explosive mixtures
Sucrose: violent or explosive reaction
Sugars: forms sensitive explosive mixtures
Sugar + sulfuric acid: violent or explosive reaction
Sulfur: forms sensitive explosive mixtures
Sulfur + metal derivatives: forms sensitive explosive mixtures
Sulfur dioxide: violent or explosive reaction
Sulfur dioxide solutions in ether or ethanol: explosive reaction
Sulfuric acid: violent or explosive reaction
Tannic acid: forms sensitive explosive mixtures
Thiocyanates: violent or explosive reaction
Thiuram: forms sensitive explosive mixtures
Thorium dicarbide: violent or explosive reaction
Titanium dust: forms sensitive explosive mixtures
Tricopper diphosphide: forms sensitive explosive mixtures
Trimercury tetraphosphide: forms sensitive explosive mixtures
Zinc dust: forms sensitive explosive mixtures
Zirconium dust: forms sensitive explosive mixtures

POTASSIUM CYANIDE

CAS #151-50-8 mf: CN·K
DOT #1680 mw: 65.12

NFPA LABEL

SYNONYMS
Cyanide of potassium (solution)
Cyanure de potassium (French)
Hydrocyanic acid, potassium salt (solution)
Potassium cyanide solution (DOT)

PROPERTIES   The solution in water is colorless; slight odor of bitter almonds

EPA TSCA Inventory

OSHA PEL   TWA 5 mg(CN)/m^3 (skin)

ACGIH TLV   TWA 5 mg(CN)/m^3

NIOSH REL   ceiling 5 mg(CN)/m^3/10 min

DOT Classification   Poison B; Label: Poison

TOXIC EFFECTS   A deadly human poison by ingestion. An experimental poison by ocular, subcutaneous, intravenous, intramuscular, and intraperitoneal routes. Experimental teratogenic and reproductive effects.

HAZARDOUS REACTIONS

Acids or acid fumes: reaction forms deadly HCN gas

When heated to decomposition, it emits very toxic fumes of CN$^-$ and NO$_x$.

POTASSIUM FLUORIDE

CAS #7789-23-3   mf: FK
DOT #1812   mw: 58.10

SYNONYMS
Fluorure de potassium (French)
Potassium fluoride (DOT)
Potassium fluorure (French)

PROPERTIES   White, crystalline, deliquescent powder; sharp saline taste; melting point 859.9°C, boiling point 1500°C; very soluble in boiling water density 2.48; vapor pressure 1 mm Hg at 885°C

OSHA PEL   TWA 2500 μg(F)/m^3

NIOSH REL   TWA 2.5 mg(F)/m^3

DOT Classification   Solid: Corrosive material; Label: Corrosive; Solution: ORM-B; Label: None

EPA TSCA Inventory

TOXIC EFFECTS   Poisonous by ingestion and some other routes. A corrosive irritant to the eyes, skin, and mucous membranes. A teratogen.

HAZARDOUS REACTIONS   A very reactive and irritating material.
When heated to decomposition, it emits toxic fumes of F$^-$.

POTASSIUM HYDROXIDE

CAS #1310-58-3   mf: KOH
DOT #1813/1814   mw: 56.11

NFPA LABEL

SYNONYMS
Caustic potash
Hydroxyde de potassium (French)
Kaliumhydroxid (German)
Kaliumhydroxyde (Dutch)
Lye
Potassa
Potasse caustique (French)
Potassio (idrossido di) (Italian)
Potassium hydrate

PROPERTIES   White, deliquescent pieces, lumps or sticks having crystalline fracture; melting point 360°C ± 7°C, boiling point 1320°C; density 2.044; violent exothermic reaction with water

ACGIH TLV   Ceiling 2 mg/m^3

DOT Classification   Corrosive material; Label: Corrosive; Corrosive material; Label: Corrosive, solution

EPA TSCA Inventory

TOXIC EFFECTS   Poisonous by ingestion. A severe eye and skin irritant. Very corrosive to the eyes, skin, and mucous membranes. Ingestion may cause violent pain in throat and epigastrium, hematemesis, collapse. Stricture of esophagus may result if not immediately fatal.

HAZARDOUS REACTIONS

Acids: violent reaction or ignition under some conditions
Alcohols: violent reaction or ignition under some conditions
Ammonium hexachloroplatinate (2$^-$); when heated, reaction forms a heat-sensitive explosive product
p-Bis (1,3-dibromoethyl) benzene: violent reaction or ignition under some conditions
Bromoform + crown ethers: potentially explosive reaction
Chlorine dioxide: potentially explosive reaction

...olent reaction or ignition under
...reaction or ignition under some
...lent reaction or ignition under
...conditions
Maleic anhydride: violent reaction or ignition under some conditions
Nitroalkanes: violent reaction or ignition under some conditions
Nitrobenzene: potentially explosive reaction
Nitrogen trichloride: potentially explosive reaction
Nitromethane: potentially explosive reaction
2-Nitrophenol: violent reaction or ignition under some conditions
Potassium peroxodisulphate: violent reaction or ignition under some conditions
Sugars: violent reaction or ignition under some conditions
Sugars (reducing): reaction above 84°C forms carbon monoxide gas
Tetrahydrofuran (peroxidized): potentially explosive reaction
2,2,3,3-Tetrafluoropropanol: violent reaction or ignition under some conditions
Thorium dicarbide: violent reaction or ignition under some conditions
2,4,6-Trinitrotoluene: potentially explosive reaction
Water: violent, exothermic reaction

POTASSIUM PERCHLORATE

CAS #7778-74-7   mf: $ClO_4 \cdot K$
DOT #1489   mw: 138.55

NFPA LABEL

SYNONYMS
Periodin
Potassium hyperchloride

PROPERTIES   Colorless crystals or white powder; insoluble in alcohol; decomposes at 400°C and with organic matter; density 2.52; melting point 610°C ± 10°C

DOT Classification   Label: Oxidizer

EPA TSCA Inventory

TOXIC EFFECTS   A severe irritant to skin, eyes, and mucous membranes. Absorption can cause methemoglobinemia and kidney injury.

HAZARDOUS REACTIONS   A powerful oxidizer. Has been involved in many industrial explosions.

Aluminum + aluminum fluoride: violent reaction or ignition under some conditions
Aluminum + barium nitrate + potassium nitrate + water: explodes on contact
Aluminum + magnesium: violent reaction or ignition under some conditions
Aluminum powder: mixture is explosive
Aluminum powder + titanium dioxide: mixture is explosive
Barium chromate + tungsten or titanium: violent reaction or ignition under some conditions
Boron + magnesium + silicone rubber: violent reaction or ignition under some conditions
Charcoal: violent reaction or ignition under some conditions
Cotton lint: mixture is explosive above 245°C
Ethanol: heated reaction forms the explosive ethyl perchlorate
Ethylene glycol: mixture is explosive above 240°C
Ferrocenium diamminetetrakis (thiocyanato-N) chromate (1^-): violent reaction or ignition under some conditions
Fluorine: violent reaction or ignition under some conditions
Furfural: mixture is explosive above 270°C
Iron powder: mixture is explosive
Lactose: mixture is explosive
Magnesium powder: mixture is explosive
Metal powders: mixture is explosive
Molybdenum powder: mixture is explosive
Nickel + titanium: violent reaction or ignition under some conditions
Nickel powder: mixture is explosive
Potassium hexacyanocobaltate (3^-): violent reaction or ignition under some conditions
Reducing agents: violent reaction or ignition under some conditions
Sulfur: mixture is explosive
Tantalum powder: mixture is explosive
Titanium hydride: mixture is explosive
Titanium powder: mixture is explosive

POTASSIUM PERMANGANATE

CAS #7722-64-7
DOT #1490
mf: $MnO_4 \cdot K$
mw: 158.04

NFPA LABEL

SYNONYMS
Chameleon mineral
Condy's crystals
Kaliumpermanganaat (Dutch)
Kaliumpermanganat (German)
Permanganate de potassium (French)
Permanganate of potash
Potassio (permanganato di) (Italian)
Potassium (permanganate de) (French)
Potassium permanganate (DOT)

PROPERTIES   Dark purple crystals with a blue metallic sheen; sweetish astringent taste; decomposes before melting at <240°C; density 2.703

ACGIH TLV   Ceiling 5 mg(Mn)/m³

DOT Classification   Oxidizer; Label: Oxidizer

EPA TSCA Inventory and Community Right to Know List (manganese and its compounds)

TOXIC EFFECTS   A poison by ingestion. A strong irritant by chemical reaction.

Symptoms (ingestion)   Dyspnea, nausea, other gastrointestinal effects

HAZARDOUS REACTIONS   A powerful oxidizer. A dangerous explosion hazard.

Acetaldehyde: ignites on contact
Acetic acid: explodes on contact
Acetic anhydride: explodes on contact
Acetone + *tert*-butylamine: violent reaction or ignition under some conditions
Acetylacetone: ignites on contact
Alcohols + nitric acid: violent reaction or ignition under some conditions
Aluminum carbide: ignites on contact
Aluminum powder + ammonium nitrate + glyceryl nitrate + nitrocellulose: mixture is explosive
Ammonia: ignites on contact
Ammonia + sulfuric acid: violent reaction or ignition under some conditions
Ammonium hydroxide: ignites on contact
Ammonium nitrate: explodes on contact
Ammonium perchlorate: ignites on contact
Antimony: violent reaction or ignition under some conditions
Arsenic: mixture is explosive
Benzaldehyde: ignites on contact
Carbon: violent reaction or ignition under some conditions
3-Chloropropane-1,2-diol: ignites on contact
Coal + peroxomonosulfuric acid: violent reaction or ignition under some conditions
Dichloromethylosilane: violent reaction or ignition under some conditions
Dimethyl sulfoxide: ignites on contact
Dimethylformamide: explodes on contact
Erythritol: ignites on contact
Ethylene glycol: ignites on contact
Ethylene glycol esters: ignites on contact
Ethanol + sulfuric acid: violent reaction or ignition under some conditions
Formaldehyde: explodes on contact
Glycerol: violent reaction or ignition under some conditions
H_2S_3: ignites on contact
Hydrochloric acid (concentrated): explodes on contact
Hydrofluoric acid (concentrated): violent reaction or ignition under some conditions
Hydrogen peroxide: violent reaction or ignition under some conditions
Hydrogen trisulfide: violent reaction or ignition under some conditions
Hydroxylamine: violent reaction or ignition under some conditions
Isobutyraldehyde: ignites on contact
Lactic acid: ignites on contact
Mannitol: ignites on contact
Nitrates: ignites on contact
Nitro compounds (organic): violent reaction or ignition under some conditions
Organic materials (dry or in solution): may explode on contact
Oxalic acid: ignites on contact
Oxidizable materials (dry or in solution): may explode on contact
Oxygenated organic compounds: ignites on contact
Phosphorus: mixture is explosive
Polypropylene: violent reaction or ignition under some conditions
Potassium chloride + sulfuric acid: explodes on contact

Propane-1,2-diol: ignites on contact
Slag wool: mixture is explosive
Sulfur: mixture is explosive
Sulfuric acid: ignites on contact
Sulfuric acid + organic matter: ignites on contact
Sulfuric acid + potassium chloride: ignites on contact
Sulfuric acid + water: explodes on contact
Titanium: mixture is explosive
Triethanolamine: ignites on contact
3,4,4'-Trimethyldiphenyl sulfone: violent reaction or ignition under some conditions
Wood: ignites on contact

FIRE HAZARD   Flammable by chemical reaction.

PROPANE

CAS #74-98-6   mf: C_3H_8
DOT #1075; 1978   mw: 44.11

STRUCTURE   NFPA LABEL
$CH_3CH_2CH_3$

SYNONYMS
Dimethylmethane
Propyl hydride

PROPERTIES   Colorless, odorless gas (a foul-smelling odorant is often added); insoluble in water; soluble in alcohol, ether; freezing point $-187.1°C$, boiling point $-42.1°C$; lower explosive limit 2.2%, upper explosive limit 9.5%; flash point $-104.4°C$; density 0.5852 at $-44.5°C/4°C$; autoignition temperature 450°C; vapor density 1.56

OSHA PEL   TWA 1000 ppm; IDLH 20,000 ppm

ACGIH TLV   Asphyxiant

DOT Classification   Flammable gas; Label: Flammable gas

EPA TSCA Inventory

TOXIC EFFECTS   An asphyxiant. Central nervous system effects at high concentrations.

Routes   Inhalation, contact

Symptoms   Dizziness, disorientation, excitation, frostbite

Target Organs   Central nervous system

HARZARDOUS REACTIONS
Barium peroxide: violent exothermic reaction when heated
Chlorine dioxide: explosive reaction
Oxidizing materials: vigorous reaction

FIRST AID   In case of contact with the eyes, immediately wash the eyes with large amounts of water, occasionally lifting the lower and upper lids. Get medical attention immediately. Contact lenses should not be worn when working with this chemical.

In case of contact with the skin, immediately flush the contaminated skin with water. If this chemical penetrates the clothing, immediately remove the clothing and flush the skin with water. Get medical attention promptly.

If a person breathes in large amounts of this chemical, move the victim to fresh air at once. If breathing has stopped, perform artificial respiration. Keep the affected person warm and at rest. Get medical attention as soon as possible.

FIRE HAZARD   A highly dangerous fire hazard when exposed to heat, flame, or oxidizers. Explosive in the form of vapor when exposed to heat or flame.

To extinguish fires, stop flow of gas.

HANDLING   When using propane, the following must be borne in mind:

1. Mixtures of pure propane and air containing from about 2% to 10% of propane by volume are inflammable. However, commercial propane often contains other products which may extend the above range of flammability. Detonation may occur throughout the whole range. When working with propane, the most obvious precaution is to avoid the formation of inflammable mixtures.

2. The ignition temperature of a propane-air mixture can be as low as 481°C.

3. One liter of liquid propane evaporates to about 280 L of propane gas at normal pressure and temperature.

4. Propane is 1.5 times as heavy as air.

5. At 20°C the vapor pressure of propane is 120 psi; at the critical point the temperature is 97°C and the pressure 650 psi.

Under no circumstance should propane be mixed with air inside buildings in appreciable quantities.

To render safe an apparatus which has contained propane, the residual propane vapor must be removed by some safe procedure, for example, by exhausting or displacing the propane (without oxygen) into a properly vented system. Any operation, such as heating, which results in release of additional propane vapor from porous or absorbent parts of the apparatus must be carried out in a safe manner so that no dangerous mixture with oxygen results.

Wear appropriate clothing and equipment to prevent skin freezing. Immediately remove any clothing that becomes wet to avoid flammability hazard.

Wear eye protection to reduce probability of eye contact.

PROPYL ALCOHOL

CAS #71-23-8   mf: C_3H_8O
DOT #1274   mw: 60.11

NFPA LABEL

SYNONYMS

Alcool propilico (Italian)	Propanolen (Dutch)
Alcool propylique (French)	Propanoli (Italian)
Ethyl carbinol	n-Propyl alcohol
n-Propanol	1-Propyl alcohol
Propanol-1	n-Propyl alkohol (German)
1-Propanol	Propylic alcohol
Propanole (German)	Propylowy alkohol (Polish)

PROPERTIES   Clear liquid with a mild, nonresidual alcoholic odor; miscible in water, alcohol, ether; melting point −127°C, boiling point 97.19°C; flash point 15°C (closed cup); Underwriters Laboratory code 55–60; density 0.8044 at 20°C/4°C; lower explosive limit 2%, upper explosive limit 14%; autoignition temperature 440°C; vapor pressure 10 mm Hg at 14.7°C; vapor density 2.07

OSHA PEL   TWA 200 ppm; IDLH 4000 ppm

ACGIH TLV   TWA 200 ppm (skin); STEL 250 ppm

DOT Classification   Label: Flammable liquid

EPA TSCA Inventory

TOXIC EFFECTS   Moderately toxic by inhalation, ingestion, and several other routes. Mildly toxic by skin contact. A skin and severe eye irritant. A carcinogen.

Routes   Inhalation, contact, ingestion

Symptoms   Mild irritation of the eyes, nose, and throat; dry, cracking skin; drowsiness, headache; ataxia; gastrointestinal pain; abdominal cramps; nausea, vomiting, diarrhea

Target Organs   Skin, eyes, respiratory system, gastrointestinal tract

HAZARDOUS REACTIONS
Oxidizing agents: incompatible
Potassium-*tert*-butoxide: violent reaction

FIRST AID   In case of contact with the eyes, immediately wash the eyes with large amounts of water, occasionally lifting the lower and upper lids. Get medical attention immediately. Contact lenses should not be worn when working with this chemical.

If this chemical comes in contact with the skin, immediately flush the contaminated skin with water. If it penetrates the clothing, immediately remove the clothing and flush the skin with water. Get medical attention immediately.

If a person breathes in large amounts of this chemical, move the victim to fresh air at once. If breathing has stopped, perform artificial respiration. Keep the affected person warm and at rest. Get medical attention as soon as possible.

If this chemical has been swallowed, get medical attention immediately.

FIRE HAZARD   Dangerous fire hazard when exposed to heat, flame, or oxidizers. Moderate explosion hazard when exposed to flame.

To fight a fire, use alcohol foam, CO_2, or dry chemical extinguishers.

HANDLING   Wear appropriate clothing and equipment to prevent repeated or prolonged skin contact. Wash promptly if skin becomes wet. Immediately remove any clothing that becomes wet.

Wear eye protection to reduce probability of eye contact.

See Chap. 3 for information on handling carcinogens.

PYRIDINE

CAS #110-86-1 mf: C_5H_5N
DOT #1282 mw: 79.11

STRUCTURE

NFPA LABEL

SYNONYMS
Azabenzene
Azine

PROPERTIES   Colorless to yellow, basic liquid with a sharp, unpleasant, penetrating odor, threshold 1 ppm; burning taste; miscible with water and most organic solvents; flammable; boiling point 115.6°C, freezing point −42°C; lower explosive limit 1.8%, upper explosive limit 12.4%; flash point 20°C (closed cup); density 0.986 at 15.5°C; autoignition temperature 482.2°C; vapor pressure 10 mm Hg at 13.2°C, 20 mm Hg at 25°C; vapor density 2.73; volatile with steam

OSHA PEL   TWA 5 ppm; IDLH 3600 ppm

ACGIH TLV   TWA 5 ppm

DOT Classification   Flammable liquid; Label: Flammable liquid; Flammable liquid; Label: Flammable liquid, poison

EPA TSCA Inventory and Community Right to Know List

TOXIC HAZARD   Moderately toxic by ingestion, skin contact, and some other routes. Mildly toxic by inhalation. The liquid and vapor are irritating to the eyes, nose, and throat.

Chronic exposure has produced serious liver and kidney damage. Ingestion of 2 mL/day, prescribed therapeutically, has caused death. Continuing skin contact with the liquid can produce a dry, scaly inflammation.

Exposure to more than 100 ppm on a relatively regular basis has caused transient headaches, dizziness or lightheadedness, insomnia, mental dullness, nausea, and anorexia.

Routes   Inhalation, absorption, ingestion, contact

Symptoms   Eye irritation; dermatitis; headache; nervousness; dizziness; insomnia; nausea, anorexia, frequent urination; liver and kidney damage; central nervous system depression, gastrointestinal upset

Target Organs   Central nervous system, liver, kidneys, skin, gastrointestinal tract

HAZARDOUS REACTIONS
Acids (strong): incompatible
Bromine trifluoride: reaction forms pyrophoric or explosive products
Chlorosulfonic acid: violent reaction
Chromic acid: violent reaction
Chromium trioxide: violent reaction
Dinitrogen tetraoxide: violent reaction
Fluorine: incandescent reaction
Formamide + iodine + sulfur trioxide: storage hazard, mixtures release carbon dioxide and sulfuric acid
Maleic anhydride: reaction evolves carbon dioxide above 150°C
Nitric acid: violent reaction
Oleum: violent reaction
Oxidizing materials: incompatible
Perchromates: violent reaction
β-Propiolactone: violent reaction
Silver perchlorate: violent reaction
Sulfuric acid: violent reaction
Trifluoromethyl hypofluorite: reaction forms pyrophoric or explosive products

When heated to decomposition, it emits highly toxic fumes of NO_x and possibly cyanides.

FIRST AID   In case of contact with the eyes, immediately wash the eyes with large amounts of water, occasionally lifting the lower and upper lids. Get medical attention immediately. Contact lenses should not be worn when working with this chemical.

In case of contact with the skin, immediately flush the contaminated skin with water. If this chemical penetrates the clothing, immediately remove the clothing and flush the skin with water. Get medical attention promptly.

If a person breathes in large amounts of this chemical, move the victim to fresh air at once. If breathing has stopped, perform artificial respiration. Keep the affected person warm and at rest. Get medical attention as soon as possible.

If this chemical has been swallowed, get medical attention immediately.

FIRE HAZARD   Dangerous fire hazard when exposed to heat, flame, or oxidizers. Severe explosion hazard in the form of vapor when exposed to flame or spark.

To extinguish fires, use alcohol foam.

HANDLING   The odor threshold of approximately 1 ppm requires that pyridine be handled in a laboratory hood or with the use of some other local-exhaust ventilation.

Wear appropriate clothing and equipment to prevent skin contact. Wash immediately if skin becomes contaminated. Immediately remove any clothing that becomes wet.

Wear eye protection to prevent any possibility of eye contact.

The following equipment should be available: an eyewash station and a quick-drench shower.

SELENIUM

CAS #7782-49-2 mf: Se
DOT #2658 mw: 78.96

SYNONYMS
Colloidal selenium
Elemental selenium
Selenium alloy
Selenium base
Selenium dust
Selenium elemental
Selenium homopolymer
Selenium metal powder, nonpyrophoric (DOT)
Vandex

PROPERTIES   Gray, nonmetallic element; insoluble in water and alcohol, slightly soluble in ether; melting point 170–217°C, boiling point 690°C; density 4.81–4.26; vapor pressure 1 mm Hg at 356°C

OSHA PEL   TWA 0.2 mg(Se)/m^3; IDLH 100 mg/m^3

ACGIH TLV   TWA 0.2 mg/m^3

DOT Classification   Poison B; Label: St. Andrew's cross

EPA TSCA Inventory and Community Right to Know List (selenium and its compounds)

TOXIC EFFECTS   Poisonous by ingestion, inhalation, and some other routes. A teratogen.

Chronic ingestion of 5 mg of selenium per day resulted in 49% morbidity in five Chinese villages. Selenosis in humans has occurred from ingestion of 3.2 mg of selenium per day and in populations with selenium blood levels of 800 µg/L. In cattle, "alkali disease" is associated with consumption of grain or plants containing 5–25 mg/kg of selenium. Consumption of plants grown in areas with high selenium concentrations in the soil or water can cause effects in humans and animals. Selenium is an essential trace element for many species.

Routes   Inhalation, absorption, ingestion, contact

Symptoms   Irritated eyes, nose, and throat; blurred vision; headache; chills, fever, dyspnea, bronchitis; metallic taste, gastrointestinal disturbances; dermatitis; pallor, nervousness, depression, garlic odor of breath and sweat, liver damage, brittle hair, new hair with no pigment, brittle nails with spots and streaks, skin lesions, peripheral anesthesia, acroparaesthesia, pain, and hyperreflexia

Target Organs   Upper respiratory system, eyes, skin, liver, kidneys, blood

HAZARDOUS REACTIONS
Acids: incompatible
Barium carbide: violent reaction
Bromine pentafluoride: violent reaction
Calcium carbide: violent reaction
Chlorates: violent reaction
Chlorine trifluoride: violent reaction
Chromium trioxide: violent reaction
Fluorine: violent reaction
Lithium carbide: violent reaction
Lithium silicon (Li$_6$Si$_2$): violent reaction
Metal amides: reaction forms explosive products
Nickel: violent reaction
Nitric acid: violent reaction
Nitrogen trichloride: violent reaction
Oxidizing agents (strong): incompatible
Oxygen: violent reaction
Potassium: violent reaction
Potassium bromate: violent reaction
Rubidium carbide: violent reaction
Silver bromate: violent reaction
Sodium: violent reaction
Strontium carbide: violent reaction
Thorium carbide: violent reaction
Uranium: violent reaction
Zinc: violent reaction

When heated to decomposition, it emits toxic fumes of Se.

FIRST AID   In case of contact with the eyes, immediately wash the eyes with large amounts of water, occasionally lifting the lower and upper lids. Get medical attention immediately. Contact lenses should not be worn when working with this chemical.

In case of contact with the skin, immediately wash the contaminated skin with soap and water. If this chemical penetrates through the clothing, immediately remove the clothing and wash the skin with soap and water. Get medical attention promptly.

If a person breathes in large amounts of this chemical, move the victim to fresh air at once. If breathing has stopped, perform artificial respiration. Keep the

affected person warm and at rest. Get medical attention as soon as possible.

If this chemical has been swallowed, get medical attention immediately.

SILVER AND ITS COMPOUNDS

CAS #7440-22-4 mf: Ag
mw: 107.868

SYNONYMS
Argentum
Shell silver
Silver atom

PROPERTIES   Soft, ductile, malleable, lustrous, white metal; melting point 961.93°C, boiling point 2212°C; density 10.50 at 20°C

OSHA PEL   TWA 10 µg/m³

ACGIH TLV   TWA (metal) 0.1 mg/m³ (soluble compounds as Ag) 0.01 mg/m³

Community Right to Know List (silver and its compounds) and EPA TSCA Inventory

TOXIC EFFECTS   An experimental tumorigen. Human systemic effects by inhalation; skin effects. Inhalation of dusts can cause argyrosis (generalized grayish pigmentation of the skin and mucous membranes).

Silver is not expelled efficiently from the body. Chronic exposure to silver dust at 1 mg/m³ for years forms deposits throughout the body. There are no constitutional symptoms or physical disability.

Routes   Inhalation, ingestion, contact

Symptoms   Blue-gray eyes, nasal septum, throat, skin; skin irritation and ulceration; gastrointestinal disturbances

Target Organs   nasal septum, skin, eyes

Water-soluble silver compounds   Poisonous by ingestion. Skin and mucous membrane irritants.

Symptoms   Pulmonary edema, hemorrhage, and necrosis of the bone marrow, liver, and kidney

HAZARDOUS REACTIONS
3-Bromopropyne: incompatible
Acetylene: ignites on contact
Acetylene compounds: incompatible
Ammonia: ignites on contact
Aziridine: incompatible
Bromine azide: incompatible
Bromoazide: ignites on contact
Carboxylic acids: incompatible
Chlorine trifluoride: ignites on contact
Copper + ethylene glycol: incompatible
Electrolytes + zinc: incompatible
Ethanol + nitric acid: incompatible
Ethyl hydroperoxide: incompatible
Ethylene imine: ignites on contact
Ethylene oxide: incompatible
Ethyleneimine: incompatible
Hydrogen peroxide: ignites on contact
Iodoform: incompatible
Nitric acid: incompatible
Oxalic acid: ignites on contact
Ozonides: incompatible
Peroxomonosulfuric acid: incompatible
Peroxyformic acid: incompatible
Sulfuric acid: ignites on contact
Tartaric acid: ignites on contact

FIRST AID   For soluble silver compounds: In case of contact with the eyes, immediately wash the eyes with large amounts of water, occasionally lifting the lower and upper lids. Get medical attention immediately. Contact lenses should not be worn when working with this chemical.

If this chemical comes in contact with the skin, immediately flush the contaminated skin with water. If this chemical penetrates the clothing, immediately remove the clothing and flush the skin with water. Get medical attention immediately.

If a person breathes in large amounts of this chemical, move the victim to fresh air at once. If breathing has stopped, perform artificial respiration. Keep the affected person warm and at rest. Get medical attention as soon as possible.

If this chemical has been swallowed, get medical attention immediately.

FIRE HAZARD   Flammable in the form of dust when exposed to flame or by chemical reaction (see hazardous reactions).

HANDLING   For soluble silver compounds: Wear appropriate clothing and equipment to prevent skin contact. Wash promptly if the skin becomes contaminated. Promptly remove any nonimpervious clothing that becomes contaminated. Work clothing should be changed daily if it is probable that the clothing has been contaminated.

Wear eye protection to prevent any possibility of eye contact.

The following equipment should be available: an eyewash station.

SODIUM

CAS #7440-23-5 mf: Na
DOT #1428; 1429 mw: 22.9898

NFPA LABEL

SYNONYMS
Natrium
Sodium metal (DOT)

PROPERTIES   Light, soft, ductile, malleable, silver-white metal; melting point 97.81°C, boiling point 881.4°C; density 0.9710 at 20°C; autoignition temperature >115°C in dry air; vapor pressure 1.2 mm Hg at 400°C

EPA TSCA Inventory

DOT Classification   Label: Flammable solid and dangerous when wet

TOXIC EFFECTS   Sodium is highly reactive and reacts exothermally with the moisture in body tissues to form sodium hydroxide. Thermal burns are caused by the reaction, and chemical burns are caused by the sodium hydroxide.

HAZARDOUS REACTIONS
Acids: exothermic reaction
Air: incompatible
Aluminum tribromide: incompatible
Aluminum trichloride: incompatible
Aluminum trifluoride: incompatible
Ammonium chlorocuprate: incompatible
Ammonium nitrate: incompatible
Antimony tribromide: incompatible
Antimony trichloride: incompatible
Antimony triiodide: incompatible
Arsenic trichloride: incompatible
Arsenic triodide: incompatible
Bismuth oxide: incompatible
Bismuth tribromide: incompatible
Bismuth trichloride: incompatible
Bismuth triiodide: incompatible
Boron tribromide: incompatible
Bromine: incompatible
Bromoazide: incompatible
Carbon dioxide: incompatible
Carbon monoxide + ammonia: incompatible
Carbon tetrachloride: incompatible
Chlorine: incompatible
Chlorine trifluoride: incompatible
Chromium tetrachloride: incompatible
Chromium trioxide: incompatible
Cobalt bromide: incompatible
Cobalt chloride: incompatible
Copper chloride: incompatible
Copper oxide: incompatible
1,2-Dichloroethylene: incompatible
Dichloromethane: incompatible
Fluorine: incompatible
Halogenated hydrocarbons: exothermic reaction
Halogens: exothermic reaction
Hydrazine hydrate: incompatible
Hydrogen chloride: incompatible
Hydrogen fluoride: incompatible
Hydrogen peroxide: incompatible
Hydrogen sulfide: incompatible
Hydroxylamine: incompatible
Iodine: incompatible
Iodine monochloride: incompatible
Iodine pentafluoride: incompatible
Iron dibromide: incompatible
Iron diiodide: incompatible
Iron trichloride: incompatible
Lead oxide: incompatible
Maleic anhydride: incompatible
Manganous chloride: incompatible
Mercuric bromide: incompatible
Mercuric chloride: incompatible
Mercuric fluoride: incompatible
Mercuric iodide: incompatible
Mercurous chloride: incompatible
Mercurous oxide: incompatible
Methyl chloride: incompatible
Moisture: incompatible
Molybdenum trioxide: incompatible
Monoammonium phosphate: incompatible
Nitric acid: incompatible
Nitrogen peroxide: incompatible
Nitrosyl fluoride: incompatible
Nitrous oxide: incompatible
Oxidizing materials: incompatible
Phosgene: incompatible
Phosphorus: incompatible
Phosphorus pentafluoride: incompatible
Phosphorus pentoxide: incompatible
Phosphorus tribromide: incompatible
Phosphorus trichloride: incompatible
Phosphoryl chloride: incompatible
Potassium oxides: incompatible
Potassium ozonide: incompatible

Potassium superoxide: incompatible
Selenium: incompatible
Silicon tetrachloride: incompatible
Silver bromide: incompatible
Silver chloride: incompatible
Silver fluoride: incompatible
Silver iodide: incompatible
Sulfuric acid: incompatible
Sodium peroxide: incompatible
Stannic chloride: incompatible
Stannic iodide + sulfur: incompatible
Stannic oxide: incompatible
Stannous chloride: incompatible
Sulfur: incompatible
Sulfur dibromide: incompatible
Sulfur dichloride: incompatible
Sulfur dioxide: incompatible
Tellurium: incompatible
Tetrachloroethane: incompatible
Thallous bromide: incompatible
Thiophosphoryl bromide: incompatible
Trichlorethylene: incompatible
Vanadium pentachloride: incompatible
Vanadyl chloride: incompatible
Water: reaction evolves hydrogen and heat
Zinc bromide: incompatible

When heated to decomposition, it emits toxic fumes of Na_2O.

FIRE HAZARD   Heated sodium ignites spontaneously in air. A very dangerous fire hazard when exposed to heat and moisture. Heated sodium is spontaneously flammable in air. Dangerous explosion hazard in reaction with water in any form.

To extinguish fires, use soda ash, dry sodium chloride, or graphite, in order of preference. Attempts to put out sodium fires using water, foam, or halon will convert what might be a minor fire into a major explosion.

HANDLING   There are three main hazards from sodium.

1. It always reacts violently, and usually explosively, with water.
2. At temperatures just over its melting point it easily ignites in air, producing highly irritant sodium oxide smoke.
3. Sodium, or the oxide coating on sodium, in contact with bare skin, can cause caustic burns.

Sodium is stored dry in metal containers. These containers should be kept in a dry, fireproof building which should not contain any water pipes and whose floor should be sufficiently high that no reasonable accident could lead to water leaking in. Sodium can be safely stored under liquid hydrocarbons.

Cleaning sodium-contaminated systems after use: There are four main techniques for removing residual sodium:

1. Washing out with alcohol
2. Blowing through with steam
3. Washing with liquid ammonia
4. Weathering

Small quantities (on the order of a few grams) of waste sodium can be safely disposed of by dissolving in alcohol followed by mixing of the alcohol with a large volume of water.

Anyone handling equipment containing sodium during operations where spillage might reasonably be anticipated (i.e., loading, unloading, cleaning, and maintenance) should always wear, at the very least, tightly fitting eye goggles and rubber or plastic gloves, and preferably a face mask and a fireproofed coverall in addition. Under conditions where exposure to sodium is likely, i.e., when burning waste sodium or dealing with sodium fires, more extensive protective clothing should be worn. This should include a flameproof hood covering the entire head and shoulders, a flameproof coverall with the trousers pulled outside rubber boots, and thick leather gloves.

SODIUM ACETATE

CAS #127-09-3   mf: $C_2H_3O_2 \cdot Na$
mw: 82.04

SYNONYMS
Acetic acid, sodium salt
Anhydrous sodium acetate

PROPERTIES   White crystals, soluble in water; melting point 58°C, decomposes at higher temperatures; density 1.45; autoignition temperature 607.2°C

EPA TSCA Inventory and EPA Genetic Toxicology Program

TOXIC EFFECTS   Moderately toxic by ingestion. A skin and eye irritant.

HAZARDOUS REACTIONS
Fluorine: violent reaction

Diketene: violent reaction
Potassium nitrate: violent reaction

When heated to decomposition, it emits toxic fumes of Na$_2$O.

SODIUM AZIDE

CAS #26628-22-8 mf: N$_3$Na
DOT #1687 mw: 65.02

STRUCTURE
NaN$_3$

SYNONYMS
Azide
Azium
Kazoe

PROPERTIES   Colorless, hexagonal crystals; insoluble in ether, soluble in liquid ammonia; decomposes before melting; density 1.846

ACGIH TLV   Ceiling 0.3 mg/m^3

DOT Classification   Poison B; Label: Poison

EPA TSCA Inventory, EPA Genetic Toxicology Program, and EPA Extremely Hazardous Substances List

TOXIC EFFECTS   Poisonous by ingestion, skin contact, and many other routes.

It is highly soluble in water and rapidly converted to hydrazoic acid, which may be the ultimate toxic agent.

Acute exposure to hydrazoic acid vapor can produce irritation of the eyes, tracheal bronchitis, headache, possibly a dramatic decrease in blood pressure, weakness, pulmonary edema, and collapse.

Accidental ingestion of 50–60 mg of sodium azide has resulted in brief loss of consciousness, nausea, and severe headache, but recovery was rapid. In another incident, while acidifying 10 g of sodium azide, a chemist complained of dizziness, blurred vision, shortness of breathing, and faintness following a few minutes of exposure. Hypotension and bradycardia were seen but, again, recovery was complete in 1 h.

Sodium azide has been used to control blood pressure therapeutically. Doses of 0.65–3.9 mg by mouth daily for up to 2.5 years lowered the blood pressure and reduced the transient pounding sensation of the headache with no evidence of organic damage.

Symptoms by Ingestion   General anesthesia, somnolence, and kidney changes

HAZARDOUS REACTIONS   An unstable explosive, sensitive to impact.

Acids: incompatible
Ammonium chloride + trichloroacetonitrile: incompatible
Barium carbonate: violent reaction
Benzoyl chloride + potassium hydroxide: violent reaction
Brass: reaction forms explosive azides
Bromine: violent reaction
Carbon disulfide: violent reaction
Chromium oxychloride: violent reaction
Copper or its salts: reaction forms explosive azides
Cyanuric chloride: incompatible
Dibromomalononitrile: violent reaction
Dimethylsulfate: violent reaction
2,5-Dinitro-3-methylbenzoic acid + oleum: incompatible
Lead or its salts: reaction forms explosive azides
Metals (heavy) or their salts: reaction forms explosive azides
Nitric acid: violent reaction
Phosgene: incompatible
Sulfuric acid: violent reaction
Trifluororacryloyl chloride: incompatible
Water (hot): violent reaction

When heated to decomposition, it emits very toxic fumes of NO$_x$ and Na$_2$O.

HANDLING   Sodium azide is one of the azides that is not explosive. It is especially important that sodium azide not be allowed to come in contact with heavy metals (for example, by being poured into a lead or copper drain) or their salts; heavy metal azides detonate with notorious ease.

The ACGIH limit is also recommended as a ceiling that should not be exceeded even instantaneously during the day. Use of a laboratory hood that has a protection factor adequate to prevent significant worker exposure or a closed system is recommended for operations using sodium azide.

SODIUM BISULFITE

CAS #7631-90-5 mf: HO$_3$S·Na
DOT #2693 mw: 104.06

SYNONYMS
Hydrogen sulfite sodium
Sodium acid sulfite
Sodium bisulfite, solid (DOT)

Sodium bisulfite, solution (DOT)
Sodium bisulphite
Sodium hydrogen sulfite
Sodium hydrogen sulfite, solid (DOT)
Sodium hydrogen sulfite, solution (DOT)
Sodium sulhydrate
Sulfurous acid, monosodium salt

PROPERTIES   White, crystalline powder with an odor of SO_2 and a disagreeable taste; very soluble in hot or cold water, slightly soluble in alcohol; decomposes before melting; density 1.48

EPA TSCA Inventory and EPA Genetic Toxicology Program

ACGIH TLV   TWA 5 mg/m^3

DOT Classification   ORM-B; Label: None; Corrosive material; Label: Corrosive

TOXIC EFFECTS   Moderately toxic by ingestion. The solid and concentrated solutions are corrosive irritants to the skin, eyes, and mucous membranes. An allergen.

HAZARDOUS REACTIONS   When heated to decomposition, it emits toxic fumes of SO_x and Na_2O.

SODIUM CYANIDE

CAS #143-33-9   mf: CNNa
DOT #1689   mw: 49.01

STRUCTURE   NFPA LABEL
NaC≡N

SYNONYMS
Cyanide of sodium   Hydrocyanic acid,
Cyanobrik   sodium salt
Cyanogran   Sodium cyanide, solid
Cymag   and solution (DOT)

PROPERTIES   White, deliquescent, crystalline powder; readily soluble in water, slightly soluble in alcohol; melting point 563.7°C, boiling point 1496°C; vapor pressure 1 mm Hg at 817°C, 10 mm Hg at 983°C

EPA TSCA Inventory and Community Right to Know List (cyanide and its compounds)

OSHA PEL   TWA 5 mg(CN)/m^3 (skin)
ACGIH TLV   TWA 5 mg(CN)/m^3 (skin)
NIOSH REL   Ceiling 5 mg(CN)/m^3/10M

DOT Classification   Poison B; Label: Poison, solid and solution

TOXIC EFFECTS   A deadly poison by ingestion and many other routes. Can be adsorbed through the skin. A teratogen. Solutions irritate the skin, nose, and eyes. Inhibits the enzymes responsible for oxidative phosphorylation.

A very fast-acting poison. The lethal oral dose for humans is 200 mg. Higher doses may cause immediate death.

People who work with cyanide solutions may develop a cyanide rash, characterized by itching and skin eruptions. Chronic exposure to low levels of cyanide compounds can cause loss of appetite, headache, weakness, nausea, dizziness, and irritation of the upper respiratory tract and eyes.

Symptoms by Ingestion   Headache, confusion, hallucinations, distorted perceptions, muscle weakness, gastritis, nausea, and vomiting

HAZARDOUS REACTIONS
Acids or acid fumes: reaction forms hydrogen cyanide gas
Air: reaction of carbon dioxide with cyanide solutions forms hydrocyanic acid
Chlorate: mixture explodes at about 450°C
Fluorine: violent reaction
Magnesium: violent reaction
Nitrates: violent reaction
Nitric acid: violent reaction
Nitrites: violent reaction
Water or steam: reaction forms hydrogen cyanide gas

FIRE HAZARD   Flammable by chemical reaction with heat, moisture, acid

HANDLING   The OSHA limits include a warning of the potential contribution of skin absorption to the overall exposure.

Dry cotton gloves should be worn when handling dry sodium cyanide. Rubber gloves and splash-proof goggles should be worn when substantial amounts of sodium cyanide solution are used.

Acid must not be allowed to come in contact with sodium cyanide, as gaseous hydrogen cyanide will be liberated.

SODIUM FLUORIDE

CAS #7681-49-4   mf: FNa
DOT #1690   mw: 41.99

NFPA LABEL

SYNONYMS

Alcoa sodium fluoride	Luride
Antibulit	Nafeen
Cavi-trol	Nafpak
Chemifluor	Na frinse
Credo	Natrium fluoride
Disodium difluoride	Nufluor
FDA 0101	Ossalin
F1-tabs	Ossin
Floridine	Pediaflor
Florocid	Pedident
Flozenges	Pennwhite
Fluoral	Pergantene
Fluorident	Phos-flur
Fluorigard	Point two
Fluorineed	Predent
Fluorinse	Rafluor
Fluoritab	Rescue squad
Fluor-o-kote	Roach salt
Flura-gel	Sodium fluoride, solid
Flurcare	and solution (DOT)
Fungol b	Sodium hydrofluoride
Gel ii	Sodium monofluoride
Gelution	So-flo
Gleem	Stay-flo
Iradicav	Studafluor
Karidium	Super-dent
Karigel	T-fluoride
Kari-rinse	Thera-flur-n
Lea-cov	Trisodium trifluoride
Lemoflur	Villiaumite

PROPERTIES   Clear, lustrous crystals or white powder or balls; melting point 993°C, boiling point 1700°C; density 2 at 41°C; vapor pressure 1 mm Hg at 1077°C

EPA TSCA Inventory and EPA Genetic Toxicology Program

OSHA PEL   TWA 2500 µg(F)/m³

ACGIH TLV   TWA 2.5 mg(F)/m³

NIOSH REL   TWA (inorganic fluorides) 2.5 mg (F)/m³

DOT Classification   ORM-B; Label: None; Corrosive material; Label: Corrosive, solution; Poison B; Label: St. Andrew's cross

TOXIC EFFECTS   A poison by ingestion, skin contact, and many other routes. A teratogen. A corrosive irritant to skin, eyes, and mucous membranes.

Symptoms by Ingestion   Paresthesia, ptosis (drooping of the eyelid), tremors, fluid intake, muscle weakness, headache, EKG changes, cyanosis, respiratory depression, hypermotility, diarrhea, nausea or vomiting, salivary gland changes, changes in teeth and supporting structures and other musculoskeletal changes, and increased immune response

USES   Used in chemical cleaning, for fluoridation of drinking water, as a dental treatment, and as a fungicide and insecticide

HAZARDOUS REACTIONS   When heated to decomposition, it emits toxic fumes of F^- and Na_2O.

SODIUM HYDROXIDE

CAS #1310-73-2   mf: HNaO
DOT #1823; 1824   mw: 40.00

NFPA LABEL

SYNONYMS
Caustic soda
Caustic soda, bead (DOT)
Caustic soda, dry (DOT)
Caustic soda, flake (DOT)
Caustic soda, granular (DOT)
Caustic soda, liquid (DOT)
Caustic soda, solid (DOT)
Caustic soda, solution (DOT)
Lewis-red devil lye
Lye (DOT)
Soda lye
Sodium hydrate (DOT)
Sodium hydroxide, bead (DOT)
Sodium hydroxide, dry (DOT)

258 SODIUM HYDROXIDE

Sodium hydroxide, flake (DOT)
Sodium hydroxide, granular (DOT)
Sodium hydroxide, solid (DOT)
White caustic

PROPERTIES   Colorless, odorless solid as white deliquescent pieces, lumps or sticks; solubility in water 50%; melting point 318.4°C, boiling point 1390°C, density 2.120 at 20°C/4°C, vapor pressure 1 mm Hg at 739°C

EPA TSCA Inventory and EPA Genetic Toxicology Program

OSHA PEL   TWA 2 mg/m^3; IDLH 250 mg/m^3

ACGIH TLV   Ceiling 2 mg/m^3

NIOSH REL   Ceiling (sodium hydroxide) 2 mg/m^3/15 min

DOT Classification   Corrosive material; Label: Corrosive

TOXIC EFFECTS   Moderately toxic by ingestion. All forms are very corrosive to skin, eyes, and mucous membranes.

Routes   Inhalation, ingestion, contact

Symptoms   Irritation to eyes, nose, and respiratory system; burns of the eyes and skin; pneumonitis, damage to upper respiratory tract and lung tissue; temporary loss of hair

Target Organs   Eyes, respiratory system, skin

HAZARDOUS REACTIONS   A strong base.

Acetaldehyde: violent reaction or ignition under some conditions
Acetic acid: violent reaction or ignition under some conditions
Acetic anhydride: violent reaction or ignition under some conditions
Acids: incompatible
Acrolein: violent reaction or ignition under some conditions
Acrylonitrile: violent reaction or ignition under some conditions
Allyl alcohol: violent reaction or ignition under some conditions
Allyl chloride: violent reaction or ignition under some conditions
Aluminum: violent reaction or ignition under some conditions
Aluminum + arsenic compounds: reaction forms poisonous arsine
Ammonia + silver nitrate: reaction forms explosive silver nitride
Benzene-1,4-diol: violent reaction or ignition under some conditions
N,N'-bis (trinitroethyl) urea: reaction forms explosive products in storage
Bromine: potentially explosive reaction
Chlorine trifluoride: violent reaction or ignition under some conditions
4-Chlorobutyronitrile: potentially explosive reaction
Chloroform + methanol: violent reaction or ignition under some conditions
Chlorohydrin: violent reaction or ignition under some conditions
4-Chloro-2-methylphenol: potentially explosive reaction in storage
Chloronitrotoluenes: violent reaction or ignition under some conditions
Chlorosulfonic acid: violent reaction or ignition under some conditions
Cinnamaldehyde: violent reaction or ignition under some conditions
Cyanogen azide: reaction forms explosive products
Diborane + octyl oxime: violent reaction or ignition under some conditions
2,2-Dichloro-3,3-dimethylbutane: violent reaction or ignition under some conditions
1,2-Dichloroethylene: violent reaction or ignition under some conditions
Diethylene glycol: reaction forms explosive products above 230°C
Ethylene cyanhydrin: violent reaction or ignition under some conditions
Ethylene glycol: reaction forms explosive products above 230°C
Flammable liquids: incompatible
Glycols: reaction forms explosive products above 230°C
Glyoxal: violent reaction or ignition under some conditions
Halogens (organic): incompatible
Hydrogen chloride: violent reaction or ignition under some conditions
Hydrogen fluoride: violent reaction or ignition under some conditions
Hydroquinone: violent reaction or ignition under some conditions
Maleic anhydride: violent reaction or ignition under some conditions
Metals: incompatible
Methanol + tetrachloro-benzene: violent reaction or ignition under some conditions
4-Methyl-2-nitrophenol: violent reaction or ignition under some conditions

3-Methyl-2-penten-4-yn-1-ol: reaction forms explosive products
Nitric acid: violent reaction or ignition under some conditions
Nitro compounds: incompatible
Nitrobenzene: potentially explosive reaction when heated
Nitroethane: violent reaction or ignition under some conditions
Nitromethane: violent reaction or ignition under some conditions
Nitroparaffins: violent reaction or ignition under some conditions
Nitropropane: violent reaction or ignition under some conditions
Oleum: violent reaction or ignition under some conditions
Pentol: violent reaction or ignition under some conditions
Phosphorus: violent reaction or ignition under some conditions
Phosphorus pentoxide: violent reaction or ignition under some conditions
β-Propiolactone: violent reaction or ignition under some conditions
Sodium tetrahydroborate: potentially explosive reaction
Sulfuric acid: violent reaction or ignition under some conditions
1,2,4,5-Tetrachlorobenzene: dangerously violent reaction forms the extremely toxic 2,3,7,8-tetrachlorodibenzodioxin
Tetrahydrofuran: violent reaction or ignition under some conditions
Tin: incompatible
1,1,1-Trichloroethanol: violent reaction or ignition under some conditions
2,2,2-Trichloroethanol: potentially explosive reaction
Trichloroethylene: reaction forms explosive dichloroacetylene
Trichloronitromethane: violent reaction or ignition under some conditions
Water: violent reaction or ignition under some conditions
Zinc: violent reaction or ignition under some conditions
Zirconium: potentially explosive reaction when heated

When heated to decomposition, it emits toxic fumes of Na_2O.

FIRST AID   In case of contact with the eyes, immediately wash the eyes with large amounts of water, occasionally lifting the lower and upper lids. Get medical attention immediately. Contact lenses should not be worn when working with this chemical.

In case of contact with the skin, immediately flush the contaminated skin with water. If this chemical penetrates the clothing, immediately remove the clothing and flush the skin with water. Get medical attention promptly.

If a person breathes in large amounts of this chemical, move the victim to fresh air at once. If breathing has stopped, perform artificial respiration. Keep the affected person warm and at rest. Get medical attention as soon as possible.

If this chemical has been swallowed, get medical attention immediately.

HANDLING   A dangerous material to handle. Wear appropriate clothing and equipment to prevent any possibility of skin contact. Wash immediately if skin becomes contaminated. Immediately remove any nonimpervious clothing that becomes contaminated. Work clothing should be changed daily if it is possible that the clothing has been contaminated.

Wear eye protection to prevent any possibility of eye contact.

The following equipment should be available: an eyewash station and a quick-drench shower.

SODIUM IODIDE

CAS #7681-82-5   mf: INa
mw: 149.89

SYNONYMS
Anayodin
Ioduril
Jodid sodny
Sodium iodine
Sodium monoiodide

PROPERTIES   Cubic, colorless crystals; melting point 651°C, boiling point 1300°C; density 3.667; vapor pressure 1 mm Hg at 767°C

EPA TSCA Inventory

TOXIC EFFECTS   Moderately toxic by ingestion. A teratogen by ingestion which causes developmental abnormalities of the endocrine system. A skin and eye irritant.

HAZARDOUS REACTIONS
Bromine trifluoride: violent reaction
Oxidants: violent reaction
Perchloric acid: violent reaction

When heated to decomposition, it emits toxic fumes of I⁻ and Na_2O.

SODIUM NITRATE

CAS #7631-99-4   mf: $NaNO_3$
DOT #1498   mw: 85.00

NFPA LABEL

SYNONYMS
Chile saltpeter
Cubic niter
Nitratine
Nitric acid, sodium salt
Soda niter
Sodium nitrate (DOT)

PROPERTIES   Colorless, transparent, odorless crystals; saline, slightly bitter taste; melting point 306.8°C, decomposes at 380°C; density 2.261

EPA TSCA Inventory and EPA Genetic Toxicology Program

DOT Classification   Oxidizer; Label: Oxidizer

TOXIC EFFECTS   Moderately toxic by ingestion.

HAZARDOUS REACTIONS   A powerful oxidizer. Explodes when heated to over 537.8°C.

Acetic anhydride: incompatible
Aluminum powder: mixture is explosive
Antimony powder: mixture is explosive
Barium thiocyanate: mixture is explosive
Bitumen: potentially violent reaction or ignition
Boron phosphide: explodes when mixed
Calcium-silicon alloy: potentially violent reaction or ignition
Cyanides: explodes when mixed
Jute + magnesium chloride: potentially violent reaction or ignition
Magnesium: potentially violent reaction or ignition
Metal amidosulfates: mixture is explosive
Metal cyanides: potentially violent reaction or ignition
Nonmetals: potentially violent reaction or ignition
Organic matter: potentially violent reaction or ignition
Peroxyformic acid: potentially violent reaction or ignition
Phenol + trifluoroacetic acid: potentially violent reaction or ignition
Sodium: mixture is explosive
Sodium hypophosphite: explodes when mixed
Sodium phosphinate: mixture is explosive
Sodium thiosulfate: mixture is explosive
Sulfur + charcoal (gunpowder): mixture is explosive
Wood: incompatible

When heated to decomposition, it emits toxic fumes of Na_2O.

FIRE HAZARD   This chemical will ignite with heat or friction.

SODIUM NITRITE

CAS #7632-00-0   mf: $NaNO_2$
DOT #1500   mw: 69.00

SYNONYMS
Antirust
Diazotizing salts
Erinitrit
Filmerine
Nitrous acid, sodium salt

PROPERTIES   Slightly yellowish or white crystals, sticks or powder; melting point 271°C, decomposes at 320°C; density 2.168

EPA TSCA Inventory and EPA Genetic Toxicology Program

DOT Classification   Oxidizer; Label: Oxidizer

TOXIC EFFECTS   A poison by ingestion and several other routes. A carcinogen and teratogen. Mutagenic data. An eye irritant.

Nitrites may react with organic amines in the body to form carcinogenic nitrosamines.

Symptoms (ingestion): motor activity changes, coma, decreased blood pressure, possible pulse-rate increase without fall in blood pressure, arteriolar or venous dilation, nausea or vomiting, and blood methemoglobinemia-carboxhemoglobinemia.

HAZARDOUS REACTIONS   Explodes when heated to above 537°C.

Aminoguanidine salts: incompatible
Ammonium salts: explodes on contact
Butadiene: incompatible
Cellulose: explodes on contact
Cyanides: explodes on contact
Lithium: explodes on contact
$Na_2S_2O_3$: explodes on contact
Phthalic acid: incompatible
Phthalic anhydride: incompatible
Potassium + ammonia: explodes on contact

Reducants: incompatible
Sodium amide: incompatible
Sodium disulphite: incompatible
Sodium thiocyanate: incompatible
Urea: incompatible
Wood: incompatible

When heated to decomposition, it emits toxic fumes of NO_x and Na_2O.

FIRE HAZARD   Flammable; a strong oxidizing agent. In contact with organic matter, it will ignite by friction.

HANDLING   See Chap. 3 for information on handling carcinogens.

SODIUM PEROXIDE

CAS #1313-60-6   mf: Na_2O_2
DOT #1504   mw: 77.98

NFPA LABEL

SYNONYMS
Disodium dioxide   Sodium oxide (Na2-02)
Disodium peroxide   Sodium peroxide
Flocool 180   (DOT)
Sodium dioxide   Solozone

PROPERTIES   White powder which turns yellow when heated; decomposes at 460°C; density 2.805

EPA TSCA Inventory

DOT Classification   Oxidizer; Label: Oxidizer

TOXIC EFFECTS   A severe irritant to skin, eyes, and mucous membranes.

HAZARDOUS REACTIONS   Reacts explosively or violently under the appropriate conditions with the following:

Acetic acid: explosive or violent reaction under some conditions
Acetic anhydride: explosive or violent reaction under some conditions
Acids: explosive or violent reaction under some conditions
Almond oil: explosive or violent reaction under some conditions
Aluminum: explosive or violent reaction under some conditions
Aluminum + aluminum chloride: explosive or violent reaction under some conditions
Aluminum + carbon dioxide: explosive or violent reaction under some conditions
Ammonium sulfate: explosive or violent reaction under some conditions
Aniline: explosive or violent reaction under some conditions
Antimony: explosive or violent reaction under some conditions
Arsenic: explosive or violent reaction under some conditions
Benzene: explosive or violent reaction under some conditions
Boron: explosive or violent reaction under some conditions
Boron nitride: explosive or violent reaction under some conditions
Calcium acetylide: explosive or violent reaction under some conditions
Carbon: explosive or violent reaction under some conditions
Charcoal: explosive or violent reaction under some conditions
Copper: explosive or violent reaction under some conditions
Cotton: explosive or violent reaction under some conditions
Diselenium dichloride: explosive or violent reaction under some conditions
Disulfur dichloride: explosive or violent reaction under some conditions
Ethanol: explosive or violent reaction under some conditions
Ethyl ether: explosive or violent reaction under some conditions
Ethylene glycol: explosive or violent reaction under some conditions
Fibrous materials + water: explosive or violent reaction under some conditions
Glucose + potassium nitrate: explosive or violent reaction under some conditions
Glycerol: explosive or violent reaction under some conditions
Hexamethylene-tetramine: explosive or violent reaction under some conditions
Hydrogen sulfide: explosive or violent reaction under some conditions
Hydroxy compounds: explosive or violent reaction under some conditions

Magnesium: explosive or violent reaction under some conditions
Magnesium + carbon dioxide: explosive or violent reaction under some conditions
Manganese dioxide: explosive or violent reaction under some conditions
Metals: explosive or violent reaction under some conditions
Metals + carbon dioxide + water: explosive or violent reaction under some conditions
Nonmetal halides: explosive or violent reaction under some conditions
Nonmetals: explosive or violent reaction under some conditions
Organic matter: explosive or violent reaction under some conditions
Paraffin: explosive or violent reaction under some conditions
Peroxyformic acid: explosive or violent reaction under some conditions
Phosphorus: explosive or violent reaction under some conditions
Phosphorus trichloride: explosive or violent reaction under some conditions
Potassium: explosive or violent reaction under some conditions
Potassium nitrate + dextrose: explosive or violent reaction under some conditions
Reducing materials: explosive or violent reaction under some conditions
Selenium: explosive or violent reaction under some conditions
Silver chloride + charcoal: explosive or violent reaction under some conditions
Soap: explosive or violent reaction under some conditions
Sodium: explosive or violent reaction under some conditions
Sodium dioxide: explosive or violent reaction under some conditions
Sulfur: explosive or violent reaction under some conditions
Sulfur chloride: explosive or violent reaction under some conditions
Tin: explosive or violent reaction under some conditions
Water: explosive or violent reaction under some conditions
Wood: explosive or violent reaction under some conditions
Wool: explosive or violent reaction under some conditions
Zinc: explosive or violent reaction under some conditions

When heated to decomposition, it emits toxic fumes of Na_2O.

FIRE HAZARD   Dangerous fire hazard by chemical reaction; a powerful oxidizing agent.

To fight a small fire, use carbon dioxide or dry chemical extinguishers. Combustible materials ignited by contact with sodium peroxide should be smothered with soda ash, salt, or dolomite mixtures. Chemical fire extinguishers should not be used for large fires. If the fire cannot be smothered, it should be flooded with large quantities of water from a hose.

SODIUM PHOSPHATE TRIBASIC DODECAHYDRATE

CAS #10101-89-0   mf: $O_4P \cdot 3Na \cdot 12H_2O$
                       mw: 380.18

SYNONYMS
Phosphoric acid, trisodium salt, dodecahydrate
Sodium phosphate dodecahydrate

TOXIC EFFECTS   Mildly toxic by ingestion. Irritating to skin, eyes, and mucous membranes.

FIRE HAZARD   When heated to decomposition, it emits toxic fumes of PO_x and Na_2O.

SODIUM SULFATE

CAS #7757-82-6   mf: $O_4S \cdot 2Na$
                       mw: 142.04

SYNONYMS
Disodium sulfate
Salt cake
Sodium sulfate anhydrous
Sodium sulphate
Sulfuric acid, disodium salt
Thenardite
Trona

PROPERTIES   Odorless, white crystals or powder; soluble in water, glycerin; insoluble in alcohol; melting point 888°C; density 2.671

EPA TSCA Inventory and EPA Genetic Toxicology Program

TOXIC EFFECTS   Moderately toxic by intravenous route. Mildly toxic by ingestion. A teratogen. Experimental reproductive effects.

HAZARDOUS REACTIONS
Aluminum: violent reaction

When heated to decomposition, it emits toxic fumes of SO_x and Na_2O.

SODIUM SULFIDE (ANHYDROUS)

CAS #1313-82-2 mf: Na_2S
DOT #1385 mw: 78.04

NFPA LABEL

SYNONYMS
Sodium monosulfide
Sodium sulphide

PROPERTIES   Amorphous, yellow-pink or white, deliquescent crystals; melting point 1180°C; density 1.856 at 14°C

EPA TSCA Inventory

DOT Classification   Label: Flammable solid

HAZARDOUS REACTIONS   This material is unstable and can explode on rapid heating or percussion.

Carbon: violent reaction
Diazonium salts: violent reaction
N,N-Dichloromethylamine: violent reaction
o-Nitroaniline diazonium salt: violent reaction
Water: violent reaction

When heated to decomposition, it emits toxic fumes of SO_x and Na_2O.

SULFAMIC ACID

CAS #5329-14-6 mf: H_3NO_3S
DOT #2967 mw: 97.10

SYNONYMS
Amidosulfonic acid   Aminosulfonic acid
Amidosulfuric acid   Sulfamidic acid

PROPERTIES   White crystals; decomposes at 200°C; density 203 at 12°C

EPA TSCA Inventory

TOXIC EFFECTS   Moderately toxic by ingestion. A corrosive irritant to skin, eyes, and mucous membranes.

HAZARDOUS REACTIONS
Chlorine: violent or explosive reaction
Metal nitrates: violent or explosive reaction when heated
Metal nitrites: violent or explosive reaction when heated
Nitric acid (fuming): violent or explosive reaction

When heated to decomposition, it emits very toxic fumes of SO_x and NO_x.

SULFUR

CAS #7704-34-9 mf: S
DOT #1350; 2448 mw: 32.06

NFPA LABEL

SYNONYMS
Brimstone   Precipitated sulfur
Colloidal sulfur   Sublimed sulfur
Flowers of sulfur   Sulfur, solid

PROPERTIES   Rhombic, yellow crystals or yellow powder; insoluble in water; slightly soluble in alcohol, ether, soluble in carbon disulfide, benzene, toluene; melting point 119°C, boiling point 444.6°C; density 2.07; density (liquid) 1.803; flash point 207.2°C (closed cup); autoignition temperature 232.2°C; vapor pressure 1 mm Hg at 183.8°C

EPA TSCA Inventory

DOT Classification   ORM-C; Label: None; Flammable solid; Label: Flammable solid

TOXIC EFFECTS   Poisonous by ingestion. An eye and mucous membrane irritant.

HAZARDOUS REACTIONS   Explosive in the form of dust when exposed to flame.

Aluminum: violent reaction
Aluminum + niobium pentoxide: violent reaction
Ammonia: violent reaction
Ammonium nitrate: violent reaction
Ammonium perchlorate: violent reaction
Boron: violent reaction
Bromine pentafluoride: violent reaction
Bromine trifluoride: violent reaction
Calcium: violent reaction

Calcium + vanadium oxide + water: violent reaction
Calcium hypochlorite: violent reaction
Calcium phosphide: violent reaction
Carbides: violent reaction
Cesium nitride: violent reaction
Charcoal: violent reaction
Chlorine dioxide: violent reaction
Chlorine monoxide: violent reaction
Chlorine trifluoride: violent reaction
Chromium hypochlorite: violent reaction
Chromium trioxide: violent reaction
Copper + chlorates: violent reaction
Fluorine: violent reaction
Halogenates: violent reaction
Halogenites: violent reaction
Halogens: violent reaction
Hydrocarbons: violent reaction
Indium: violent reaction
Iodine pentafluoride: violent reaction
Iodine pentoxide: violent reaction
Lead dioxide: violent reaction
Lithium: violent reaction
Mercuric oxide: violent reaction
Mercurous oxide: violent reaction
Mercury nitrate: violent reaction
Nickel: violent reaction
Nitrogen dioxide: violent reaction
Oxidizing materials: incompatible
Palladium: violent reaction
Phosphorus: violent reaction
P_2O_3: violent reaction
Potassium: violent reaction
Potassium nitrate + arsenic trisulfide: violent reaction
Potassium nitride: violent reaction
Potassium permanganate: violent reaction
Silver nitrate: violent reaction
Silver oxide: violent reaction
Sodium: violent reaction
Sodium + tin tetraiodide: violent reaction
Sodium hydride: violent reaction
Sodium nitrate + charcoal: violent reaction
Sulfur dichloride: violent reaction
Thallium trioxide: violent reaction
Tin: violent reaction
Uranium: violent reaction
Zinc: violent reaction

Dangerous; when heated, sulfur burns and emits highly toxic fumes of SO_x.

FIRE HAZARD   Combustible when exposed to heat or flame or by chemical reaction with oxidizers.

To fight a fire, use water or special mixtures of dry chemical extinguishers.

SULFUR DIOXIDE

CAS #7446-09-5   mf: O_2S
DOT #1079   mw: 64.06

NFPA LABEL

SYNONYMS
Bisulfite
Fermenicide liquid
Fermenicide powder
Sulfur oxide
Sulfurous acid
  anhydride
Sulfurous anhydride
Sulfurous oxide

PROPERTIES   Colorless liquid or gas with a characteristic, pungent odor; solubility in water 10%; melting point −75.5°C, boiling point −10.0°C; density (liquid) 1.434 at 0°C; vapor pressure 2538 mm Hg at 21.1°C; vapor density 2.264 at 0°C

OSHA PEL   TWA 5 ppm; IDLH 100 ppm

ACGIH TLV   TWA 2 ppm; STEL 5 ppm

NIOSH REL   TWA 0.5 ppm

DOT Classification   Nonflammable gas; Label: Nonflammable gas

IMO Classification   Poison A; Label: Poison gas

EPA Extremely Hazardous Substances List

TOXIC EFFECTS   Mildly toxic to humans by inhalation. A teratogen. A corrosive irritant to eyes, skin, and mucous membranes.

Irritation occurs at levels which supply good warning of toxic exposures. Exposure to 400–500 ppm is immediately dangerous. Inhalation of high concentrations can be fatal.

Routes   Inhalation, contact

Symptoms   Irritated eyes, nose, and throat, eye and skin burns, rhinitis; choking, cough; reflex bronchoconstriction, pulmonary edema, respiratory paralysis, bronchitis, asphyxia, conjunctivitis

Target Organs   Respiratory system, skin, eyes

HAZARDOUS REACTIONS
Acrolein: violent reaction
Alkali metal powders: incompatible
Aluminum: violent reaction
Cesium oxide: violent reaction

Chlorates: violent reaction
Chlorine trifluoride: violent reaction
Chromium: violent reaction
$CsHC_2$: violent reaction
Fluorine: violent reaction
Halogens: incompatible
Interhalogens: incompatible
Iron monoxide: violent reaction
KHC_2: violent reaction
Lithium acetylene carbide diammino: violent reaction
Lithium nitrate: incompatible
Manganese: violent reaction
Metal acetylides: incompatible
Metal oxides: incompatible
Metals: incompatible
Polymeric tubing: incompatible
Potassium chlorate: violent reaction
Potassium powder: incompatible
Rubidium carbide: violent reaction
Sodium: violent reaction
Sodium carbide: violent reaction
Sodium hydride: incompatible
Tin oxide: violent reaction
Water or steam: reaction produces toxic and corrosive fumes

When heated to decomposition, it emits toxic fumes of SO_x.

FIRST AID   In case of contact with the eyes, immediately wash the eyes with large amounts of water, occasionally lifting the lower and upper lids. Get medical attention immediately. Contact lenses should not be worn when working with this chemical.

In case of contact with the skin, immediately flush the contaminated skin with water. If this chemical penetrates the clothing, immediately remove the clothing and flush the skin with water. Get medical attention promptly.

If a person breathes in large amounts of this chemical, move the victim to fresh air at once. If breathing has stopped, perform artificial respiration. Keep the affected person warm and at rest. Get medical attention as soon as possible.

HANDLING   Usually supplied in a liquid form in small metal cylinders or in glass siphons. Like ammonia, sulfur dioxide is soluble in water, and care must be taken to avoid water being drawn back into the cylinder.

Respiratory protective equipment of a type approved by the MSHA for sulfur dioxide service should always be readily available in a place where this substance is used and so located as to be easily reached in case of need. When liquid sulfur dioxide is used, the eyes should be protected by the use of goggles or large-lens spectacles.

Leaks of sulfur dioxide may be detected by passing a rag dampened with aqueous ammonia over the suspected valve or fitting. White fumes indicate escaping sulfur dioxide gas.

Wear appropriate clothing and equipment to prevent skin freezing. Immediately remove any clothing that becomes wet.

Wear eye protection to prevent any possibility of eye contact.

The following equipment should be available: an eyewash station.

SPILL CLEANUP   If a cylinder valve is leaking, eliminate all sources of ignition and bubble the gas through calcium hypochlorite solution. Include a trap in the line to prevent suck-back.

SULFURIC ACID

CAS #7664-93-9   mf: H_2O_4S
DOT #1830; 1832   mw: 98.08

NFPA LABEL

SYNONYMS

Dipping acid   Oil of vitriol
Nordhausen acid   Sulphuric acid
  (DOT)   Vitriol brown oil

PROPERTIES   Colorless to dark brown, oily liquid; odorless; miscible with water and alcohol; solution releases large amounts of heat; melting point 10.49°C, boiling point 290°C; decomposes at 340°C; density 1.834; vapor press 1 mm Hg at 145.8°C

EPA TSCA Inventory

OSHA PEL   TWA 1 mg/m³; IDLH 80 mg/m³

ACGIH TLV   TWA 1 mg/m³

NIOSH REL   TWA 1 mg/m³

DOT Classification   Corrosive material; Label: Corrosive

TOXIC EFFECTS   A poison by inhalation. Moderately toxic by ingestion. Corrosive and toxic, causing rapid destruction of tissue and severe burns. Contact with large areas of skin can cause shock, collapse, and other symptoms of severe burns. Repeated exposure to dilute solutions causes dermatitis.

Concentrations in air of 0.125–0.50 ppm may be mildly annoying, and 1.5–2.5 ppm can be unpleasant; 10–20 ppm is unbearable. Repeated or prolonged inhalation of sulfuric acid mists can cause an inflammation of the upper respiratory tract leading to chronic bronchitis. Chronic exposure to low levels of sulfuric acid fumes results in gradual desensitization to the irritation. Inhalation of concentrated vapor or mists from hot acid or oleum can cause rapid loss of consciousness with serious damage to lung tissue.

Routes   Inhalation, ingestion, contact

Symptoms   Eye, nose, and throat irritation; skin and eye burns; dermatitis, pulmonary edema, bronchial emphysema; conjunctivitis; stomatitis; dental erosion; tracheobronchitis

Target Organs   Respiratory system, eyes, skin, teeth

HAZARDOUS REACTIONS   A very powerful, acidic oxidizer. Reaction with metals produces hydrogen gas.

Acetic acid: ignition or explosion on contact
Acetone + nitric acid: ignition or explosion on contact
Acetone + potassium dichromate: ignition or explosion on contact
Acetone cyanhydrin: ignition or explosion on contact
Acetonitrile: ignition or explosion on contact
Acrolein: ignition or explosion on contact
Acrylonitrile: ignition or explosion on contact
Acrylonitrile + water: ignition or explosion on contact
Alcohols + hydrogen peroxide: ignition or explosion on contact
Allyl alcohol: ignition or explosion on contact
Allyl chloride: ignition or explosion on contact
2-Amino ethanol: ignition or explosion on contact
Ammonia: ignition or explosion on contact
Ammonium hydroxide: ignition or explosion on contact
Aniline: ignition or explosion on contact
Benzene + permanganates: ignition or explosion on contact
Bromates + metals: ignition or explosion on contact
Bromine pentafluoride: ignition or explosion on contact
n-Butyraldehyde: ignition or explosion on contact
Carbides: ignition or explosion on contact
Chlorates: ignition or explosion on contact
Chlorine trifluoride: ignition or explosion on contact
Chlorosulfonic acid: ignition or explosion on contact
$CoHC_2$: ignition or explosion on contact
Copper nitride: ignition or explosion on contact
Diisobutylene: ignition or explosion on contact
Dimethyl benzylcarbinol + hydrogen peroxide: ignition or explosion on contact
Epichlorohydrin: ignition or explosion on contact
Ethylene cyanhydrin: ignition or explosion on contact
Ethylene diamine: ignition or explosion on contact
Ethylene glycol: ignition or explosion on contact
Ethylene imine: ignition or explosion on contact
Fulminates: ignition or explosion on contact
Hexalithium disilicide: ignition or explosion on contact
Hydrogen: ignition or explosion on contact
Hydrogen chloride: ignition or explosion on contact
Indene + nitric acid: ignition or explosion on contact
Iodine heptafluoride: ignition or explosion on contact
Iron: ignition or explosion on contact
Isoprene: ignition or explosion on contact
Mercury nitride: ignition or explosion on contact
Mesityl oxide: ignition or explosion on contact
Metals: ignition or explosion on contact
Metals + chlorates: ignition or explosion on contact
Nitric acid + glycerides: ignition or explosion on contact
Nitric acid + toluene: ignition or explosion on contact
p-Nitrotoluene: ignition or explosion on contact
Organics: incompatible
Oxidizing materials: incompatible
$P(OCN)_3$: ignition or explosion on contact
Pentasilver trihydroxydiamino phosphate: ignition or explosion on contact
Perchlorates: ignition or explosion on contact
Perchloric acid: ignition or explosion on contact
1-Phenyl-2-methyl propyl alcohol + hydrogen peroxide: ignition or explosion on contact
Phosphorus: ignition or explosion on contact
Picrates: ignition or explosion on contact
Potassium chlorate: ignition or explosion on contact
Potassium permanganate: ignition or explosion on contact
Potassium permanganate + potassium chloride: ignition or explosion on contact
Potassium permanganate + water: ignition or explosion on contact
Potassium-*tert*-butoxide: ignition or explosion on contact
β-Propiolactone: ignition or explosion on contact
Propylene oxide: ignition or explosion on contact
Pyridine: ignition or explosion on contact
$RbHC_2$: ignition or explosion on contact
Reducing materials: incompatible
Sodium: ignition or explosion on contact

Sodium carbonate: ignition or explosion on contact
Sodium hydroxide: ignition or explosion on contact
Steel: ignition or explosion on contact
Styrene monomer: ignition or explosion on contact
Triperchromate: ignition or explosion on contact
Vinyl acetate: ignition or explosion on contact
Water: ignition or explosion on contact

FIRST AID   In case of contact with the eyes, immediately wash the eyes with large amounts of water, occasionally lifting the lower and upper lids. Get medical attention immediately. Contact lenses should not be worn when working with this chemical.

In case of contact with the skin, immediately flush the contaminated skin with water. If this chemical penetrates the clothing, immediately remove the clothing and flush the skin with water. Get medical attention promptly.

If a person breathes in large amounts of this chemical, move the victim to fresh air at once. If breathing has stopped, perform artificial respiration. Keep the affected person warm and at rest. Get medical attention as soon as possible.

If this chemical has been swallowed, get medical attention immediately.

FIRE HAZARD   When heated to decomposition, it emits toxic fumes of SO_x.

HANDLING   Wear appropriate equipment to prevent any possibility of skin contact with liquids containing >1% of the acid, and repeated or prolonged skin contact with liquids containing <1%. Wash immediately if skin becomes contaminated. Immediately remove any nonimpervious clothing that becomes contaminated.

Wear eye protection to prevent any possibility of eye contact.

Provide an eyewash station and a quick-drench shower if liquids containing >1% acid are present.

TANNIC ACID

CAS #1401-55-4   mf: $C_{76}H_{52}O_{46}$
                       mw: 1701.28

SYNONYMS
Gallotannic acid   Glycerite
Gallotannin   Tannin

PROPERTIES   Yellowish-white or brown, bulky powder or flakes; very soluble in alcohol, acetone; almost insoluble in benzene, chloroform, ether, petroleum ether, carbon disulfide; melting point 200°C; flash point 198.9°C (open cup); autoignition temperature 526.7°C

IARC Cancer Review: Animal Sufficient Evidence and EPA TSCA Inventory

TOXIC EFFECTS   Poisonous by ingestion and several other routes. A carcinogen and tumorigen.

HAZARDOUS REACTIONS
Albumin: incompatible
Alkaloids: incompatible
Gelatin: incompatible
Lime water: incompatible
Metal salts (heavy): incompatible
Oxidizing materials: incompatible
Spirit nitrous ether: incompatible
Starch: incompatible

When heated to decomposition, it emits acrid smoke and irritating fumes.

FIRE HAZARD   Combustible when exposed to heat or flame.

To extinguish fires, use water.

HANDLING   See Chap. 3 for information on handling carcinogens.

TETRAHYDROFURAN

CAS #109-99-9   mf: C_4H_8O
DOT #2056   mw: 72.12

NFPA LABEL

SYNONYMS
Butylene oxide   Hydrofuran
Cyclotetramethylene   Oxacyclopentane
  oxide   Oxolane
Diethylene oxide   Tetramethylene oxide
1,4-Epoxybutane   THF
Furanidine

PROPERTIES   Colorless, mobile liquid with an etherlike odor; miscible with water, alcohol, ketones, esters, ethers, and hydrocarbons; freezing point −108.5°C, boiling point 65.4°C; density 0.888 at 20°C/4°C; flash point −17°C (TCC); lower explosive limit

2%, upper explosive limit 11.8%; autoignition temperature 321°C; vapor pressure 114 mm Hg at 15°C; vapor density 2.5

OSHA PEL   TWA 200 ppm; IDLH 20,000 ppm

ACGIH TLV   TWA 200 ppm; STEL 250 ppm

DOT Classification   Flammable liquid; Label: Flammable liquid

EPA TSCA Inventory

TOXIC EFFECTS   Moderately toxic by ingestion. Mildly toxic by inhalation. Irritant to eyes and mucous membranes. Narcotic in high concentrations. May cause liver and kidney damage.

Routes   Inhalation, contact, ingestion

Symptoms   Irritated eyes and upper respiratory tract; nausea; dizziness; headache, general anesthesia

Target Organs   Eyes, skin, respiratory system, central nervous system

HAZARDOUS REACTIONS   Like other ethers, unstabilized tetrahydrofuran forms sensitive, explosive peroxides on exposure to air. Alkalies can deplete the inhibitor in THF and cause an unexpected buildup of peroxides.

Air: forms explosive peroxides
2-Aminophenol + potassium dioxide: reaction forms an explosive product
Borane: reaction forms hydrogen gas
Bromine: vigorous reaction
Calcium hydride: vigorous reaction when heated
Hafnium tetrachloride: violent reaction
Lithium tetrahydroaluminate: reaction forms hydrogen gas
Metal halides: violent reaction
Oxidizing materials: incompatible
Potassium hydroxide: explosive reaction
Sodium aluminum hydride: explosive reaction
Sodium hydroxide: explosive reaction
Sodium tetrahydroaluminate: explosive reaction
Titanium tetrachloride: violent reaction
Zirconium tetrachloride: violent reaction

When heated to decomposition, it emits acrid smoke and irritating fumes.

FIRST AID   In case of contact with the eyes, immediately wash the eyes with large amounts of water, occasionally lifting the lower and upper lids. Get medical attention immediately. Contact lenses should not be worn when working with this chemical.

In case of contact with the skin, immediately flush the contaminated skin with water promptly. If this chemical penetrates the clothing, immediately remove the clothing and flush the skin with water promptly. If irritation persists after washing, get medical attention.

If a person breathes in large amounts of this chemical, move the victim to fresh air at once. If breathing has stopped, perform artificial respiration. Keep the affected person warm and at rest. Get medical attention as soon as possible.

If this chemical has been swallowed, get medical attention immediately.

FIRE HAZARD   A very dangerous fire hazard when exposed to heat, flames, or oxidizers. Vapor may explode if ignited in an enclosed area. Flashback may occur along vapor trail.

To fight a fire, use foam, dry chemical or CO_2 extinguishers.

HANDLING   Stored ether samples must always be tested for peroxide prior to distillation. Peroxides can be removed by treatment with strong ferrous sulfate solution made slightly acidic with sodium bisulfate.

Wear appropriate clothing and equipment to prevent repeated or prolonged skin contact. Wash promptly if skin becomes wet. Immediately remove any clothing that becomes wet.

Wear eye protection to reduce probability of eye contact.

TITANIUM CHLORIDE

CAS #7705-07-9   mf: $TiCl_3$
DOT #2441   mw: 154.25

SYNONYMS
Tac 121
Tac 131
Titanium(III) chloride
Titanium trichloride
Titanium trichloride, pyrophoric (DOT)
Titanous chloride
Trichloro titanium

PROPERTIES   Colorless to light yellow liquid, fumes in moist air; melting point −30°C, boiling point 136.4°C; density 1.772 at 25°C/25°C; vapor pressure 10 mm Hg at 21.3°C

EPA TSCA Inventory

DOT Classification   Flammable solid; Label: Spontaneously combustible, corrosive

TOXIC EFFECTS   A corrosive irritant to skin, eyes, and mucous membranes. Reacts exothermally with moisture in body tissues to form hydrochloric acid. Both the acid and the released heat can cause burns.

HAZARDOUS REACTIONS
Air: may ignite spontaneously
Hydrogen fluoride: violent reaction
Potassium: violent reaction

When heated to decomposition, it emits toxic fumes of Cl^-.

FIRE HAZARD   Flammable when exposed to heat or flame.

HANDLING   If spilled on skin, wipe off with dry cloth before applying water.

TITANIUM OXIDE

CAS #13463-67-7   mf: TiO_2
DOT #2546   mw: 79.90

SYNONYMS
A-fil cream
Anatase
Atlas white titanium dioxide
Austiox
Bayeritian
Bayertitan
Baytitan
Brookite
Calcotone white T
Cosmetic white C47-5175
Flamenco
Hombitan
Horse headache A-410
KH 360
Kronos titanium dioxide
Levanox white RKB
Rayox
Runa RH20
Rutile
Tiofine
Tioxide
Titanium dioxide
Trioxide(s)
Tronox
Unitane O-110
1700 White
Zopaque

PROPERTIES   White powder; insoluble in water; decomposes at 1640°C; density 4.26

OSHA PEL   TWA 15 mg/m³

ACGIH TLV   TWA 10 mg/m³ of total dust when toxic impurities are not present, for example, when quartz dust is <1%

EPA TSCA Inventory and Community Right to Know List

TOXIC EFFECTS   A carcinogen. A skin irritant.

Routes   Inhalation

Symptoms   Slight lung fibrosis
Target Organs   Lungs

HAZARDOUS REACTIONS
Metals: violent or incandescent reaction
Aluminum: violent or incandescent reaction
Calcium: violent or incandescent reaction
Lithium: violent reaction
Magnesium: violent or incandescent reaction
Potassium: violent or incandescent reaction
Sodium: violent or incandescent reaction
Zinc: violent or incandescent reaction
Lithium: violent or incandescent reaction

FIRST AID   If a person breathes in large amounts of this chemical, move the victim to fresh air at once. If breathing has stopped, perform artificial respiration. Keep the affected person warm and at rest. Get medical attention as soon as possible.

HANDLING   See Chap. 3 for information on handling carcinogens.

TOLUENE

CAS #108-88-3   mf: C_7H_8
DOT #1294   mw: 92.15

NFPA LABEL

SYNONYMS
Antisal 1a
Methacide
Methylbenzene
Methylbenzol
Phenylmethane
Toluol (DOT)

PROPERTIES   Colorless liquid with an aromatic, benzenelike odor; solubility in water 0.05%, soluble in acetone; miscible in absolute alcohol, ether, chloroform; melting point −95°C to −94.5°C, boiling point 110.4°C; density 0.866 at 20°C/4°C; flash point 4.4°C (closed cup); Underwriters Laboratory Classification 75–80; lower explosive limit 1.3%, upper explosive limit 7.1%; autoignition temperature 535.5°C; vapor pressure 36.7 mm Hg at 30°C; vapor density 3.14

OSHA PEL   TWA 200 ppm; ceiling 300 ppm; peak 500/10 min; IDLH 2000 ppm

ACGIH TLV   TWA 100 ppm; STEL 150 ppm

NIOSH REL   TWA 100 ppm; ceiling 200 ppm/10 min

DOT Classification   Flammable liquid; Label: Flammable liquid

Community Right to Know List and EPA TSCA Inventory

TOXIC EFFECTS   Mildly toxic by inhalation. A teratogen. A human eye irritant. An experimental skin and severe eye irritant.

Toluene is derived from coal tar, and commercial grades usually contain small amounts of benzene as an impurity.

Inhalation of 200 ppm of toluene for 8 h may cause impairment of coordination and reaction time; with higher concentrations (up to 800 ppm) these effects are increased and are observed in a shorter time.

Routes   Inhalation, absorption, ingestion, contact

Symptoms   Eye irritation, dilated pupils, lacrimation, photophobia; dermatitis, headache, dizziness, nervousness, confusion, euphoria, hallucinations or distorted perceptions, nausea, loss of appetite, a bad taste, lassitude, impairment of coordination and reaction time, fatigue, weakness, insomnia, paresthesia, bone marrow changes

Target Organs   Central nervous system, liver, kidneys, skin

HAZARDOUS REACTIONS

Bromine trifluoride: explosive reaction
1,3-Dichloro-5,5-dimethyl-2,4-imidazolididione: explosive reaction
Dinitrogen tetroxide: explosive reaction
Nitric acid (conc.): explosive reaction
Oxidizing materials: vigorous reaction
Silver perchlorate: explosive reaction
Sulfuric acid + nitric acid: explosive reaction
Tetranitromethane: forms an explosive mixture
Uranium hexafluoride: explosive reaction

When heated to decomposition, it emits acrid smoke and irritating fumes.

FIRST AID   In case of contact with the eyes, immediately wash the eyes with large amounts of water, occasionally lifting the lower and upper lids. Get medical attention immediately. Contact lenses should not be worn when working with this chemical.

In case of contact with the skin, promptly wash the contaminated skin with soap and water. If this chemical penetrates through the clothing, promptly remove the clothing and wash the skin with soap and water. Get medical attention promptly.

If a person breathes in large amounts of this chemical, move the victim to fresh air at once. If breathing has stopped, perform artificial respiration. Keep the affected person warm and at rest. Get medical attention as soon as possible.

If this chemical has been swallowed, get medical attention immediately.

FIRE HAZARD   A very dangerous fire hazard when exposed to heat, flame, or oxidizers. Explosive in the form of vapor when exposed to heat or flame.

To fight a fire, use foam, CO_2, or dry chemical extinguishers.

HANDLING   Wear appropriate clothing and equipment to prevent repeated or prolonged skin contact. Wash promptly if skin becomes wet. Immediately remove any clothing that becomes wet.

Wear eye protection to reduce probability of eye contact.

TRICHLOROETHYLENE

CAS #79-01-6   mf: C_2HCl_3
DOT #1710   mw: 131.38

STRUCTURE   NFPA LABEL

H
C=CCl$_2$
Cl

SYNONYMS

Acetylene trichloride	Dukeron
Algylen	Ethene
Anamenth	Ethinyl trichloride
Benzinol	Ethylene trichloride
Blacosolv	Trichloran
Cecolene	Trichloro
1-Chloro-2,2-dichloroethylene	Trichloroethene
	1,2,2-Trichloroethylene
Chlorylea	Triclene
Chorylen	Trilene
Circosolv	Trimar
Crawhaspol	Tri-plus
Densinfluat	Vestrol
1,1-Dichloro-2-chloro-ethylene	Vitran
	Westrosol
Dow-tri	

TRICHLOROETHYLENE

PROPERTIES   Colorless, mobile liquid; sweet odor of chloroform, threshold 21.4 ppm; solubility in water 0.1%, soluble in ethanol and ethyl ether; melting point: −84.8°C, boiling point 86.7°C; density 1.4649 at 20°C/4°C; flash point 32°C; lower explosive limit 11%, upper explosive limit 41%, 90% at >30°C; autoignition temperature 420°C; vapor pressure 19.9 mm Hg at 0°C, 57.8 mm Hg at 20°C, 100 mm Hg at 32°C; vapor density 4.53

OSHA PEL   TWA 100 ppm; ceiling 1200; peak 300/5 min/2 h

ACGIH TLV   TWA 50 ppm; STEL 200 ppm

NIOSH REL   TWA (trichloroethylene) 250 ppm; ceiling (waste anesthetic gases) 2 ppm/1h

DOT Classification   ORM-A; Label: None; Poison B; Label: St. Andrew's cross

IARC Cancer Review: Animal Sufficient Evidence, EPA TSCA Inventory, EPA Genetic Toxicology Program, and Community Right to Know List

TOXIC EFFECTS   Moderately toxic by ingestion. Mildly toxic by inhalation. A carcinogen and teratogen. An eye and severe skin irritant.

A form of addiction has been observed in exposed workers. Fatalities following severe, acute exposure have been attributed to ventricular fibrillation resulting in cardiac failure.

In humans, acute exposure affects primarily the central nervous system. Common symptoms are headache, dizziness, nausea, fatigue, and drunkenness. Coma and sudden death have been reported in severe intoxication. Exposure for 2 h to 1000 ppm caused adverse affects to performance of steadiness and manual dexterity tests. Prolonged inhalation of moderate concentrations causes headache and drowsiness.

Acute skin exposure has caused erythema, burning, and inflammation of the skin. With repeated exposure, the liquid can produce inflammation and skin vesicles; in the eye, it produces pain and inflammation. There is damage to liver and other organs from chronic exposure.

Female mice, however, exhibited fatty degeneration of the liver after exposure to 1600 ppm for up to 8 weeks. After inhaling 500–750 ppm for 3–8 weeks, dogs exhibited lethargy, anorexia, nausea, vomiting, weight loss, and liver dysfunction.

Routes   Inhalation, ingestion, contact

Symptoms   Irritated eyes, dermatitis; headache, vertigo, visual disturbances, tremors, somnolence, hallucinations or distorted perceptions; nausea, vomiting; cardiac arrhythmias; paresthesia; gastrointestinal changes and jaundice; a carcinogen

Target Organs   Respiratory system, heart, liver, kidneys, central nervous system, skin

HAZARDOUS REACTIONS   Commercial-grade trichloroethylene often contains an alkyl amine stabilizer, which, in contact with concentrated sodium hydroxide, can produce the spontaneously inflammable gas chloroacetylene.

Alkali: reaction forms spontaneously flammable gas dichloroacetylene
Aluminum: reaction occurs under acid conditions
Barium: can react violently
2,2-Bis((4(2′,3′-epoxypropoxy)phenyl)propane: reaction forms spontaneously flammable gas dichloroacetylene
1,4-Butanediol mono-2,3-epoxypropylether: reaction forms spontaneously flammable gas dichloroacetylene
Caustics: incompatible
1-Chloro-2,3-epoxypropane: reaction forms spontaneously flammable gas dichloroacetylene
Dinitrogen tetroxide: can react violently
Epoxides: reaction forms spontaneously flammable gas dichloroacetylene
Lithium: can react violently
Magnesium: incompatible
Metals (chemically active): incompatible
Oxygen (liquid): can react violently
Ozone: can react violently
Potassium hydroxide: can react violently
Potassium nitrate: can react violently
Sodium: can react violently
Sodium hydroxide: can react violently
Titanium: can react violently
Water: reacts with heat and pressure to form HCl gas

When heated to decomposition, it emits toxic fumes of Cl^-.

FIRST AID   In case of contact with the eyes, immediately wash the eyes with large amounts of water, occasionally lifting the lower and upper lids. Get medical attention immediately. Contact lenses should not be worn when working with this chemical.

In case of contact with the skin, promptly wash the contaminated skin with soap and water. If this chemical penetrates through the clothing, promptly remove the clothing and wash the skin with soap and water. Get medical attention promptly.

If a person breathes in large amounts of this chemical, move the victim to fresh air at once. If breathing

has stopped, perform artificial respiration. Keep the affected person warm and at rest. Get medical attention as soon as possible.

If this chemical has been swallowed, get medical attention immediately.

FIRE HAZARD   Occasionally, high concentrations of trichloroethylene vapor in air at high temperature can burn if exposed to flame. Flames or arcs should not be used in closed areas which contain any solvent residue or vapor. Explosive in the form of vapor when exposed to heat or flame.

HANDLING   Although trichloroethylene has caused cancer in mice, its carcinogenic potency is so low that no special precautions are needed for laboratory work with it beyond normal good practices. This includes the use of a hood for most operations. Gloves should be used to prevent these solvents from damaging the skin.

If degreasing with trichloroethylene cannot be carried out without release of vapor, it should be done in a well-ventilated area. Operators should wear masks, gloves, and protective clothing to prevent repeated or prolonged skin contact. Wash promptly if skin becomes wet. Promptly remove any nonimpervious clothing that becomes wet.

Wear eye protection to reduce probability of eye contact.

Disposal should be done by a commercial disposal company.

See Chap. 3 for information on handling carcinogens.

TRIETHYLAMINE

CAS #121-44-8
DOT #1296
mf: $C_6H_{15}N$
mw: 101

STRUCTURE
$(C_2H_5)_3N$

NFPA LABEL

SYNONYMS
(Diethylamino)ethane
N,N-Diethylethanamine
TEN

PROPERTIES   Colorless liquid with a fishy, ammonialike odor; solubility in water 0.03%; miscible in alcohol and ether; melting point −167.82°C, boiling point −57.8°C; flash point −6.7°C (open cup); lower explosive limit 1.2%, upper explosive limit 8%; density 0.7255 at 25°C/4°C; vapor density 3.48

OSHA PEL   TWA 25 ppm; IDLH 1000 ppm

ACGIH TLV   TWA 10 ppm

DOT Classification   Flammable liquid; Label: Flammable liquid

TOXIC EFFECTS   Moderately toxic by ingestion and skin contact. Mildly toxic by inhalation. A skin and severe eye irritant.

Routes   Inhalation, ingestion, absorption, contact

Symptoms   Irritated eyes, respiratory system, and skin; kidney and liver damage

Target Organs   Respiratory system, eyes, skin, liver, kidneys

HAZARDOUS REACTIONS
Acids: incompatible
Dinitrogen tetroxide: undiluted complex explodes below 0°C
Maleic anhydride: exothermic reaction above 150°C
Oxidizing materials: incompatible

FIRST AID   In case of contact with the eyes, immediately wash the eyes with large amounts of water, occasionally lifting the lower and upper lids. Get medical attention immediately. Contact lenses should not be worn when working with this chemical.

In case of contact with the skin, immediately wash the contaminated skin with soap and water. If this chemical penetrates through the clothing, immediately remove the clothing and wash the skin with soap and water. Get medical attention promptly.

If a person breathes in large amounts of this chemical, move the victim to fresh air at once. If breathing has stopped, perform artificial respiration. Keep the affected person warm and at rest. Get medical attention as soon as possible.

If this chemical has been swallowed, get medical attention immediately.

FIRE HAZARD   A very dangerous fire hazard when exposed to heat, flame, or oxidizers. Explosive in the form of vapor when exposed to heat or flame.

To fight a fire, use CO_2, dry chemical, or alcohol foam extinguishers.

HANDLING   Wear appropriate clothing and equipment to prevent repeated or prolonged skin contact. Wash immediately if skin becomes contaminated. Immediately remove any clothing that becomes wet.

Wear eye protection to prevent any possibility of eye contact.

The following equipment should be available if liquids containing more than 1% TEA are present: an eyewash station and a quick-drench shower.

TRIFLUOROACETIC ACID

CAS #76-05-1 mf: $C_2HF_3O_2$
DOT #2699 mw: 114.03

SYNONYMS
Perfluoroacetic
Trifluoroethanoic acid

PROPERTIES   Colorless liquid with a strong, pungent odor; melting point −15.25°C; boiling point 71.1°C at 734 mm Hg; density 1.535 at 0°C

DOT Classification   Corrosive material; Label: Corrosive

EPA TSCA Inventory

TOXIC EFFECTS   Poisonous by ingestion. Mildly toxic by inhalation. A corrosive irritant to skin, eyes, and mucous membranes.

2,2,4-TRIMETHYLPENTANE

CAS #540-84-1 mf: C_8H_{18}
DOT #1262 mw: 114.26

SYNONYMS
Isobutyltrimethylethane
Isooctane

PROPERTIES   Clear liquid; odor of gasoline; boiling point 99.2°C, freezing point −116°C; flash point −12.2°C; density 0.692 at 20°C/4°C; lower explosive limit 1.1%, upper explosive limit 6.0%, autoignition temperature 415°C; vapor pressure 40.6 mm Hg at 21°C; vapor density 3.93

EPA TSCA Inventory

NIOSH REL   TWA 350 mg/m³

DOT Classification   Flammable liquid; Label: Flammable liquid

TOXIC EFFECTS   High concentrations can cause narcosis.

HAZARDOUS REACTIONS
Reducing materials: vigorous reaction

When heated to decomposition, it emits acrid smoke and irritating fumes.

FIRE HAZARD   A very dangerous fire hazard when exposed to heat, flame, or oxidizers. Explosive in the form of vapor when exposed to heat or flame.

To fight a fire, use CO_2 or dry chemical extinguishers.

TURPENTINE

CAS #8006-64-2 mf: $C_{10}H_{16}$
DOT #1299 mw: ≈136

NFPA LABEL

SYNONYMS
Gum spirits
Gum turpentine
Oil of turpentine
Oil of turpentine, rectified
Spirits of turpentine
Steam-distilled turpentine
Sulfate wood turpentine
Turpentine oil, rectified
Turpentine, steam distilled
Turps
Wood turpentine

PROPERTIES   Colorless liquid with a characteristic paint odor; insoluble in water; boiling point 154–170°C; density 0.854–0.868 at 25°C/25°C; flash point 35°C (closed cup); lower explosive limit 0.8%; autoignition temperature 253.3°C; vapor density 4.84; vapor pressure 5 mm Hg; Underwriters Laboratory classification 40–50, as a natural product it is a mixture of components

OSHA PEL   TWA 100 ppm; IDLH 1900 ppm

ACGIH TLV   TWA 100 ppm

DOT Classification   Flammable liquid; Label: Flammable liquid; Combustible liquid; Label: None; Flammable or combustible liquid; Label: Flammable liquid

EPA TSCA Inventory

TOXIC EFFECTS   Moderately toxic by ingestion and possibly other routes. Mildly toxic by inhalation. A human eye irritant. Irritating to eyes, skin, and respiratory system. A systemic kidney irritant.

Routes   Inhalation, absorption, ingestion, contact

Symptoms   Irritation of the eyes, nose, and throat; sensitization, headache, vertigo; hematuria, albuminuria, hallucinations or distorted perceptions; antipsychotic, pulmonary effects

Target Organs   Skin, eyes, kidneys, respiratory system

HAZARDOUS REACTIONS   Spontaneous heating is possible.

Calcium hypochlorite: can react violently
Chlorine: can react violently
Chromium hypochlorite: can react violently
Chromium trioxide: can react violently
Hexachloromelamine: can react violently
Oxidizing materials: incompatible
Tin tetrachloride: can react violently
Trichloromelamine: can react violently

When heated to decomposition, it emits acrid smoke and irritating fumes.

FIRST AID   In case of contact with the eyes, immediately wash the eyes with large amounts of water, occasionally lifting the lower and upper lids. Get medical attention immediately. Contact lenses should not be worn when working with this chemical.

In case of contact with the skin, promptly wash the contaminated skin with soap and water. If this chemical penetrates through the clothing, promptly remove the clothing and wash the skin with soap and water. Get medical attention promptly.

If a person breathes in large amounts of this chemical, move the victim to fresh air at once. If breathing has stopped, perform artificial respiration. Keep the affected person warm and at rest. Get medical attention as soon as possible.

If this chemical has been swallowed, get medical attention immediately.

FIRE HAZARD   A very dangerous fire hazard when exposed to heat or flame. Moderate explosion hazard in the form of vapor when exposed to flame. Turpentine-soaked combustible materials (e.g., rags, cardboard) may ignite spontaneously.

To fight a fire, use foam, CO_2, or dry chemical extinguishers.

HANDLING   Keep cool and ventilated.

Wear appropriate clothing and equipment to prevent repeated or prolonged skin contact. Wash promptly if the skin becomes contaminated. Immediately remove any clothing that becomes wet.

Wear eye protection to reduce probability of eye contact.

URANIUM AND ITS INSOLUBLE COMPOUNDS

CAS #7440-61-1   mf: U
DOT #9175   mw: 238.00

PROPERTIES   Heavy, silvery metal; melting point 1132°C, boiling point 3818°C; density 18.95; radioactive

OSHA PEL   TWA 0.25 mg/m^3; IDLH 30 mg/m^3

ACGIH TLV   TWA 0.2 mg/m^3

DOT Classification   Radioactive material; Label: Radioactive and flammable

TOXIC EFFECTS   A poison. Uranium can cause irreversible kidney damage.

Routes   Inhalation, contact, ingestion

Symptoms   Dermatitis, kidney damage

Target Organs   Skin, bone marrow, lymphatic system, kidneys

HAZARDOUS REACTIONS

Uranium
Air: incompatible
Ammonia: potentially violent reaction
Bromine trifluoride: potentially violent reaction
Carbon dioxide: incompatible
Carbon tetrachloride: incompatible
Chlorine: potentially violent reaction
Fluorine: potentially violent reaction
Nitric acid: incompatible
Nitrogen monoxide: potentially violent reaction
Nitryl fluoride: potentially violent reaction
Selenium: potentially violent reaction
Sulfur: potentially violent reaction
Trichloroethylene: potentially violent reaction
Water: potentially violent reaction

Uranium hydride
Halogenated hydrocarbons: incompatible
Strong oxidizers: incompatible
Water: incompatible

FIRST AID   In case of contact with the eyes, immediately wash the eyes with large amounts of water, occasionally lifting the lower and upper lids. Get medical attention immediately. Contact lenses should not be worn when working with this chemical.

In case of contact with the skin, promptly wash the contaminated skin with soap and water. If this chemical penetrates through the clothing, promptly remove the clothing and wash the skin with soap and water. Get medical attention promptly.

If a person breathes in large amounts of this chemical, move the victim to fresh air at once. If breathing has stopped, perform artificial respiration. Keep the affected person warm and at rest. Get medical attention as soon as possible.

If this chemical has been swallowed, get medical attention immediately.

FIRE HAZARD   Very dangerous fire hazard as the solid or dust when exposed to heat or flame.

HANDLING   Wear appropriate clothing and equipment to prevent repeated or prolonged skin contact. Wash promptly if the skin becomes contaminated. Promptly remove any nonimpervious clothing that becomes contaminated. Work clothing should be changed daily if it is probable that the clothing has been contaminated.

Wear eye protection to prevent any possibility of eye contact.

The following equipment should be available: an eyewash station.

URANIUM COMPOUNDS, SOLUBLE

OSHA PEL   TWA 0.05 mg/m^3; IDLH 20 mg/m^3

ACGIH TLV   TWA 0.2 mg/m^3

TOXIC EFFECTS   The toxicity of uranium compounds is directly proportional to their solubility. The most soluble uranium compounds are uranium hexafluoride, uranium nitrate, uranium hypochlorite, and the uranyl acetates, sulfates, and carbonates.

Routes   Inhalation, contact, ingestion

Symptoms   Lacrymation, conjunctivitis; short breath, cough, chest rales; nausea, vomiting; skin burns; casts in urine, albuminuria, high BUN

Target Organs   Respiratory system, blood, liver, lymphatic system, kidneys, skin, bone marrow

HAZARDOUS REACTIONS
Uranyl nitrates
Combustibles: incompatible

Uranium hexafluoride
water: incompatible

FIRST AID   In case of contact with the eyes, immediately wash the eyes with large amounts of water, occasionally lifting the lower and upper lids. Get medical attention immediately. Contact lenses should not be worn when working with this chemical.

In case of contact with the skin, immediately flush the contaminated skin with water. If this chemical penetrates the clothing, immediately remove the clothing and flush the skin with water. Get medical attention promptly.

If a person breathes in large amounts of this chemical, move the victim to fresh air at once. If breathing has stopped, perform artificial respiration. Keep the affected person warm and at rest. Get medical attention as soon as possible.

If this chemical has been swallowed, get medical attention immediately.

HANDLING   Wear appropriate equipment to prevent any possibility of skin contact with UF$_6$. Employees should wash immediately if skin has come in contact with UF$_6$ and daily when working with UF$_6$. Work clothing should be changed daily if it is probable that the clothing has been contaminated, or immediately if there is any possibility that it may have been contaminated with UF$_6$.

Wear eye protection to prevent any possibility of eye contact.

The following equipment should be available: an eyewash station and a quick-drench shower.

URETHANE

CAS #51-79-6   mf: $C_3H_7NO_2$
mw: 89.11

SYNONYMS
A 11032   Leucothane
Estane 5703   NSC 746
Ethyl carbamate   Pracarbamin
Ethylurethan   Pracarbamine
Ethyl urethane   U-compound
o-Ethylurethane   Urethan
Leucethane

PROPERTIES   Colorless, odorless crystals; very soluble in water, alcohol, ether; melting point 49°C, boiling point 184°C; density 0.9862; vapor pressure 10 mm Hg at 77.8°C; vapor density 3.07

IARC Cancer Review: Animal Sufficient Evidence, EPA TSCA Inventory, EPA Genetic Toxicology Program, and Community Right to Know List

TOXIC EFFECTS   Moderately toxic by ingestion and several other routes. A carcinogen. A powerful teratogen in mice.

Small amounts are found as a side product of processing in many alcoholic drinks. The beverages with the greatest concentrations are bourbons, sherries, and fruit brandies (some have 1,000 to 12,000 ppb urethane). Many whiskeys, table and dessert wines, brandies, and liqueurs contain potentially hazardous amounts of urethane. The allowable limit for urethane in alcoholic beverages became 125 ppb on January 1, 1989.

Symptoms   Depression of bone marrow, brain damage, central nervous system depression, nausea, vomiting

HAZARDOUS REACTIONS

Acids: reaction with heat forms ethanol, carbon dioxide, and ammonia
Alkalies: reaction with heat forms ethanol, carbon dioxide, and ammonia
Phosphorus pentachloride: reaction forms an explosive product

When heated, it emits toxic fumes of NO_x.

HANDLING   See Chap. 3 for information on handling carcinogens.

VINYL CHLORIDE

CAS #75-01-4 mf: C_2H_3Cl
DOT #1086 mw: 62.50

NFPA LABEL

SYNONYMS

Chlorethene
Chlorethylene
Chloroethene
Chloroethylene
Ethylene monochloride
Monochloroethene
Monochloroethylene (DOT)
Trovidur
VC
VCM
Vinyl chloride monomer
Vinyl C monomer

PROPERTIES   Colorless gas which is usually cooled and handled as a liquid; faintly sweet odor; slightly soluble in water, soluble in alcohol, very soluble in ether; melting point −153.7°C, freezing point −159.7°C, boiling point −13.8°C; density (liquid) 0.9195 at 15°C/4°C; flash point −77°C; lower explosive limit 3.6%, upper explosive limit 33%; autoignition temperature 472°C; vapor pressure 2600 mm Hg at 25°C; vapor density 2.15

OSHA PEL   TWA 1 ppm; ceiling 5 ppm/15 min; cancer suspect agent

ACGIH TLV   TWA 5 ppm; human carcinogen

NIOSH REL   TWA (vinyl chloride), lowest detectable level

DOT Classification   Flammable gas; Label: Flammable gas

EPA TSCA Inventory, EPA Genetic Toxicology Program, Community Right to Know List, and IARC Cancer Review: Animal Sufficient Evidence, Human Sufficient Evidence

TOXIC EFFECTS   Poisonous by inhalation. Moderately toxic by ingestion. A carcinogen which causes liver and blood tumors. A teratogen which when inhaled causes defects in spermatogenesis. A severe irritant to skin, eyes, and mucous membranes.

Causes skin burns by rapid evaporation and consequent freezing. In high concentration it acts as an anesthetic.

Routes   Inhalation

Symptoms   Weakness, abdominal pain, gastrointestinal bleeding, hematomegaly, pallor or cyanosis of extremities, liver damage; a carcinogen

Target Organs   Liver, central nervous system, blood, respiratory system, lymphatic system

HAZARDOUS REACTIONS   Long-term exposure to air may result in formation of peroxides which can initiate explosive polymerization of the chloride.

Air: peroxidation may lead to explosive polymerization reaction
Oxides of nitrogen: explodes on contact
Oxidizing materials: can react vigorously

When heated to decomposition, it emits highly toxic fumes of Cl^-.

FIRST AID   If a person breathes in large amounts of this chemical, move the victim to fresh air at once. If breathing has stopped, perform artificial respiration. Keep the affected person warm and at rest. Get medical attention as soon as possible.

FIRE HAZARD   A very dangerous fire hazard when exposed to heat, flame, or oxidizers. Large fires of this material are practically inextinguishable. A severe explosion hazard in the form of vapor when exposed to heat or flame.

To extinguish fires, stop flow of gas.

HANDLING   Before storing or handling this material, instructions for its use should be obtained from the supplier.

Vinyl chloride is regulated by OSHA as a human carcinogen. The regulations, which call for more stringent precautions than Procedure A (see Chap. 3), should be consulted before work with vinyl chloride is begun. Contact with liquid vinyl chloride is prohibited. A monitoring program is required for all vinyl chloride operations. Exposures to vinyl chloride must be controlled by feasible engineering controls or work practices. Use of closed systems or laboratory hoods that have protection factors adequate to prevent significant worker exposure are recommended. Whenever respirators are required, they must be used in accordance with a standard respirator program.

Wear appropriate clothing and equipment to prevent any possibility of skin contact. Wash immediately if skin becomes contaminated, and at the end of each work shift. Immediately remove any clothing that becomes contaminated. Work clothing should be changed daily if there is any possibility that the clothing has been contaminated.

Wear eye protection to prevent any possibility of eye contact.

The following equipment should be available: an eyewash station and a quick-drench shower.

See Chap. 3 for information on handling carcinogens.

VINYLACETYLENE

CAS #689-97-4   mf: C_4H_4
   mw: 52.08

STRUCTURE
HC≡CCH=CH$_2$

SYNONYM
Buten-3-yne

HAZARDOUS REACTIONS   Very exothermic decomposition when heated.

Air: forms explosive peroxides
1,3-Butadiene: explosive reaction when heated
Oxygen: explosive reaction when heated; forms explosive peroxides

Silver nitrate: reaction forms explosive silver buten-3-ynide

When heated to decomposition, it emits acrid smoke and irritating fumes.

XYLENE

CAS #1330-20-7   mf: C_8H_{10}
DOT #1307   mw: 106.18

NFPA LABEL

SYNONYMS
Dimethylbenzene   Violet 3
Methyl toluene   Xylol (DOT)

PROPERTIES   Colorless liquid; solubility in water 0.00003%; melting points (o-, m-, p-) −24.4/−47.8/55°C, boiling point 144.4/138.9/138.3°C; density 0.864 at 20°C/4°C; flash point 32.2/28.9/27.2°C; lower explosive limits 1/1.1/1.1%, upper explosive limits 6/7/7%; vapor pressure 7/9/9 mm Hg at 21°C

OSHA PEL   TWA 100 ppm; IDLH 1000 ppm

ACGIH TLV   TWA 100 ppm; STEL 150 ppm

NIOSH REL   TWA (xylene) 100 ppm; ceiling 200 ppm/10 min

DOT Classification   Flammable liquid; Label: Flammable liquid; Flammable or combustible liquid; Label: Flammable liquid

EPA TSCA Inventory, EPA Genetic Toxicology Program, and Community Right to Know List

TOXIC EFFECTS   Mildly toxic by ingestion and inhalation. A teratogen. A skin and severe eye irritant. Irritation can start at 200 ppm.

Routes   Inhalation

Symptoms   Irritated eyes, nose, and throat; corneal vacuolization; dizziness, excitement, drowsiness, incoherence, staggering gait; anorexia, nausea, vomiting, abdominal pain; dermatitis

Target Organs   Central nervous system, eyes, gastrointestinal tract, blood, liver, kidneys, skin

HAZARDOUS REACTIONS
Acetic acid + air: potentially explosive reaction
1,3-Dichloro-5,5-dimethyl-2,4-imidazolidindione: explosive reaction

Nitric acid: potentially explosive reaction under pressure
Oxidizing materials: incompatible

When heated to decomposition, it emits acrid smoke and irritating fumes.

FIRST AID   In case of contact with the eyes, immediately wash the eyes with large amounts of water, occasionally lifting the lower and upper lids. Get medical attention immediately. Contact lenses should not be worn when working with this chemical.

In case of contact with the skin, promptly wash the contaminated skin with soap and water. If this chemical penetrates through the clothing, promptly remove the clothing and wash the skin with soap and water. Get medical attention promptly.

If a person breathes in large amounts of this chemical, move the victim to fresh air at once. If breathing has stopped, perform artificial respiration. Keep the affected person warm and at rest. Get medical attention as soon as possible.

If this chemical has been swallowed, get medical attention immediately.

FIRE HAZARD   A very dangerous fire hazard when exposed to heat or flame.

To fight a fire, use foam, CO_2, or dry chemical extinguishers.

HANDLING   Wear appropriate clothing and equipment to prevent repeated or prolonged skin contact. Wash promptly if skin becomes contaminated. Immediately remove any clothing that becomes wet.

Wear eye protection to reduce probability of eye contact.

m-XYLENE

CAS #108-38-3 mf: C_8H_{10}
DOT #1307 mw: 106.18

NFPA LABEL

SYNONYMS
m-Dimethylbenzene   1,3-Xylene
1,3-Dimethylbenzene   m-Xylol (DOT)

PROPERTIES   Colorless liquid; insoluble in water; miscible with alcohol, ether, and some organic solvents; melting point −47.9°C, boiling point 139°C; density 0.864 at 20°C/4°C; flash point 25°C; lower explosive limit 1.1%, upper explosive limit 7.0%; autoignition temperature 530°C; vapor pressure 10 mm Hg at 28.3°C; vapor density 3.66

For further information, see xylene.

o-XYLENE

CAS #95-47-6 mf: C_8H_{10}
DOT #1307 mw: 106.18

NFPA LABEL

SYNONYMS
o-Dimethylbenzene   1,2-Xylene
1,2-Dimethylbenzene   o-Xylol (DOT)
o-Methyltoluene

PROPERTIES   Colorless liquid; insoluble in water; miscible in absolute alcohol, ether; density 0.880 at 20°C/4°C; melting point 25.2°C, boiling point 144.4°C; flash point 17°C; lower explosive limit 1.0%, upper explosive limit 6.0%

For further information, see xylene.

p-XYLENE

CAS #106-42-3 mf: C_8H_{10}
DOT #1307 mw: 106.18

NFPA LABEL

SYNONYMS
Chromar   Scintillar
p-Dimethylbenzene   1,4-Xylene
1,4-Dimethylbenzene   p-Xylol (DOT)
p-Methyltoluene

PROPERTIES   Colorless liquid which crystallizes when cooled; insoluble in water; soluble in alcohol, ether, organic solvents; melting point 13–14°C, boiling point 138.3°C; density 0.8611 at 20°C/4°C; flash point 25°C (closed cup); lower explosive limit 1.1%, upper explosive limit 7.0%; autoignition temperature 530°C;

vapor pressure 10 mm Hg at 27.3°C; vapor density 3.66

For further information, see xylene.

ZINC ACETATE

CAS #557-34-6 mf: $C_4H_6O_4 \cdot Zn$
DOT #9153 mw: 183.47

SYNONYMS
Acetic acid, zinc salt
Dicarbomethoxyzinc
Zinc diacetate

PROPERTIES   Crystals, astringent taste; very soluble in water, somewhat soluble in alcohol; melting point 237°C; density 1.735

DOT Classification   ORM-E; Label: None

EPA TSCA Inventory and Community Right to Know List (zinc and its compounds)

TOXIC EFFECTS   Moderately toxic by ingestion.

HAZARDOUS REACTIONS
Alkalies and their carbonates: incompatible
Oxalates: incompatible
Phosphates: incompatible
Sulfides: incompatible
Zinc salts: incompatible

When heated to decomposition, it emits toxic fumes of ZnO.

ZINC CHLORIDE

CAS #17646-85-7 mf: Cl_2Zn
DOT #1840; 2331 mw: 136.27

NFPA LABEL

SYNONYMS
Butter of zinc
Tinning flux (DOT)
Zinc chloride, anhydrous (DOT)
Zinc chloride, solid (DOT)
Zinc chloride, solution (DOT)
Zinc dichloride
Zinc muriate, solution (DOT)

PROPERTIES   Odorless, cubic white, deliquescent crystals; solubility in water 81%; melting point 290°C, boiling point 732°C; density 2.91 at 25°C; vapor pressure 1 mm Hg at 428°C

OSHA PEL   TWA 1 mg(fume)/m³; IDLH 2000 mg(fume)/m³

ACGIH TLV   TWA 1 mg/m³; STEL 2 mg/m³ (fume)

DOT Classification   Corrosive material; Label: Corrosive and corrosive, solution; ORM-E; Label: None, solid

EPA TSCA Inventory, EPA Genetic Toxicology Program, and Community Right to Know List (zinc and its compounds)

TOXIC EFFECTS   Poisonous by inhalation and ingestion. A tumorigen and teratogen. A corrosive irritant to skin, eyes, and mucous membranes.

The fumes or dusts can cause dermatitis, boils, conjunctivitis, and gastrointestinal tract upsets.

Routes   Inhalation

Symptoms   Irritated nose and throat; burning skin and eyes; conjunctivitis; cough, copious sputum, dyspnea, chest pain, pulmonary edema, bronchial pneumonia; pulmonary fibrosis, acute right heart strain or chronic right ventricular hypertrophy; fever; cyanosis; tachypnea

Target Organs   Respiratory system, skin, eyes

HAZARDOUS REACTIONS
Potassium: incompatible
Zinc powder: mixtures with the powdered chloride are flammable

When heated to decomposition, it emits toxic fumes of Cl⁻ and ZnO.

FIRST AID   If a person breathes in large amounts of this chemical, move the victim to fresh air at once. If breathing has stopped, perform artificial respiration. Keep the affected person warm and at rest. Get medical attention as soon as possible.

ZINC NITRATE

CAS #7779-88-6 mf: $N_2O_6 \cdot Zn$
DOT #1514 mw: 189.39

NFPA LABEL

SYNONYMS Nitric acid, zinc salt

PROPERTIES There are two forms, both very soluble in alcohol and soluble in water.

Trihydrate Needle-shaped crystals; melting point 42.5°C

Hexahydrate Tetragonal, colorless crystals; melting point 36.4°C; loses $6H_2O$ at 105–131°C; density 2.065 at 14°C

DOT Classification Oxidizer; Label: Oxidizer

EPA TSCA Inventory and Community Right to Know List (zinc and its compounds)

HAZARDOUS REACTIONS A powerful oxidizer.

Carbon: violent reaction
Copper: violent reaction
Metal sulfides: violent reaction
Organic matter: violent reaction
Phosphorus: violent reaction
Sulfur: violent reaction

When heated to decomposition, it emits toxic fumes of NO_x and ZnO.

ZINC SULFATE

CAS #7733-02-0 mf: $O_4S \cdot Zn$
DOT #9161 mw: 161.43

SYNONYMS
Bonazen
Bufopto zinc sulfate
Op-thal-zin
Sulfuric acid, zinc salt (1:1)
White copperas
White vitriol
Verazinc
Zinc sulphate
Zinc vitriol
Zinkosite

PROPERTIES Rhombic colorless crystals; soluble in water, almost insoluble in alcohol; decomposes at 740°C; density 3.74 at 15°C

DOT Classification ORM-E; Label: None

EPA TSCA Inventory, EPA Genetic Toxicology Program, and Community Right to Know List (zinc and its compounds)

TOXIC EFFECTS Moderately toxic by ingestion. A tumorigen and teratogen. An eye irritant.

Symptoms Increased pulse rate without blood pressure decrease, blood pressure decrease, acute pulmonary edema, normocytic anemia, hypermotility, diarrhea and other gastrointestinal changes

HAZARDOUS REACTIONS When heated to decomposition, it emits toxic fumes of SO_x and ZnO.

ZIRCONIUM NITRATE

CAS #13746-89-9 mf: $N_4O_{12} \cdot Zr$
DOT #2728 mw: 339.26

NFPA LABEL

PROPERTIES White crystals.

OSHA PEL TWA 5 mg(Zr)/m³

ACGIH TLV TWA 5 mg(Zr)/m³; STEL 10 mg(Zr)/m³

DOT Classification Oxidizer; Label: Oxidizer

EPA TSCA Inventory

TOXIC EFFECTS Moderately toxic by inhalation and ingestion.

HAZARDOUS REACTIONS When heated to decomposition, it emits toxic fumes of NO_x.

ZIRCONIUM OXYCHLORIDE

CAS #7699-43-6 mf: $ZrCl_2O$
mw: 178.12

SYNONYMS
Basic zirconium chloride
Chlorozirconyl
Dichlorooxozirconium
Zirconyl chloride

PROPERTIES Crystals; very soluble in water, alcohol; density 1.91

OSHA PEL TWA 5 mg(Zr)/m³

ACGIH TLV TWA 5 mg(Zr)/m³; STEL 10 mg(Zr)/m³

EPA TSCA Inventory

TOXIC EFFECTS Moderately toxic by ingestion.

HAZARDOUS REACTIONS When heated to decomposition, it emits toxic fumes of Cl^-.

Part 3
APPENDICES

APPENDIX A: HAZARD RATINGS AND CLASSIFICATIONS

TOXICITY RATING SCALE

Rating	Description	Rat Oral LD_{50} (mg/kg)	Rat 4-h LC_{50} (ppm)[a]	Probable Lethal Dose for Adult
1	Extremely toxic	<1	<10	taste (1 grain)
2	Highly toxic	1–50	10–100	4 cc (teaspoon)
3	Moderately toxic	50–500	100–1,000	30 g (1 oz)
4	Slightly toxic	500–5,000	1,000–10,000	250 g (1 pt)
5	Practically nontoxic	5,000–15,000	10,000–100,000	1 kg (1 qt)
6	Relatively harmless	>15,000	>100,000	>1 kg (1 qt)

[a]Concentration in air that kills 50% of test animals in 4 h.
Source: Reprinted from H. C. Hodge and J. H. Sterner, "Tabulation of Toxicity Classes," *American Industrial Hygiene Association Quarterly*, Vol. 10, No. 4, Dec. 1949, p. 93, by permission of the American Industrial Hygiene Association.

FIRE HAZARD RATINGS AND CLASSES

Class	Criteria	Fire Hazard
IA	Boiling point <100°F Flash point ≥73°F	4
IB	Boiling point at or above 100°F	3
IC	Boiling point not considered Flash point 73–100°F	3
II	Flash point 100–140°F	2
IIIA	Flash point 140–200°F	2
IIIB	Flash point >200°F	1
	No flash point or nonflammable	0

UNDERWRITERS LABORATORIES SCALE OF RELATIVE HAZARD OF VARIOUS FLAMMABLE LIQUIDS[a]

Ether class	100
Gasoline class	90–100
Alcohol (ethyl) class	60–70
Kerosene class	30–40
Paraffin oil class	10–20

[a]A flammable liquid is any liquid having a flash point below 100°F and a vapor pressure not exceeding 40 lb/in.^{2} (absolute) at 100°F. A combustible liquid is any liquid having a flash point at or above 100°F.

APPENDIX B: NATIONAL FIRE PROTECTION ASSOCIATION (NFPA) LABELS

	Identification of Health Hazard Color Code: Blue		Identification of Flammability Color Code: Red		Identification of Reactivity (Stability) Color Code: Yellow
	Type of Possible Injury		**Susceptibility of Materials to Burning**		**Susceptibility to Release of Energy**
4	Materials which on very short exposure could cause death or major residual injury even though prompt medical treatment were given.	4	Materials which will rapidly or completely vaporize at atmospheric pressure and normal ambient temperature, or which are readily dispersed in air and which will burn readily.	4	Materials which in themselves are readily capable of detonation or of explosive decomposition or reaction at normal temperatures and pressures.
3	Materials which on short exposure could cause serious temporary or residual injury even though prompt medical treatment were given.	3	Liquids and solids that can be ignited under almost all ambient temperature conditions.	3	Materials which in themselves are capable of detonation or explosive reaction but require a strong initiating source or which must be heated under confinement before initiation or which react explosively with water.
2	Materials which on intense or continuous exposure could cause temporary incapacitation or possible residual injury unless prompt medical treatment is given.	2	Materials that must be moderately heated or exposed to relatively high ambient temperatures before ignition can occur.	2	Materials which in themselves are normally unstable and readily undergo violent chemical change but do not detonate. Also materials which may react violently with water or which may form potentially explosive mixtures with water.
1	Materials which on exposure would cause irritation but only minor residual injury even if no treatment is given.	1	Materials that must be preheated before ignition can occur.	1	Materials which in themselves are normally stable, but which can become unstable at elevated temperatures and pressures or which may react with water with some release of energy but not violently.

Identification of Health Hazard Color Code: Blue		Identification of Flammability Color Code: Red		Identification of Reactivity (Stability) Color Code: Yellow	
Type of Possible Injury		**Susceptibility of Materials to Burning**		**Susceptibility to Release of Energy**	
0	Materials which on exposure under fire conditions would offer no hazard beyond that of ordinary combustible material.	0	Materials that will not burn.	0	Materials which in themselves are normally stable, even under fire exposure conditions, and which are not reactive with water.

```
         FIRE
         (red)
HEALTH        SAFETY
(blue)       (yellow)
```

Source: Reprinted with permission from *Identification of the Fire Hazards of Materials,* NFPA 704-1990, copyright 1990, National Fire Protection Association, Quincy, Mass. This reprinted material is not the complete and official position of the National Fire Protection Association on the referenced subject, which is represented only by the standard in its entirety.

DIMENSIONS OF NFPA LABELS

Distances at Which Signals Must Be Legible (ft)	Size of Signals Required (in.)
50	1
75	2
100	3
200	4
300	6

Source: Reprinted with permission from *Identification of the Fire Hazards of Materials,* NFPA 704-1990, copyright 1990, National Fire Protection Association, Quincy, Mass. This reprinted material is not the complete and official position of the National Fire Protection Association on the referenced subject, which is represented only by the standard in its entirety.

NFPA HAZARD RATINGS AND SIGNAL WORDS

Rating	Health Hazards	Flammability	Reactivity
4	Extreme health hazard	Extremely flammable	Extremely reactive
3	High health hazard	Highly flammable	Highly reactive
2	Moderate health hazard	Moderately flammable	Moderately reactive
1	Slight health hazard	Slightly flammable	Slightly reactive
0	No significant health hazard	Noncombustible	Nonreactive

APPENDIX C: HAZARDOUS MATERIALS WARNING LABELS

The following information is provided to assist those involved in the storage, handling, and transportation of hazardous materials by presenting potential hazards and precautions associated with warning labels.

The presence of a warning label on a package affords the same quick and easy recognition of hazards to warehouse personnel as is offered to the shippers. For complete details, refer to one or more of the following:

Code of Federal Regulations (CFR), Title 49, Transportation, Parts 100–199 (all modes of transport). Section numbers listed in this appendix refer to this CFR.

International Civil Aviation Organization (ICAO) Technical Instructions for the Safe Transport of Dangerous Goods by Air (air transport).

International Maritime Organization (IMO) Dangerous Goods Code (water transport).

Canadian Transport Commission (CTC) Regulations (rail transport).

GENERAL GUIDELINES ON THE USE OF LABELS

Labels described in this appendix are used on domestic shipments. Domestic warning labels may display UN class number, division number (and compatibility group for explosives only) [49 CFR, Section 172.407(g)].

Any person who offers a hazardous material for transportation MUST label the package, if required [49 CFR, Section 172.400(A)].

Label(s), when required, must be printed on or affixed to the surface of the package near the proper shipping name [49 CFR, Section 172.406(a)].

When two or more different labels are required, they must be displayed next to each other [49 CFR, Section 172.406(c)].

Labels may be affixed to packages (even when not required by regulations), provided each label represents a hazard of the material in the package [49 CFR, Section 172.401].

The Hazardous Materials Tables [49 CFR, Sections 172.101 and 172.102] identify the proper label(s) for the hazardous materials listed.

UN CLASS NUMBERS

Hazardous materials class numbers are as follows:

Class 1 Explosives

Class 2 Gases (compressed, liquefied, or dissolved under pressure)

Class 3 Flammable liquids

Class 4 Flammable solids or substances

Class 5 Oxidizing substances

Class 6 Poisonous and infectious substances

Class 7 Radioactive substances

Class 8 Corrosives

Class 9 Miscellaneous dangerous substances

UN Class Numbers appear at the bottom of most diamond labels (domestic and international) and placards. See Fig. C-1.

FIGURE C-1. Location of the UN class numbers. (From Department of Defense, *Hazardous Materials Storage and Handling Handbook,* U.S. Government Printing Office, Washington, D.C., July 1987, p. 82.)

Explosives (UN Class 1)

Examples:

A Dynamite
B Propellants or flares
C Common fireworks

Explosive A: capable of exploding with a small spark, shock, or flame and spreading the explosion hazard to other packages.

Explosive B: very rapidly combustible.

Explosive C: low hazard but may explode under high heat when many are tightly packed together.

Hazards/Precautions

No flares, smoking, flames, or sparks in the hazard area.

May explode if dropped, heated, or sparked.

See Fig. C-2.

Compressed Gases (UN Class 2)

Examples:

Acetylene
Chlorine

Items in this class require storage and handling under pressure in compressed gas cylinders.

Hazards/Precautions

Container may explode in heat or fire.

Contact with liquid may cause frostbite.

May be flammable, poisonous, explosive, irritating, corrosive, or suffocating.

FIGURE C-2. Label for explosives, UN class 1. (From Department of Defense, *Hazardous Materials Storage and Handling Handbook,* U.S. Government Printing Office, Washington, D.C., July 1987, p. 83.)

May be *extremely hazardous.*

See Fig. C-3.

Flammable Liquids (UN Class 3)

Examples:

Ether
Acetone
Gasoline
Toluene
Pentane

This class includes liquids with a flash point less than 100 ° F.

FIGURE C-3. Label for compressed gases, UN class 2. (From Department of Defense, *Hazardous Materials Storage and Handling Handbook,* U.S. Government Printing Office, Washington, D.C., July 1987, p. 84.)

Hazards/Precautions

No flares, smoking, flames, or sparks in the hazard area.

Vapors are an explosion hazard.

Can be poisonous; check labels.

If it is poisonous, it can cause death when inhaled, swallowed, or touched.

See Fig. C-4.

Flammable Solids (UN Class 4)

Examples:

Calcium resinate
Potassium metal
Sodium amide

This class includes any solid material which, under certain conditions, might cause a fire or which can be ignited readily and burns vigorously.

Hazards/Precautions

May ignite when exposed to air or moisture.

May reignite after being extinguished.

Fires may produce irritating or poisonous gases.

Contact may cause burns to skin or eyes.

See Fig. C-5.

Dangerous When Wet (UN Class 4)

Examples:

Magnesium metal
Aluminum phosphide

FIGURE C-4. Label for flammable liquids, UN class 3. (From Department of Defense, *Hazardous Materials Storage and Handling Handbook,* U.S. Government Printing Office, Washington, D.C., July 1987, p. 85.)

FIGURE C-5. Label for flammable solids, UN class 4. (From Department of Defense, *Hazardous Materials Storage and Handling Handbook,* U.S. Government Printing Office, Washington, D.C., July 1987, p. 86.)

Lithium hydride
Calcium carbide

These items include flammable solids that react with water.

Hazards/Precautions

May ignite in the presence of moisture.

May reignite after fire has been extinguished.

Contact may cause burns to skin and eyes.

Skin contact may be poisonous.

Inhalation of vapors may be harmful.

Flames or smoking should not be permitted in the vicinity.

See Fig. C-6.

Oxidizing Materials (UN Class 5)

Examples:

Calcium permanganate
Calcium hypochlorite
Barium perchlorate

FIGURE C-6. Label for materials that are dangerous when wet, UN class 4. (From Department of Defense, *Hazardous Materials Storage and Handling Handbook,* U.S. Government Printing Office, Washington, D.C., July 1987, p. 87.)

Hydrogen peroxide

Ammonium nitrate

Items in this class are chemically reactive and will provide both heat and oxygen to support a fire.

Hazards/Precautions

May ignite combustibles (wood, paper, etc.).

Reaction with fuels may be violent.

Fires may produce poisonous fumes.

Vapors and dusts may be irritating.

Contact may burn skin and eyes.

Peroxides may explode from heat or contamination.

See Fig. C-7.

Poisonous Materials (Class B Poisons, UN Class 6)

Examples:

Cyanogen gas

Lead cyanide

Parathion

These poisons are extremely toxic to man and animals.

Hazards/Precautions

May cause death quickly if breathed, swallowed, or touched.

May be flammable, explosive, corrosive, or irritating.

May be *extremely hazardous.*

Look for the "skull and crossbones" on the label.

Degree-of-hazard key words:

"Poison"   Highly toxic
"Danger"   Moderately toxic
"Warning" Least toxic

Read label carefully for storage and safety information.

See Fig. C-8.

Irritating Materials (UN Class 6)

Examples:

Tear gas

Riot-control agents

This class includes materials capable of causing discomfort such as tearing, choking, vomiting, or skin irritation.

Hazards/Precautions

May cause difficulty in breathing.

May burn, but does not ignite readily.

Exposure in enclosed areas may be harmful.

May cause tearing of the eyes, choking, nausea, or skin irritation.

See Fig. C-9.

Biomedical Materials (UN Class 6)

Examples:

Live virus vaccines

Etiologic agents

FIGURE C-7. Label for oxidizing materials, UN class 5. (From Department of Defense, *Hazardous Materials Storage and Handling Handbook,* U.S. Government Printing Office, Washington, D.C., July 1987, p. 88.)

FIGURE C-8. Label for poisonous materials, class B poisons, UN class 6. (From Department of Defense, *Hazardous Materials Storage and Handling Handbook,* U.S. Government Printing Office, Washington, D.C., July 1987, p. 89.)

FIGURE C-9. Label for irritating materials, UN class 6. (From Department of Defense, *Hazardous Materials Storage and Handling Handbook,* U.S. Government Printing Office, Washington, D.C., July 1987, p. 90.)

Materials in this class can cause human disease (infectious/etiological agents).

Hazards/Precautions

May be ignited if carrier is flammable.

Contact may cause infection/disease.

Damage to outer container may not affect inner container.

See Fig. C-10.

Radioactive Materials (UN Class 7)

Examples:

Thorium-232

Carbon-14

Radium-226

The degree of hazard will vary depending on the type and quantity of material.

Hazards/Precautions

Avoid touching broken or damaged radioactive items.

Persons handling damaged items must wear rubber or plastic gloves.

Damaged items are to be monitored and safely packaged under the surveillance of the radiological monitor.

Persons having come in direct contact with damaged or broken radioactive items are to move away from the spill site (but stay in the area) to be monitored and decontaminated.

See Fig. C-11.

Corrosives (UN Class 8)

Examples:

Sodium hydroxide

Hydrochloric acid

Alkaline liquids

This class includes materials that cause destruction to human tissue and corrode metal (i.e., steel), and thus many packaging materials, upon contact.

Hazards/Precautions

Contact causes burns to skin and eyes.

May be harmful if breathed.

Fire may produce poisonous fumes.

May react violently with water.

May ignite combustibles.

Explosive gases may accumulate.

See Fig. C-12.

FIGURE C-10. Label for biological hazards, UN class 6. (From Department of Defense, *Hazardous Materials Storage and Handling Handbook,* U.S. Government Printing Office, Washington, D.C., July 1987, p. 91.)

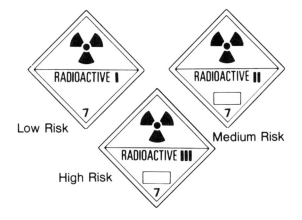

FIGURE C-11. Label for radioactive materials, UN class 7. (From Department of Defense, *Hazardous Materials Storage and Handling Handbook,* U.S. Government Printing Office, Washington, D.C., July 1987, p. 92.)

FIGURE C-12. Label for corrosives, UN class 8. (From Department of Defense, *Hazardous Materials Storage and Handling Handbook,* U.S. Government Printing Office, Washington, D.C., July 1987, p. 93.)

Other Regulated Materials (ORMs) (49 CFR, Part 172.316)

This classification applies to materials that may pose an unreasonable risk to health and safety or property, and are not covered under the hazardous materials warning label requirements.

ORM-A. ORM-A includes materials with an anesthetic, irritating, noxious, toxic, or other property that can cause discomfort to persons in the event of leakage.

See Fig. C-13.

Examples include trichloroethylene, 1,1,1-trichloroethane, dry ice, chloroform, and carbon tetrachloride.

ORM-B. ORM-B includes materials specifically named or capable of causing significant corrosion damage form leakage.

See Fig. C-14.

Examples include lead chloride, quicklime, metallic mercury, and barium oxide.

ORM-C. ORM-C covers materials specifically named and with characteristics which make them unsuitable for shipment unless properly packaged.

FIGURE C-13. ORM-A label. (From Department of Defense, *Hazardous Materials Storage and Handling Handbook,* U.S. Government Printing Office, Washington, D.C., July 1987, p. 94.)

FIGURE C-14. ORM-B label. (From Department of Defense, *Hazardous Materials Storage and Handling Handbook,* U.S. Government Printing Office, Washington, D.C., July 1987, p. 95.)

See Fig. C-15.

Examples include bleaching powder, lithium batteries (for disposal), magnetized materials, sawdust, and asbestos.

ORM-D. ORM-D covers materials such as consumer commodities which present a limited hazard due to form, quantity, or packaging. They must be materials for which exceptions are provided.

See Fig. C-16.

Examples include chemical consumer commodities (e.g., hairspray and shaving lotion) and small-arms ammunition (reclassified because of packaging).

ORM-E. ORM-E covers materials that are not included in any other hazard class, but are regulated as ORM.

See Fig. C-17.

Examples include hazardous waste and hazardous substances.

Special Handling

Additional labeling is to be used with DOT hazardous materials warning labels, as required (49 CFR, Part 172.402).

See Fig. C-18.

FIGURE C-15. ORM-C label. (From Department of Defense, *Hazardous Materials Storage and Handling Handbook,* U.S. Government Printing Office, Washington, D.C., July 1987, p. 95.)

FIGURE C-16. ORM-D label. (From Department of Defense, *Hazardous Materials Storage and Handling Handbook,* U.S. Government Printing Office, Washington, D.C., July 1987, p. 96.)

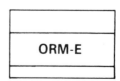

FIGURE C-17. ORM-E label. (From Department of Defense, *Hazardous Materials Storage and Handling Handbook,* U.S. Government Printing Office, Washington, D.C., July 1987, p. 96.)

Examples include asbestos (29 CFR, Part 1910.1001), for which the label would read:

CAUTION: Contains Asbestos Fibers.
Avoid Creating Dust.
Breathing Asbestos Dust May Cause
Serious Bodily Harm.

Special-Item Markings

Examples of items requiring special markings include:

PCBs (40 CFR, Part 761.40)

Hazardous waste (40 CFR, Part 262.32)

See Fig. C-19.

Manufacturer's Labels

Hazard Communication label (29 CFR, Part 1910.1200)

- Applies to the hazards of all chemicals manufactured or imported.
- Ensures that information on chemical hazards is properly transmitted.

FIGURE C-19. Special-item labels. (From Department of Defense, *Hazardous Materials Storage and Handling Handbook,* U.S. Government Printing Office, Washington, D.C., July 1987, p. 98.)

Cargo Aircraft only Label | Magnetized Material Label

Bung Label | Empty Label

FIGURE C-18. Examples of labels describing special handling. (From Department of Defense, *Hazardous Materials Storage and Handling Handbook,* U.S. Government Printing Office, Washington, D.C., July 1987, p. 97.)

Substance liable to Spontaneous Combustion | Poisonous Substance

Poisonous Substance | Infectious Substance

FIGURE C-20. Examples of international labels. (From Department of Defense, *Hazardous Materials Storage and Handling Handbook,* U.S. Government Printing Office, Washington, D.C., July 1987, p. 100.)

Label contents: product identity, hazard warnings, and name and address of the manufacturer, importer, or other responsible party.

Pesticides (40 CFR, Part 162.10)

Applies to all pesticide products.

Label contents: product identity, producer/registrant data, net contents, product registration number, ingredients statement, warning or precautionary statement, directions for use, and use classification.

International Labeling

Examples of international labels are shown in Fig. C-20. These are examples of international labels that are not presently used for domestic shipments. Most of the domestic labels may be used internationally.

Text, when used internationally, may be in the language of the country of origin.

Text is mandatory on radioactive materials, materials marked with St. Andrew's cross, and infectious substance labels.

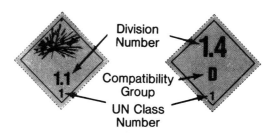

FIGURE C-21. International explosives labels. (From Department of Defense, *Hazardous Materials Storage and Handling Handbook,* U.S. Government Printing Office, Washington, D.C., July 1987, p. 101.)

International Labeling for Explosives

Examples of explosives labels are shown in Fig. C-21.

USEFUL REFERENCE

Hazardous Materials Storage and Handling Handbook, U.S. Government Printing Office, Washington, D.C., 1987.

APPENDIX D: HAZARDOUS LABORATORY SUBSTANCES

Assessment of the chemicals in this list indicated that their hazardous nature is greater than their potential usefulness in many secondary school science programs. Evaluation included toxicity, carcinogenicity, teratogenicity, flammability, and explosive propensity. These chemicals should be removed from schools if alternatives can be used. For those that must be retained, amounts should be kept to a minimum. Irritants and corrosives can be used with caution and approved safety procedures.

Chemical Name	CAS Number	Hazardous Properties
Acetaldehyde	75-07-0	IARC 3; irritant
Acetamide		Remove if feasible; animal carcinogen
Acetic acid		Corrosive
Acetic anhydride		Irritant
N-Acetoxy-2-acetylaminofluorene (N-AcO-AAF)		Potential occupational carcinogenic risk
2-Acetylaminofluorene	53-96-3	NTPAHC
Acid green		Remove if feasible
Acrylamide	79-06-1	IARC 3
Acrylonitrile	107-13-1	IARC 2A; NTPAHC; too hazardous
Actinolite	13768-00-8	IARC 1; NTPHC
Actinomycin D	50-76-0	IARC 2B
Adrenaline		Toxic
Adriamycin	23214-92-8	IARC 2B; NTPAHC
Aflatoxins	1402-68-2	IARC 2A; NTPAHC
Aflatoxin B_1	1162-65-8	IARC 2A
Aflatoxin B_2		Potential occupational carcinogenic risk
Aflatoxin G_1		Potential occupational carcinogenic risk
Aflatoxin G_2		Potential occupational carcinogenic risk
Aflatoxin M_1		Potential occupational carcinogenic risk
Aflatoxin M_2		Potential occupational carcinogenic risk
Aflatoxin P_1		Potential occupational carcinogenic risk
AF-2	3688-53-7	IARC 3
Aldrin		Suspected carcinogen
Aliphatic halides		Potential occupational carcinogenic risk
Aluminum chloride		Remove if feasible; fire hazard, corrosive
2-Aminoanthraquinone	117-79-3	NTPAHC
o-Aminoazotoluene	97-56-3	IARC 2Bs; PAFAR
4-Aminobiphenyl	92-67-1	IARC 1; NTPHC
2-Aminodipyrido[1,2-a:3′,2′-d]dimidazole	67730-10-3	IARC 3
2-Aminofluorene (AF)		Potential occupational carcinogenic risk
Aminofluorene derivatives		Potential occupational caracinogenic risk
2-Amino-6-methyldipyrido[1,2-a:3′,2′-d]imidazole	67730-11-4	IARC 3
2-Amino-3-methyl-9H-pyrido[2,3-b]indole	68006-83-7	IARC 3

Chemical Name	CAS Number	Hazardous Properties
2-Amino-5-(5-nitro-2-furyl)-1,3,4-thiadiazole	712-68-5	IARC 2Bs
2-Amino-9H-pyrido[2,3-b]indole	26148-68-5	IARC 3
Amitrole	61-82-5	IARC 2B; IARC 3; NTPAHC
Ammonia		Irritant
Ammonium bichromate corrosive		Remove if feasible; toxic; fire hazard
Ammonium chromate		Too hazardous
Ammonium oxalate		Remove if feasible; corrosive
Ammonium vanadate		Remove if feasible; toxic
Amosite	12172-73-5	IARC 1; NTPHC
Aniline		Too hazardous; toxic
Aniline hydrochloride		Too hazardous; toxic
o-Anisidine	90-04-0	IARC 2Bs; NTPAHC
o-Anisidine hydrochloride	134-29-2	NTPAHC
Anthophyllite	17068-78-9	IARC 1; NTPHC
Anthracene		Too hazardous; carcinogen
Antimony		Remove if feasible; toxic; irritant
Antimony chloride		Toxic, corrosive, reactive hazard
Antimony oxide		Remove if feasible; toxic
Antimony pentachloride		Corrosive
Antimony potassium tartrate		Remove if feasible; toxic
Antimony trichloride		Too hazardous; corrosive
Antimony trioxide		Suspected carcinogen
Antracene oils		SEQ NO-2-1 IARC 3
Aramite	140-57-8	IARC 2Bs; NTPAHC
Aroclor 1260	11096-82-5	NTPAHC
Aromatic amines and related compounds		Potential occupational carcinogenic risk
Arsenic	7440-38-2	IARC 1; NTPHC; too hazardous
Arsenic chloride		Too hazardous; toxic
Arsenic compounds	7440-38-2	NTPHC
Arsenic pentoxide	1303-28-2	IARC 1; too hazardous; toxic
Arsenic trioxide	1327-53-3	IARC 1; NTPHC; too hazardous
Arsenic, inorganic compounds	7440-38-2	IARC 1; NTPHC
Aryl halides		Potential occupational carcinogenic risk
Asbestos	1332-21-4	IARC 1; NTPHC; too hazardous
Asbestos, amosite		Carcinogen
Asbestos, chrysotile		Carcinogen
Asbestos, crocidolite	12001-28-4	IARC 1; NTPHC
Ascarite		Too hazardous; carcinogen
Auramine	492-80-8	IARC 2B; NTPHC
Auramine, technical grade	492-80-8	IARC 2B
Azaserine	115-02-6	IARC 2Bs
Azathioprine	446-86-6	IARC 1; NTPHC
Aziridines		Potential occupational carcinogenic risk
Azo and azoxy derivatives		Potential occupational carcinogenic risk
Barium chloride		Remove if feasible; toxic
Barium chromate	10294-40-3	IARC 1
Barium compounds		Toxic
Barium hydroxide		Toxic
Barium oxalate		Remove if feasible
Benz(a)anthracene	56-55-3	IARC 2Bs; NTPAHC
Benz(c)acridine		Suspected carcinogen
Benzene	71-43-2	IARC 1; NTPHC; too hazardous
Benzidine	92-87-5	IARC 1; NTPHC
Benzo(a)pyrene	50-32-8	IARC 2A; NTPAHC
Benzo(b)fluoranthene	205-99-2	IARC 2Bs; NTPAHC
Benzo(j)fluoranthene	205-82-3	IARC 3; PAFAR
Benzo(k)fluoranthene	207-08-9	IARC 3; PAFAR
Benzone		Remove if feasible
Benzotrichloride	98-07-7	IARC 2B; NTPAHC
Benzoyl peroxide		Too hazardous; toxic, reactive hazard, explosive
Benzyl violet	1694-09-3	IARC 2Bs
Beryl	1302-52-9	IARC 2A; NTPAHC

Chemical Name	CAS Number	Hazardous Properties
Beryllium	7440-41-7	IARC 2A; NTPAHC
Beryllium-aluminum alloy	12770-50-2	NTPAHC; IARC 2A
Beryllium carbonate	66104-24-3	IARC 2A; NTPAHC; remove if feasible; toxic
Beryllium chloride	7787-47-5	IARC 2A; NTPAHC
Beryllium compounds	7440-41-7	IARC 2A; NTPAHC
Beryllium fluoride	7787-49-7	IARC 2A; NTPAHC
Beryllium hydrogen phosphate	13598-15-7	IARC 2A; NTPAHC
Beryllium hydroxide	13327-32-7	IARC 2A; NTPAHC
Beryllium oxide	1304-56-9	IARC 2A; NTPAHC
Beryllium silicate	13598-00-0	IARC 2A; NTPAHC
Beryllium sulfate	13510-49-1	IARC 2A; NTPAHC
Beryllium sulfate tetrahydrate	7787-56-6	IARC 2A; NTPAHC
N,N-Bis(2-chloroethyl)-2-naphthylamine		Suspected carcinogen
Bis-(chloromethyl) ether	542-88-1	IARC 1; NTPHC
Bis-chloroethyl nitrosourea	154-93-8	IARC 2B; NTPAHC
Bismuth trichloride		Corrosive
Bitumens		SEQ NO-3-9-IARC 3
Bonine fluid		Remove if feasible
Bromine		Remove if feasible; corrosive; toxic
Bromkal 80	61288-13-9	IARC 3
Bromoethyl methanesulfonate (BEMS)		Potential occupational carcinogenic risk
1,3-Butadiene	106-99-0	IARC 3; PAFAR
1,4-Butanediol dimethanesulphonate	55-98-1	IARC 1; NTPHC
Butanol		Toxic
Butylated hydroxyanisole	25013-16-5	IARC 3
β-butyrolactone	3068-88-0	IARC 2Bs
C.I. basic red 9 (p-rosaniline)	569-61-9	PAFAR
C.I. direct black 38	1937-37-7	IARC 2B; NTPAHC
C.I. direct blue 6	2602-46-2	NTPAHC
C.I. direct blue 14	72-57-1	IARC 2Bs
C.I. direct disperse orange 11	82-28-0	NTPAHC
C.I. solvent orange 2	2646-17-5	IARC 2Bs
Cadmium	7440-43-9	NTPAHC; IARC 2B; remove if feasible
Cadmium acetate		Remove if feasible; toxic
Cadmium bromide		Remove if feasible; toxic
Cadmium carbonate		Remove if feasible; toxic
Cadmium chloride	10108-64-2	IARC 2B; NTPAHC; remove if feasible
Cadmium compounds	7440-43-9	IARC 2B; NTPAHC
Cadmium dust	7440-43-9	IARC 2B
Cadmium oxide	1306-19-0	IARC 2B; NTPAHC
Cadmium oxide, fume, as Cd	1306-19-0	NTPAHC
Cadmium sulfate	10124-36-4	IARC 2B; NTPAHC; remove if feasible
Cadmium sulfide	1306-23-6	IARC 2B; NTPAHC
Calcium carbide		Corrosive
Calcium chromate, sintered	13765-19-0	IARC 1; NTPHC
Calcium cyanide		Too hazardous; toxic
Calcium fluoride		Too hazardous; irritant
Carbon blacks, solvent (benzene) extracts	SEQ NO-40-6	IARC 3
Carbon disulfide		Explosive
Carbon tetrachloride	56-23-5	IARC 2B; NTPAHC; too hazardous
Carboxylic acid derivatives		Potential occupational carcinogenic risk
Carmine		Remove if feasible
Carrageenan (degraded)	9000-07-1	IARC 3
Catechol		Remove if feasible; toxic, irritant
Chloral hydrate		Too hazardous; toxic
Chlorambucil	305-03-3	IARC 1; NTPHC
Chloramphenicol	56-75-7	IARC 2B
Chlorendic acid	115-28-6	PAFAR
Chloretone		Too hazardous; toxic
Chlorinated ethanes		Suspected carcinogen
Chlorinated toluenes, production of	SEQ NO-39-8	IARC 2B
Chlorine		Too hazardous; toxic, corrosive

APPENDIX D 297

Chemical Name	CAS Number	Hazardous Properties
Chlormethyl methyl ether		Suspected carcinogen
Chlornaphazine	494-03-1	IARC 1; 494-03-1 NTPHC
1-(2-Chloroethyl)-3-cyclohexyl-1-nitrosourea	13010-47-4	IARC 2B; NTPAHC
Chlorodiphenyl (54% chlorine)	11097-69-1	NTPAHC
Chloroform	67-66-3	IARC 2B; NTPAHC; too hazardous
Chloromethyl ethers		Potential occupational carcinogenic risk
Chloromethyl methyl ether, technical grade	107-30-2	IARC 1; NTPHC
4-Chloro-2-methyl-phenoxy acetic acid (occupational exposure)	94-74-6	IARC 2Bs
2-(4-Chloro-2-methyl-phenoxy) propanoic acid (occupational exposure)	93-65-2	IARC 2Bs
3-Chloro-2-methylpropene	563-47-3	PAFAR
Chlorophenols (occupational exposure)	SEQ NO-5-8	IARC 2B; IARC 2Bs
4-Chloro-o-phenylenediamine	95-83-0	IARC 2Bs; NTPAHC
Chloropromazine		Too hazardous; toxic
p-Chloro-o-toluidine	95-69-2	IARC 3
Chromates		Suspected carcinogen
Chromic acid		Remove if feasible; corrosive, reactive hazard
Chromic acid, calcium salt (1:1)	13765-19-0	IARC 1; NTPHC
Chromic acid, calcium salt (1:1), dihydrate	8012-75-7	IARC 1
Chromium	7440-47-3	IARC 1; NTPHC; too hazardous
Chromium acetate		Remove if feasible
Chromium compounds	7440-47-3	IARC 1; NTPHC
Chromium(III) oxide		Suspected carcinogen
Chromium oxide		Too hazardous
Chromium potassium sulfate		Too hazardous
Chromium trioxide	1333-82-0	IARC 1; NTPHC; too hazardous
Chromium trioxide, sintered	1333-82-0	IARC 1; NTPHC
Chromium VI oxide		Suspected carcinogen
Chrysene		Suspected carcinogen
Chrysotile asbestos	12001-29-5	IARC 1; NTPHC
Cisplatin	15663-27-1	IARC 2B
Citrus red no. 2	6358-53-8	IARC 2Bs
Coal gasification	SEQ NO-40-3	IARC 2Bs
Coal soot (occupational exposure)	SEQ NO-7-0	IARC 1; NTPHC
Coal soot extracts	SEQ NO-7-0	IARC 3
Coal tar pitches	65996-93-2	IARC 2Bs
Coal tar pitches (occupational exposure)	65996-93-2	IARC 1
Coal tars	8007-45-2	IARC 3
Coal tars (occupational exposure)	807-45-2	IARC 1
Cobalt		Suspected carcinogen; remove if feasible
Cobalt nitrate		Remove if feasible; toxic, reactive hazard
Cobalt-chromium alloy	11114-92-4	IARC 1; NTPHC
Coke oven emissions	SEQ NO- 7-5	NTPHC
Coke production	SEQ NO-40-4	IARC 2Bs
Colchicine		Suspected carcinogen; too hazardous; toxic
Creosote(s)	8001-58-9	IARC 2Bs
Creosote(s) (occupational exposure)	8001-58-9	IARC 1; NTPHC
p-Cresidine	120-71-8	IARC 2Bs; NTPAHC
Cummingtonite-grunerite	SEQ NO- 37-3	NTPHC
Cupferron	135-20-6	NTPAHC
Cupric bromide		Corrosive
Cupric chloride		Irritant
Cupric nitrate		Irritant
Cupric sulfate		Irritant
Cutting oils (occupational exposure)	SEQ NO-36-1	IARC 1; NTPHC
Cycasin	14901-08-7	IARC 2Bs; NTPAHC
Cyclohexane		Remove if feasible; toxic; fire hazard
Cyclohexene		Remove if feasible; toxic; fire hazard
Cyclophosphamide	50-18-0	IARC 1; NTPHC
Dacarbazine	4342-03-4	IARC 2B; NTPAHC
Daunomycin	20830-81-3	IARC 2Bs
DDT	50-29-3	IARC 2B; NTPAHC

Chemical Name	CAS Number	Hazardous Properties
Decabromobiphenyl	13654-09-6	IARC3
Dehydrosafrole		Suspected carcinogen
Dethylene oxide	75-21-8	NTPAHC
N,N'-Diacetylbenzidine	613-35-4	IARC 2Bs
Diallate		Suspected carcinogen
2,4-Diaminoanisole sulfate	39156-41-7	IARC 2Bs; NTPAHC
4,4'-Diaminodiphenyl ether	101-80-4	IARC 2Bs; PAFAR
2,4-Diaminotoluene	95-80-7	IARC 2Bs; NTPAHC
Diazomethane		Potential occupational carcinogenic risk
Dibenz(a,h)acridine	226-36-8	IARC 2Bs; NTPAHC
Dibenz[a,j]acridine	224-42-0	IARC 2Bs; NTPAHC
Dibenz[a,h]anthracene	53-70-3	IARC 2Bs; NTPAHC
Dibenzo(a,e)pyrene	192-65-4	IARC 2Bs; PAFAR
Dibenzo[a,h]pyrene	189-64-0	NTPAHC; IARC 2Bs
Dibenzo[a,i]pyrene	189-55-9	IARC 2Bs
Dibenzo(a,l)pyrene	191-30-0	IARC 3; PAFAR
Dibenzo[q,i]pyrene	189-55-9	NTPAHC
7H-Dibenzo(c,g)carbazole	194-59-2	IARC 2Bs; NTPAHC
1,2-Dibromo-3-chloropropane (DBCP)	96-12-8	IARC 2Bs; NTPAHC
1,2-Dibromoethane		Suspected carcinogen
p-Dichlorobenzene	106-46-7	PAFAR; irritant
3,3'-Dichlorobenzidine	91-94-1	IARC 2B; NTPAHC
3,3'-Dichloro-4,4'-diaminodiphenyl ether	28434-86-8	IARC 2Bs
1,2-dichloroethane		Suspected carcinogen
2,4-Dichlorophenol (occupational exposure)	120-83-2	IARC 2Bs
2,4-Dichlorophenoxy acetic acid (occupational exposure)	94-75-7	IARC 2Bs
2-(2,4-Dichlorophenoxy)-propanoic acid (occupational exposure)	120-36-5	IARC 2Bs
1,3-Dichloropropene	542-75-6	IARC 3
1,2-Diethylhydrazine	1615-80-1	IARC 2Bs
Dichlorobenzene		Too hazardous
Dichloroethane		Too hazardous
Dichloroindophenol, sodium salt		Remove if feasible
Dichloropropene	542-75-6	PAFAR
Dieldrin		Suspected carcinogen
Dienestrol	84-17-3	IARC 2B
Diepoxybutane (DEB)	1464-53-5	IARC 2Bs; NTPAHC
Diethyl phthalate		Irritant
Diethyl sulfate	64-67-5	IARC 2A; NTPAHC
Diethylstilbestrol	56-53-1	IARC 1; NTPHC
Diglycidyl resorcinol ether	101-90-6	IARC 3; PAFAR
Dihydrosafrole	94-58-6	IARC 2Bs
Diisopropyl ether		Explosive
3,3'-Dimethoxybenzidine	119-90-4	NTPAHC; IARC 2B
4-Dimethylaminoazobenzene		Potential occupational carcinogenic risk
trans-2((Dimethylamino)methylimino)-5-(2-(5-nitro-2-furyl) vinyl)-1,3,4-oxadiazole	55738-54-0	IARC 2Bs
N,N-Dimethylaniline		Toxic
7,12-Dimethylbenz[a]anthracene (DMBA)		Potential occupational carcinogenic risk
3,3'-Dimethylbenzidine		Potential occupational carcinogenic risk
1,1-Dimethylethylenimine		Potential occupational carcinogenic risk
1,1-Dimethylhydrazine (UDMH)	57-14-7	IARC 2Bs; NTPAHC
1,2-Dimethylhydrazine (SDMH)	540-73-8	IARC 2Bs
Dimethyl sulfate	77-78-1	IARC 2A; NTPAHC
Dimethylaminoazobenzene	60-11-7	IARC 2Bs; NTPAHC
Dimethylaniline		Too hazardous
Dimethylcarbamoyl chloride	79-44-7	IARC 2B; NTPAHC
Dimethylvinyl chloride	513-37-1	PAFAR
2,4-Dinitrophenol		Remove if feasible; toxic, explosion hazard
1,4-Dinitrosopiperzine (DNP)		Potential occupational carcinogenic risk
2,4-Dinitrotoluene		Suspected carcinogen
Dioxane	123-91-1	IARC 2B; NTPAHC
p-Dioxane		Suspected carcinogen; too hazardous

Chemical Name	CAS Number	Hazardous Properties
Diphenyl ester carbonic acid		Suspected carcinogen; too hazardous
1,2-Diphenylhydrazine	122-66-7	NTPAHC
Direct black 38, technical grade	1937-37-7	IARC 2B
Direct blue 6, technical grade	2602-46-2	IARC 2B
Direct brown 95, technical grade	16071-86-6	IARC 2B
Epichlorohydrin	106-89-8	IARC 2B; NTPAHC
Estradiol	50-28-2	IARC 2B; NTPAHC
Estrogens, conjugated	SEQ NO-24-0	IARC 1; NTPHC
Estrone	53-16-7	IARC 2B; NTPAHC
Ethenylamine, N-methyl-N-nitroso	4549-40-0	IARC 2Bs; NTPAHC
Ethers, oxides and epoxides		Potential occupational carcinogenic risk
Ethinylestradiol	57-63-6	IARC 2B; NTPAHC
Ethionine		Potential occupational carcinogenic risk
Ethyl acrylate	140-88-5	IARC 3; PAFAR
Ethyl carbamate	51-79-6	IARC 2Bs; NTPAHC
Ethyl ether		Fire hazard, explosion hazard
Ethyl methacrylate		Irritant
Ethyl methanesulfonate (EMS)	62-50-0	IARC 2Bs
Ethylene bis dithiocarbamate		Suspected carcinogen
Ethylene dibromide (EDB)	106-93-4	IARC 2B; NTPAHC
Ethylene dichloride	107-06-2	IARC 2Bs; NTPAHC; too hazardous
Ethylenimine (EI)		Potential occupational carcinogenic risk
Ethylene oxide	75-21-8	IARC 2B; too hazardous; toxic, fire hazard
Ethylene thiourea	96-45-7	IARC 2B; NTPAHC
Ferric chloride		Irritant
Ferrous sulfate		Remove if feasible
Firemaster BP-6	59536-65-1	IARC 3; NTPAHC
Firemaster FF-1	67774-32-7	IARC 3; NTPAHC
Formaldehyde	50-00-0	Suspected carcinogen; remove if feasible; toxic, fire hazard
Formaldehyde (gas)	50-00-0	IARC 2B; NTPAHC
Formalin		Remove if feasible
2-(2-Formylhydrazino)-4-(5-nitro-2-furyl)thiazole	3570-75-0	IARC 2Bs
Fuchsin		Remove if feasible
Furniture manufacture	SEQ NO-40-0	IARC 1
Gasoline		Remove if feasible; fire hazard
Glycidylaldehyde	765-34-4	IARC 2Bs
Gunpowder		Too hazardous; explosion hazard
Gyromitrin	16568-02-8	IARC 3
Hematite, underground mining with exposure to radon	1317-60-8	IARC 1; NTPHC
Hematoxylin		Remove if feasible
Heptachlor		Suspected carcinogen
2,2',4,4',5,5'-Hexabromo-1,1'-biphenyl	59080-40-9	IARC 3
Hexachlorobenzene	118-74-1	IARC 2Bs; NTPAHC
Hexachlorobutadiene		Suspected carcinogen
Hexachlorocyclohexane		Suspected carcinogen
Hexachlorophene		Too hazardous; toxic, irritant
Hexamethyl phosphoramide	680-31-9	NTPAHC; IARC 2Bs
Hydrazine (HZ)	302-01-2	IARC 2B; NTPAHC
Hydrazines		Potential occupational carcinogenic risk
Hydrazine, sulfate (1:1)	10034-93-2	NTPAHC
Hydriodic acid		Too hazardous
Hydrobromic acid		Too hazardous; toxic, corrosive
Hydrochloric acid		Corrosive; too hazardous
Hydrofluoric acid		Corrosive
Hydrogen		Too hazardous
Hydrogen peroxide (30%)		Irritant
Hydrogen sulfide		Remove if feasible; irritant, fire hazard
Hydroiodic acid		Corrosive
Hydroquinone		Remove if feasible; irritant
N-Hydroxy-2-acetylaminofluorene (N-HO-AAF)		Potential occupational carcinogenic risk
Indeno(1,2,3-cd)pyrene	193-39-5	IARC 2Bs; NTPAHC

Chemical Name	CAS Number	Hazardous Properties
Indigo carmine		Too hazardous
Iodine crystals		Toxic, irritant
2-amino-3-methylimidazo[4,5-f]quinoline	76180-96-6	IARC 3
Iron and steel founding	SEQ NO-40-5	IARC 2Bs
Iron dextran	9004-66-4	IARC 3; NTPAHC
Iso-amyl alcohol		Remove if feasible
Iso-butyl alcohol		Remove if feasible
Iso-pentyl alcohol		Remove if feasible
Isopropyl alcohol manufacture (strong acid process)	67-63-0	IARC 1; NTPHC
Isosafrole	120-58-1	IARC 2Bs
Kepone	143-50-0	IARC 2Bs; NTPAHC
Lasiocarpine	303-34-4	IARC 2Bs
Lead acetate	301-04-2	IARC 3; NTPAHC
Lead arsenate		Suspected carcinogen; too hazardous
Lead carbonate		Too hazardous; toxic, irritant
Lead chromate	7758-97-6	IARC 1; NTPHC
Lead chromate(VI) oxide	18454-12-1	IARC 1; 18454-12-1 NTPHC
Lead compounds		Toxic
Lead phosphate	7446-27-7	IARC 3; NTPAHC
Lead subacetate	1335-32-6	IARC 3
Lead(VI) chromate		Too hazardous
Lindane and other hexaclorocyclohexane isomers	58-89-9	NTPAHC
α-Lindane	319-84-6	NTPAHC
β-Lindane	319-85-7	NTPAHC
γ-Lindane	58-89-9	NTPAHC
Lindane-mixed isomers	608-73-1	NTPAHC
Lithium		Too hazardous; toxic, fire hazard, explosion hazard, corrosive
Lithium nitrate		Too hazardous
Magenta, manufacture of	632-99-5	IARC 2A
Magnesium		Fire hazard
Magnesium, metal (powder)		Too hazardous
Magnesium chlorate		Remove if feasible; toxic, reactive hazard, fire hazard
Melphalan	148-82-3	IARC 1; NTPHC
Mercuric sulfide		Toxic
Mercuric bichloride		Remove if feasible
Mercuric chloride		Toxic; too hazardous
Mercuric iodide		Remove if feasible; toxic
Mercuric nitrate		Toxic, fire hazard
Mercuric oxide		Remove if feasible; toxic, fire hazard
Mercuric sulfate		Remove if feasible; toxic, fire hazard
Mercuric sulfide		Remove if feasible
Mercurous chloride		Remove if feasible; toxic
Mercurous nitrate		Remove if feasible; toxic
Mercurous oxide		Remove if feasible; toxic
Mercury		Too hazardous; toxic
Merphalan	531-76-0	IARC 2Bs
Mesitylene		Too hazardous; toxic
Mestranol	72-33-3	IARC 2B; NTPAHC
5-methoxypsoralen	484-20-8	IARC 3
3'-Methyl-4-aminoazobenzene		Potential occupational carcinogenic risk
3-Methylcholanthrene (MC)		Potential occupational carcinogenic risk
Methyl chloromethyl ether	107-30-2	NTPHC
5-Methylcrysene	3697-24-3	IARC 3; PAFAR
4,4'-Methylene bis-(2-chloroaniline)	101-14-4	IARC 2Bs; NTPAHC
4,4'-Methylene bis(N,N-dimethyl)benzeneamine	101-61-1	NTPAHC
4,4'-Methylene bis(2-methylaniline)	838-88-0	IARC 2Bs
4,4'-Methylenedianiline	101-77-9	IARC 3; NTPAHC
4,4'-Methylenedianiline dihydrochloride	13552-44-8	IARC 3; NTPAHC
Methyl ethyl ketone		Remove if feasible; toxic, irritant, fire hazard, explosion hazard
Methyl hydrazine (NMH)		Potential occupational carcinogenic risk

Chemical Name	CAS Number	Hazardous Properties
Methyl iodide	74-88-4	IARC 2Bs; NTPAHC
Methyl iodine		Too hazardous
Methyl methacrylate		Too hazardous; toxic, fire hazard, explosion hazard, irritant
Methyl methanesulfonate (MMS)	66-27-3	IARC 2Bs
1-Methyl-3-nitro-1-nitrosoguanidine (NNMG)		Potential occupational carcinogenic risk
2-Methyl-1-nitroanthraquinone	129-15-7	IARC 2Bs
N-Methyl-N'-nitro-N-nitrosoguanidine	70-25-7	IARC 2Bs
Methyl oleate		Remove if feasible
Methyl orange[a]		Too hazardous; toxic
Methyl red[a]		Too hazardous
Methyl salicylate		Irritant
Methylazoxymethanol acetate	592-62-1	IARC 2Bs
Methylene chloride	75-09-2	IARC 3; PAFAR
Methylthiouracil	56-04-2	IARC 2Bs
Metronidazole	443-48-1	IARC 2B; NTPAHC
Michler's ketone	90-94-8	NTPAHC
Mirex	2385-85-5	IARC 2Bs; NTPAHC
Mitomycin C	50-07-7	IARC 2Bs
Monocrotaline	315-22-0	IARC 2Bs
5-(Morpholinomethyl)-3-((5-nitrofurfurylidene)amino)-2-oxazolidinone (dl-form)	139-91-3	IARC 2Bs
5-(Morpholinomethyl)-3-((5-nitrofurfurylidene)amino)-2-oxazolidinone (dl-form hydrochloride)	13146-28-6	IARC 2Bs
5-(Morpholinomethyl)-3-((5-nitrofurfurylidene)amino)-2-oxazolidinone (l-form)	3795-88-8	IARC 2Bs
Mustard gas	505-60-2	IARC 1; NTPHC
1-Naphthylamine(1-NA)		Potential occupational carcinogenic risk
2-Naphthylamine (2-NA)	91-59-8	IARC 1; NTPHC
5-Nitroacenaphthene	602-87-9	IARC 2Bs
5-Nitro-o-anisidine	99-59-2	NTPAHC
4-Nitrodiphenyl		Suspected carcinogen
1-((5-Nitrofurfurylidene)amino)-2-imidazolidinone	555-84-0	IARC 2Bs
N-(4-(5-Nitro-2-furyl)-2-thiazolyl)acetamide	531-82-8	IARC 2Bs
N-[4-(5-Nitro-2-furyl)-2-thiazolyl]formamide (FANFT)		Potential occupational carcinogenic risk
Nitrofuran derivatives		Potential occupational carcinogenic risk
Nitrogen mustards		Potential occupational carcinogenic risk
2-Nitropropane	79-46-9	IARC 2Bs; NTPAHC
4-Nitroquinoline-1-oxide		Potential occupational carcinogenic risk
Nitrosamides		Potential occupational carcinogenic risk
Nitrosamines		Potential occupational carcinogenic risk
N-Nitroso-N-ethylurea	759-73-9	IARC 2Bs; NTPAHC
N-Nitroso-N-ethylurethane (ENUT)		Potential occupational carcinogenic risk
p-Nitrosodiphenylamine	156-10-5	NTPAHC
N-Nitroso-N-methylurea	684-93-5	IARC 2Bs; NTPAHC
N-Nitroso-N-methylurethane (MNUT)		Potential occupational carcinogenic risk
N-Nitrosodi-n-butylamine	924-16-3	IARC 2Bs; NTPAHC
N-Nitrosodi-n-propylamine	621-64-7	IARC 2Bs; NTPAHC
N-Nitrosodiethanolamine	1116-54-7	IARC 2Bs; NTPAHC
N-Nitrosodiethylamine	55-18-5	IARC 2Bs; NTPAHC
N-Nitrosodimethylamine	62-75-9	IARC 2Bs; NTPACH
N-Nitrosomethylethylamine	10595-95-6	IARC 2Bs
N-Nitrosomorpholine	59-89-2	IARC 2Bs; NTPAHC
N-Nitrosonornicotine	16543-55-8	IARC 2Bs; NTPAHC
N-Nitrosopiperidine	100-75-4	IARC 2Bs; NTPAHC
N-Nitrosopyrrolidine	930-55-2	IARC 2Bs; NTPAHC
N-Nitrososarcosine	13256-22-9	IARC 2Bs; NTPAHC
5-Nitro-o-toluidine		Suspected carcinogen
N-Phenyl-β-naphthylamine		Suspected carcinogen
Nafenopin	3771-19-5	IARC 2Bs
Naphthalene		Irritant
Nickel	7440-02-0	IARC 2A; NTPAHC

302 APPENDICES

Chemical Name	CAS Number	Hazardous Properties
Nickel dusts, powders, and fumes	7440-02-0	IARC 2A; NTPAHC
Nickel metal/nickel oxide		Too hazardous
Nickel carbonate	3333-67-3	IARC 2A; NTPAHC; remove if feasible
Nickel carbonyl	13463-39-3	IARC 2A; NTPAHC
Nickel compounds	7440-02-0	IARC 2A; NTPAHC
Nickel hydroxide [$Ni(OH)_2$]	12054-48-7	IARC 2A
Nickel(II) acetate		Suspected carcinogen
Nickel(II) oxide		Suspected carcinogen
Nickel oxide	1313-99-1	IARC 2A; NTPAHC
Nickel refining	7440-02-0	IARC 1; NTPHC
Nickel subsulfide	12035-72-2	IARC 2A; NTPAHC
Nickel sulfide fumes and dusts		Suspected carcinogen
Nickelocene	1271-28-9	IARC 2A; NTPAHC
Nickelous acetate		Remove if feasible
Nicotine		Too hazardous; toxic, fire hazard
Niridazole	61-57-4	IARC 2Bs
Nitric acid		Corrosive
Nitrilotriacetic acid	139-13-9	NTPAHC
Nitrofen	1836-75-5	IARC 3; NTPAHC
Nitrogen mustard	51-75-2	IARC 2A; NTPAHC
Nitrogen mustard hydrochloride		Suspected carcinogen
Nitrogen mustard N-oxide	126-85-2	IARC 2Bs
Nitrogen mustard N-oxide hydrochloride		Suspected carcinogen
Norethisterone	68-22-4	IARC 2B; NTPAHC
Octyl phthalate, di-sec	117-81-7	IARC 2Bs, NTPAHC
Oil shale soot extracts	SEQ NO-24-1	IARC 2Bs
Oral contraceptives, combined	SEQ NO-24-2	IARC 2A
Oral contraceptives, sequential	SEQ NO-24-3	IARC 2B
Osmium tetroxide		Too hazardous; toxic, irritant
Oxalic acid		Corrosive
Oxygen, tank		Too hazardous
Oxymetholone	434-07-1	IARC 2A; NTPAHC
Panfuran S	794-93-4	IARC 2Bs
Paradichlorobenzene		Remove if feasible
Paraffin waxes and hydrocarbon waxes, chlorinated	63449-39-8	PAFAR
Paris green		Too hazardous; toxic
Pentachloronitrobenzene		Suspected carcinogen
Pentachlorophenol (occupational exposure)	87-86-5	IARC 2Bs
Pentane		Remove if feasible; fire hazard, explosion hazard
Perchloric acid		Irritant, reactive hazard, explosion hazard
Perchloroethylene		Suspected carcinogen
Petroleum ether		Remove if feasible; toxic, fire hazard, explosion hazard
Phenacetin	62-44-2	IARC 2A; NTPAHC
Phenacetin, analgesic mixtures containing	62-44-2	IARC 1; NTPHC
Phenazopyridine	94-78-0	IARC 2B
Phenazopyridine hydrochloride	136-40-3	NTPAHC
Phenol		Suspected carcinogen; too hazardous
Phenoxyacetic acid herbicides (occupational exposure)	SEQ NO-26-0	IARC 2B
Phenoxybenzamine	59-96-1	IARC 2Bs
Phenoxybenzamine hydrochloride	63-92-3	IARC 2Bs; PAFAR
Phenylhydrazine		Suspected carcinogen
Phenylthiocarbamide		Remove if feasible
1-Phenyl-2-thiourea		Remove if feasible; toxic
Phenytoin	57-41-0	IARC 2B; NTPAHC
Phenytoin, sodium salt	630-93-3	NTPAHC
Phosphorus pentoxide		Too hazardous; irritant, reactive hazard, fire hazard, corrosive
Phosphorus, red		Too hazardous; fire hazard, explosion hazard, reactive hazard
Phosphorus, white		Too hazardous; toxic, fire hazard, corrosive
Phthalic anhydride		Too hazardous; irritant
Picric acid		Too hazardous; toxic, explosion hazard

Chemical Name	CAS Number	Hazardous Properties
Polychlorinated biphenyls	1336-36-3	IARC 2B; NTPAHC
Polycyclic aromatic hydrocarbons		Potential occupational carcinogenic risk
Ponceau 3R	3564-09-8	IARC 2Bs
Ponceau MX	3761-53-3	IARC 2Bs
Potassium		Fire hazard, explosion hazard, corrosive; too hazardous
Potassium bromate	7758-01-2	IARC 3
Potassium chlorate		Remove if feasible; toxic, fire hazard, reaction hazard, explosion hazard
Potassium chromate		Suspected carcinogen; remove if feasible; corrosive
Potassium cyanide		Toxic, corrosive
Potassium fluoride		Corrosive
Potassium hydroxide		Corrosive
Potassium oxalate		Too hazardous; toxic, corrosive
Potassium periodate		Remove if feasible; toxic, irritant, fire hazard
Potassium permanganate		Remove if feasible; irritant, fire hazard, reactive hazard
Potassium sulfide		Too hazardous; toxic, fire hazard, explosion hazard
Procarbazine (MIH)	671-16-9	IARC 2A; NTPAHC
Procarbazine hydrochloride	366-70-1	NTPAHC
Progesterone	57-83-0	IARC 2B; NTPAHC
Pronamine		Suspected carcinogen
Propane sultone		Suspected carcinogen
1,3-Propane sultone	1120-71-4	IARC 2Bs; NTPAHC
β-Propiolactone (BPL)	57-57-8	NTPAHC; IARC 2Bs
Propyleneimine	75-55-8	IARC 2Bs; NTPAHC
Propylene oxide	75-56-9	PAFAR; IARC 3
Propylthiouracil	51-52-5	IARC 2B; NTPAHC; suspected carcinogen
Pyridine		Too hazardous; toxic, fire hazard, explosion hazard
Pyrogallic acid		Too hazardous; toxic
Reserpine	50-55-5	NTPAHC
Rubber industry (certain occupations)	SEQ NO-40-1	IARC 1; NTPHC
Saccharin	81-07-2	NTPAHC; too hazardous
Safrole	94-59-7	IARC 2Bs; NTPAHC
Salol		Remove if feasible; toxic
Sodium chromate		Too hazardous
Selenium		Suspected carcinogen; too hazardous
Selenium sulfide (SeS)	7446-34-6	NTPAHC
Silver compounds		Irritant, toxic
Silver cyanide		Too hazardous; toxic
Silver nitrate		Too hazardous; toxic, irritant
Silver oxide		Too hazardous; toxic, fire hazard, explosion hazard
Sodium		Too hazardous; irritant, fire hazard, reactive hazard, corrosive
Sodium arsenate	7631-89-2	IARC 1; too hazardous
Sodium arsenite	7784-46-5	IARC 1; too hazardous
Sodium azide		Too hazardous; toxic, explosion hazard
Sodium bromate		Remove if feasible
Sodium chlorate		Remove if feasible; irritant, reactive hazard, fire hazard
Sodium chromate		Suspected carcinogen
Sodium cyanide		Too hazardous; toxic, corrosive
Sodium dichloroindophenol		Too hazardous
Sodium dichromate		Suspected carcinogen; too hazardous
Sodium ferrocyanide		Too hazardous; corrosive
Sodium fluoride		Remove if feasible; toxic
Sodium hydroxide		Corrosive
Sodium nitrate		Remove if feasible
Sodium nitrite		Suspected carcinogen; too hazardous
Sodium oxalate		Remove if feasible
Sodium permanganate		Toxic, irritant, reactive hazard, fire hazard
Sodium saccharin	128-44-9	IARC 2Bs
Sodium silicofluoride		Remove if feasible; toxic, corrosive
Sodium sulfide		Too hazardous; toxic, irritant, fire hazard
Sodium thiocyanate		Too hazardous; toxic

Chemical Name	CAS Number	Hazardous Properties
Soots, tars and mineral oils (occupational exposure)	SEQ NO-29-1	IARC 1; NTPHC
Stannic chloride		Too hazardous; corrosive, reactive hazard
Stearic acid		Suspected carcinogen; too hazardous
Sterigmatocystin	10048-13-2	IARC 2Bs
Streptozotocin	18883-66-4	IARC 2Bs; NTPAHC
Strontium		Too hazardous; fire hazard
Strontium chromate	7789-06-2	IARC 1; NTPHC
Strontium nitrate		Too hazardous; toxic, reactive hazard, fire hazard
Styrene oxide	96-09-3	IARC 3
Sudan III		Remove if feasible
Sudan IV		Suspected carcinogen; too hazardous
Sulfallate	95-06-7	IARC 3; NTPAHC
Sulfamethazine		Remove if feasible
Sulfonic acid derivatives		Potential occupational carcinogenic risk
Sulfuric acid		Toxic, irritant, reactive hazard, corrosive
Sulfuric acid, fuming		Too hazardous; corrosive
Talc[a]		Too hazardous
Tannic acid		Suspected carcinogen; too hazardous
Testosterone	58-22-0	IARC 2Bs
Testosterone oenanthate	315-37-7	IARC 2Bs
Testosterone propionate	57-85-2	IARC 2Bs
Tetrabromoethane		Too hazardous; toxic, irritant
Tetrachloroethylene	127-18-4	PAFAR
2,3,7,8-Tetrachlorodibenzo-p-dioxin	1746-01-6	IARC 2B; NTPAHC
2,3,4,6-Tetrachlorophenol (occupational exposure)	58-90-2	IARC 2Bs
Thermite and compounds		Too hazardous; fire hazard
Thioacetamide	62-55-5	IARC 2Bs; NTPAHC; too hazardous
4,4'-thiodianiline	136-65-1	IARC 2Bs
Thiourea	62-56-6	IARC 2Bs; NTPAHC; too hazardous
Thorium dioxide	1314-20-1	NTPHC
Titanium trichloride		Too hazardous; fire hazard, irritant
o-Tolidine	119-93-7	IARC 2Bs; NTPAHC
Toluene		Remove if feasible; toxic, fire hazard, explosion hazard, irritant
Toluene diisocyanate	26471-62-5	IARC 3
Toluene-2,4-diisocyanate	584-84-9	IARC 3; NTPAHC
Toluene-2,6-diisocyanate	91-08-7	IARC 3
m-Toluenediamine (TDA)		Potential occupational carcinogenic risk
o-Toluidine	95-53-4	IARC 2A; NTPAHC; too hazardous
o-Toluidine hydrochloride	636-21-5	NTPAHC
p-Toluidine		Suspected carcinogen
Toxaphene	8001-35-2	IARC 2Bs; NTPAHC
Tremolite	14567-73-8	IARC 1; NTPHC
Treosulphan	299-75-2	IARC 1
Trichloroethylene		Suspected carcinogen; remove if feasible
2,2,4-Trimethylpentane		Toxic, fire hazard
2,4,5-Trichlorophenol (occupational exposure)	95-95-4	IARC 2Bs
2,4,6-Trichlorophenol	88-06-2	IARC 2B; NTPAHC
2,4,6-Trichlorophenol (occupational exposure)	88-06-2	IARC 2Bs
2,4,5-Trichlorophenoxy acetic acid (occupational exposure)	93-76-5	IARC 2Bs
2-(2,4,5-Trichlorophenoxy)propionic acid (occupational exposure)	93-72-1	IARC 2Bs
Trichlorotrifluoroethane		Irritant
Tris(1-aziridinyl)phosphate		Suspected carcinogen
Tris(2,3-dibromopropyl) phosphate	126-72-7	IARC 2Bs; NTPAHC
Tris(aziridinyl)-p-benzoquinone	68-76-8	IARC 2B
Tris(aziridinyl)-phosphine sulfide	52-24-4	IARC 2B; NTPAHC
Trp-P-1	62450-06-0	IARC 3
Trp-P-1 acetate	75104-43-7	IARC 3
Trp-P-1 monoacetate	68808-54-8	IARC 3
Trp-P-2	62450-07-1	IARC 3
Trp-P-2 acetate	72254-58-1	IARC 3

Chemical Name	CAS Number	Hazardous Properties
Trypan blue		Suspected carcinogen
Turpentine		Irritant
Uracil mustard	66-75-1	IARC 2B
Uranium		Suspected carcinogen; too hazardous
Uranyl acetate		Suspected carcinogen; too hazardous
Uranyl nitrate		Suspected carcinogen; too hazardous
Urethane		Carcinogen; remove if feasible; too hazardous
Vinyl bromide	593-60-2	IARC 3
Vinyl chloride (VC)	75-01-4	IARC 1; NTPHC
Vinyl cyclohexene dioxide		Suspected carcinogen
Vinylidene chloride		Suspected carcinogen
Vinylite		Suspected carcinogen; too hazardous
Wood's metal		Too hazardous
Xylene		Remove if feasible
Zinc beryllium silicate	39413-47-3	IARC 2A; NTPAHC
Zinc chromate	13530-65-9	IARC 1; NTPHC
Zinc chromate hydroxide	15930-94-6	IARC 1
Zinc chromates		Suspected carcinogen

[a]Suggested alternatives: For methyl orange and methyl red, bromophenol blue and bromothymol blue; for talc, starch talc.

IARC CATEGORIES

Group	Hazardous Properties
1	= Causally associated with human cancer.
2A	= Probably carcinogenic to humans—higher degree of evidence.
2B	= Probably carcinogenic to humans—lower degree of evidence.
2Bs	= Probably carcinogenic to humans—lower degree of evidence: evaluated subsequent to IARC Supplement 4.
2B[a]	= IARC animal carcinogen for which human data is not available. Considered by OSHA to correspond to Group 2B.
3	= IARC animal carcinogen for which human data is not available. To be safe, these should be treated as carcinogens.

NTP Human Carcinogens:
 NTPHC=National Toxicology Program Human Carcinogens.
 NTPAHC=National Toxicology Program Anticipated Human Carcinogens.
 PAFAR=Proposed for addition in NTP's Fifth Annual Report on Carcinogens.

APPENDIX E: STANDARD CHEMICAL LABEL STATEMENTS

Label statements are intermediate in information content between the hazard symbol and detailed material safety data sheets. It is stressed that the requirements detailed in this section are minimal because of the complexity and diversity of hazardous materials. Additional warnings and detailed medical and other information is desirable on labels attached to containers of new or unusual materials. In addition, the manufacturer or user may supply additional cautions on ecological or other matters, as appropriate.

Suggested label statements are supplied for guidance only. Label wording should be tailored specifically for each material or combination of materials.

For health hazards, the major considerations are modes of entry, speed of attack, and whether the effects are acute or chronic.

For fire hazards, considerations include vapor pressure and vapor density, autoignition temperature, explosive limits, viscosity, products of combustion, and extinguishing media.

Reactivity hazards require knowledge of sensitivity to detonation by shock or heat, tendency to rapid polymerization, reactivity with common substances, ability to supply oxygen in a fire situation, and other harmful properties.

The number of statements used will depend on the hazard involved. Extremely dangerous materials may require extensive warnings and detailed instructions for safe use and disposal. Minimally hazardous substances may require little more than the hazard statement. Specific and more detailed first-aid statements, including notes to physicians, may be necessary for extremely hazardous materials. These statements are best formulated by physicians familiar with the hazards of the specific material and aware of the capabilities of industrial paramedical personnel and facilities.

Health-Related Statements

1. Fatal if swallowed.
2. Fatal if inhaled.
3. Fatal if absorbed through the skin.
4. Harmful if swallowed.
5. Harmful if inhaled.
6. Harmful if absorbed through the skin.
7. Can cause allergic respiratory reaction.
8. Can cause allergic skin reaction.
9. Vapor (gas) may cause suffocation.
10. Causes (severe) eye burns.
11. Causes eye irritation.
12. Causes burns.
13. Causes irritation.
14. Can be fatal or cause blindness if swallowed.
15. Cannot be made nonpoisonous.
16. Repeated absorption can cause bladder tumors.
17. Rapidly absorbed through skin.
18. Inhalation can be fatal or cause delayed lung damage.
19. Harmful if inhaled and can cause delayed lung damage.
20. Can cause delayed effect.
21. Vapor extremely irritating.
22. Extremely irritating gas and liquid under pressure.
23. Gas extremely irritating.

24. Lung injury and burns may be delayed.
25. Contact with water or moist air liberates irritating gas.
26. Contact with acid liberates poisonous gas.
27. Contact with water or acid slowly liberates poisonous and flammable hydrogen sulfide gas.
28. Liberates gas which may cause suffocation.
29. Repeated inhalation or skin contact can, without symptoms, increase hazard.
30. Causes severe burns which may not be immediately painful or visible.
31. Can cause rash or external sores.
32. Can cause burns or external sores.
33. Liquid or vapor causes burns which may be delayed.
34. May cause eye injury—effects may be delayed.
35. Liquid penetrates shoes and leather, causing delayed burns.
36. May cause sterility.
37. May affect unborn children.
38. Cancer-suspect agent.

Example: Fatal if swallowed, inhaled, or absorbed through the skin.

Health Hazard Action Statements

1. Do not breathe dust.
2. Do not breathe vapor.
3. Do not breathe mist.
4. Do not breathe gas.
5. Do not get in eyes, on skin, on clothing.
6. Do not breathe vapor or get in eyes, on skin, on clothing.
7. Prevent contact with food, chewing, or smoking materials.
8. Wash thoroughly after handling.
9. Use only in well-ventilated area.
10. Keep container closed.
11. Avoid prolonged or repeated contact with skin.
12. Do not enter storage areas unless they are well ventilated.
13. Avoid breathing dust or solution spray or vapor.
14. Avoid prolonged or repeated breathing of vapor.
15. Use special protective clothing and gloves.
16. Wear goggles; neoprene, butyl rubber, or vinyl gloves; neoprene shoes or boots; and clean protective outer clothing.
17. Wear goggles; neoprene, butyl rubber, or vinyl gloves.
18. Always wear a self-contained breathing apparatus or full-face, airline respirator when using this product.
19. Have available emergency self-contained breathing apparatus or full-face airline respirator when using this product.
20. Wear respirator approved by NIOSH or the U.S. Bureau of Mines for organic vapor, dust, etc.
21. Wear goggles or face shield, rubber gloves, and protective clothing when handling.
22. Do not wear ordinary rubber protective clothing, including gloves or boots.
23. Wear full neoprene suit, rubber boots, rubber gloves, and self-contained breathing apparatus.
24. Do not taste.
25. This gas deadens the sense of smell. Do not depend on odor to detect presence of gas.
26. Use fresh clothing daily. Take hot shower at the end of work shift using plenty of soap.
27. POISON (with skull-and-crossbones symbol).
28. Avoid exposing women of child-bearing age.

Example: Do not breathe vapor or get in eyes, on skin, or on clothing.

Fire Hazard Amplifying Statements

1. Strong oxidizer—contact with other materials may cause fire.
2. Catches fire if exposed to air.
3. Spillage may cause fire or liberate dangerous gas.
4. Highly volatile.
5. Contact with water or acid slowly liberates flammable gas.
6. Contact with water may cause flash fire.
7. May ignite if allowed to become damp.

8. Heat, shock, or contact with other materials may cause fire or explosive decomposition.
9. Contact with other materials may cause fire or explosion, especially if heated.

Fire Hazard Action Statements

1. Keep away from fire, sparks, and open flames.
2. Keep from contact with clothing and other combustible materials to avoid fire.
3. Drying of this product on clothing or combustible materials may cause fire.
4. Spills on clothing or combustible materials may cause fire.
5. Contents packed under water will ignite if water is removed.
6. Avoid friction or rough handling because of fire hazard.
7. Keep wet in storage—dry powder may ignite by friction, static electricity, or heat.
8. Wear goggles or face shield and fire-retardant clothing when handling.
9. Clothing and vegetation contaminated with chlorate or its solutions are *dangerously flammable*. Remove clothing and wash thoroughly in water. Keep persons and animals off treated areas.
10. Store in cool place.
11. Keep container tightly closed.
12. Loosen closure cautiously before opening.
13. Store in cool, dry place.
14. Store in a cool place in original container and protect from direct sunlight.
15. In case of fire, stop flow of gas. Use dry chemical or carbon dioxide when necessary to gain access to valve.
16. Avoid spillage and contact with moisture or combustion.
17. In case of spillage, flush with plenty of water and remove contaminated articles.
18. Flush area with water spray.
19. In case of fire, smother with dry sand, dry ground limestone, or dry powder-type materials specially designed for metal powder fires.
20. Spillage may cause fire. Do not get on floor. Sweep up and remove immediately.

Reactivity Hazard Amplifying Statements

1. Powerful oxidizer.
2. Strong oxidizer.
3. Strong acid.
4. Strong caustic (alkali).
5. Causes severe burns which may not be immediately painful or visible.
6. Heat, shock, or contact with other materials may cause fire or explosion, especially if heated.
7. Contact with other materials may cause fire or explosion, especially if heated.
8. Reacts violently with water, liberating and igniting hydrogen.
9. May form explosive peroxides.
10. Forms shock-sensitive mixtures with certain other materials.
11. May explode if water content is 10% or below.
12. Contamination may result in dangerous pressure.
13. Liquid and gas under pressure.
14. Extremely hazardous liquid and vapor under pressure.
15. Extremely cold (give degrees F or C below zero).
16. High explosive.
17. Explosive.
18. Inhibited monomer, subject to violent polymerization.
19. Liquid and gas under pressure.
20. Gas under pressure.

Reactivity Hazard Action Statements

1. Keep from contact with oxidizing materials, highly oxygenated or halogenated solvents, organic compounds containing reducible functional groups, or aqueous ammonia.
2. Keep from contact with oxidizing materials.
3. Keep from contamination from any source, including metals, dust, and organic materials. Such contamination can cause rapid decomposition, generation of high pressures, or formation of explosive mixtures.

4. Solidifies at about ____°F (____°C) and may break container. Store in moderately warm place.
5. Keep from any contact with water.
6. Use only dry, clean utensils in handling.
7. While making solutions, add slowly to surface to avoid violent splattering.
8. Keep wet in storage—dry powder may ignite by friction, static electricity, or heat.
9. Do not add to hot materials; do not grind or subject to frictional heat or shock—explosive decomposition may result.
10. Prevent contamination with readily oxidizable materials and polymerization accelerators.
11. Do not allow to evaporate to near dryness. Addition of water or appropriate reducing materials will lessen peroxide formation.
12. Do not add water to contents while in a container because of violent reaction and possible flash fire.
13. Do not attempt to loosen or remove material from container with any tool.
14. Wear goggles and *dry* gloves when handling.
15. Put nothing else in this container.
16. Keep dry and handle only in suitable equipment to prevent metallic contamination. Consult manufacturer.
17. Keep container tightly closed and away from water or acids.
18. Keep container tightly closed; flush container clean before discarding.
19. Do not put in stoppered or closed container.
20. Note: Suck-back into cylinder may cause explosion. Under no circumstances should the cylinder entry tube be inserted in a liquid or gas without a vacuum break or other protective apparatus in the line to prevent suck-back.
21. Store in original vented container.
22. Store in cool place.
23. Keep drum in upright position. Do not roll drum on side.
24. Handle under inert gas atmosphere in *dry* equipment.
25. Keep from freezing.
26. Loosen closure cautiously before opening.
27. Store separately from, and avoid contact with, dehydrating materials and other materials.
28. Keep away from fire.
29. Open container cautiously and only under a dry, oxygen-free or inert atmosphere.
30. Store in cool, dry place.
31. Store in cool place in original container and protect from direct sunlight.
32. Keep container closed to prevent drying out.
33. Do not heat cylinders.
34. Keep away from acids and heat.
35. Never return unused hydrogen peroxide to container. Dilute with plenty of water.
36. Avoid spillage and contact with moisture or combustibles.
37. Fire or high temperatures may cause explosive decomposition if confined.
38. In case of fire, smother with dry sand, dry ground limestone, or dry powder-type materials specially designed for metal powder fires. Do not use carbon tetrachloride, carbon dioxide extinguishers, or water.
39. Do not use air pressure to transfer.

First-Aid Statements

1. First Aid: *Call a physician as soon as possible.* If swallowed, induce vomiting by sticking finger down throat or by giving soapy or strong salty water to drink. Repeat until vomit is clear. Never give anything by mouth to an unconscious person.
2. First Aid: *Call a physician as soon as possible.* In case of contact, immediately flush eyes or skin with plenty of water for at least 15 minutes while removing contaminated clothing and shoes. Wash clothing before reuse. (Discard contaminated shoes.)
3. First Aid: *Call a physician as soon as possible.* If inhaled, remove to fresh air. If not breathing, give artificial respiration, preferably mouth-to-mouth. If breathing is difficult, give oxygen.
4. First Aid: *Call a physician as soon as possible.* In case of eye contact, immediately flush eyes with plenty of water for at least 15 minutes. Remove contact lenses if worn.

5. First Aid: *Call a physician as soon as possible.* In case of contact, immediately flush eyes with plenty of water for at least 15 minutes. Flush skin with water. (Wash clothing before reuse.)
6. First Aid: In case of contact, immediately wash skin with soap and plenty of water.
7. First Aid: Do *not* induce vomiting. Call a physician as soon as possible.
8. Antidote: (Indicate commonly available antidote.)
9. Note to Physicians: (Give detailed, specific treatment, including drug dosage.)
10. Call the local emergency unit.

Statements Specifying Specific Disposal Instructions

1. Flush spill area with water spray.
2. Soak up spill with sand or earth. Do not use water.
3. Flush away spill by flooding with water applied quickly to entire spill.
4. Keep upwind of leak. Evacuate enclosed places until gas has dispersed.
5. Dike spill and decontaminate by. . . .
6. Do not flush into sewers.
7. Dispose of sodium by burning carefully in an open fire.
8. Sweep up spillage with strong calcium hypochlorite solution.
9. Treat spillage with strong calcium hypochlorite solution and flush to sewer.
10. In case of spillage, keep wet and remove carefully.
11. Soak up with rags and dispose in covered metal containers.
12. Consult local solid waste regulations for safe disposal.
13. Do not sweep. Use vacuum cleaning equipment only.

Sample Label

4	Extreme Health Hazard
4	Extremely Flammable
2	Moderately Reactive

Fatal if swallowed, inhaled, or absorbed through the skin.

Causes severe eye burns.

Protect from all sources of ignition.

Subject to violent polymerization.

Do not breathe vapor or get in eyes, on skin, on clothing.

When possibility of contact exists, wear full neoprene suit, rubber boots, rubber gloves, and self-contained breathing apparatus.

Avoid contact with acid, organic compounds, or water.

First Aid: *Call a physician as soon as possible.* Immediately upon exposure, flush skin and eyes with water for 15 minutes while removing contaminated clothing and shoes. Wash clothing before reuse. Discard contaminated shoes.

Refer to the Material Safety Data Sheet (MSDS) on file.

APPENDIX F: TOXICOLOGY GLOSSARY

AAPCC	American Association of Poison Control Centers.
ABIH	American Board of Industrial Hygiene.
ACGIH	American Conference of Governmental Industrial Hygienists.
ACS	American Chemical Society.
Acute	A short time frame of action—measurable in seconds, minutes, hours, days. Used to describe brief exposures and effects which appear promptly after exposure.
Adsorption	The process by which chemicals are held on the surface of a material.
Aerosol	A suspension of fine liquid particles in the air.
AIHA	American Industrial Hygiene Association.
ALC	The approximate lethal concentration in air for experimental animals: The test animal and the test condition should be specified; the value is expressed in mg/L, mg/m^3, or ppm.
ANSI	American National Standards Institute.
Antagonism	Interference or inhibition of the effect of one chemical by the action of another chemical.
Autoignition temperature	Minimum temperature that a substance must be to ignite spontaneously in air.
Bioaccumulation	The retention and concentration of a substance by an organism.
Caustic	A substance that attacks tissue by direct chemical reaction. Often alkali.
Cancer	A disease characterized by the rapid and uncontrolled growth of aberrant cells into malignant tumors.
Carcinogenic	Causing malignant (cancerous) tumors (OSHA, NIOSH, and FDA consider any tumor to be either a cancer or a precursor of a cancer). Because of the difficulty of proving that a material is a human carcinogen, very few materials are labeled as such. However, many chemicals are suspected or possible human carcinogens and should be treated as true carcinogens in the school laboratory.
CAS	Chemical Abstract Service.
CAS Registration Number	A number assigned by the Chemical Abstracts Service to identify a chemical.
CAS RN	Chemical Abstracts Registration Number.
CDC	Centers for Disease Control.
Ceiling concentration (CL)	Maximum concentration of a toxic substance at any one time.
CERCLA	Comprehensive Environmental Response, Compensation and Liability Act of 1980.
CFR	Code of Federal Regulations.
Chronic	Long period of action—weeks, years. Used to describe ongoing exposures and effects that develop only after long exposure.

CNS	Central nervous system. The portion of the nervous system which consists of the brain and spinal cord.
Corrosive	A chemical that attacks metals and body tissue, usually an acid or material that reacts with water to produce an acid.
Danger	Highest degree of hazard.
DOT	U.S. Department of Transportation. Provides information (classifications and labels) which indicates hazards expected in transportation of materials.
Dust	Airborne fine solid material.
EL	Exposure level.
Embryotoxic	Poisonous to an embryo (without necessarily poisoning the mother).
EPA	Environmental Protection Agency.
Etiologic	Pertaining to the factors that cause disease and the methods of their introduction to the host.
Experimental carcinogen	A substance that has been shown by valid, statistically significant experimental evidence to induce cancer in animals.
Explosive limits	The lower explosive limit (LEL) is the minimum concentration in air below which contact with a source of ignition does not result in explosion. The upper explosive limit (UEL) is the maximum concentration in air above which contact with a source of ignition does not result in explosion.
Explosive range	The difference between the lower and upper explosive limits, expressed in percentage of vapor or gas in air by volume.
FDA	U.S. Food and Drug Administration.
Flash point	Lowest temperature at which a flammable liquid will ignite when a source of ignition is applied.
Hazard	A condition of danger either potential or actual that will interrupt or interfere with a normal, orderly process.
Hazardous chemical	A chemical having one or more of the OSHA hazardous chemical properties.
HEPA	High-efficiency particulate air filter. Used to capture toxic particles from the air or vacuum cleaner exhaust.
Human carcinogen	A substance that has been shown by valid, statistically significant epidemiological evidence to be carcinogenic to humans.
IARC	International Agency for Research on Cancer.
IDLH	Immediately dangerous to life and health. A level defined by the Standards Completion Program (a joint effort of NIOSH and OSHA) only for the purpose of respirator selection. Represents the maximum concentration from which, in the event of respirator failure, one could escape within 30 min without experiencing any escape-impairing or irreversible health effects.
Ignition temperature	Minimum temperature required to initiate or cause self-sustained combustion independently of the heating or heated element.
IMO	International Maritime Organization. Provides classifications of hazardous materials to improve safety during their transport. This material is similar to that provided by DOT.
Insidious material	Materials possessing a great potential for harmful effects upon people.
Insidious material chemical	A chemical acting in imperceptible steps, having a cumulative chronic effect.
Irreversible effect	Effect characterized by the inability of the body to partially or fully repair injury caused by a toxic agent.
Irritant	A substance producing local inflammation on contact.
Latency	Time from the first exposure to a chemical until the appearance of a toxic effect.

LC_{50}	The concentration in air of a toxicant that kills 50% of the test animals in a specific time. The test animal and the test conditions should be specified; the value is expressed mg/L, mg/m^3, or ppm.
LD_{50}	The quantity of material that when ingested, injected, or applied to the skin as a single dose will cause death of 50% of the test animals: The test conditions should be specified; the value is expressed in g/kg or mg/kg of body weight.
LEL	Lower explosive limit. The minimum concentration in air below which contact with a source of ignition does not result in explosion.
lfm	Linear feet per minute. Used as a measure of air velocity through the face of a fume hood or other local ventilation device.
LOAEL	Lowest observed adverse effect level.
LOEL	Lowest observed effect level.
MAWP	Maximum allowable working pressure.
mg/m^3	Milligrams per cubic meter. Usually used in describing the concentration of fumes, mists, and dusts in air; i.e., milligrams (dust) per cubic meter (air).
MSDS	Material Safety Data Sheet.
MSHA	U.S. Mine Safety and Health Administration.
Mutagenic	Causing a heritable change in the gene structure.
NFPA	National Fire Protection Association.
NIOSH	National Institute for Occupational Safety and Health.
Oncogenic	Causing tumors.
OSHA TWA	OSHA occupational standard for air concentration.
Pathogen	Any disease-causing agent, usually applied to living agents.
Pathogenic	Causing or capable of causing disease.
PEL	Permissible exposure limits for the workplace, set by regulation and enforced by OSHA; most of these limit values were originally set, by consensus, by the ACGIH to assist industrial hygienists in implementing exposure control programs. As law, these are listed in 29 CFR 1910.1000 and subject to revision through the regulatory process.
Poison	A substance with an LD_{50} rat oral dose of 50 mg/kg or less.
Potentiation	The effect of one chemical to increase the effect of another chemical.
ppm	Parts per million. Usually used in describing the concentration of gases and vapors in air by volume; i.e., microliters/liter (air), cubic meters/10^6 (air).
rad	Unit of the absorbed dose of ionizing radiation in any material. The rad can be related to any type of radiation. For gamma rays or x-rays, one rad (tissue) is nearly equal to one roentgen.
RCRA	Resource Conservation and Recovery Act of 1976.
Registry	Registry of Toxic Effects of Chemical Substances, published by NIOSH.
rem	Unit which expresses the biological effect or biological dose in humans for any type of radiation. The biological dose in rems for x-rays or gamma rays is equal to the absorbed dose in rads or the exposure in roentgens: (1 gamma-ray roentgen $\approx$ 1 rad = 1 rem).
Reversible effect	An effect which is not permanent, especially an adverse effect which diminishes when exposure to a toxic chemical ceases.
Roentgen	The basic unit of measurement of gamma or x-radiation based on the ionizing effects of radiation in air.
SARA	Superfund Amendments and Reauthorization Act of 1986.

SEQ NO	An arbitrary sequence number assigned to entries when a CAS RN is not available.
STEL	Short-term exposure limit. Maximum concentration to which workers can be exposed for periods up to 15 min. Such exposures should be limited to no more than four per day with periods of at least 60 min each between exposures; the total time-weighted exposure per day should not exceed the TLV value.
Systemic	Relating to the whole body, rather than its individual parts.
TC	Toxic concentration.
TD	Toxic dose.
Teratogenic	Producing a malformation of the embryo (birth defect).
Threshold	The lowest dose of a chemical at which a specified measurable effect is observed and below which it is not observed.
TLV-TWA	Threshold limit value established by the ACGIH: The time-weighted average concentration for a normal 8-h workday or 40-h work week to which nearly all workers may be repeatedly exposed, day after day, without adverse affect.
Toxicant	A substance capable of producing injury, illness, or death to humans as a result of a dermal absorption, ingestion, or inhalation.
Tumorigenic	Causing tumors.
TWA	Time-weighted average.

$$TWA = \sum_{i=1}^{i=n} \frac{t_i C_i}{t_{total}} = \frac{t_1 C_1 + t_2 C_2 + \cdots + t_n C_n}{t_1 + t_2 + \cdots + t_n}$$

where:

t = exposure time in minutes
C = concentration in ppm or mg/m^3
n = total number of exposures
i = exposure number

UEL	Upper explosive limit. The maximum concentration in air above which contact with a source of ignition does not result in explosion.
Vapor density	Relative density of the vapor as compared to air.
WHO	World Health Organization.
Zoonose	Any disease of animals which may be transmitted to humans under natural conditions (e.g., brucellosis, rabies).

APPENDIX G: SCIENCE INVENTORY AND STORAGE HAZARDS

The following table contains most of the chemicals found in school teaching laboratories. Two complementary organization schemes are included under the "Organization" heading. The shelf number and organic/inorganic classification are based on the Flinn Scientific scheme. The color coding refers to the J. T. Baker system. Each organic and inorganic storage cabinet should have shelves numbered from the top down. See Chap. 4 for more information on chemical storage systems. The suggested joint use of these two systems is as follows:

1. Separate the chemicals by their color coding.
2. Separate chemicals within colors according to the inorganic and organic numerical classifications.
4. Store chemicals in the proper color area and on the correct shelf as diagrammed at the end of this appendix.
5. Store all compressed gases separately by color code.

Materials not in this table may be categorized by flammability (red and red stripe), acids (white), bases (white stripe), oxidizers (yellow), reducing agents (red), poisons and carcinogens (blue). If a material may fit into more than one category (e.g., flammable and an acid), put it in the category with the greatest risk (e.g., flammable).

The National Fire Protection Association flammability classifications are one of the criteria used to designate the "Storage Hazards." They are based on the flash points of materials and are the following: If the flash point is less than 73°F, the material is highly flammable; between 73°F and 100°F the material is flammable; and between 100°F and 200°F the material is combustible.

Poisons can be determined by LD_{50} values and the toxicity rating scale in Appendix A.

Tabulated below are storage and use hazards for each chemical. The absence of a hazard or toxicity designation is not meant to imply safety. Chemical carcinogens should be clearly designated.

316 APPENDICES

Substance	Organization	Storage Hazards
Acetaldehyde	Organic #3, red	Oxidizes readily in air to form unstable peroxides.
Acetamide	Organic #2, orange	
Acetanilide	Organic #2, orange	
Acetic acid	Organic #1, red	Combustible; above 103°F, explosive vapor-air mixture (fireproof storage).
Acetic anhydride	Organic #1, red stripe	Combustible; above 120°F, explosive potential (fireproof storage).
Aceto carmine (Natural Red 4)	Miscellaneous, dye	
Acetone	Organic #4, red	Highly flammable, vapor-air mixture explosive (fireproof, cool storage).
Aceto-orcein (orcinol)	Miscellaneous, orange	
Acetylcholine (as bromide or chloride)	Organic #3	
Acridine orange	Miscellaneous, dye	
Acrylonitrile (Inhibited)	Organic #2, red	Flammable, explosive (fireproof storage).
Adenine	Organic #2, orange	
Adrenaline	Organic #2, orange	
Agar	Miscellaneous, orange	
Alanine	Organic #2, orange	
Albumin	Miscellaneous, orange	
Alizarin yellow	Miscellaneous, dye	
Alizarin red (Red #1)	Miscellaneous, dye	
Alum	Inorganic #2, orange	See aluminum ammonium sulfate, aluminum potassium sulfate.
Aluminum ammonium sulfate	Inorganic #2, orange	
Aluminum chloride, hydrate	Inorganic #2, orange	
Aluminum chloride, anhydrous	Inorganic #2, yellow	Store separately from strong bases. Reacts violently with water.
Aluminum hydroxide	Inorganic #4, orange	
Aluminum metal	Inorganic #1, red	
Aluminum nitrate	Inorganic #3, yellow	Enhances combustion of other materials (avoid contamination).
Aluminum oxide	Inorganic #2, orange	
Aluminum potassium sulfate	Inorganic #2, orange	
Aluminum sodium sulfate	Inorganic #2, orange	
Aluminum sulfate	Inorganic #2, orange	
Ammonia, liquid	Inorg. #4, white stripe	
Ammonium acetate	Inorganic #2, orange	
Ammonium bicarbonate	Inorganic #4, orange	
Ammonium bichromate	Inorganic #8, yellow	Many reactions may cause fire and explosion (fireproof storage).
Ammonium bromide	Inorganic #2, orange	
Ammonium carbonate	Inorganic #4, orange	
Ammonium chloride	Inorganic #2, orange	
Ammonium chromate	Inorganic #8, blue	
Ammonium citrate	Inorganic #8, orange	
Ammonium dichromate	Inorganic #8, yellow	See ammonium bichromate.
Ammonium hydroxide	Inorg. #4, white stripe	
Ammonium iodide	Inorganic #2, orange	
Ammonium metavanadate	Inorganic #2, blue	
Ammonium molybdate	Inorganic #8, orange	
Ammonium nitrate	Yellow, store separately	Enhances combustion of other substances. Strong oxidant (fireproof storage).
Ammonium oxalate	Inorganic #2, white	
Ammonium persulfate	Inorganic #6, yellow	Enhances combustion of other substances. Explosive reaction with reducing agents, metals.
Ammonium phosphate	Inorganic #2, orange	
Ammonium sulfate	Inorganic #2, orange	
Ammonium sulfide	Inorganic #5, red	
Ammonium sulfite	Inorganic #2, orange	
Ammonium tartrate	Inorganic #2, orange	
Ammonium thiocyanate	Inorganic #7, orange	
Amyl acetate	Organic #3, red	Flammable, explosive (fireproof storage).

Substance	Organization	Storage Hazards
n-Amyl alcohol	Organic #2, red	Combustible.
Aniline	Organic #2, red	Combustible; above 160°F, explosive air-vapor mixtures (fireproof storage, away from acids, oxidants).
Aniline blue	Miscellaneous, dye	
Aniline hydrochloride	Organic #2, orange	
Aniline violet	Miscellaneous, dye	
Anthracene	Organic #3, orange	
Antimony	Inorganic #1, blue	
Antimony oxide (trioxide)	Inorganic #4, blue	
Antimony pentachloride	Inorganic #2, white	
Antimony potassium tartrate	Inorganic #2, white	
Antimony trichloride	Inorganic #2, white	
Antimony trisulfide	Inorganic #5, blue	May enhance combustion of other substances.
Arabinose	Organic #2, orange	
Arsenic	Inorganic #10, blue	
Arsenic chloride (trichloride)	Inorganic #10, blue	
Arsenic pentoxide	Inorganic #10, blue	
Arsenic trioxide (arsenous acid)	Inorganic #7, blue	
Asbestos	Inorganic #4, blue	
Ascorbic acid	Organic #1, orange	
Balsam	Organic #2, orange	
Barford reagent	Organic #1, orange	Contains cupric acetate, acetic acid, and water.
Barium acetate	Inorganic #2, blue	
Barium carbonate	Inorganic #4, orange	
Barium chlorate	Inorganic #6, yellow	Enhances combustion of other substances. Explosive, oxidant.
Barium chloride	Inorganic #2, blue	
Barium hydroxide	Inorganic #4, blue	
Barium nitrate	Inorganic #3, yellow	Enhances combustion of other substances. Explosive, oxidant.
Barium oxalate	Inorganic #2, blue	
Barium oxide	Inorganic #4, blue	Oxidant.
Barium peroxide	Inorganic #6, yellow	Enhances the combustion of other substances. Many reactions cause fire or explosion.
Barium sulfate	Inorganic #2, orange	
Barium sulfide	Inorganic #5, blue	
Beal orcinol reagent	Organic #2, red	Contains resorcinol, ethyl alcohol, and ferric chloride.
Beeswax	Miscellaneous, orange	
Benedict's solution	Inorganic #2, orange	
Benzaldehyde	Organic #3, red	Combustible; above 145°F, explosive air-vapor mixtures (fireproof storage).
Benzene	Organic #3, red stripe	Highly flammable (fireproof storage).
Benzidine	Organic #2, blue	
Benzoic acid	Organic #1, orange	
Benzoyl peroxide	Organic #6, yellow stripe	Contamination or heating can cause violent decomposition.
Beryllium carbonate	Inorganic #4, blue	
Biphenyl (diphenyl)	Organic #3, red	
Bismuth nitrate	Inorganic #3, yellow	Oxidant.
Bismuth trichloride	Inorganic #2	
Boric acid	Inorganic #9, orange	
Bouin's fluid	Organic #1, white	Saturated picric acid solution, formalin, and acetic acid.
Brilliant green	Organic #3, dye	
Bromine	Inorganic #2, yellow	Many reactions may cause fire and explosion. Oxidant.
Bromine water	Inorganic #2, yellow	Oxidant.
Bromocresol green	Miscellaneous, dye	
Bromocresol purple	Miscellaneous, dye	
Bromophenol blue	Miscellaneous, dye	
Bromothymol blue	Miscellaneous, dye	
Butanol (*n*-butyl alcohol)	Organic #2, red	Flammable, explosive (fireproof storage).
Butyric acid	Organic #1, white	Explosive above 161°F in air-vapor mixtures (fireproof storage).

Substance	Organization	Storage Hazards
Cadmium acetate	Inorganic #2, blue	
Cadmium carbonate	Inorganic #4, blue	
Cadmium chloride	Inorganic #2, blue	
Cadmium, metal	Inorganic #1, blue	
Cadmium nitrate	Inorg. #3, yellow stripe	Oxidant.
Cadmium oxide	Inorganic #4, blue	
Cadmium sulfate	Inorganic #2, white	
Calcium	Inorganic #1, red	Many reactions may cause fire or explosion.
Calcium acetate	Inorganic #2, orange	
Calcium bromide	Inorganic #2, orange	
Calcium carbide	Inorganic #5, red	Reaction with water may cause fire and explosion.
Calcium carbonate	Inorganic #4, orange	
Calcium chloride	Inorganic #2, orange	
Calcium dioxide	Inorganic #4, yellow	
Calcium fluoride	Inorganic #2, orange	
Calcium hydroxide	Inorganic #4, orange	
Calcium hypochlorite	Inorganic #6, yellow	Enhances combustion of other substances. Oxidant.
Calcium nitrate	Inorganic #3, yellow	Enhances combustion of other substances. Oxidant.
Calcium oxide	Inorganic #4, orange	
Calcium phosphate	Inorganic #2, orange	
Calcium sulfate	Inorganic #2, orange	
Camphor	Organic #4, red	Combustible; above 150° F, explosive vapor air mixtures.
Carbolfuchsin (Ziehl's stain)	Organic #2	
Carbolic acid (phenol)	Organic #8, blue	
Carbon	Inorganic #10, orange	
Carbon dioxide	Miscellaneous	Solid can cause frostbite.
Carbon disulfide	Organic #7, red	Highly flammable, explosive (fireproof storage under water or inert gas).
Carbon tetrachloride	Organic #4, blue	
Carborundum	Inorganic #4, blue	
Carmine	Miscellaneous, dye	
Carnoy fixative (mixture of alcohol, acetic acid, and chloroform)	Organic #2, red	Flammable.
Casein	Miscellaneous, orange	
Catechol (1,2-dihydroxybenzene)	Organic #8, red	Combustible.
Ceric sulfate	Inorganic #2, yellow	Fire risk in presence of organic substances.
Charcoal	Inorganic #10, red	
Chloral hydrate	Controlled substance, blue	Should not be stored on school premises.
Chloretone (chlorobutanol)	Organic #2, blue	
Chlorine	Bottled gas, yellow	Many reactions may cause fire and explosion.
Chlorine water	Inorganic #2, yellow	
Chlorobenzene	Organic #4, red	Combustible; above 84°F, explosive vapor-air mixtures (fireproof storage).
Chloroform	Organic #4, blue	
Chorionic gonadotropin	Miscellaneous, orange	
Chromium	Inorganic #1, blue	
Chromium acetate	Inorganic #2, blue	
Chromium chloride	Inorganic #2, orange	
Chromium nitrate	Inorganic #3, yellow	Strong oxidant.
Chromium(VI) oxide	Inorganic #4, blue	Oxidant.
Chromium(III) potassium sulfate	Inorganic #2, orange	
Chromium trioxide	Inorganic #4, yellow	Many reactions may cause fire and explosion; strong oxidant.
Cobalt	Inorganic #1, orange	Dust is flammable.
Cobalt chloride	Inorganic #2, blue	
Cobalt nitrate	Inorganic #3, yellow	Enhances the combustion of other substances; oxidant.
Cobalt sulfate	Inorganic #2, orange	
Colchicine	Organic #8, blue	
Cupric acetate	Inorganic #2, orange	
Cupric bromide	Inorganic #2, orange	
Cupric carbonate	Inorganic #4, orange	
Cupric chloride	Inorganic #2, orange	

Substance	Organization	Storage Hazards
Cupric nitrate	Inorganic #3, yellow	Strong oxidant.
Cupric oxide	Inorganic #4, orange	
Cupric sulfate	Inorganic #2, orange	
Cyclohexane	Organic #3, red	Highly flammable (fireproof storage).
Cyclohexene	Organic #3, red	Highly flammable (fireproof storage, add inhibitor).
Deoxyribonucleic acid	Organic #1, orange	
Dextrin starch	Miscellaneous, orange	
Dextrose	Miscellaneous, orange	
Diastase of malt	Miscellaneous, orange	
p-Dichlorobenzene	Organic #4, red	Combustible; above 150°F, explosive air-vapor mixtures (fireproof storage).
Dichloroethane	Organic #4, red	See ethylene dichloride.
Dichloroindophenol sodium salt	Organic #8	
Dichloromethane	Organic #4, blue	See methylene chloride.
Dichlorophenol	Organic #8, blue	
Diethyl phthalate	Organic #4, red	Combustible.
Digitonin	Organic #3	
Diisopropyl ether	Organic #4, red	Explosive.
N,N-Dimethylaniline	Organic #2, red	Combustible; above 145°F, explosive vapor-air mixtures (fireproof storage).
Dimethylglyoxime	Organic #2	
1,4-Dioxane (p-dioxane)	Organic #4, red	Flammable; may develop explosive peroxides (fireproof storage).
Diphenylamine	Organic #2, orange	
Dipotassium chromate	Inorganic #8, yellow	Oxidant.
EDTA	Organic #1, orange	
Eosin	Miscellaneous	
Epinephrine	Organic #2, orange	See adrenaline.
Epsom salt	Inorganic #2, orange	See magnesium sulfate.
Erythrosine	Miscellaneous, dye	
Ether, ethyl	Organic #4, red	Highly flammable, explosive, forms peroxides.
Ethyl acetate	Organic #4, red	Flammable, explosive (fireproof storage).
Ethyl alcohol	Organic #2, red	Flammables
Ethyl methacrylate	Organic #3, red	Flammable.
Ethylene dichloride	Organic #4, red	Flammables
Ethylene glycol	Organic #2, orange	
Ethylene oxide	Organic #5, red	Highly flammable (fireproof storage).
F.A.A. solution	Organic #2, red	Contains formaldehyde, ethyl alcohol, and acetic acid.
Fehling's solution A	Inorganic #2, orange	
Fehling's solution B	Inorganic #4	
Ferric acetate	Inorganic #2, orange	Combustible.
Ferric ammonium acetate	Inorganic #2, orange	
Ferric ammonium citrate	Inorganic #2, orange	
Ferric ammonium sulfate	Inorganic #2, orange	
Ferric chloride	Inorganic #2, orange	
Ferric nitrate	Inorganic #3, yellow	Oxidant.
Ferric oxide	Inorganic #4, orange	
Ferric phosphate	Inorganic #2, orange	
Ferric sulfate	Inorganic #2, orange	
Ferrous ammonium sulfate	Inorganic #2, orange	
Ferrous chloride	Inorganic #2, orange	
Ferrous nitrate	Inorganic #3, orange	
Ferrous oxide	Inorganic #4, orange	
Ferrous sulfate	Inorganic #2, orange	
Ferrous sulfide	Inorganic #5, orange	
Feulgen stain	Miscellaneous, dye	See Schiff reagent.
Flagella stain	Miscellaneous, dye	See Loeffler's stain.
Fluorescein	Organic #8	
Formaldehyde	Organic #3, red	
Formalin	Organic #3	37–50% solution of formaldehyde.
Formic acid	Organic #1, red	Above 156°F, explosive vapor-air mixtures.
Fructose	Miscellaneous, orange	

Substance	Organization	Storage Hazards
Fuchsin	Miscellaneous, orange	
Fumaric acid	Organic #1	Combustible.
Gasoline	Organic #3, red	Highly flammable.
Gelatin	Miscellaneous, orange	
Gentian violet	Miscellaneous, orange	See methyl violet.
Gibberellic acid	Organic #1, orange	
Giemsa stain	Organic #2, dye	
Gilson fluid	Organic #2	Contains acetic acid, nitric acid, ethyl alcohol, and zinc chloride.
Glucose	Organic #2, orange	
Glycerine	Organic #2, orange	See glycerol.
Glycerol	Organic #2, orange	
Gold foil	Inorganic #1, orange	
Gram's iodine stain	Miscellaneous, dye	
Graphite	Inorganic #1, red	
Gum arabic	Organic #1, orange	
Gum tragacanth	Organic #2, orange	
Gypsum	Inorganic #2, orange	See calcium sulfate.
Hayem's solution	Inorganic #2, blue	Contains mercuric chloride, sodium chloride, and sodium sulfate.
Helium	Bottled gas	
Hematoxylin	Organic #2, blue	
Heptane	Organic #3, red	Flammable; explosive vapor-air mixtures.
Hexachlorophene [2,2-methylene-bis(3,4,6-trichlorophenol)]	Organic #8, blue	
Hexane	Organic #3, red	
Holtfreter's solution	Inorganic #2, orange	Contains sodium chloride, potassium chloride, calcium chloride, sodium bicarbonate.
Hydroiodic acid	Inorganic #9, white	
Hydrochloric acid	Inorganic #9, white	
Hydrofluoric acid	Inorganic #9, white	
Hydrogen	Bottled gas, red	Highly flammable, explosive.
Hydrogen peroxide, 30%	Inorganic #6, yellow	Enhances combustion of other substances, possible explosive mixed with other substances.
Hydrogen sulfide	Inorganic #5, red	Highly flammable, explosive gas.
Hydroquinone	Organic #3, red	
Indigo	Miscellaneous, dye	
Indigo carmine	Miscellaneous, dye	
Indolacetic acid	Organic #1	
Indolphenol sodium salt	Inorganic #8	
Iodine	Inorganic #2, white	
Iron acetate	Inorganic #2, orange	See ferric acetate.
Iron metal	Inorganic #1, orange	
Iron pyrite	Inorganic #2, orange	See ferrous sulfide.
Isoamyl alcohol	Organic #2, red	Combustible.
Isobutyl alcohol	Organic #2, red	Combustilbe, explosive above 82°F.
Isopentyl alcohol	Organic #2, red	See isoamyl alcohol.
Isopropyl alcohol	Organic #2, red	Flammable.
Janus green B	Miscellaneous, dye	
Kaolin	Inorganic #4, orange	
Kerosene	Organic #3, red	Combustible; above 110°F, vapor-air mixtures are explosive.
Lactic acid	Organic #1, white	
Lactose	Miscellaneous, orange	
Lauric acid	Organic #1	Combustible.
Lead acetate	Inorganic #2, blue	
Lead arsenate	Inorganic #7, blue	
Lead carbonate	Inorganic #4, blue	
Lead chloride	Inorganic #2, blue	Oxidant.
Lead dioxide	Inorganic #4, yellow	Enhances the combustion of other substances; oxidant; reacts violently.
Lead iodide	Inorganic #2, blue	

Substance	Organization	Storage Hazards
Lead metal	Inorganic #1, orange	
Lead monoxide (litharge)	Inorganic #4, blue	
Lead nitrate	Inorganic #3, yellow	Enhances combustion of other substances; oxidant.
Lead oxide	Inorganic #4, blue	Oxidant; strong reactant.
Lead peroxide	Inorganic #4, yellow	See lead dioxide.
Lead sulfate	Inorganic #2, white	
Lead sulfide (galena)	Inorganic #5, blue	
Lead tetraoxide	Inorganic #4, blue	See lead oxide.
Lime water	Inorganic #4, orange	See calcium hydroxide.
Linseed oil	Organic #2, red	
Lithium carbonate	Inorganic #4, white	
Lithium chloride	Inorganic #2, orange	
Lithium hydroxide	Inorganic #4, white	Reacts violently with acids.
Lithium metal	Inorganic #1, red stripe	Flammable, reacts violently with water, oxidants (fireproof storage).
Lithium nitrate	Inorganic #3, yellow	Oxidant.
Lithium sulfate	Inorganic #2, orange	
Litmus	Miscellaneous	
Loeffler's flagella stain	Organic #2	Contains fuchsin, ethyl alcohol, and aniline.
Logwood extract (hematin)	Organic #2	
Luminol	Miscellaneous	
Lugol's iodine	Inorganic #2, blue	
Lycopodium powder	Miscellaneous	Explosive dust.
Lye	Inorg. #4, white stripe	See sodium hydroxide.
Magnesium acetate	Inorganic #2, orange	
Magnesium bromide	Inorganic #2, orange	
Magnesium carbonate	Inorganic #4, orange	
Magnesium chloride	Inorganic #2, orange	
Magnesium metal	Inorganic #1, red	Highly flammable in powder form; explosive.
Magnesium nitrate	Inorganic #3, yellow	Enhances combustion of other substances; oxidant.
Magnesium oxide	Inorganic #4, orange	
Magnesium sulfate	Inorganic #2, orange	
Magnesium trisilicate	Inorganic #4, orange	
Malachite green	Miscellaneous	
Maleic acid	Organic #1, red	Combustible.
Malonic acid	Organic #1, white	
Maltose	Miscellaneous, orange	
Manganese bromide (manganous bromide)	Inorganic #2, orange	
Manganese chloride (manganous chloride)	Inorganic #2, orange	
Manganese carbonate	Inorganic #4, orange	
Manganese dioxide	Inorganic #4, yellow	Enhances combustion of other substances; many reactions may cause fire and explosion.
Manganese metal	Inorganic #1, red stripe	Dust is flammable.
Manganese nitrate (manganous nitrate)	Inorganic #3, yellow	Oxidant.
Manganese oxide (manganous oxide)	Inorganic #4, orange	
Manganese sulfate (manganous sulfate)	Inorganic #2, orange	
Mayer's fluid	Inorganic #2	Contains potassium phosphate, magnesium sulfate, ammonium nitrate, calcium phosphate.
Mercuric chloride	Inorganic #2, blue	
Mercuric iodide	Inorganic #2, blue	
Mercuric nitrate	Inorg. #3, yellow stripe	Enhances combustion of other substances, strong oxidant.
Mercuric oxide	Inorganic #4, blue	
Mercuric sulfate	Inorganic #2, blue	Decomposes on exposure to light.
Mercuric sulfide	Inorganic #5	
Mercurous chloride	Inorganic #2, blue	
Mercurous nitrate	Inorg. #3, yellow stripe	

Substance	Organization	Storage Hazards
Mercurous oxide	Inorganic #4, yellow	Oxidant.
Mercury bichloride	Inorganic #2, blue	See mercuric chloride.
Mercury metal	Inorganic #1, blue	Toxic vapors.
Methanol, methyl alcohol	Organic #2, red	Flammable; vapor-air mixture explosive (fireproof storage, separate from oxidants).
Methylcellulose	Miscellaneous, blue	
Methylene blue	Miscellaneous, orange	
Methylene chloride	Organic #4, blue	
Methyl ethyl ketone	Organic #2, red	Highly flammable; vapor-air mixtures explosive (fireproof storage, separate from oxidants).
Methyl iodide	Organic #4, blue	
Methyl methacrylate (inhibited)	Organic #3, red stripe	Flammable; vapor-air mixture explosive (fireproof storage, cool).
Methyl orange	Miscellaneous	
Methyl red	Miscellaneous, orange	
Methyl salicylate	Organic #3, orange	
Methyl sulfoxide (dimethyl sulfoxide)	Organic #4, orange	
Methyl violet	Miscellaneous	
Mineral oil	Organic #3, red	
Molasses	Miscellaneous, orange	
Monochloroacetic acid	Organic #1, white	
Naphthalene	Organic #2, red stripe	
2-Naphthol (β-naphthol)	Organic #2	Combustible.
Nessler's reagent	Inorganic #2, blue	
Nickel(II) acetate	Inorganic #2, blue	
Nickel(II) ammonium sulfate	Inorganic #2, blue	
Nickel(II) carbonate	Inorganic #6, blue	
Nickel chloride	Inorganic #2, blue	
Nickel hydroxide	Inorganic #4, blue	
Nickel metal	Inorganic #1, orange	
Nickel nitrate	Inorg. #3, yellow stripe	Oxidant.
Nickel oxide	Inorganic #4, blue	
Nickel sulfate	Inorganic #2, blue	
Nicotine sulfate	Organic #2, blue	
Nicotinic acid (niacin)	Organic #1, blue	
Nigrosine black	Miscellaneous	
Ninhydrin	Organic #2, blue	
Nitric acid	Inorganic #3, yellow	Many reactions may cause explosion; oxidant.
Nitrobenzene	Inorganic #4, red	
Nitrobenzeneazoresorcinol	Organic #8	Combustible.
Nitrogen	Bottled gas	
p-Nitrophenol	Organic #8, yellow	Strong oxidant.
Nucleic acid	Organic #1, orange	
Oleic acid	Organic #1, orange	
Olive oil	Miscellaneous, orange	
Orange IV (Torpeolin 00)	Miscellaneous	
Orcein staining solution	Miscellaneous	Contains orcein, hydrochloric acid, and ethylanol. Flammable liquid.
Osmium tetroxide	Inorganic #4, blue	Vapors are highly irritant.
Oxalic acid	Organic #1, white	Separate from oxidants and strong bases.
Oxygen	Bottled gas	Fire and explosion risk.
Pancreatin	Miscellaneous, orange	
Paraffin	Miscellaneous, orange	
Peanut oil	Miscellaneous, orange	
Pentane	Organic #3, red	Highly flammable; vapor-air mixture explosive (fireproof storage).
Perchloric acid	Inorganic #6, yellow	
Petroleum ether	Organic #4, red	Highly flammable.
Phenolphthalein	Miscellaneous, orange	
Phenyl salicylate (salol)	Organic #3	Combustible.
Phosphoric acid	Organic #1, white	

Substance	Organization	Storage Hazards
Phosphorus (red)	Inorg. #10, red stripe	Separate from oxidants.
Phosphorus (white)	Inorg. #10, red stripe	Flammable; ignites upon contact with air.
Phosphorus pentoxide	Inorganic #10, yellow	Many reactions may cause fire or explosion.
Phthalic anhydride	Organic #1, white	
Picric acid	Organic #8, red	Explosive if dry.
Potassium bicarbonate	Inorganic #4, orange	
Potassium bisulfate	Inorganic #2, orange	
Potassium bitartrate	Inorganic #2, orange	
Potassium bromate	Inorganic #2, yellow	
Potassium bromide	Inorganic #2, orange	
Potassium carbonate	Inorganic #4, orange	
Potassium chlorate	Inorganic #6, yellow	Enhances combustion of other substances; if contaminated, may explode from shock or mechanical friction.
Potassium chloride	Inorganic #2, orange	
Potassium chromate	Inorganic #8, blue	
Potassium cyanide	Inorg. #7, white stripe	
Potassium dichromate	Inorganic #8, yellow	Strong oxidant.
Potassium ferricyanide	Inorganic #7, orange	
Potassium ferrocyanide	Inorganic #7, orange	
Potassium fluoride	Inorganic #2, blue	
Potassium hydroxide	Inorg. #4, white stripe	Reacts violently with acids.
Potassium iodate	Inorganic #8, yellow	Enhances combustion of other substances; strong oxidant.
Potassium iodide	Inorganic #2, orange	
Potassium metal	Inorg. #1, red stripe	Combustible; many reactions may cause fire and explosion; reacts violently with water (fireproof storage separately under paraffin or oil).
Potassium nitrate	Inorganic #3, yellow	Enhances combustion of other substances; oxidant; violent reactant.
Potassium oxalate	Inorganic #2, blue	
Potassium oxide	Inorganic #4, white	
Potassium periodate, meta	Inorganic #6, yellow	Enhances combustion of other substances; many reactions may cause fire or explosion.
Potassium permanganate	Inorganic #8, yellow	Enhances combustion of other substances; many reactions may cause fire and explosion; powerful oxidant; violent reactant.
Potassium phosphate	Inorganic #2, orange	
Potassium pyrosulfate	Inorganic #2, orange	
Potassium sodium tartrate	Inorganic #2, orange	
Potassium sulfate	Inorganic #2, orange	
Potassium sulfide	Inorganic #5, red	May ignite spontaneously on contact with air; flammable; explosive on heating (fireproof storage).
Potassium tartrate	Inorganic #2, orange	
Potassium thiocyanate	Inorg. #2, yellow stripe	
Propane	Bottled gas, red	Highly flammable; explosive air-vapor mixtures.
Propionic acid	Organic #1, red	Combustible.
Propyl alcohol	Organic #2, red	Flammable; vapor-air mixtures explosive.
Pyridine	Organic #2, red	Flammable; vapor-air mixtures explosive (fireproof storage separate from oxidants).
Pyrogallic acid	Organic #4, blue	
Quinine sulfate	Organic #2	
Resorcinol	Organic #2, red	
Ringer's solution	Miscellaneous	
Rosin	Miscellaneous, red	
Safranine	Miscellaneous, dye	
Salicylic acid	Organic #1, orange	Dust explosive.
Sand	Miscellaneous, orange	
Schiff reagent	Organic #2	Contains fuchsin, sodium bisulfite, and hydrochloric acid.
Selenium	Inorganic #1, orange	
Sesame oil	Organic #4, orange	

Substance	Organization	Storage Hazards
Silicic acid	Inorganic #9, orange	
Silica gel	Miscellaneous, orange	
Silicon metal	Inorganic #1, orange	
Silver acetate	Inorganic #2, blue	
Silver chloride	Inorganic #2, blue	
Silver cyanide	Inorganic #7, blue	
Silver iodide	Inorganic #2, blue	
Silver metal	Inorganic #1, blue	
Silver nitrate	Inorganic #3, yellow	Many reactions may cause fire and explosion; violent reaction with organic substances.
Silver oxide	Inorganic #4, orange	Oxidant.
Silver sulfate	Inorganic #2, blue	
Sodium acetate	Inorganic #2, orange	
Sodium arsenate	Inorganic #7, blue	
Sodium arsenite	Inorganic #7, blue	
Sodium azide	Inorganic #3, blue	Explosion possible from concussion, friction (fireproof storage, mix with water, 20%).
Sodium bicarbonate	Inorganic #4, orange	
Sodium bismuthate	Inorganic #7, orange	
Sodium bisulfate	Inorganic #2, orange	
Sodium bisulfite	Inorganic #2, orange	
Sodium borate	Inorganic #8, orange	
Sodium bromide	Inorganic #2, orange	
Sodium carbonate	Inorganic #4, orange	
Sodium chlorate	Inorganic #6, yellow	Many reactions may cause fire and explosion; strong oxidant.
Sodium chloride	Inorganic #2, orange	
Sodium chromate	Inorganic #8, yellow	Oxidant.
Sodium citrate	Inorganic #8, orange	
Sodium cyanide	Inorganic #7, blue	
Sodium dichromate	Inorganic #8, yellow	Many reactions may cause fire and explosion; oxidant.
Sodium dithionite	Inorganic #2, red stripe	Oxidant.
Sodium ferrocyanide	Inorganic #7, orange	
Sodium fluoride	Inorganic #2, blue	
Sodium hydroxide	Inorg. #4, white stripe	Reacts violently with acid.
Sodium hydrosulfite	Inorganic #2, red stripe	See sodium dithionite.
Sodium hypochlorite	Inorganic #6, orange	Reacts violently with acids; forms toxic fumes in presence of ammonia.
Sodium hyposulfate	Inorganic #2, orange	See sodium dithionite.
Sodium iodate	Inorganic #2, yellow	
Sodium iodide	Inorganic #2, orange	
Sodium lauryl sulfate	Inorganic #2	
Sodium metabisulfite	Inorganic #2, orange	
Sodium metal	Inorganic #1, red stripe	Combustible; many reactions may cause fire and explosion; violent reaction with water (fireproof storage; separate under paraffin oil or kerosene from all substances).
Sodium metaphosphate	Inorganic #2, orange	
Sodium molybdate	Inorganic #2, orange	
Sodium nitrate	Inorganic #3, yellow	Enhances the combustion of other substances; oxidant.
Sodium nitrite	Inorganic #3, yellow	Many reactions may cause fire and explosion.
Sodium oxalate	Inorganic #2, blue	
Sodium perborate	Inorganic #8, orange	Oxidant.
Sodium permanganate	Inorganic #8, yellow	Oxidant.
Sodium peroxide	Inorg. #6, yellow stripe	Many reactions may cause fire and explosion; reacts violently with water.
Sodium phosphate	Inorganic #2, orange	
Sodium pyrophosphate	Inorganic #2, orange	
Sodium salicylate	Organic #1, orange	
Sodium silicate	Inorganic #2, orange	
Sodium silicofluoride (disodium hexafluorosilicate)	Inorganic #4	

APPENDIX G 325

Substance	Organization	Storage Hazards
Sodium sulfate	Inorganic #2, orange	
Sodium sulfide (anhydrous)	Inorganic #5, red	Store separately from acids, oxidants, dry.
Sodium sulfite	Inorganic #2, orange	
Sodium tartrate	Inorganic #2, orange	
Sodium tetraborate	Inorganic #8, orange	See sodium borate.
Sodium thiocyanate	Inorganic #7, orange	
Sodium thiosulfate	Inorganic #2, orange	
Sodium tungstate	Inorganic #2, orange	
Stannic chloride	Inorganic #2, white	
Stannic oxide	Inorganic #2, orange	
Stannous chloride	Inorganic #2, orange	
Starch	Miscellaneous, orange	
Stearic acid	Organic #1, orange	
Strontium	Inorganic #1, red stripe	
Strontium bromide	Inorganic #2, orange	
Strontium chloride	Inorganic #2, orange	
Strontium nitrate	Inorganic #3, yellow	
Succinic acid	Organic #1, orange	
Sucrose	Miscellaneous, orange	
Sudan black B	Miscellaneous, dye	
Sudan III	Miscellaneous, dye	
Sudan IV	Organic #2, dye	
Sugar	Miscellaneous, orange	
Sulfamic acid	Organic #1, white	Separate from strong bases.
Sulfanilic acid	Organic #1, white	
Sulfur	Inorganic #10, orange	
Sulfur black dye	Inorganic #10, dye	
Sulfur blue dye	Inorganic #10, dye	
Sulfur yellow dye (napthol yellow, citronin)	Inorganic #10, dye	
Sulfuric acid	Inorganic #9, white	Many reactions may cause fire and explosion; water-reactive.
Talc	Miscellaneous, orange	
Tannic acid	Organic #1, orange	
Tartaric acid	Organic #1, orange	
Terpineol	Organic #2, orange	
Testosterone	Miscellaneous, blue	
Tetrahydrofuran	Organic #4, red	Highly flammable, vapor-air mixtures are explosive; also forms explosive peroxides.
Thermite igniting mixture	Inorganic #4, red	Contains FeO_2, and Al; flammable; burning difficult to stop once started (fireproof storage).
Thioacetamide	Organic #2, blue	
Thiourea	Organic #2, blue	
Thymol blue	Miscellaneous	
Thyroxine	Miscellaneous, orange	
Tin metal	Inorganic #1, orange	Combustible as dust.
Titanium metal	Inorganic #1, red	Combustible as dust.
Titanium dioxide (titanium oxide)	Inorganic #4, orange	Combustible; many reactions may cause fire and explosion; store under inert gas.
Titanium trichloride	Inorganic #2, red	
Toluene	Inorganic #3, red	Flammable; vapor-air mixtures explosive (fireproof storage, separate from H_3SO_4).
o-toluidine	Organic #2, blue	Separate from acids.
Tricans methane sulfonate	Organic #2	
Trichloroacetic acid	Organic #1, white	
Trichlorotrifluoroethane	Organic #4, orange	
Triethanolamine	Organic #2, orange	
Trimethylpentane	Organic #3, red	Flammable.
Triphenyl tetrazolium chloride	Miscellaneous	
Trisodium phosphate	Inorganic #2, white	Separate from strong acids.
Tumeric powder	Organic #2, orange	
Tungsten metal	Inorganic #1, orange	Dust is flammable.

Substance	Organization	Storage Hazards
Turpentine	Organic #2, red	Combustible (fireproof storage, separate from oxidants).
Ultramarine blue	Miscellaneous	
Uranyl nitrate	Inorganic #3, yellow	Strong oxidant.
Urea	Organic #2, orange	
Urethane	Organic #2, orange	
Vegetable oil	Organic #2, orange	
Wood's metal	Inorganic #1, orange	Contains bismuth, lead, tin, cadmium.
Wright's stain solution	Miscellaneous, red	Flammable.
Xylene	Organic #3, red	Combustible; above 81°F, explosive vapor-air mixtures (fireproof storage).
Yeast	Miscellaneous, orange	
Zenker's fluid	Inorganic #2, blue	Contains mercuric chloride, potassium dichromate, sodium sulfate, and acetic acid.
Zeolite	Inorganic #4, orange	
Zinc acetate	Inorganic #2, orange	
Zinc carbonate	Inorganic #2	
Zinc chloride	Inorganic #2, white	
Zinc metal	Inorganic #1, red stripe	Combustible as dust (fireproof storage, separated from oxidants).
Zinc nitrate	Inorganic #3, yellow	Enhances combustion of other substances.
Zinc oxide	Inorganic #4, orange	
Zinc stearate	Inorganic #2, orange	
Zinc sulfate	Inorganic #2, orange	
Zinc sulfide	Inorganic #5, orange	
Zirconium nitrate	Inorganic #3, yellow	

SUGGESTED SHELF STORAGE PATTERNS

Organics

Organic: Orange—No Special Hazard

```
Organic shelf 1

Organic shelf 2

Organic shelf 3

Organic shelf 4

Miscellaneous organics shelf
```

Organic: Red—Flammables

```
Organic shelf 1

Organic shelf 2

Organic shelf 3

Organic shelf 4

Organic shelf 5

Organic shelf 7

Organic shelf 8

Miscellaneous shelf
```

Approved flammables cabinet

Organic: Red Stripe—Flammables

Organic shelf 1
Organic shelf 2
Organic shelf 3

Approved flammables cabinet

Organic: Yellow—Reactivity Hazard

Organic shelf 8

Store separately from other chemicals.

Organic: Yellow Stripe—Reactivity Hazard

Organic shelf 6

Store separately from other chemicals.

Organic: White—Contact Hazard

Organic shelf 1

Store separately in a corrosion-proof cabinet or locker.

Organic: Blue—Poisons and Controlled Substances

Organic shelf 2
Organic shelf 4
Organic shelf 8
Miscellaneous shelf

Lockable poisons cabinet

Inorganics

Inorganic: Orange—No Special Hazard

Inorganic shelf 1
Inorganic shelf 2
Inorganic shelf 3
Inorganic shelf 4
Inorganic shelf 5
Inorganic shelf 6
Inorganic shelf 7
Inorganic shelf 8
Inorganic shelf 9
Inorganic shelf 10
Miscellaneous shelf

Standard shelf or chemical storage cabinet

Inorganic: Red—Flammables

Inorganic shelf 1
Inorganic shelf 2
Inorganic shelf 3
Inorganic shelf 4
Inorganic shelf 5
Inorganic shelf 10
Miscellaneous shelf

Approved flammables cabinet

Inorganic: Red Stripe—Flammables

Inorganic shelf 1
Inorganic shelf 2
Inorganic shelf 10

Approved flammables cabinet

Inorganic: Yellow—Reactivity Hazard

Inorganic shelf 2
Inorganic shelf 3
Inorganic shelf 4
Inorganic shelf 6
Inorganic shelf 8
Inorganic shelf 10

Store separately from other chemicals.

Inorganic: Yellow Stripe—Reactivity Hazard

Inorganic shelf 2
Inorganic shelf 3
Inorganic shelf 6

Store separately from other chemicals.

Inorganic: White—Contact Hazard

Inorganic shelf 2
Inorganic shelf 4
Inorganic shelf 7
Inorganic shelf 9

Store separately in a corrosion-proof cabinet or locker.

Inorganic: White Stripe—Contact Hazard

Inorganic shelf 4
Inorganic shelf 7

Store separately in a corrosion-proof cabinet or locker.

Compressed Gases

No Color—No Special Hazard

Red—Flammable Compressed Gas

Yellow—Corrosive Compressed Gas

Store separately from other gases.

SUBJECT INDEX

Italicized page numbers indicate pages on which definitions appear.

Acetylation, 39
Acetylcholine esterase inhibitors, 43
Acid cabinets, 57
Acids, 7, 24, 27
 disposal, 72
ACS-certified grade, chemical, 55
Acute, *312*
 toxicity, 20, 44
Acutely hazardous solid waste, 67
Additive effects of gases, 49
Additive toxic effects, 40
Adsorption, *312*
Aerosols, 45, *312*
ALC, *312*
Alkalies, 24, 27
Allergens, 29
American Association of Poison Control Centers (AAPCC), *312*
American Board of Industrial Hygiene (ABIH), *312*
American Chemical Society (ACS), *312*
American Conference of Governmental Industrial Hygienists (ACGIH), 23, *312*
American Industrial Hygiene Association (AIHA), *312*
American National Standards Institute (ANSI), 52, *312*
Ames test, 31, 34
Anesthetics, 28
Antagonism, *312*
Antioxidant stabilizers, 9
Approximate lethal concentration, *312*
Asphyxiants, 28
Autoignition temperature, *4, 312*
Autotoxics, 26

Barricades, 14
Bases, disposal, 72
Behavioral toxicity, 32
Bioaccumulation, *312*
Biological waste disposal, 75

Cancer, 30, *312*
Carcinogenic, *312*
Carcinogenicity, 29, 32
Carcinogens, 35, 44, 45
 central storage, 56
 laboratory storage, 62
 procedures for handling, 44, 46
CAS registration number (RN), *312*
Cats, 38
Catalysts, 3

Catalytic reactions, *18*
Caustic, *312*
Ceiling concentration, *312*
Centers for Disease Control (CDC), *312*
Central Nervous System (CNS), *313*
CERCLA, *312*
CFR, *312*
Chemical Abstract Service, *312*
Chemical abstracts registration number, *312*
Chemical conversion of wastes, 73
Chemical disposal, 70
Chemical storage, 57
 in the laboratory, 61
Chemical waste
 storage, 76
 transportation, 76
Chinese hamster ovary test, 34
Chlorinated hydrocarbon waste, 71
CHO test, 34
Chronic, *312*
Chronic toxicity, 20, 44
CL, *312*
Cleveland open-cup (COC), 4
Code of Federal Regulations, *312*
Cold traps, 12
Color code system for chemical storage, 59
Combined effects of gases, 49
Combustible liquids, storage, 60
Community Right to Know Laws, 66
Comprehensive Environmental Response, Compensation and Liability Act of 1980, *312*
Compressed gases
 central storage, 60
 laboratory storage, 62
 transport, 61
Concentration effects, 3
Consumer Product Safety Commission, 51
Controlled areas, 46
Corrosive, *313*
Corrosive chemicals, *15*, 27
 handling, 48
 laboratory storage, 64
 wastes, 69
Cutoff area, 59

Danger, *313*
Dated receiving system, 56
Deflagration, 7
Dehydrating agents, 28
Dermal route, 22

Dermatitis, 24, 49
Detonation, 7
Detonators, *11*
Detoxification, 39
Diet, effect on toxicity, 39
Differential thermal analysis, 8
Disposal, chemical, 47, 54, 66
Dog, 38
Dose-response relationship, 34
Dose sensitivity, 29
Drain disposal of chemicals, 72
Dust, *313*

EL, *313*
Embryotoxic, *313*
Embryotoxins, 28
Endpoints, toxicological, 29
Environmental Protection Agency (EPA), *313*
Enzymes, 25, 33
Etiologic, *313*
Exothermic reaction, 3, 8
Experimental carcinogen, *313*
Explosions, 7
Explosive gases, 7
Explosive limits, *313*
Explosive range, *313*
Explosives
 handling, 14
 powerful group, 7
 sensitive group, 7
 wastes, 70
Exposure control, 50
Exposure level, *313*

Face shields, 14
Faculty, 20
FCC grade, chemical, 55
Federal Hazardous Substances Act, 51
Fetotoxicity, 29
Finely divided materials, 12
Fire doors, 59
Flammable gases, 7
Flammable limits, 5
Flammable liquids, 16
 central storage, 59
 laboratory storage, 62
Flammable reagents, storage in refrigerators, 63
Flammables, 4
Flash point, 4, 6, *313*
Fume hoods, 45, 63

329

Gas cylinders, central storage, 60
Gases, handling, 49
Genotoxicity, 29
Glove box, 47
Gloves, 14
Guards, 14
Guinea pigs, 38

Hazard, *313*
Hazardous chemical, *313*
Hazardous waste, 68
  definitions, 67
Heat of formation, 8
Hematologic toxicity, 32
Hepatotoxicity, 29, 32
High-efficiency particulate air (HEPA) filter, 46–48, *313*
Hoods, chemical storage, 61
Human carcinogen, *313*
Human equivalent dose, 39
Hydrophilic, 24
Hyperinsulinism, 39
Hyperthyroidism, 39

IDLH, *313*
Ignitable wastes, 69
Ignition sources, 6
Ignition temperature, *4, 313*
Immediately dangerous to life and health, *313*
Incineration of chemicals, 71
Incompatible chemicals, 15, *16*
Ingestion, 24
Inhalation, 22, 23
Inhalation toxicity, 32
Inhibitors, 17
Insecticides, 43
Insidious materials, *313*
International Agency for Research on Cancer (IARC), *313*
International Maritime Organization (IMO), *313*
Intramuscular route, 25
Intraperitoneal route, 25
Intravenous route, 25
Intumescent materials, 62
Inventory records, 56
Irreversible effect, *313*
Irritant(s), 26, *313*
  delayed, 26

Kidney damage, 33
Kreb's cycle inhibitors, 43

Lab packs, 43, *75*
Labeling, chemical, 51
Laboratory coats, 14
Lactational toxicity, 33
Landfills, 73
Latency, *313*
LC_{50}, 22, *314*
LD_{50}, *21, 314*
LEL, *4, 6, 314*
lfm, *314*
Lipophilic, 24
Liquefied gases, 7
Liver damage, 33
Local toxicity, 20
Lower explosive limit, *314*
Lowest observed adverse effect level (LOAEL), 30, *314*
Lowest observed effect level (LOEL), 30, *314*
Lungs, 22

Manufacturing Chemists Association (MCA), 51
Margin of error, toxicity assessment, 40
Material Safety Data Sheets (MSDSs), 42, 50, 71, *314*
Maximum allowable working pressure (MAWP), *314*
Maximum tolerated dose, 30
Metabolism, 38
Metals, handling, 49
Methane, 6
Methylation, 38
mg/m^3, *314*
Mixtures, toxic effects, 40
Monkey, 38
Mouse, 37, 38
MTD, 30
Mutagenesis, 33
Mutagenic, *314*
Mutagenicity, 32
  tests, 31

Nasopharyngeal effects, 23
National Fire Protection Association (NFPA), 53, *314*
National Institute for Occupational Safety and Health (NIOSH), 52, *314*
Nervous system damage, 32
Neurotoxicity, 32
NF grade, chemical, 55
No observed adverse effect level (NOAEL), 30
Nonhazardous waste, 68

Occupational Safety and Health Act, 51
Occupational Safety and Health Administration (OSHA), 22
Occupational toxicology, 20
Oncogenic, *314*
Optimal dose, 20
Organophosphate insecticides, 39
OSHA, 22, 44
OSHA TWA, *314*
Oxidizing materials, laboratory storage, 63
Oxygen balance, 8

Paracelsus, 20
Parenteral route, 25
Particulates, toxic effects, 29
Pathogen, *314*
Pathogenic, *314*
Perinatal toxicity, 33
Permissible exposure levels (PEL) limits, 23, *314*
Peroxide, tests, 10
Peroxide formers, laboratory storage, 64
Peroxides
  disposal, 10
  maximum levels, 10
Phase separations, 18
Poison, *314*
Potentiation, *314*
Powders, 12
  laboratory storage, 64
ppm, *314*
Practical grade, chemical, 55
Primary standard grade, chemical, 55
Purchasing chemicals, 54
Purified grade, chemical, 55
Pyrophoric chemicals, laboratory storage, 64, 65

Pyrophoric metals, 6
Pyrophoric wastes, 70

Quality control, safety, 50

Rabbits, 33, 38
rad, *314*
Radioactive materials, laboratory storage, 63
Rats, 22, 38
Reactive wastes, 70
Reagent grade, chemical, 55
Receiving ordered chemicals, 55
Recordkeeping, 45, 56
  for waste chemicals, 76
Recycling, 71
Refrigerators, 61, 63
Registry, *314*
Relief valves, 19
rem, *314*
Renal toxicity, 32
Reproductive
  system effects, 33
  toxicity, 32
Reserve capacity, 33, 34
Resources Conservation and Recovery Act (RCRA), 66, *314*
Reversible effect, *314*
Roentgen, *314*
Routine chemical reaction, 17
Rules for
  chemical storage, 57
  handling chemicals, 4, 6, 43
  handling peroxides, 8
  stockroom personnel, 55
Rupture disks, 19

Safety factor, toxicity, 39
Safety glasses, 14
Safety officer, 8, 11
Safety precautions, 42
Salmonella typhimurium, 34
Saponification, 24
Sensitizers, 27
SEQ NO, *315*
Sewer systems, chemical disposal, 72
Shields, 14
Short-term exposure limit (STEL), 23, *315*
Side reactions, 18
Silicosis, 29
Skin contact, 23
Skin irritation, 24
Solid waste, definitions, 67
Solvent waste disposal, 74
Sources of ignition, 59
Specific half-life, 8
Spills, 5, 45
Spontaneous ignition, 6
Static electricity, 6
Storage cabinets, flammables, 62
Storerooms, central, 58
Students, 20
Subcutaneous route, 25
Superfund Amendments and Reauthorization Act (SARA), 66, *314*
Surface area in toxicology, 36
Synergism, 40
Systemic, *315*
Systemic toxicity, 20

Tagliabue closed-cup (TCC), 4
Tagliabue open-cup (TOC), 4
TC, *315*

SUBJECT INDEX

TD, *315*
Technical grade, chemical, 55
Temperature effects, 3
Teratogenic, *315*
Teratogenicity, 29, 32
Teratogens, 28
Teratology, 33
Threshold, *315*
Threshold Limit Value (TLV), 22, 23, 49, *315*
Time-Weighted Average (TWA), 22, *315*
TLV-TWA, *315*
TOC, 4
Toxic chemicals
  central storage, 59
  laboratory storage, 62, 64
Toxic concentration, *315*
Toxic dose, *315*

Toxic wastes, 69
Toxicant, *315*
Toxicity assessment, 30
Toxicokinetic differences, 36
Toxicology, *20*
Toxophores, 25
Tracheobronchiolar effects, 23
Transformation tests, 31
Transporting chemicals, 60
Tumorigenic, *315*

U.S. Department of Transportation (DOT), *313*
U.S. Food and Drug Administration (FDA), *313*
U.S. Mine Safety and Health Administration (MSHA), *314*
Uncertainty factor in toxicity assessment, 40

Unlabeled chemicals, disposal, 73
Upper explosive limit (UEL), *315*
USP grade, chemical, 55

Vacuum distillation, 13
Vapor density, 5, *315*
Vapors, 45
Vermiculite, 45
Volatile substances, 45

Waste chemicals, definitions, 66
Waste minimization, 76
Weight equivalence, 36
Welding, 49
World Health Organization (WHO), *315*

Zoonose, *315*

CHEMICAL INDEX

*A boldfaced term indicates the prime name used in the Chemical Entries section.
Italicized page numbers indicate that the reference is a chemical entry under the listed name.*

A 11032, 275
Absolute ethanol, 162
Acetaldehyde, 5, *81*
   waste, 70
7-Acetamido-6,7-dihydro-
   1,2,3,10-tetramethoxy-
   benzo(a)heptalen-
   9(5H)-one, 144
Acetdimethylamide, 154
Acetic acid, 5, 16, 27, *82*
   anhydride, 83
Acetic acid-n-butyl ester, 116
Acetic acid dimethylamide, 154
Acetic acid
   glacial, 82
   laboratory storage, 64
   sodium salt, 254
   zinc salt, 278
Acetic aldehyde, 81
Acetic anhydride, 27, *83*
   laboratory storage, 64
Acetic ether, 161
Acetic oxide, 83
Acetidin, 161
Acetone, 5, 16, *84*
Acetonitrile, *86*
Acetoxyethane, 161
Acetoxyl, 107
2-Acetylaminofluorene, 46
Acetyl anhydride, 83
Acetyl chloride,
   laboratory storage, 64
Acetylen, 87
Acetylene, 7, 8, 12, 16, *87*
   storage, 60
Acetylene trichloride, 270
Acetylenic compounds, 12
Acetyl ether, 83
Acetyl hydroperoxide, 230
Acetylides, 7, 8, 25
Acetyl nitrate, 11
Acetyl peroxide, 8, *87*
   waste, 70
Acetyl oxide, 83
N-Acetyltransferase, 39
N-Acetyl trimethylcolchicinic
   acid methylether, 144
Acid anhydrides, 8
Acid chlorides, 8
Acide nitrique (French), 214
Acide oxalique (French), 224

Acido nitrico (Italian), 214
Acido ossalico (Italian), 224
Acids, 7, 24, 27
Acnegel, 107
Acquinite, 89
Acraldehyde, 89
Acrolein, 7, 27, *89*
   waste, 67
Acryaldehyde, 89
Acrylamide, 29, *90*
   monomer, 90
Acrylic aldehyde, 89
Acrylic amide, 90
Acrylonitrile (inhibited), *91*
   monomer, 91
Actinolite, 103
Adronal, 147
Aero liquid HCN, 187
Aerothene MM, 210
Aether, 163
A-fil cream, 269
Aflatoxin B_1, 46
Agricultural limestone, 120
Agstone, 120
Air, liquid, 7
Alabaster, 123
Albone, 192
Alcide, 134
Alcoa sodium fluoride, 257
Alcohol, 26, 28, 162
   anhydrous, 162
   dehydrated, 162
Alcohol isobutylique (French),
   199
Alcohols, 27
Alcool propilico (Italian), 249
Alcool propylique (French), 249
Aldehydes, 9, 25, 26
Aldifen, 158
Algrain, 162
Algylen, 270
Alkali metals, 6
Alkalies, 24, 27
Alkaline earth metals, 16
Alkaline sulfides, 27
Alkaloids, 25
Alkenes, 26
Alkyl aluminum compounds,
   waste, 70
Alkyl zinc, waste, 70
Allomaleic acid, 176

Allyl alcohol, waste, 67
Allyl aldehyde, 89
Allyl compounds, 9
Almond artificial essential oil,
   105
Alum, 94
Alumina fiber, 92
Aluminum, *92*
Aluminum alkyls
   laboratory storage, 64
   waste, 70
Aluminum dehydrated, 92
Aluminum dust, laboratory stor-
   age, 64
Aluminum flake, 92
Aluminum powder, 12, 16, 92
Aluminum phosphide, waste, 67
Aluminum sulfate (2:3), *94*
Aluminum trisulfate, 94
Amianthus, 103
Amidosulfonic acid, 263
Amines, 23, 26
Aminic acid, 176
Aminobenzene, 100
2-Aminoethanol, 160
β-Aminoethyl alcohol, 160
Aminophen, 100
Aminosulfonic acid, 263
Ammonia, 7, 12, 26–28, 40,
   94
Ammonia
   anhydrous, 16, 94
   aqueous, 97
   gas, 94
   laboratory storage, 64
   solution, 97
Ammoniacal silver nitrate, 11
Ammonium carbonate, 27
Ammonium chloride, *97*
Ammonium hydroxide, 43, *97*
Ammonium muriate, 97
Ammonium nitrate, 7, 16, *98*
   waste, 70
Ammonium perbromate, 10
Ammonium perchlorate, 10
Ammonium periodate, 10
Ammonium peroxydisulfate, 99
Ammonium persulfate, *99*
Amosite, 103
Amphibole, 103
Amprolene, 167

Amyl acetate, 5
AN, 91
Anamenth, 270
Anatase, 269
Anayodin, 259
Anesthesia ether, 163
Anhydrol, 162
Anhydrone, 11
Anhydrous ammonia, 94
Anhydrous dextrose, 177
Anhydrous hydrazine, 182
Anhydrous hydrobromic acid,
   186
Anhydrous hydrofluoric acid,
   190
Anhydrous hydrogen chloride,
   186
Anhydrous sodium acetate, 254
Aniline, 16, 23, *100*
Aniline oil, 100
Annaline, 123
Anol, 147
Anproline, 167
Anthium dioxide, 134
Anthophylite, 103
Antibulit, 257
Antimony, *101*
Antimony and its salts, 27
Antimony black, 101
Antimony compounds, 25
Antimony regulus, 101
Antirust, 260
Antisal 1a, 269
Aqua ammonia, 97
Aqua fortis, 214
Aquacat, 143
Aquachloral, 130
Aqualine, 89
Aragonite, 120
Arctuvin, 196
Argentum, 252
Arsenic, *102*
   wastes, 72
Arsenic and its salts, 27
Arsenic black, 102
Arsenic compounds, 16, 25
Arsenic oxide, 103
Arsenic sesquioxide, 103
Arsenic trichloride, 27
Arsenic trioxide, *103*
Arsenic(III) oxide, 103

CHEMICAL INDEX

Arsenicals, 102
Arsenious acid, 103
Arsenious oxide, 103
Arsenious trioxide, 103
Arsenous acid, 103
Arsenous acid anhydride, 103
Arsenous anhydride, 103
Arsenous oxide, 103
Arsenous oxide anhydride, 103
Arsines, 26
Artificial almond oil, 105
Asbestos, 25, 43, *103*
Asbestos fiber, 103
Asymmetrical dichloroethane, 151
Atlas white titanium dioxide, 269
Atomit, 120
Atropine, 39
Atropine esterase, 39
Austiox, 269
Azabenzene, 250
Azide, 255
Azides, 7, 8, 16, *104*
Azine, 250
Azium, 255
Azotic acid, 214
Azotowy kwas (Polish), 214
Aztec BPO, 107

B-K powder, 121
Baker's P and S liquid and ointment, 232
Barium, 39, *104*
Barium carbonate, 27
Barium chloride, *104*
Barium cyanide, waste, 67
Barium dichloride, 104
Barium hydroxide, 27
Barium peroxide, 8
Bases, 7, 28
Basic zirconium chloride, 280
Bayeritian, 269
BCME, 109
BCS copper fungicide, 145
Bell mine, 121
Bell mine pulverized limestone, 120
Benoxyl, 107
Benzac, 107
Benzaknew, 107
Benzaldehyde, *105*
Benzenamine, 100
Benzene, 5, 6, 16, 21, 22, 26, 39, *105*
Benzene chloride, 137
Benzenecarbaldehyde, 105
Benzenecarbonal, 105
1,2-Benzenediol, 130
1,4-Benzenediol, 196
o-Benzenediol, 130
p-Benzenediol, 196
Benzenetetrahydride, 149
Benzenol, 232
Benzinoform, 128
Benzinol, 270
Benzo(def)chrysene, 106
Benzohydroquinone, 196
Benzoic acid, peroxide, 107
Benzoic aldehyde, 105
Benzol, 105
Benzole, 105

Benzolene, 105
Benzoperoxide, 107
Benzo[a]pyrene, 46, 47, *106*
3,4-Benzopyrene, 106
6,7-Benzopyrene, 106
Benzoquinol, 196
Benzoyl, 107
Benzoyl peroxide, 8, 12, *107*
waste, 70
Benzoyl superoxide, 107
Benz(a)pyrene, 106
3,4-Benzpyrene, 106
3,4-Benz(a)pyrene, 106
3,4-Benzpyrene, 106
handling, 47
Berelex, 177
Bertholite, 131
Beryllium, 43, *108*
handling, 48
Beryllium compounds, 25
BFV, 174
BHT, 9
Bibenzene, 109
1,2-Bichloroethane, 166
Bichromate of potash, 243
Biocide, 89
Biphenyl, *109*
1,1'-Biphenyl, *109*
Bis(chloromethyl)ether, 40, 46, *109*
Bis-CME, 109
3,3-Bis(p-hydroxyphenyl)-phthalide, 233
Bisulfite, 264
Black and white bleaching cream, 196
Black pearls, 124
Blacosolv, 270
Bleaching powder, 49, 121
Blue copper, 145
Blue oil, 100
Blue stone, 145
Blue vitriol, 145
Bluestone, 145
Boletic acid, 176
Bonazen, 280
Boracic acid, 110
Borer sol, 166
Boric acid, *110*
2-Bornanone, 123
Borofax, 110
Boron chloride, 110
Boron fluoride, 111
Boron trichloride, *110*
Boron trifluoride, *111*
Brellin, 177
Brimstone, 263
Brocide, 166
Bromates, 7
laboratory storage, 63
Bromine, 16, 27, *112*
laboratory storage, 64
Bromine solution, 112
Bromofume, 165
Brookite, 269
Brown acetate, 120
Bufopto zinc sulfate, 280
Burnt lime, 122
Butane, 16
Butanoic acid, 117
Butan-1-ol, 113
Butan-2-ol, 114

Butanol, 113
1-Butanol, *113*
2-Butanol, *114*
n-Butanol, 113
sec-Butanol, 114
tert-Butanol, *115*
2-Butanone, 209
(E)-Butenedioic acid, 176
trans-Butenedioic acid, 176
Buten-3-yne, 277
Butter of zinc, 279
Butyl acetate, 116
1-Butyl acetate, 116
n-Butyl acetate, *116*
Butyl alcohol, 5, 113
2-Butyl alcohol, 114
n-Butyl alcohol, 113
tert-Butyl alcohol, 115
Butyl chloride, 116
n-Butyl chloride, *116*
Butyl ethanoate, 116
Butyl hydroxide, 113
Butyl methyl ketone, 181
tert-Butyl hydroperoxide, *117*
tert-Butyl hydroxide, 115
Butylated hydroxytoluene, 9
Butylene hydrate, 114
Butylene oxide, 267
n-Butyric acid, *117*
Butyric or normal primary butyl alcohol, 113
BZF-60, 107

Caddy, 118
Cadet, 107
Cadmium, 39, 43, *118*
Cadmium chloride, *118*
Cadmium compounds, 25
Cadmium dichloride, 118
Cadmium dinitrate, 119
Cadmium nitrate, *119*
Cadmium nitrate tetrahydrate, *119*
Cadmium oxide, *119*
Cadox, 107
Cake alum, 94
Calcia, 122
Calcicat, 120
Calcite, 120
Calcium, *120*
crystalline, 120
metal, 120
nonpyrophoric, 120
powdered, 16
pyrophoric, 120
Calcium acetate, *120*
Calcium carbide, waste, 70
Calcium carbonate, *120*
Calcium chloride, *121*
Calcium chlorohydrochlorite, 121
Calcium cyanamide, 27
Calcium cyanide, waste, 67
Calcium diacetate, 120
Calcium hydrate, 121
Calcium hydroxide, *121*
Calcium hypochloride, 121
Calcium hypochlorite, *121*
Calcium oxide, 16, 28, *122*
Calcium oxychloride, 121
Calcium phosphide, waste, 70
Calcium sulfate dihydrate, *123*

Calcotone white T, 269
Calplus, 121
Caltac, 121
Calx, 122
2-Camphanone, 123
Camphor, *123*
natural, 123
synthetic, 123
Camphor tar, 212
Caporit, 121
Carbacryl, 91
Carbazoic acid, 240
Carbinol, 207
Carbolic acid, 232
Carbon, *124*
activated, 16
Carbon bisulfide, 125
Carbon black, 124
Carbon chloride, 128
Carbon dioxide, *124*
Carbon disulfide, 4, 5, 27, *125*
waste, 67
Carbona, 128
Carbonic acid, calcium salt (1:1), 120
Carbonic acid gas, 124
Carbonic anhydride, 124
Carbonic oxide, 127
Carbon monoxide, 7, 26, 28, *127*
cryogenic liquid, 127
Carbon oil, 105
Carbon oxide (CO), 127
Carbon oxychloride, 234
Carbon sulfide, 125
Carbon tet, 128
Carbon tetrachloride, 5, 12, 16, 23, 26, 29, *128*
Carbonyl chloride, 234
Carbonyls, 25
handling, 48
Caro's acid, 8
Cardox TBH, 117
Cartose, 177
Catalysts, 8, 13
Catechin, 130
Catechol, *130*
Caustic potash, 245
Caustic soda, 257
bead (DOT), 257
dry (DOT), 257
flake (DOT), 257
granular (DOT), 257
liquid (DOT), 257
solid (DOT), 257
solution (DOT), 257
Cavi-trol, 257
CCH, 121
Cecolene, 270
Cekugib, 177
Cellosolve, 161
Cellosolve solvent, 161
Cellosolves, 9
laboratory storage, 64
Cellulose nitrate, 43
Celmide, 165
Cerelose, 177
Chalk, 120
Chameleon mineral, 247
Charcoal black, 124
Chemifluor, 257
Chemox PE, 158

334 CHEMICAL INDEX

Chile saltpeter, 260
Chloral hydrate, *130*
Chloraldurat, 130
Chlorate de potassium (French), 243
Chlorate of potash, 243
Chlorates, 6, 7, 16, 17, 25
  laboratory storage, 63
Chlorbenzene, 137
Chlorbenzol, 137
Chlorethene, 276
Chlorethylene, 276
Chloride of lime, 121
Chlorinated hydrocarbon solvents, 27
Chlorinated lime, 121
Chlorine, 7, 12, 16, 26, 27, *131*
Chlorine dioxide, 16, *134*
Chlorine fluoride, 135
Chlorine mol., 131
Chlorine oxide, 134
Chlorine(IV) oxide, 134
Chlorine peroxide, 134
Chlorine trifluoride, *135*
Chlorites, 7
Chloroacetaldehyde, waste, 67
Chloroacetic acids, 27
Chloroben, 150
Chlorobenzene, *137*
Chlorobenzol, 137
1-Chlorobutane, 116
Chloro(chloromethoxy)methane, 109
Chloro compounds, 26
Chloroden, 150
1-Chloro-2,2-dichloroethylene, 270
Chloroethene, 276
Chloroethylene, 276
Chloroform, 12, 22, 39, *137*
Chloroformyl chloride, 234
Chloromethane, *139*
Chloromethyl ether, 109
Chloroperoxyl, 134
Chloropicrin, 27
Chlorosulfonic acid, laboratory storage, 64
Chlorotrifluoride, 135
Chlorozirconyl, 280
Chloryl radical, 134
Chlorylea, 270
Chorylen, 270
Chromar, 278
Chrome, 140
Chromic acid, 10, 16, *140*, 141
  handling, 48
  solid, 141
  solution, 141
Chromic anhydride, 141
Chromic chloride, 141
Chromic(VI) acid, 140
Chromium, 8, 43, *140*
Chromium and the alkaline chromates, 27
Chromium chloride, *141*
  anhydrous, 141
Chromium compounds, 25
  handling, 48
Chromium oxide, 141
Chromium(VI) oxide, 141
Chromium(VI) oxide (1:3), 141

Chromium trichloride, 141
Chromium trioxide, 16, *141*
  anhydrous, 141
  laboratory storage, 63
Chromium(6+) trioxide, 141
Chromium trioxide-pyridine complex, 12
Chrysotile, 103
Circosolv, 270
Clearasil benzoyl peroxide lotion, 107
Cloroben, 150
Coal naphtha, 105
Coal tar solvents, 27
Cobalt, *143*
Cobalt-59, 143
Cobalt dichloride, 143
Cobalt dinitrate, 144
Cobalt muriate, 143
Cobalt sulfate, 144
Cobalt(II) chloride, *143*
Cobalt(II) nitrate, *144*
Cobalt(2+) sulfate, 144
Cobalt(III) sulfate (1:1), *144*
Cobaltous chloride, 143
Cobaltous dichloride, 143
Cobaltous nitrate, 144
Cobaltous sulfate, 144
Colchicine, *144*
7-α-H-colchicine, 144
Colchineos, 144
Colchisol, 144
Colcin, 144
Colloidal arsenic, 102
Colloidal mercury, 204
Colloidal selenium, 251
Colloidal sulfur, 263
Cologne spirits (alcohol), 162
Colonial spirit, 207
Colsaloid, 144
Columbian carbon, 124
Columbian spirits, 207
Common-sense cockroach and rat preparations, 237
Condy's crystals, 247
Condylon, 144
Copper, 8, 16
Copper compounds, 25
Copper cyanide, 27
Copper dinitrate, 145
Copper monosulfate, 145
Copper sulfate, 27
Copper sulfate pentahydrate, *145*
Copper(2+) nitrate, 145
Copper(II) nitrate, *145*
Copper(II) sulfate, *145*
Copper(II) sulfate (1:1), 145
Copperas, 169
Copperfine-zinc, 145
Corn sugar, 177
Cosmetic white C47-5175, 269
CP Basic sulfate, 145
Crawhaspol, 270
Credo, 257
3-Cresol, 146
***m*-Cresol**, *146*
Cresylic acid, 27
m-Cresylic acid, 146
Crocidolite, 103
Crolean, 89
Crude arsenic, 103

CSP, 145
Cubic niter, 260
Cumene, 9
Cumene hydroperoxide, 16
Cupric dinitrate, 145
Cupric nitrate, 145
Cupric sulfate, 145
Cupric sulfate pentahydrate, 145
Curare, 43
Cuticura acne cream, 107
Cyanide of potassium (solution), 244
Cyanide of sodium, 256
Cyanides, 16, 28, 43
  handling, 48
Cyanobrik, 256
Cyanoethylene, 91
Cyanogen, 25
Cyanogran, 256
Cyanomethane, 86
Cyanure de potassium (French), 244
Cyclohexane,, 5, *146*
Cyclohexanol, *147*
Cyclohexanone, *148*
Cyclohexatriene, 105
Cyclohexene, 9, *149*
  laboratory storage, 64
Cyclohexyl alcohol, 147
Cyclon, 187
Cyclone B, 187
Cyclooctene, 9
  laboratory storage, 64
Cyclopentane, *149*
Cyclotetramethylene oxide, 267
Cymag, 256

DANA, 152
DBE, 165
DCB, 150
1,2-DCE, 166
DCM, 210
DCP, 152
2,4-DCP, 152
Debroxide, 107
Decahydronaphthalene, 9
Decalin, 9
  laboratory storage, 64
Deltan, 157
Demasorb, 157
Demavet, 157
Demeso, 157
Demsodrox, 157
DEN, 152
DENA, 152
Densinfluat, 270
Dermasorb, 157
Destruxol borer-sol, 166
Dextropur, 177
Dextrose, 177
  anhydrous, 177
Dextrosol, 177
DFP, 153
Diacetyl peroxide, 8, 87
Diacyl peroxides, 9
Dialuminum sulphate, 94
Dialuminum trisulfate, 94
Diamide, 182
Diamine, 182
Diamond, 124
Diarsenic trioxide, 103
Diazo compounds, 7, 8

Diazomethane, 7, 12
Diazotizing salts, 260
Dibenzoyl peroxide, 107
Diborane waste, 70
Dibromoethane, 165
1,2-Dibromoethane, 165
sym-Dibromoethane, 165
Dibutyl ether, 9
  laboratory storage, 64
Dicarbomethoxyzinc, 278
Dichloremulsion, 166
1,2-Dichlorobenzene, 150
o-Dichlorobenzene, 150
o-Dichlorobenzol, 150
1,1-Dichloro-2-chloroethylene, 270
Dichlorodifluoromethane, *150*
sym-Dichlorodimethyl ether, 109
Dichloroethane, *150*
1,2-Dichloroethane, 166
α,β-Dichloroethane, 166
sym-Dichloroethane, 166
Dichlorethyl sulfide, 27
Dichloroethylene, 166
Dichloromethane, 210
Dichloromethyl ether, 27
sym-Dichloromethyl ether, 109
Dichloroöxozirconium, 280
2,4-Dichlorophenol, *152*
Dieldrin, waste, 67
(Diethylamino)ethane, 272
Diethylene dioxide, 159
Diethylene ether, 159
1,4-Diethylene ether, 159
Diethylene imidoxide, 211
Diethylene oxide, 267
Diethyleneimide oxide, 211
N,N-Diethylethanamine, 272
Diethyl ether, 4, 5, 9, *163*
  disposal, 72
  laboratory storage, 64
  waste, 70
Diethylnitrosamine, *152*
N,N-Diethylnitrosamine, 152
Diethyl phthalate, 4
Difluorodichloromethane, 150
Diflupyl, 153
Dihydrogen dioxide, 192
Dihydrooxirene, 167
Dihydroxybenzene, 196
1,2-Dihydroxybenzene, 130
1,4-Dihydroxybenzene, 196
1,2-Dihydroxyethane, 167
o-Dihydroxybenzene, 130
p-Dihydroxybenzene, 196
Diisopropoxyphosphoryl fluoride, 153
Diisopropyl ether, 9, *153*
  laboratory storage, 64
Diisopropyl fluorophosphate, *153*
O,O-Diisopropyl fluorophosphate, 153
Diisopropyl fluorophosphonate, 153
Diisopropylfluorophosphoric acid ester, 153
Diisopropyl oxide, 153
Diisopropyl phosphofluoridate, 153
Diisopropyl phosphorofluoridate, 153

CHEMICAL INDEX

Dimethoxyethane, 178
1,2-Dimethoxyethane, 178
α,β-Dimethoxyethane, 178
Dimethyl acetamide, *154*
Dimethylacetone amide, 154
Dimethylamide acetate, 154
7,12-Dimethylbenz[a]anthracene, 46
Dimethylbenzene, 277
1,3-Dimethylbenzene, 278
1,4-Dimethylbenzene, 278
m-Dimethylbenzene, 278
p-Dimethylbenzene, 278
Dimethylcarbamoyl chloride, 46
Dimethylcarbinol, 200
Dimethylcellosolve, 178
Dimethyl-1,1'-dichloroether, 109
Dimethylene oxide, 167
1,1-Dimethylethanol, 115
1,1-Dimethylethyl hydroperoxide, 117
Dimethylformaldehyde, 85
Dimethylformamide, *155*
N,N-Dimethylformamide, 155
Dimethyl ketone, 85
Dimethyl mercury, 46
Dimethyl monosulfate, 156
Dimethyl sulfate, 27, *156*
  laboratory storage, 64
Dimethyl sulfoxide, 12, *157*
Dimethyl sulphoxide, 157
Dimethylketal, 85
Dimethylmethane, 248
Dimexide, 157
Dinitrogen tetroxide, 222
2,4-Dinitrophenol, *158*
  waste, 67
α-Dinitrophenol, 158
Diokan, 159
1,4-Dioxacyclohexane, 159
Dioxane, 5, 8, *159*
  storage, 56
  waste, 70
1,4-Dioxane, 159
p-Dioxane, 9, 159
  laboratory storage, 64
2,5-Dioxahexane, 178
p-Dioxobenzene, 196
o-Dioxybenzene, 130
Dioxyethylene ether, 159
o-Diphenol, 130
Diphenyl, 109
Diphenylglyoxal peroxide, 107
Diphosgene, 234
Diphosphorus pentoxide, 239
Dipirartril-tropico, 157
Dipotassium dichromate, 243
Dipping acid, 265
Dipropyl methane, 179
Disodium difluoride, 257
Disodium dioxide, 261
Disodium peroxide, 261
Disodium sulfate, 262
Dithiocarbonic anhydride, 125
DMA, 154
DMAC, 154
DMF, 155
DMFA, 155
DMS (methyl sulfate), 156
DMS-70, 157
DMS-90, 157

DMSO, 157
2,4-DNP, 158
Dolicur, 157
Doligur, 157
Dolomite, 120
Domoso, 157
Dormal, 130
Dow-tri, 270
Dowanol EE, 161
Dowflake, 121
Dowfume 40, 165
Dowfume EDB, 165
Dromisol, 157
Dry and Clear, 107
Dry box, 15
Dry ice, 12
Dukeron, 270
Durafur developer C, 130
Durasorb, 157
Duretter, 169
Duroferon, 169
Dutch liquid, 166
Dutch oil, 166
Dyflos, 153

EDB, 165
EDC, 166
EDGME, 178
EGM, 208
EGME, 208
Ektasolve, 208
Ektasolve EE, 161
Eldopaque, 196
Eldoquin, 196
Elemental selenium, 251
Emanay atomized aluminum powder, 92
Epi-clar, 107
1,4-Epoxybutane, 267
Epoxyethane, 167
1,2-Epoxyethane, 167
Epoxy resins, handling, 49
Erinitrit, 260
Essence of mirbane, 220
Essence of myrbane, 220
Estane 5703, 275
Esters, 27
Ethanal, 81
Ethane, 28
Ethane dichloride, 166
Ethanedioic acid, *224*
1,2-Ethanediol, 167
Ethanenitrile, 86
Ethaneperoxoic acid, 230
Ethanoic acid, 82
Ethanoic anhydrate, 83
Ethanol, 28, 162
Ethanol-200 proof, 162
Ethanol solution, 162
Ethanolamine, *160*
β-Ethanolamine, 160
Ethene, 270
Ethene oxide, 167
Ether, 28, 163
Ethers, 9
Ethine, 87
Ethinyl trichloride, 270
Ethoxyethane, 163
2-Ethoxyethanol, *161*
Ethyl acetate, 5, *161*
Ethyl acetic ester, 161
Ethyl acetone, 229

Ethyl alcohol, 5, *162*
  anhydrous, 162
Ethyl aldehyde, 81
Ethyl carbamate, 275
Ethyl carbinol, 249
Ethyl cellosolve, 161
Ethyl chlorosulfonate, 27
Ethyl compounds, 26
Ethyl ethanoate, 161
Ethyl ether, 4, 5, 9, *163*
  storage, 56
Ethyl hydrate, 162
Ethyl hydroxide, 162
Ethyl methyl ketone, 209
Ethyl nitrile, 86
Ethyl urethane, 275
Ethylacetic acid, 117
1,1-Ethylidene dichloride, 151
Ethylene, 22
Ethylene alcohol, 167
Ethylene aldehyde, 89
Ethylene bromide, 165
Ethylene chloride, 166
Ethylene dibromide, *165*
Ethylene dichloride, 5, *166*
1,2-Ethylene dichloride, 166
Ethylene dihydrate, 167
Ethylene dimethyl ether, 178
Ethylene glycol, *167*
Ethylene glycol ethyl ether, 161
Ethylene monochloride, 276
Ethylene oxide, 12, *167*
Ethylene trichloride, 270
Ethylenecarboxamide, 90
(E)1,2-Ethylenedicarboxylic acid, 176
trans-1,2-Ethylenedicarboxylic acid, 176
Ethyleneglycol diethyl ether, 9
Ethyleneglycol dimethyl ether, 9, 178
Ethyleneglycol monoethyl ether, 9, 161
Ethyleneglycol monomethyl ether, 9, 208
Ethyleneimine, waste, 67
Ethylic acid, 82
Ethylidene chloride, 151
Ethylmethyl carbinol, 114
N-Ethyl-N-nitroso-ethan-amine, 152
Ethylolamine, 160
Ethylurethan, 275
o-Ethylurethane, 275
Ethyne, 87
ETO, 167
Exhaust gas, 127
Exsiccated ferrous sulfate, 169

F12, 150
F1-tabs, 257
Fannoform, 174
Fasciolin, 128
FDA 0101, 257
Fekabit, 243
Felsules, 130
Fenoxyl carbon N, 158
Feosol, 169
Feospan, 169
Fer-in-sol, 169
Fermenicide liquid, 264

Fermenicide powder, 264
Fermentation alcohol, 162
Fermentation butyl alcohol, 199
Fero-gradumet, 169
Ferralyn, 169
Ferric chloride, 169
  solid, anhydrous, 169
Ferro-theron, 169
Ferrous sulfate, *169*
Fersolate, 169
Fibrous grunerite, 103
Filmerine, 260
Flamenco, 269
Flocool 180, 261
Floraltone, 177
Flores martis, 169
Floridine, 257
Florocid, 257
Floropryl, 153
Flowers of sulfur, 263
Flozenges, 257
Flue gas, 127
Flukoids, 128
Fluohydric acid gas, 187
Fluophosphoric acid, diisopropyl ester, 153
Fluoral, 257
Fluorident, 257
Fluorides, *169*
Fluorigard, 257
Fluorine, 16, *170*
  compressed, 170
  waste, 67
Fluorineed, 257
Fluorinse, 257
Fluoritab, 257
Fluorodiisopropyl phosphate, 153
Fluor-o-kote, 257
Fluoropryl, 153
Fluorotrichloromethane, *173*
Fluorure de potassium (French), 245
Fluostigmine, 153
Flura-gel, 257
Flurcare, 257
Formaldehyde, 27, 40, *174*
  solution, 174
Formalin, 174
Formalith, 174
Formamide, 28
Formic acid, 27, *176*
  solution, 176
Formic aldehyde, 174
Formol, 174
Formonitrile, 187
Formosa camphor, 123
Formyl trichloride, 137
N-Formyldimethylamine, 155
Formylic acid, 176
Fosforo bianco (Italian), 237
Fostex, 107
Fouramine PCH, 130
Fourrine 68, 130
Freon 11, 173
Freon 12, 150
Freon 30, 210
Fumaric acid, *176*
Fumigants, 25
Fumigrain, 91
Fumo-gas, 165
Fungol B, 257

336 CHEMICAL INDEX

Furanidine, 267
Fyde, 174

GA, 177
Gallic acid, *177*
  wastes, 72
Gallotannic acid, 267
Gallotannin, 267
Gamasol 90, 157
Garox, 107
Gel II, 257
Gelber phosphor (German), 237
Gelution, 257
Gettysolve-B, 180
Gettysolve-C, 179
Gib-sol, 177
Gib-tabs, 177
Gibberellic acid, *177*
Gibberellin, 177
Gibbrel, 177
Glacial acetic acid, 82
Gleem, 257
Glover, 201
Glucolin, 177
Glucose, 177
d-**Glucose,** *177*
  anhydrous, 177
Glucose liquid, 177
Glucosides, 25
Glutathione, 35
Glycerite, 267
Glycerol, *178*
Glycinol, 160
Glycol, 167
Glycol alcohol, 167
Glycol bromide, 165
Glycol dichloride, 166
Glycol dimethyl ether, 178
Glycol ether EE, 161
Glycol ether EM, 208
Glycol ethyl ether, 161
Glycol methyl ether, 208
Glycol monoethyl ether, 161
Glycol monomethyl ether, 208
Glyme, *178*
Grain alcohol, 162
Grape sugar, 177
Gray acetate, 120
Green vitriol, 169
Grey arsenic, 102
Grocel, 177
GSH, 35
Gum camphor, 123
Gum spirits, 273
Gum turpentine, 273
Gypsum, 123
Gypsum stone, 123

Hafnium and compounds, *178*
Halamines, 7
Halides, 25
Halogenated compounds, 12
Halogenated hydrocarbons, 25
Halogens, 25
Halon 104, 128
Halon 122, 150
HCN, 187
Heavy metals, 25
Hemoglobin, 34
Hepatic gas, 194
Heptachlor, waste, 67
Heptane, *179*

n-Heptane, 5, 179
Heptyl hydride, 179
Hexahydrobenzene, 147
Hexahydrophenol, 147
Hexalin, 147
Hexamethylene, 147
Hexamethylphosphoramide, 46
Hexanaphthene, 147
Hexane, *180*
n-Hexane, 5, 180
Hexanon, 148
Hexanone-2, 181
2-Hexanone, *181*
Hexone, 181
Hexyl hydride, 180
Hioxyl, 192
Hoch, 174
Hombitan, 269
Horse headache A-410, 269
HTH, 121
Hyadur, 157
Hydral, 130
Hydralin, 147
Hydrated lime, 121
Hydrazine and its salts, 40,
  182
  aqueous solution, 182
Hydrazine base, 182
Hydrobromic acid, 186
  anhydrous, 186
Hydrocarbons, 16, 28
Hydrochloric acid, 27, 186
  laboratory storage, 64
Hydrocyanic acid, 16, 187
  liquefied, 187
  potassium salt (solution), 244
  prussic, unstabilized, 187
  sodium salt, 256
  waste, 67
Hydrofluoric acid, 27, 28, 190
  anhydrous, 16
  laboratory storage, 64
Hydrofluoride, 190
Hydrofuran, 267
Hydrogen, 7, *184*
  compressed, 184
  refrigerated liquid, 184
Hydrogen bromide, *186*
Hydrogen carboxylic acid, 176
Hydrogen chloride, 26, 40, *186*
Hydrogen cyanide, 21, *187*
  anhydrous, stabilized, 187
Hydrogen dioxide, 192
Hydrogen fluoride, *190*
Hydrogen nitrate, 214
Hydrogen peroxide, 7, 8, 16,
  192
  laboratory storage, 63
  stabilized (over 60% peroxide), 192
Hydrogen peroxide solution
  (over 52% peroxide), *192*
Hydrogen sulfide, 7, 16, 21,
  26, *194*
Hydrogen sulfite sodium, 255
Hydrogen sulfuric acid, 194
Hydroperoxide, 192
  acetyl, 230
Hydroperoxides, 9
2 Hydroperoxy 2 methylpropane,
  117
Hydrophenol, 147

Hydroquinol, 196
Hydroquinone, *196*
α-Hydroquinone, 196
o-Hydroquinone, 130
p-Hydroquinone, 196
Hydrosulfuric acid, 194
Hydroxyde de potassium
  (French), 245
Hydroxybenzene, 232
1-Hydroxybutane, 113
2-Hydroxybutane, 114
Hydroxycyclohexane, 147
1-Hydroxy-2,4-dinitrobenzene,
  158
Hydroxy ether, 161
2-Hydroxyethylamine, 160
β-Hydroxyethylamine, 160
1-Hydroxy-3-methylbenzene, 146
1-Hydroxymethylpropane, 199
2-Hydroxyphenol, 130
o-Hydroxyphenol, 130
p-Hydroxyphenol, 196
β-hydroxypropionic acid, 47
m-Hydroxytoluene, 146
2-Hydroxy-1,3,5-trinitrobenzene,
  240
Hypochlorites, 16, *197*
Hypochlorous acid, calcium salt,
  121

Incidol, 107
Infiltrina, 157
Inhibine, *192*
Iodates, 7
  laboratory storage, 63
Iodine, 16, 27, *197*
  crystals, 197
  sublimed, 197
Ioduril, 259
Iopezite, 243
IPA, 200
Iradicav, 257
Iron, 8
  waste, 70
Iron chloride, 169
Iron monosulfate, 169
Iron protosulfate, 169
Iron sesquichloride, solid, 169
Iron trichloride, 169
Iron vitriol, 169
Irospan, 169
Irosul, 169
Isoamyl methyl ketone, 211
Isobutanol, 199
Isobutyl alcohol, 199
Isobutyl methyl ketone, *181*
Isobutylalkohol (Czech), 199
Isobutyltrimethylethane, 273
Isocyanic acid and its esters, 25
Isofluorophate, 153
Isohol, 200
Isooctane, 273
Isopentyl methyl ketone, 211
Isopropanol, 200
2-Isopropoxypropane, 153
Isopropyl alcohol, 5, *200*
Isopropyl benzene, 9
Isopropyl ether, 8, 153
  storage, 57
Isopropyl fluophosphate, 153
Isopropylacetone, *181*

Isopropylcarbinol, 199
Ivalon, 174

Japan camphor, 123
Jaysol, 162
Jeffersol EE, 161
Jeffersol EM, 208
Jodid sodny, 259

Kaliumchloraat (Dutch), 243
Kaliumchlorat (German), 243
Kaliumdichromat (German), 243
Kaliumhydroxid (German), 245
Kaliumhydroxyde (Dutch), 245
Kaliumpermanganaat (Dutch),
  247
Kaliumpermanganat (German),
  247
Kari-rinse, 257
Karidium, 257
Karigel, 257
Karsan, 174
Kazoe, 255
Kemikal, 121
Kessodrate, 130
Ketohexamethylene, 148
Ketone propane, 85
Ketones, 26, 27
β-Ketopropane, 85
2-Keto-1,7,7-trimethyl-
  norcamphane, 123
KH 360, 269
Kopfume, 165
m-Kresol, 146
Kronos titanium dioxide, 269
KS-4, 201
Kwik (Dutch), 204
Kyselina stavelova (Czech),
  224

Land plaster, 123
Laurel camphor, 123
Lea-cov, 257
Lead, 21, 35, 39, 43, *201*
  waste, 70
Lead compounds, 25, 28
Lead flake, 201
Lead S2, 201
Lemoflur, 257
Lemonene, 109
Leucethane, 275
Leucothane, 275
Levanox white RKB, 269
Lewis-Red Devil lye, 257
Lichenic acid, 176
Light spar, 123
Lime, 27, 122
  burned, 122
  unslaked, 122
Lime acetate, 120
Lime chloride, 121
Lime pyrolignite, 120
Lime water, 121
Limestone, 120
Liquidow, 121
Lithium and its compounds,
  202
  powdered, 16
Lithium aluminum hydride, 13
  waste, 70
Lithographic stone, 120
Lo-bax, 121

Lorinal, 130
Loroxide, 107
Losantin, 121
Lox, 225
Lucidol, 107
Luperco, 107
Luride, 257
Lutrol-9, 167
Lye, 245
 (DOT), 257
Lysoform, 174

Macrogol 400 BPC, 167
Magnacide H, 89
Magnesia white, 123
Magnesio (Italian), 203
Magnesium, 203
Magnesium metal (DOT), 203
Magnesium nitride, waste, 70
Magnesium perchlorate, 11
Magnesium pellets, 203
Magnesium powder, 12, 16, 203
 laboratory storage, 64
Magnesium ribbons, 203
Magnesium turnings, 203
Marble, 120
Maroxol-50, 158
Matricaria camphor, 123
MBK, 181
MCB, 137
MEA, 160
MECS, 208
M.E.G., 167
MEK, 209
Melinite, 240
Mercure (French), 204
Mercuric salts, 27
Mercurio (Italian), 204
Mercury, 16, 21, 26, 43, *204*
 metallic (DOT), 204
 wastes, 72
Mercury acetyl, 8
Mercury compounds, 25, 28
Mercury fulminate, waste, 67
Merpol, 167
Metal cyanides, 25
Metal fumes, 25
Metal hydrides, laboratory storage, 64
Metallic alkyls, laboratory storage, 64
Metallic arsenic, 102
Metallic peroxides, laboratory storage, 63
Metals, handling, 49
Metana aluminum paste, 92
Metaphosphoric acid, 235
Methacide, 269
Methanal, 174
Methane, 5, 28
Methane dichloride, 210
Methane tetrachloride, 128
Methane trichloride, 137
Methanecarbonitrile, 86
Methanecarboxylic acid, 82
Methanoic acid, 176
Methanol, 43, *207*
Methenyl trichloride, 137
2-Methoxyethanol, 208
Methyl acetone, 209
Methyl alcohol, 5, 207
Methyl aldehyde, 174

Methyl butyl ketone, 181
Methyl *n*-butyl ketone, 181
Methyl cellosolve, *208*
Methyl chloride, 139
Methyl chlorosulfonate, 27
Methyl compounds, 26
Methyl cyanide, 86
Methyl ethoxol, 208
Methyl ethyl ketone, 5, *209*
Methyl ethyl ketone peroxide, *209*
Methyl glycol, 208
2-Methyl-5-hexanone, 211
5-Methyl-2-hexanone, *211*
Methyl hydrazine, waste, 67
Methyl hydroxide, 207
Methyl isoamyl ketone, 211
Methyl isobutyl ketone, *181*
Methyl ketone, 85
Methyl oxitol, 208
2-Methyl-4-pentanone, *181*
4-Methyl-2-pentanone, *181*
2-Methyl propanol, 199
2-Methyl-1-propanol, 199
2-Methyl-2,2-propanol, 115
Methyl-propyl-cetone (French), 229
Methyl propyl ketone (DOT), 229
Methyl-*n*-propyl ketone, 229
Methyl sulfate, 156
Methyl sulfoxide, 157
Methyl toluene, 277
Methyl trichloride, 137
Methylbenzene, 269
Methylbenzol, 269
Methylcarbinol, 162
3-Methylcholanthrene, 46
Methylene bichloride, 210
Methylene chloride, 5, *210*
Methylene dichloride, 210
Methylene glycol, 174
Methylene oxide, 174
Methylethylcarbinol, 114
Methylethylketonhydroperoxide, 209
Methylolpropane, 113
3-Methylphenol, 146
m-Methylphenol, 146
2-Methylpropan-1-ol, 199
2-Methylpropyl alcohol, 199
Methylopropyloketon (Polish), 229
Methylsulfinylmethane, 157
p-Methyltoluene, 278
MIAK, 211
MIBK, 181
Mighty 150, 212
MIK, 181
Mineral naphtha, 105
Mineral white, 123
Mirbane oil, 220
MNBK, 181
Molasses alcohol, 162
Molecular chlorine, 131
Monochlorbenzene, 137
Monochlorobenzene, 137
Monochloroethene, 276
Monochloroethylene (DOT), 276
Monochloromethane, 139
Monochromium oxide, 141
Monochromium trioxide, 141

Monoethanolamine, 160
Monoethylene glycol, 167
Monoethylene glycol dimethyl ether, 178
Monofluorotrichloromethane, 173
Monoglyme, 178
Monohydroxybenzene, 232
Monohydroxymethane, 207
Monomethyl ether of ethylene glycol, 208
Monopersulfuric acid, 8
Monophenol, 232
Morbocid, 174
Morpholine, *211*
 aqueous mixture, 211
Moth balls, 212
Moth flakes, 212
Motor benzol, 105
MPK, 229

Na frinse, 257
Nadone, 148
Nafeen, 257
Nafpak, 257
Naftalen (Polish), 212
Naphthalene, *212*
 crude or refined (DOT), 212
 molten (DOT), 212
Naphthalin (DOT), 212
Naphthaline, 212
Naphthene, 212
Narcylen, 87
Natrium, 253
Natrium fluoride, 257
Natural calcium carbonate, 120
NCI-C52904, 212
NCI-C55209, 224
NCI-C55447, 209
NCI-C60399, 204
NDEA, 152
Necatornia, 128
Neoglaucit, 153
Nephis, 165
Nickel carbonyl, 46, *213*
Nickel tetracarbonyl, *213*
Nitrates, 7, 16, 25, *214*
 inorganic, 13
Nitratine, 260
Nitration benzene, 105
Nitric acid, 6, 27, 28, 43, 16, *214*
 ammonium salt, 98
 cadmium salt, 119
  tetrahydrate, 119
 cobalt(2+) salt, 144
 laboratory storage, 64
 NFPA label, 54
 over 40% (DOT), 214
 sodium salt, 260
 storage, 57
 zinc salt, 279
Nitric acid-based solutions, 13
Nitric oxide, *219*
Nitrides, 25
Nitriles, 25, 26
Nitrite, 34
Nitrites, 16, *220*
Nitrito, 222
Nitro compounds, 26
Nitro kleenup, 158

Nitrobenzene, 23, *220*
 liquid, 220
Nitrobenzol, 220
 liquid, 220
Nitrocellulose, waste, 70
Nitrogen, 7, 28, *222*
 compressed (DOT), 222
 refrigerated liquid (DOT), 222
Nitrogen dioxide, 21, 26, 27, *222*
 waste, 67
Nitrogen gas, *222*
Nitrogen monoxide, 219
Nitrogen oxides, 44
Nitrogen peroxide, 222
Nitrogen tetraoxide, waste, 67
Nitrogen tetroxide, 222
Nitrogen trichloride, 13
2-Nitronaphthalene, 46
Nitroparaffins, 16
N-Nitrosamides, 46
N-Nitrosamines, 46
Nitroso compounds, 7, 8
Nitrosodiethylamine, 152
N-Nitrosodiethylamine, 152
Nitrous acid, sodium salt, 260
Nitroxanthic acid, 240
Noctec, 130
Noral ink grade aluminum, 92
Nordhausen acid (DOT), 265
Norkool, 167
Normal hexane, 180
Normal octane, 223
Normal pentane, 228
Nortec, 130
Novadelox, 107
NSC 746, 275
NTO, 222
Nufluor, 257
Nycoton, 130

Octane, *223*
ODB, 150
ODCB, 150
Oil of mirbane, 220
Oil of myrbane, 220
Oil of turpentine, 273
 rectified, 273
Oil of vitriol, 265
Olamine, 160
Oleum, laboratory storage, 64
Omaha, 201
Omaha & Grant, 201
Op-thal-zin, 280
Organic solvents, 27
ortho-**Dichlorobenzene,** *150*
Orthoboric acid, 110
Orthodichlorobenzene, 150
Orthodichlorobenzol, 150
Orthophosphoric acid, 235
Osmium tetraoxide, waste, 67
Osmium, wastes, 72
Ossalin, 257
Ossin, 257
Oxaalzuur (Dutch), 224
Oxacyclopentane, 267
Oxacyclopropane, 167
Oxalic acid, 17, 25, *224*
Oxalic acid dihydrate, 224
Oxalsaeure (German), 224
Oxane, 167
Oxides of nitrogen, 25

338 CHEMICAL INDEX

Oxidizers, 6, 10, 53
   laboratory storage, 63
Oxidoethane, 167
α,β-Oxidoethane, 167
Oxirane, 167
Oxitol, 161
1-Oxo-4-azacyclohexane, 211
2-Oxobornane, 123
Oxolane, 267
Oxomethane, 174
Oxy wash, 107
Oxy-5, 107
Oxy-10, 107
Oxybenzene, 232
Oxybis(chloromethane), 109
1,1'-Oxybisethane, 163
Oxydol, 192
Oxyfume, 167
Oxygen, 7, 17, *225*
Oxylite, 107
Oxymethylene, 174
Oxymuriate of potash, 243
Oxyphenic acid, 130
m-Oxytoluene, 146
Ozone, 13, 27, 49
Ozonides, 7

Panoxyl, 107
Pap-1, 92
Paraform, 174
Paraphenylenediamine, 44
Parathion, 39
   waste, 67
Pebble lime, 122
Pediaflor, 257
Pedident, 257
Peladow, 121
Pelagol grey C, 130
Pennwhite, 257
Pentamethylene, 149
Pentan (Polish), 228
Pentane, 5, *228*
Pentanen (Dutch), 228
2-Pentanone, *229*
Pentani (Italian), 228
Per acids, 8
Per salts, 10
Peracetic acid, *230*
Perchlorates, 7, 11, *230*
   laboratory storage, 63
Perchloric acid, 10, 11, 17, 43, *231*
   electropolishing solutions, 13
   laboratory storage, 63
Perchloromethane, 128
Perchloron, 121
Perfluoroacetic acid, 273
Pergantene, 257
Perhydrol, *192*
Periodin, 246
Permanganate de potassium (French), 247
Permanganate of potash, 247
Permanganates, 6, 13
   laboratory storage, 63
Perone, 192
Peroxan, *192*
Peroxide, *192*
Peroxide polymers, 9
Peroxides, 6–8
   organic, 17
Peroxyacetic acid, 230

Persadox, 107
Persulfates, 6, 10
   laboratory storage, 63
Pesticides, 25
Pestmaster, 165
Petrohol, 200
Petroleum ether, 4, 5
Petroleum solvents, 27
Phaldrone, 130
Phenador-x, 109
Phene, 105
Phenic acid, 232
Phenol, *232*
Phenol trinitrate, 240
Phenolics, 25, 26
Phenols, 24
Phenolphthalein, *233*
N-phenyl-1,4-benzenediamine, 44
Phenyl chloride, 137
Phenyl compounds, 26
Phenyl hydrate, 232
Phenyl hydride, 105
Phenyl hydroxide, 232
Phenyl mercaptan, waste, 67
Phenylamine, 100
Phenylbenzene, 109
o-Phenylenediol, 130
Phenylic acid, 232
Phenylic alcohol, 232
Phenylmethane, 269
Phos-flur, 257
Phosgene, 26, 27, 44, *234*
   waste, 67
Phosphoric acid, 27, *235*
   trisodium salt, dodecahydrate, 262
Phosphoric anhydride, 239
Phosphorofluoridic acid, diisopropyl ester, 153
Phosphorus, 6, 13, 27
   amorphous, red (DOT), *236*
   laboratory storage, 64
   red, *236*
   white, 17, 25, *237*
   yellow, 237
Phosphorus pentachloride, 27
Phosphorus pentoxide, 28, *239*
Phosphorus trichloride, 13, 27
   waste, 67
Phosphorus(V) oxide, 239
PHPH, 109
Picrates, 11
Picric acid, 11, *240*
   waste, 70
Picronitric acid, 240
Pimelic ketone, 148
Pittchlor, 121
Pittcide, 121
Pittclor, 121
Platinum oxide, 13
Point two, 257
Polycyclic aromatic hydrocarbons, handling, 47
Polynuclear compounds, 26
Polyoxymethylene glycols, 174
Poly-solv EE, 161
Poly-solv EM, 208
Portland stone, 120
Potash chlorate (DOT), 243
Potassa, 245
Potasse caustique (French), 245

Potassio (chlorato di) (Italian), 243
Potassio (idrossido di) (Italian), 245
Potassio (permanganato di) (Italian), 247
Potassium, 6, 13, 17, 27, *241*
   laboratory storage, 64
   powdered, 16
Potassium bichromate, *243*
Potassium carbonate, 27
Potassium chlorate (DOT), 17, *243*
Potassium (chlorate de) (French), 243
Potassium cyanide, *244*
   waste, 67
Potassium cyanide solution (DOT), 244
Potassium dichromate(VI), 243
Potassium fluoride (DOT), *245*
Potassium fluorure (French), 245
Potassium hydrate, 245
Potassium hydroxide, 28, *245*
   laboratory storage, 64
Potassium hyperchloride, 246
Potassium oxymuriate, 243
Potassium perchlorate, 17, *246*
Potassium permanganate (DOT), 17, *247*
Potassium (permanganate de) (French), 247
Potato alcohol, 162
POX, 239
Pracarbamin, 275
Pracarbamine, 275
Precipitated calcium sulfate, 123
Precipitated sulfur, 263
Predent, 257
Pro-gibb, 177
Propan-2-ol, 200
Propane, 16, *248*
Propane sultone, 46
1-Propanecarboxylic acid, 117
1,2-Propanediol, waste, 67
Propanol-1, 249
1-Propanol, 249
2-Propanol, 200
n-Propanol, 249
Propanole (German), 249
Propanolen (Dutch), 249
Propanoli (Italian), 249
Propanone, 85
Propellant 12, 150
2-Propenal, 89
Prop-2-en-1-al, 89
Propenamide, 90
2-Propenamide, 90
Propenenitrile, 91
2-Propenenitrile, 91
β-Propiolactone, 47
Propyl alcohol, *249*
1-Propyl alcohol, 249
n-Propyl alcohol, 249
sec-Propyl alcohol, 200
n-Propyl alkohol (German), 249
Propyl hydride, 248
Propylcarbinol, 113
N-Propylcarbinyl chloride, 116
Propylene aldehyde, 89
Propylformic acid, 117
Propylic alcohol, 249

Propylmethanol, 113
Propylowy alkohol (Polish), 249
Prussic acid, 187
   unstabilized, 187
Puratronic chromium chloride, 141
Puratronic chromium trioxide, 141
Purified charcoal, 124
Pyridine, 29, 38, *250*
Pyroacetic acid, 85
Pyroacetic ether, 85
Pyrobenzol, 105
Pyrobenzole, 105
Pyrocatechin, 130
Pyrocatechinic acid, 130
Pyrocatechuic acid, 130
Pyroxylic spirit, 207

Quicklime, 122
Quecksilber (German), 204
Quicksilver, 204
β-Quinol, 196
Quinolor compound, 107

R12, 150
Rafluor, 257
Raney nickel, 13
Rayox, 269
RCRA waste number U165, 212
Rectules, 130
Refrigerant 11, 173
Refrigerant 12, 150
Rescue squad, 257
Ricin, 43
Roach salt, 257
Roman vitriol, 145
Rtec (Polish), 204
Runa RH20, 269
Rutile, 269

Sal ammonia, 97
Sal ammoniac, 97
Salpetersaure (German), 214
Salpeterzuuroplossingen (Dutch), 214
Salt cake, 262
Salzburg vitriol, 145
Satin spar, 123
Satinite, 123
S.B.A., 114
Scintillar, 278
Selenides, 17
Selenium, *251*
   wastes, 72
Selenium alloy, 251
Selenium base, 251
Selenium compounds, 25
Selenium dust, 251
Selenium elemental, 251
Selenium homopolymer, 251
Selenium metal powder, nonpyrophoric (DOT), 251
Sentry, 121
Serpentine, 103
Sextone, 148
SHELL MIBK, 181
Shell silver, 252
SI, 201
Silver and its compounds, 17, *252*
Silver atom, 252

CHEMICAL INDEX

Silver compounds, 25
Silver nitrate, 27
Silver salts, 13
Simidine, 44
Sirup, 177
SK-Choral hydrate, 130
Slaked lime, 121
Slimicide, 89
Slow-Fe, 169
Snomelt, 121
SO, 201
Soda lye, 257
Soda niter, 260
Sodium, 6, 14, 17, 27, *253*
   laboratory storage, 64
   NFPA label, 54
   powdered, 16
Sodium acetate, *254*
Sodium acid sulfite, 255
Sodium amide, waste, 70
Sodium azide, 8, *255*
   waste, 67
Sodium bisulfite, *255*
   solid (DOT), 255
   solution (DOT), 255
Sodium bisulphite, 255
Sodium carbonate, 27
Sodium chloride, 28
Sodium cyanide, *256*
   solid and solution (DOT), 256
   waste, 67
Sodium dioxide, 261
Sodium fluoride, *257*
   solid and solution (DOT), 257
Sodium fluoroacetate, 43
Sodium hydrate (DOT), 257
Sodium hydrofluoride, 257
Sodium hydrogen sulfite, 255
   solid (DOT), 255
   solution (DOT), 255
Sodium hydroxide, 24, 27, 28, *257*
   bead (DOT), 257
   dry (DOT), 257
   flake (DOT), 257
   granular (DOT), 257
   laboratory storage, 64
   solid (DOT), 257
Sodium hypochlorite, 40, 49
Sodium iodide, *259*
Sodium iodine, 259
Sodium metal (DOT), 253
Sodium monofluoride, 257
Sodium monoiodide, 259
Sodium monosulfide, 263
Sodium nitrate (DOT), *260*
Sodium nitrite, 17, *260*
Sodium oxide, 261
Sodium peroxide (DOT), 8, 17, *261*
Sodium phosphate dodecahydrate, 262
Sodium phosphate tribasic dodecahydrate, *262*
Sodium silicate, 27
Sodium sulfate, *262*
Sodium sulfate anhydrous, 262
Sodium sulfide (anhydrous), *263*
   laboratory storage, 64
Sodium sulhydrate, 255
Sodium sulphate, 262

Sodium sulphide, 263
Sodium(I) nitrate, *260*
So-flo, 257
Sohnhofen stone, 120
Soilbrom-40, 165
Soilfume, 165
Solfo Black B, 158
Solmethine, 210
Solozone, 261
Solvent ether, 163
Somipront, 157
Somni sed, 130
Sontec, 130
Sorbo-calcian, 120
Sorbo-calcion, 120
Spirit, 162
Spirit of Hartshorn, 94
Spirits of turpentine, 273
Spirits of wine, 162
Stannic chloride, laboratory storage, 64
Stay-flo, 257
Steam-distilled turpentine, 273
Sterilizing gas ethylene oxide 100%, 167
Stibium, 101
Stink damp, 194
Strontium sulfide, waste, 67
Studafluor, 257
Styrene, 5
Styrene monomer, waste, 70
Sublimed sulfur, 263
Sulfamic acid, *263*
Sulfamidic acid, 263
Sulfate wood turpentine, 273
Sulferrous, 169
Sulfides, 17, 26
Sulfinylbis(methane), 157
Sulfoxyl, 107
Sulfur, *263*
   solid, 263
Sulfur dioxide, 27, *264*
   laboratory storage, 64
Sulfur hydride, 194
Sulfur monochloride, 27
Sulfur oxide, 264
Sulfur trioxide, laboratory storage, 64
Sulfurated hydrogen, 194
Sulfur(thio-) compounds, 25
Sulfuric acid, 17, 27, 28, *265*
   aluminum salt (3:2), 94
   calcium(2+) salt, dihydrate, 123
   cobalt(2+) salt (1:1), 144
   copper(2+) salt, pentahydrate, 145
   copper(2+) salt, (1:1), 145
   dimethyl ester, 156
   disodium salt, 262
   iron(2+) salt (1:1), 169
   laboratory storage, 64
   zinc salt (1:1), 280
Sulfurous acid, monosodium salt, 255
Sulfurous acid anhydride, 264
Sulfurous oxide, 264
Sulfuryl chloride, 27
Sulphocarbonic anhydride, 125
Super cobalt, 143
Super-dent, 257
Superflake anhydrous, 121

Superlysoform, 174
Superox, 107
Superoxol, 192
Syntexan, 157

Tac 121, 268
Tac 131, 268
Tannic acid, *267*
Tannin, 267
Tar camphor, 212
TCM, 137
Tecsol, 162
Telltozan, 120
Tellurides, 17
TEN, 272
Terra alba, 123
Tertrosulphur black PB, 158
Tertrosulphur PBR, 158
Tescol, 167
Tetrachlorocarbon, 128
Tetrachloroethane, 5
Tetrachloromethane, 128
Tetrafinol, 128
Tetraform, 128
Tetrafosfor (Dutch), 237
1,2,3,4-Tetrahydrobenzene, 149
Tetrahydro-1,4-dioxin, 159
Tetrahydro-p-dioxin, 159
Tetrahydrofuran, 9, *267*
   laboratory storage, 64
   waste, 70
Tetrahydro-1,4-isoxazine, 211
Tetrahydro-p-isoxazine, 211
Tetrahydronaphthalene, 9
Tetrahydro-1,4-oxazine, 211
Tetrahydro-2H-1,4-oxazine, 211
N-(5,6,7,9-tetrahydro-1,2,3,10-tetramethoxy-9-oxobenzo(α)heptalen-7-yl)-acetamide, 144
Tetralin, 9
   laboratory storage, 64
Tetramethylene oxide, 267
Tetraphosphor (German), 237
Tetrasol, 128
T-fluoride, 257
T-gas, 167
Thallium, wastes, 72
Thallium compounds, 25
Thallium(III) oxide, waste, 67
Thenardite, 262
Theraderm, 107
Thera-flur-n, 257
THF, 9, 267
Thioketones, 26
Thiols, 26
Thionyl chloride, 27
Three elephant, 110
Tinning flux (DOT), 279
Tiofine, 269
Tioxide, 269
Titanium chloride, *268*
Titanium(III) chloride, 268
Titanium dioxide, 269
Titanium oxide, *269*
Titanium tetrachloride, laboratory storage, 64
Titanium trichloride, 268
   pyrophoric (DOT), 268
Titanous chloride, 268
Toluene, 5, 9, *269*
Toluol (DOT), 269

m-Toluol, 146
Topex, 107
Topsym, 157
Tosyl, 130
Trawotox, 130
Tremolite, 103
Tri-plus, 270
Triangle, 145
Trichloran, 270
Trichloro, 270
Trichloro titanium, 268
Trichloroacetaldehyde hydrate, 130
Trichloroacetaldehyde monohydrate, 130
Trichlorochromium, 141
2,2,2-Trichloro-1,1-ethanediol, 130
Trichloroethylene, 14, *270*
Triplus, 270
1,2,2-Trichloroethylene, 270
Trichlorofluoromethane, 173
Trichloroform, 137
Trichloromethane, 137
Trichloromonofluoromethane, 173
Triclene, 270
Triethylamine, *272*
Trifluoroacetic acid, *273*
Trifluoroethanoic acid, 273
3,4,5-Trihydroxybenzoic acid, 177
2,4a,7-Trihydroxy-1-methyl-8-methylenegibb-3-ene-1,10-carboxylic acid 1-4-lactone, 177
Trilene, 270
Trimar, 270
1,7,7-trimethylbicyclo(2.2.1)-2-heptanone, 123
Trimethylcarbinol, 115
1,7,7-Trimethylnorcamphor, 123
2,2,4-Trimethylpentane, *273*
Trinagle, 145
1,3,5-Trinitrophenol, 240
2,4,6-Trinitrophenol, 11, 240
Trinitrotoluene, 7
Trioxide(s), 269
Trisodium phosphate, 27
Trisodium trifluoride, 257
Trona, 262
Tronox, 269
Trovidur, 276
T-stuff, 192
Turpentine, *273*
   steam-distilled, 273
Turpentine and terpines, 27
Turpentine oil, rectified, 273
Turps, 273

U-compound, 275
Unifume, 165
Unitane O-110, 269
Univerm, 128
Uranium and its insoluble compounds, *274*
Uranium carbide powder, 12
Uranium compounds, soluble, *275*
Uranium hexafluoride, 275
Uranium hydride powder, 12
Uranium hypochlorite, 275

Uranium nitrate, 275
Uranium powder, 12
Uranyl acetate, 275
Uranyl carbonate, 275
Uranyl sulfate, 275
Urea nitrate, waste, 70
Urethan, 275
Urethane, *275*

Vanadium pentoxide, waste, 67
Vandex, 251
Vanoxide, 107
Vaterite, 120
VC, 276
VCM, 276
VCN, 91
Ventox, 91
Verazinc, 280
Vermiculite, 9
Vermoestricid, 128
Vestrol, 270

Villiaumite, 257
Vinegar acid, 82
Vinegar naphtha, 161
Vinegar salts, 120
Vinyl acetate, 9
Vinyl C monomer, 276
Vinyl chloride, 276
  monomer, 276
Vinyl compounds, 9
Vinyl cyanide, 91
Vinylacetylene, *277*
Vinylidene chloride, 9
Vinylidene compounds, 9
Violet 3, 277
Vitran, 270
Vitriol brown oil, 265

Weevitox, 125
Weiss phosphor (German), 237
Westrosol, 270
1700 White, 269

White arsenic, 103
White caustic, 257
White copperas, 280
White phosphoric acid, 235
White phosphorus, 237
White tar, 212
White vitriol, 280
Wood alcohol, 207
Wood naphtha, 207
Wood spirit, 207
Wood turpentine, 273

Xenene, 109
Xerac, 107
Xylene, *277*
1,3-Xylene, 278
1,4-Xylene, 278
m-Xylene, *278*
p-Xylene, *278*
Xylol (DOT), 277
m-Xylol (DOT), 278
p-Xylol (DOT), 278

Yellow phosphorus, 237

Zaclon discoids, 187
Zinc acetate, *278*
Zinc chloride, *279*
  anhydrous (DOT), 279
  solid (DOT), 279
  solution (DOT), 279
Zinc diacetate, 278
Zinc dichloride, 279
Zinc muriate, solution (DOT), 279
Zinc nitrate, *279*
Zinc sulfate, *280*
Zinc sulphate, 280
Zinc vitriol, 280
Zinkosite, 280
Zirconium nitrate, *280*
Zirconium oxychloride, *280*
Zirconyl chloride, 280
Zopaque, 269

CAS NUMBER INDEX

CAS #	Chemical Name	CAS #	Chemical Name
50-00-0	Formaldehyde, 174	79-21-0	Peracetic acid, 230
50-32-8	Benzo[a]pyrene, 106	88-89-1	Picric acid, 240
50-99-7	d-Glucose, 177	91-20-3	Naphthalene, 212
51-28-5	2,4-Dinitrophenol, 158	92-52-4	Biphenyl, 109
51-79-6	Urethane, 275	94-36-0	Benzoyl peroxide, 107
55-18-5	Diethylnitrosamine, 152	95-50-1	ortho-Dichlorobenzene, 150
55-91-4	Diisopropyl fluorophosphate, 153	98-95-3	Nitrobenzene, 220
56-23-5	Carbon tetrachloride, 128	100-52-7	Benzaldehyde, 105
56-81-5	Glycerol, 178	106-42-3	p-Xylene, 278
60-29-7	Ethyl ether, 163	106-93-4	Ethylene dibromide, 165
62-53-3	Aniline, 100	107-02-8	Acrolein, 89
62-54-4	Calcium acetate, 120	107-06-2	Ethylene dichloride, 166
64-17-5	Ethyl alcohol, 162	107-13-1	Acrylonitrile (inhibited), 91
64-18-6	Formic acid, 176	107-21-1	Ethylene glycol, 167
64-19-7	Acetic acid, 82	107-87-9	2-Pentanone, 229
64-86-8	Colchicine, 144	107-92-6	n-Butyric acid, 117
67-56-1	Methanol, 207	108-10-1	Hexone, 181
67-63-0	Isopropyl alcohol, 200	108-20-3	Diisopropyl ether, 153
67-64-1	Acetone, 84	108-24-7	Acetic anhydride, 83
67-66-3	Chloroform, 137	108-38-3	m-Xylene, 278
67-68-5	Dimethyl sulfoxide, 157	108-39-4	m-Cresol, 146
68-12-2	Dimethylformamide, 155	108-88-3	Toluene, 269
71-23-8	Propyl alcohol, 249	108-90-7	Chlorobenzene, 137
71-36-3	1-Butanol, 113	108-94-1	Cyclohexanol, 147
71-43-2	Benzene, 105	108-94-1	Cyclohexanone, 148
74-86-2	Acetylene, 87	108-95-2	Phenol, 232
74-87-3	Chloromethane, 139	109-66-0	Pentane, 228
74-90-8	Hydrogen cyanide, 187	109-69-3	n-Butyl chloride, 116
74-98-6	Propane, 248	109-86-4	Methyl cellosolve, 208
75-01-4	Vinyl chloride, 276	109-99-9	Tetrahydrofuran, 267
75-05-8	Acetonitrile, 86	110-12-3	5-Methyl-2-hexanone, 211
75-07-0	Acetaldehyde, 81	110-17-8	Fumaric acid, 176
75-09-2	Methylene chloride, 210	110-22-5	Acetyl peroxide, 87
75-15-0	Carbon disulfide, 125	110-54-3	Hexane, 180
75-21-8	Ethylene oxide, 167	110-71-4	Glyme, 178
75-44-5	Phosgene, 234	110-80-5	2-Ethoxyethanol, 161
75-65-0	tert-Butanol, 115	110-82-7	Cyclohexane, 146
75-69-4	Fluorotrichloromethane, 173	110-83-8	Cyclohexene, 149
75-71-8	Dichlorodifluoromethane, 150	110-86-1	Pyridine, 250
75-91-2	tert-Butyl hydroperoxide, 117	110-91-8	Morpholine, 211
76-05-1	Trifluoroacetic acid, 273	111-65-9	Octane, 223
76-22-2	Camphor, 123	120-80-9	Catechol, 130
77-06-5	Gibberellic acid, 177	120-83-2	2,4-Dichlorophenol, 152
77-09-8	Phenolphthalein, 233	121-44-8	Triethylamine, 272
77-78-1	Dimethyl sulfate, 156	123-31-9	Hydroquinone, 196
78-83-1	Isobutyl alcohol, 199	123-86-4	n-Butyl acetate, 116
78-92-2	2-Butanol, 114	123-91-1	Dioxane, 159
78-93-3	Methyl ethyl ketone, 209	124-38-9	Carbon dioxide, 124
79-01-6	Trichloroethylene, 270	127-09-3	Sodium acetate, 254
79-06-1	Acrylamide, 90	127-19-5	Dimethyl acetamide, 154

CAS NUMBER INDEX

CAS #	Chemical Name	CAS #	Chemical Name
141-43-5	Ethanolamine, 160	7646-79-9	Cobalt(II) chloride, 143
141-78-6	Ethyl acetate, 161	7646-85-7	Zinc chloride, 279
142-82-5	Heptane, 179	7647-01-0	Hydrogen chloride, 186
143-33-9	Sodium cyanide, 256	7664-38-2	Phosphoric acid, 235
144-62-7	Oxalic acid, 224	7664-39-3	Hydrogen fluoride, 190
149-91-7	Gallic acid, 177	7664-41-7	Ammonia, 94
151-50-8	Potassium cyanide, 244	7664-93-9	Sulfuric acid, 265
287-92-3	Cyclopentane, 149	7681-49-4	Sodium fluoride, 257
302-01-2	Hydrazine and its salts, 182	7681-82-5	Sodium iodide, 259
302-17-0	Chloral hydrate, 130	7697-37-2	Nitric acid, 214
540-84-1	2,2,4-Trimethylpentane, 273	7699-43-6	Zirconium oxychloride, 280
542-88-1	Bis(chloromethyl) ether, 109	7704-34-9	Sulfur, 263
557-34-6	Zinc acetate, 278	7705-07-9	Titanium chloride, 268
591-78-6	2-Hexanone, 181	7705-08-0	Ferric chloride, 169
630-08-0	Carbon monoxide, 127	7720-78-7	Ferrous sulfate, 169
689-97-4	Vinylacetylene, 277	7722-64-7	Potassium permanganate, 247
1300-21-6	Dichloroethane, 150	7722-84-1	Hydrogen peroxide, 192
1305-62-0	Calcium hydroxide, 121	7723-14-0	Phosphorus (red), 236
1305-78-8	Calcium oxide, 122	7723-14-0	Phosphorus (white), 237
1306-19-0	Cadmium oxide, 119	7726-95-6	Bromine, 112
1310-58-3	Potassium hydroxide, 245	7727-37-9	Nitrogen, 222
1310-73-2	Sodium hydroxide, 257	7727-54-0	Ammonium persulfate, 99
1313-60-6	Sodium peroxide, 261	7733-02-0	Zinc sulfate, 280
1313-82-2	Sodium sulfide (anhydrous), 263	7738-94-5	Chromic acid, 140
1314-56-3	Phosphorus pentoxide, 239	7757-82-6	Sodium sulfate, 262
1317-65-3	Calcium carbonate, 120	7758-98-7	Copper(II) sulfate, 145
1327-53-3	Arsenic trioxide, 103	7758-99-8	Copper sulfate pentahydrate, 145
1332-21-4	Asbestos, 103	7778-50-9	Potassium bichromate, 243
1333-74-0	Hydrogen, 184	7778-54-3	Calcium hypochlorite, 121
1333-82-0	Chromium trioxide, 141	7778-74-7	Potassium perchlorate, 246
1336-21-6	Ammonium hydroxide, 97	7779-88-6	Zinc nitrate, 279
1338-23-4	Methyl ethyl ketone peroxide, 209	7782-41-4	Fluorine, 170
1401-55-4	Tannic acid, 267	7782-44-7	Oxygen, 225
3251-23-8	Copper(II) nitrate, 145	7782-49-2	Selenium, 251
3811-04-9	Potassium chlorate, 243	7782-50-5	Chlorine, 131
5329-14-6	Sulfamic acid, 263	7783-06-4	Hydrogen sulfide, 194
6484-52-2	Ammonium nitrate, 98	7789-23-3	Potassium fluoride, 245
7429-90-5	Aluminum, 92	7790-91-2	Chlorine trifluoride, 135
7439-92-1	Lead, 201	8006-64-2	Turpentine, 273
7439-93-2	Lithium and its compounds, 202	10022-68-1	Cadmium nitrate tetrahydrate, 119
7439-95-4	Magnesium, 203	10025-73-7	Chromium chloride, 141
7439-97-6	Mercury, 204	10035-10-6	Hydrogen bromide, 186
7440-09-7	Potassium, 241	10043-01-3	Aluminum sulfate, 94
7440-22-4	Silver and its compounds, 252	10043-35-3	Boric acid, 110
7440-23-5	Sodium, 253	10043-52-4	Calcium chloride, 121
7440-36-0	Antimony, 101	10049-04-4	Chlorine dioxide, 134
7440-38-2	Arsenic, 102	10101-41-4	Calcium sulfate dihydrate, 123
7440-39-3	Barium, 104	10101-89-0	Sodium phosphate tribasic dodecahydrate, 262
7440-41-7	Beryllium, 108		
7440-43-9	Cadmium, 118	10102-43-9	Nitric oxide, 219
7440-44-0	Carbon, 124	10102-44-0	Nitrogen dioxide, 222
7440-47-3	Chromium, 140	10108-64-2	Cadmium chloride, 118
7440-48-4	Cobalt, 143	10124-43-3	Cobalt(II) sulfate (1:1), 144
7440-58-6	Hafnium and compounds, 178	10141-05-6	Cobalt(II) nitrate, 144
7440-61-1	Uranium and its insoluble compounds, 274	10294-34-5	Boron trichloride, 110
		10325-94-7	Cadmium nitrate, 119
7440-70-2	Calcium, 120	10361-37-2	Barium chloride, 104
7446-09-5	Sulfur dioxide, 264	12125-02-9	Ammonium chloride, 97
7553-56-2	Iodine, 197	13300-20-7	Xylene, 277
7601-90-3	Perchloric acid, 231	13463-39-3	Nickel carbonyl, 213
7631-90-5	Sodium bisulfite, 255	13463-67-7	Titanium oxide, 269
7631-99-4	Sodium(I) nitrate, 260	13746-89-9	Zirconium nitrate, 280
7632-00-0	Sodium nitrite, 260	26628-22-8	Sodium azide, 255
7637-07-2	Boron trifluoride, 111		